Science and Technology of Photovoltaics

Science and Technology of Photovoltaics

Edited by Paul Vernon

SYRAWOOD
PUBLISHING HOUSE

New York

Published by Syrawood Publishing House,
750 Third Avenue, 9th Floor,
New York, NY 10017, USA
www.syrawoodpublishinghouse.com

Science and Technology of Photovoltaics
Edited by Paul Vernon

International Standard Book Number: 978-1-68286-531-6 (Hardback)

Cataloging-in-Publication Data

Science and technology of photovoltaics / edited by Paul Vernon.
    p. cm.
Includes bibliographical references and index.
ISBN 978-1-68286-531-6
1. Photovoltaic power generation. 2. Renewable energy sources. I. Vernon, Paul.
TK1087 .S35 2018
621.312 44--dc23

TABLE OF CONTENTS

Permissions

List of Contributors

Index

PREFACE

This book aims to highlight the current researches and provides a platform to further the scope of innovations in this area. This book is a product of the combined efforts of many researchers and scientists, after going through thorough studies and analysis from different parts of the world. The objective of this book is to provide the readers with the latest information of the field.

Photovoltaic devices help in generating electricity. This conversion is done by using semi-conducting materials which exhibit the photovoltaic effect. Photovoltaic systems are used as solar panels, which comprise of a number of solar cells. This book contains some path-breaking studies related to the science and technology of photovoltaics. For someone with an interest and eye for details, it covers the most significant topics related to the subject matter. This book is an essential guide for both academicians and those who wish to pursue this discipline further.

I would like to express my sincere thanks to the authors for their dedicated efforts in the completion of this book. I acknowledge the efforts of the publisher for providing constant support. Lastly, I would like to thank my family for their support in all academic endeavors.

Editor

Ultra-thin crystalline silicon films produced by plasma assisted epitaxial growth on silicon wafers and their transfer to foreign substrates*

M. Moreno[a] and P. Roca i Cabarrocas

Laboratoire de Physique des Interfaces et des Couches Minces, École Polytechnique, CNRS, Palaiseau, France

Abstract We have developed a new process to produce ultra-thin crystalline silicon films with thicknesses in the range of $0.1-1$ μm on flexible substrates. A crystalline silicon wafer was cleaned by SiF_4 plasma exposure and without breaking vacuum, an epitaxial film was grown from SiF_4, H_2 and Ar gas mixtures at low substrate temperature ($T_{sub} \approx 200$ °C) in a standard RF PECVD reactor. We found that H_2 dilution is a key parameter for the growth of high quality epitaxial films and modification of the structural composition of the interface with the c-Si wafer, allowing one to switch from a smooth interface at low hydrogen flow rates to a fragile one, composed of hydrogen-rich micro-cavities, at high hydrogen flow rates. This feature can be advantageously used to separate the epitaxial film from the crystalline Si wafer. As a example demonstration, we show that by depositing a metal film followed by a spin-coated polyimide layer and applying a moderate thermal treatment to the stack, the fragile interface breaks down and allows one to obtain an ultrathin crystalline wafer on the flexible polyimide support.

1 Introduction

Over the past twenty years, considerable efforts have been devoted to producing thin crystalline silicon films for micro-electronics and photovoltaic applications. The main motivation remains a reduction of the necessary amount of high grade crystalline silicon (c-Si) and therefore a reduction of device fabrication costs. For instance, very high efficiencies in heterojunction solar cells have been demonstrated ($17-22\%$), fabricated on thin c-Si substrates ($50-100$ μm) [1,2] and so the goal of further reducing the c-Si wafer thickness is still a strong driving force for cost reduction.

Various approaches have been used to produce low-cost ultra-thin c-Si wafers; however, some of them involve high temperature steps (increasing the fabrication cost). Some of the most reliable approaches are: (i) cutting thin wafers and thinning them down by chemical mechanical polishing; (ii) growing thin c-Si films by LPCVD on a seed layer; (iii) annealing macroporous arrays formed on a crystalline silicon wafer (also referred to as silicon on nothing [3,4]); and (iv) the "smart cut" [5-7] and ion cutting porcesses [8-12].

Smart cut and ion cutting processes are proven methods to transfer a thin silicon film from a wafer to another substrate, which generally is a second wafer, in which a thin SiO_2 film has been deposited (to form a silicon on insulator wafer). Those techniques employ H^+ ion implantation in a c-Si wafer (at doses between $10^{16}-10^{17}$ H/cm^2) in order to create defects (micro-cavities) at some depth from the wafer surface. In this way, it is possible to form a very thin c-Si film ($0.3-1$ μm) separated from the thick wafer by a highly defective and therefore weak layer. A hydrophilic wafer bonding is performed with a second wafer and thermal processes are applied (at $650-1000$ °C) in order to transfer the thin c-Si film from the thick wafer to the receptor substrate and to strengthen the chemical bonds.

Another approach to obtaining thin c-Si films for PV applications [13, 14] consists of producing a porous film (1 μm thick) on the surface of a c-Si wafer by electrochemical anodisation in a HF solution. An annealing is made at high temperature (1000 °C) for densification of the porous film and an epitaxial layer ($\sim$50 μm thick) is deposited by Liquid Phase Epitaxy (LPE) [13] and Vapor Phase Epitaxy (VPE) [14]. Finally the epitaxial layer is waxed to a glass substrate and is removed from the c-Si wafer by an immersion in a wet solution.

The disadvantages of these processes are the large number of steps involved, the high temperatures used ($\sim$1000 °C), the necessity of a second substrate, the wet chemical steps involved and the fact that H^+ implantation LPE and VPE are not standard techniques in a PV factory.

---

* This article has been previously published in PV Direct, the former name of EPJ Photovoltaics.

[a] *Address for correspondence:* National Institute for Astrophysics, Optics and Electronics, INAOE, Puebla, Mexico; e-mail: mmoreno@inaoep.mx

We have studied an alternative method to produce low-cost ultra-thin c-Si films without the necessity of using wet chemical steps, a second wafer, H$^+$ implantation, LPE or VPE. We have developed a completely dry process to etch the native oxide from a c-Si wafer using a SiF$_4$ plasma in a standard PECVD reactor [15], and immediately without breaking the vacuum, to grow an epitaxial silicon film at low substrate temperature ($\sim$200 °C) from a H$_2$, SiF$_4$, and Ar gas mixture.

We present a systematic study on the effect of H$_2$ dilution on the quality of the epitaxial Si films as well as on the quality of the interface with the c-Si wafer. We found that by optimizing the H$_2$ gas flow rate it is possible to obtain very highly crystalline films, and moreover, to control the interface quality from a smooth one to a highly defective interface, mainly composed of micro-cavities.

Finally, we developed a simple process to remove the epitaxial film from the c-Si wafer, which consists of the deposition of a thin metal layer plus polyimide over the epitaxial film, followed by a thermal annealing treatment (at $\sim$450 °C). As a consequence the stack composed of polyimide/metal/epitaxial film is separated from the c-Si wafer, resulting in a thin film of c-Si on a flexible substrate which functions as a mechanical support.

2 Experiments

The system used for the epitaxial growth is a standard capacitively coupled RF glow discharge PECVD reactor, and the substrates used were $\langle 100 \rangle$ FZ n-type double side polished c-Si wafers of resistivity $1-5\ \Omega$ cm. The processes were performed at a substrate temperature of 200 °C. After loading the wafers in the reactor and attaining a base pressure of 1×10^{-6} mbar, the native oxide on the c-Si wafers was removed through exposure to a SiF$_4$ plasma for 5 minutes, with the following optimized conditions: an RF power of 0.1 W/cm^2 and pressure of 30 mTorr. Immediately after this step, and without breaking the vacuum, the epitaxial film was grown from a SiF$_4$, Ar and H$_2$ gas mixture for 10 minutes, with a RF power density of 0.5 W/cm^2 and a total pressure of 2.2 Torr. H$_2$ dilution was studied in order to observe its effect on the epitaxial film crystallinity, and therefore the H$_2$/SiF$_4$ gas ratio was varied in a wide range from 0.33 to 20 (SiF$_4$ = 3 sccm, while H$_2$ = 1$-$60 sssm).

High Ar dilution (Ar = 80 sccm) was used to increase the SiF$_4$ gas dissociation and consequently the growth rate. It has been demonstrated in our previous work [16, 17], that Ar improves the dissociation in both SiF$_4$ and SiH$_4$ plasmas, and therefore stimulates nanocrystal growth in the plasma. The improvement of the crystalline structure is related to the enhancement of the contribution of nanocrystals to the film growth. Also, it has been experimentally observed that higher large grain fractions [16] are present in μc-Si films deposited from SiF$_4$+Ar+H$_2$ than in those deposited from SiH$_4$+Ar+H$_2$.

It has been demonstrated that in-situ spectroscopic UV-Visible ellipsometry is an excellent tool to characterize the composition of thin films such as microcrystalline

silicon [16–19]. An in-situ UV-Visible spectroscopic ellipsometer (Jobin Yvon – MWR UVISEL) was used to measure the imaginary part of the pseudo-dielectric function (Im[ε]) of the epitaxial films deposited on the c-Si wafer at various stages of the process. We used this technique to optimize the plasma cleaning conditions and as well the epitaxial silicon film growth. The thickness of the layers and their composition were determined by modeling the ellipsometry data using the Bruggemann effective medium approximation [16–19].

In order to study the chemical composition of the films, particularly of the interface layer, we performed Secondary Ion Mass Spectroscopy (SIMS) analysis of some samples. Moreover, a quadrupole mass spectrometer (QMS PRISMA 80) was used to measure the hydrogen effusing from the samples when submitted to an annealing in an oven under high vacuum ($\sim$10^{-7} mbar). The thermal ramp used was 10 °C/min from room temperature up to 800 °C.

A procedure to remove the epitaxial layer from the c-Si wafer was developed, which consists of depositing a thin film of chromium (150 nm thick) on the epitaxial layer by thermal evaporation, followed by the spin coating of a polyimide film and a curing step at 280 °C. The sample is then annealed for half an hour at 450 °C under vacuum ($\sim$10^{-5} mbar). This annealing step is sufficient to fragilize the porous interface layer and to delaminate the polyimide/Cr/epitaxial layer from the crystalline silicon wafer. The polyimide provides a good mechanical support for the epitaxial film. Finally, Raman spectroscopy was used to analyze the crystallinity of the epitaxial films on the foreign substrate.

3 Results and discussion

3.1 Epitaxial films

We measured in-situ the imaginary part of the pseudo-dielectric function Im[ε] of c-Si in the energy range of 1.5$-$4.7 eV, before plasma cleaning, after native SiO$_2$ etching, and after the growth of the epitaxial films with various H$_2$ flow rates. Figure 1 shows Im[ε] of four selected samples: (1) c-Si with native SiO$_2$; (2) c-Si after the native SiO$_2$ plasma etching; (3) epitaxial-Si film growth with SiF$_4$/H$_2$ = 3/1 sccm; and (4) epitaxial-Si film growth with SiF$_4$/H$_2$ = 3/3 sccm (in cases 3 and 4 Ar = 80 sccm).

As one can see in Figure 1, when the native SiO$_2$ is present on the c-Si surface, the intensity of Im[ε] at 4.2 eV (E_2) has an amplitude around $E_2 \approx 36$, but it increases to $E_2 \approx 43$ when the native SiO$_2$ is etched by the SiF$_4$ plasma. After the deposition of an epitaxial film at low H$_2$ flow rate (H$_2$ = 1 sccm), one can see that the Im[ε] spectrum has the same shape as that of the bare c-Si wafer, suggesting that the deposited film has the same structural composition and quality as the c-Si wafer (this is corroborated from the modeling of the Im[ε] spectrum). However, when the H$_2$ flow rate is increased up to 3 sccm, interference fringes appear in the Im[ε]. These interference fringes, in the energy range from 1.5$-$3 eV, provide

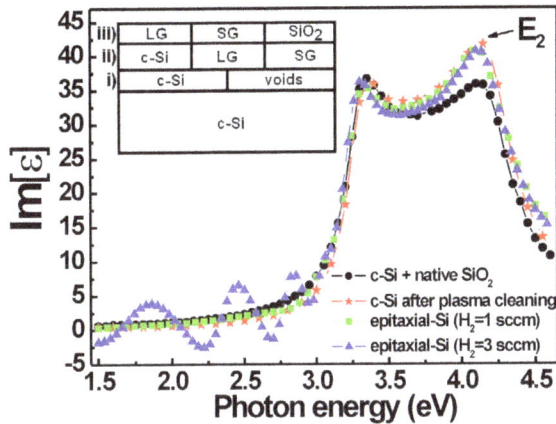

Fig. 1. (Color online) Imaginary part of the pseudo-dielectric function $Im[\varepsilon]$ of different samples: c-Si with native SiO_2, c-Si after native SiO_2 etching, and epitaxial-Si films grown with hydrogen flow rates of 1 sccm and 3 sccm. The inset shows the optical model used to analyze the epitaxial films.

a signature of the formation of a porous interface between the c-Si wafer and the epitaxial layer. This is illustrated by the insert of Figure 1, showing the optical model used to fit the experimental data. As one can see, the film is generally composed of 3 layers:

(i) A thin interface layer between the epitaxial film and the c-Si substrate composed of monocrystalline silicon (c-Si) and voids [20], of thickness in the range of $0-90$ Å (depending on H_2 dilution).

(ii) A bulk layer of thickness in the range of $900-1700$ Å (the deposition time for all the films were 10 minutes, however, as shown below, the deposition rate depends on the hydrogen flow rate). The layer is composed of a mono crystalline silicon (c-Si) fraction, a large grain (LG) fraction and a small grain (SG) fraction. The small grains have a size in the range of $1-10$ nm, while the large grains have a size in the range of $10-100$ nm.

(iii) A surface roughness characterized by a thin layer composed of a large grain (LG) fraction, a small grain (SG) fraction and SiO_2. The thickness of the surface roughness layer is in the range of $0-50$ Å (depending on the H_2 dilution). The SiO_2 fraction in the surface layers may be related to a small amount of oxygen in the deposition chamber.

The results of modeling the ellipsometric spectra of a series of films grown with a wide range of H_2 flow rates ($1-60$ sccm) are shown in Figure 2. Figure 2a shows the deposition rate and composition of the bulk epitaxial layers as a function of H_2 flow rate. At low H_2 flow rate ($H_2 = 1$ sccm) the deposition rate is around 2 Å/s, however when H_2 flow rate increases, the deposition rate also increases, with a maximum close to 3 Å/s (when $SiF_4 = H_2 = 3$ sccm). Further increase in H_2 flow rate resulted in a decrease of the deposition rate. In Figure 2a we can also see that the bulk layer has a high mono crystalline fraction (c-Si) of $80-95\%$, a small grain (SG) fraction of $5-20\%$ and large grain (LG) fraction of $1-5\%$.

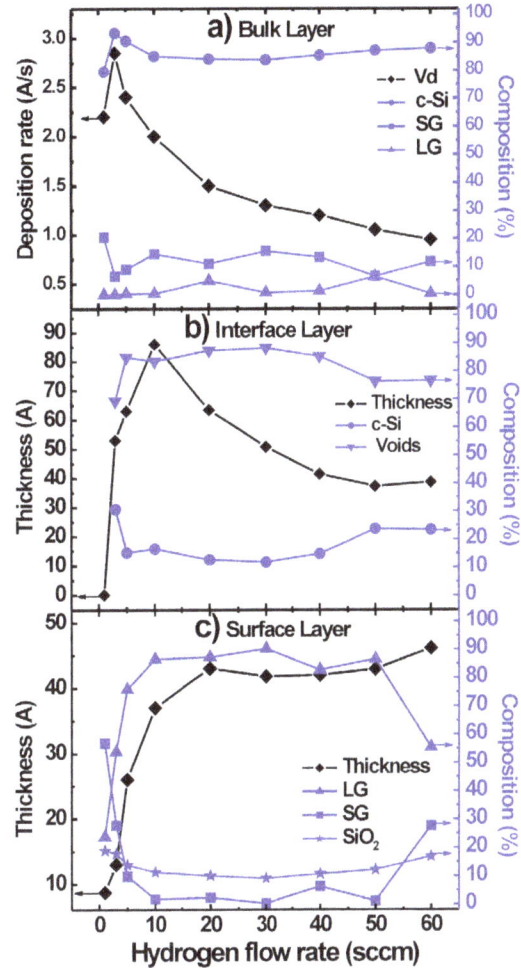

Fig. 2. (Color online) Results of modeling the ellipsometry spectra of the epitaxial films grown from SiF_4-Ar$-H_2$ mixtures as a function of the hydrogen flow rate, showing (a) the deposition rate and composition of the bulk, the thickness and composition of (b) the interface and (c) surface roughness layers.

Interestingly, the highest monocrystalline fraction (c-Si $\sim 95\%$) was obtained for the film deposited at the highest deposition rate. We believe that there is a strong relationship between the deposition rate and the crystallinity of a film. For certain deposition conditions, there is a greater dissociation of the gaseous species, improving the deposition rate. At higher deposition rate, oxygen incorporation in the film is reduced, and the film crystallinity improved.

Figure 2b shows the thickness of the interface layer as a function of H_2 flow rate. At low H_2 flow rate ($H_2 = 1$ sccm) there is no interface layer. However when the H_2 flow rate increases, an interface layer appears and reaches a maximum in thickness (~ 85 Å) when $H_2 = 10$ sccm. A further increase in H_2 dilution results in a decrease of the layer thickness. The reason for such behavior may result from a trade off between deposition rate and the hydrogen desorption from the growing film [21]. In Figure 2b also is shown the structural composition of the interface layer as a function of H_2 flow rate. As one can see, the interface

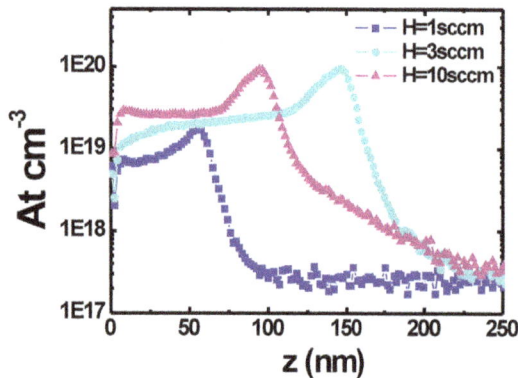

Fig. 3. (Color online) Hydrogen depth profiles measured by SIMS for epitaxial films grown with 1, 3 and 10 sccm of hydrogen in the SiF_4-Ar-H_2 gas mixture.

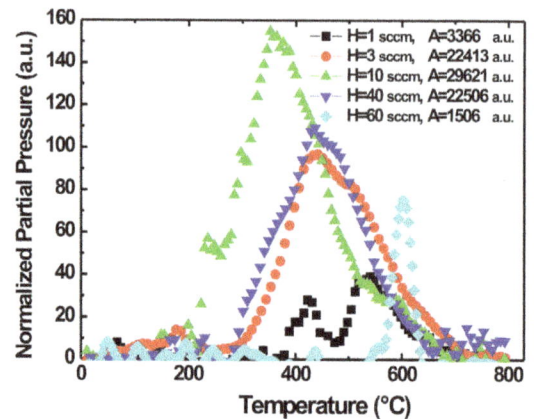

Fig. 4. (Color online) Hydrogen partial pressure measured by mass spectrometry for epitaxial layers grown with various hydrogen flow rates as a function of the annealing temperature.

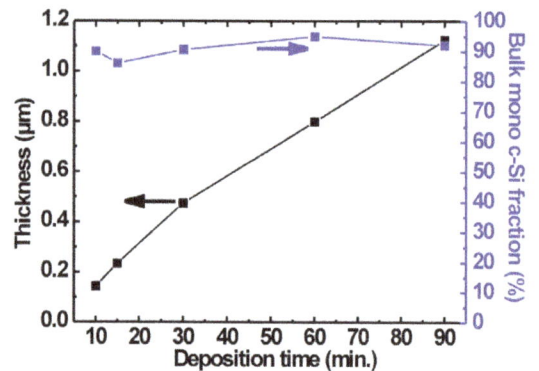

Fig. 5. (Color online) Dependence of the epitaxial film thickness and monocrystalline fraction on the deposition time.

layer is mainly composed of voids (around 80%) and a small amount of c-Si (around 20%).

Figure 2c shows the surface roughness of the epitaxial films as a function of the H_2 dilution. It is clear that the roughness of the films increases with the H_2 flow rate from $\sim$9 Å (at $H_2 = 1$ sccm) to $\sim$45 Å (at $H_2 = 60$ sccm). This increase in surface roughness can be related to the enhanced etching at high hydrogen flow rates. The structural composition of the surface roughness layer is also shown in the figure, where one can see that the SiO_2 fraction is almost independent of the H_2 flow rate.

To support the ellipsometry results indicating the presence of an interface layer, SIMS analysis were performed on the epitaxial films. Figure 3 shows the hydrogen depth profiles for the films deposited with 1, 3 and 10 sccm of hydrogen. The results show that the increase in hydrogen flow rate during deposition results in an increase of the hydrogen incorporated in the epitaxial films (from 6×10^{18} cm^{-3} for the film deposited with 1 sccm of H_2 up to 3×10^{19} cm^{-3} for the film deposited with 10 sccm of H_2), thus suggesting a better crystallinity at 1 sccm. Moreover a strong accumulation of hydrogen is observed at the interface between the epitaxial film and the crystalline substrate, with a broader peak for the samples deposited with 3 and 10 sccm, which is consistent with the thicker interface layer deduced from the ellipsometry measurements (see Fig. 2b).

Further support for the presence of a porous and hydrogen rich interface layer is given by the exodiffusion spectra shown in Figure 4. The spectra have been normalized with respect to the volume of the films in order to provide a relative comparison of the hydrogen desorbed. As one can see, the film grown with the smallest H_2 flow rate ($H_2 = 1$ sccm) shows the lowest integrated area of the curve, which indicates that the amount of H_2 is small. However, when the H_2 flow rate used for the deposition of the films is increased, the integrated area of the exodiffusion spectra increases as well and reaches a maximum for a H_2 flow rate of 10 sccm. A further increase in the H_2 flow rate results in a decrease of the integrated area. The difference in the effusing molecular H_2 from the films can be correlated to the presence of H_2 trapped at the interface,

in agreement with the SIMS results (Fig. 3). Note that the spectra of the different samples have a peak at different annealing temperatures, which is probably related to the structural composition of the films bulk (crystalline fraction) and interface layer composition [22, 23]. While a detailed analysis of these spectra is out of the scope of this paper, these results are coherent with the strong differences in the hydrogen accumulation at the interface, as suggested by the modelling of the ellipsometry spectra and the SIMS measurements.

The above results demonstrate that epitaxial films with different quality of interface layer can be produced for thicknesses up to 200 nm. However, for solar cell applications, thicker films are required. Therefore, in order to observe any dependence of the monocrystalline fraction of the epitaxial films on the layer thickness or deposition time, various epitaxial films were deposited for deposition times ranging from 10 to 90 minutes. Figure 5 shows that the mono crystalline fraction of the films is practically independent of the deposition time (and thickness) of the epitaxial films. A deposition of 10 minutes will result on a film of 1500 Å, corresponding to a deposition rate of 2.5 Å/s with a mono c-Si fraction of around 90%; while a deposition of 90 minutes will result on a film of 1.1 μm, corresponding to a deposition rate of 2 Å/s, with a mono c-Si fraction of around 92%. From the above result we conclude that is possible to growth thicker epitaxial films, without a loss of crystalline fraction.

3.2 Ultra-thin crystalline silicon film fabrication

The above results have shown that we can grow epitaxial films at low substrate temperature ($\sim$200 °C) in a standard RF glow discharge reactor. Moreover, a fine tuning of the hydrogen flow rate allows one to produce films with a thin and highly porous interface layer which mimics the hydrogen layer produced by ion implantation in the Smart Cut process [5]. In particular, a H_2 flow rate in the range of 3–10 sccm results in an interface layer with a thickness of 6–9 nm and a void fraction of $\sim$85%. Note that the void fraction deduced from spectroscopic ellipsometry is correlated with a high hydrogen content (SIMS and exodiffusion results).

Figure 6 shows the process developed to produce thin c-Si films. The c-Si wafer is placed inside the PECVD reactor without any previous cleaning. The native SiO_2 is etched by SiF_4 plasma [15] and the epitaxial c-Si film is deposited using the optimized conditions described above and in Section 3.1. The thickness of the film can be tailored according to the deposition time (as shown in Fig. 5). A chromium film of 150 nm is thermally evaporated on the epitaxial film followed by a deposition of a thin film of polymide by spin coating. A thermal annealing is performed at 450 °C under vacuum for half an hour. Finally the sample is cooled down and the vacuum is broken in order to produce the peeling of the c-Si film on a flexible substrate. This process allows to reuse the c-Si wafer, employing chemical polishing on the wafer surface, in order to remove defects from the surface and to facilitate the epitaxial grow.

Notice that the method presented above was performed with the equipment and materials available on our facilities. In fact any foreign substrate could be used for the transfer of the thin c-Si film. The only requirement is a strong adhesion between the foreign substrate and the ultrathin c-Si film.

The process described above (Fig. 6) was used to produce ultrathin crystalline silicon films as shown in Figure 7. The picture in the inset shows an ultrathin crystalline silicon film of $\sim$1 cm^2 on the polyimide foil, after being detached from the c-Si substrate. The Raman spectra of this ultrathin c-Si film (thickness of 0.5 μm) corresponds to the characteristic peak of c-Si at around 520 cm^{-1}. Its full width at half maximum is 5 cm^{-1}, which attests the high crystalline quality of the film, in good agreement with the ellipsometry spectra measured on the c-Si substrate before its detachment, from which a c-Si fraction of 95% was deduced.

This preliminary study, along with our previous work where we have demonstrated that it is possible to grow doped epitaxial films and to form a P/N junction at 175 °C, leading to a functioning solar cell [24], provides a

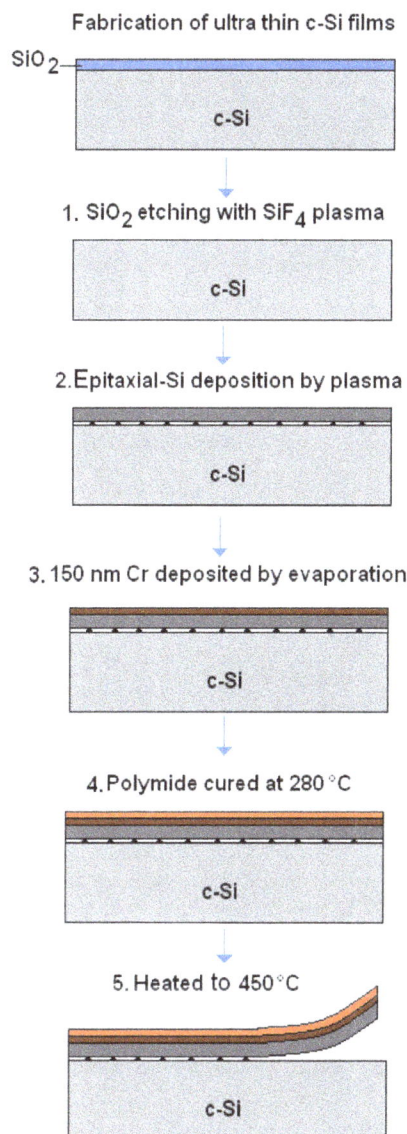

Fig. 6. (Color online) Schematic flow chart for the plasma assisted growth of ultrathin epitaxial films and their transfer to a polyimide support. Note that the maximum temperature of the process is 450 °C.

new approach for producing ultra-thin c-Si solar cells with thicknesses of a few microns.

4 Conclusions

We have developed a new process to produce ultra-thin crystalline silicon substrates for low-cost microelectronic and PV devices. Epitaxial films have been deposited on $\langle 100 \rangle$ c-Si wafers at low temperature (200 °C) using a dry process to first etch the native SiO_2, and then, without breaking the vacuum, a plasma deposition process using SiF_4-Ar-H_2 mixtures. From the film characterization, we observed that by optimizing the H_2/SiF_4 flow rate ratio while using a high Ar dilution, it is possible to deposit films with a c-Si fraction as high as 95% and having an

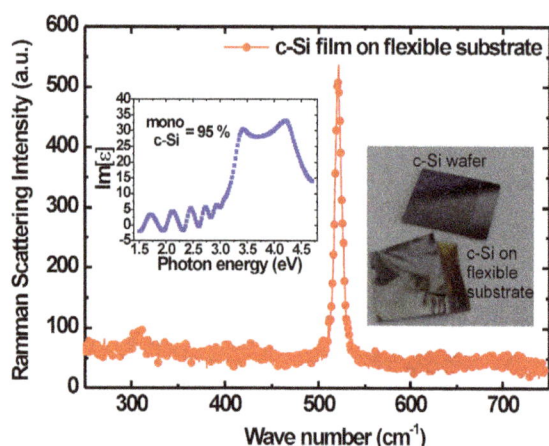

Fig. 7. (Color online) Raman spectrum of an ultra-thin c-Si film over a chromium/polyimide substrate. The left inset shows the $Im[\varepsilon]$ spectra of the film grown on c-Si before its detachment from the c-Si wafer while the right inset shows the sample after being removed from the c-Si wafer.

interface layer thickness of up to 90 Å, mainly composed of hydrogen rich micro-cavities (~85%). This fragile interface allows the separation of the epitaxial film from the crystalline silicon wafer by applying a thermal annealing step at 450 °C. As a result, crystalline silicon films were successfully obtained and transferred to chromium/polyimide flexible substrates. Finally, films up to 1 μm thick could be epitaxially grown without any decrease in their crystalline fraction. This makes this process very promising for producing ultra-thin c-Si substrates of tailored thickness for micro-electronics and PV applications.

References

1. Y. Tsunomura, Y. Yoshimine, M. Taguchi, T. Baba, T. Kinoshita, H. Kanno, H. Sakata, E. Maruyama, M. Tanaka, Sol. Energy. Mater. & Sol. Cells **93**, 670 (2009)
2. M. Reuter, W. Brendle, O. Tobail, J.H. Werner, Sol. Energy. Mater. & Sol. Cells **93**, 704 (2009)
3. I. Gordon, S. Vallon, A. Mayolet, G. Beaucarne, J. Poortmans, Sol. Energy. Mater. & Sol. Cells **94**, 381 (2010)
4. V. Depauw, I. Gordon, G. Beaucarne, J. Poortmans, R. Mertens, J.-P. Celis, J. Appl. Phys. **106**, 1 (2009)
5. M. Bruel, Mat. Res. Innovat. **3**, 9 (1999)
6. M. Bruel, Electron. Lett. **31**, 1201 (1995)
7. H. Moriceau, F. Fournel, B. Aspar, J. Electron. Mater. **32**, 829 (2003)
8. B. Terreault, Phys. Stat. Sol. (a) **204**, 2129 (2007)
9. A. Giguere, B. Terreault, J. Appl. Phys. **102**, 1 (2007)
10. O. Moutanabbir, A. Guiguere, B. Terreault, Appl. Phys. Lett. **84**, 3286 (2004)
11. S. Igarashi, A.N. Itakura, M. Kitajima, Jpn J. Appl. Phys. **46**, 7812 (2007)
12. N. Desroisers, B. Terreault, Appl. Phys. Lett. **89**, 1 (2006)
13. S. Berger, S. Quoizola, A. Fave, A. Ouldabbes, A. Kamaniski, S. Perichon, N.-E. Chabane-Sari, D. Barbier, A. Laugier, Cryst. Res. Technol. **36**, 1005 (2001)
14. J. Kraiem, O. Nichiporuk, P. Papet, A. Fave, A. Kaminski, E. Fourmond, J.-P. Boyeaux, P.-J. Ribeyron, A. Laugier, M. Lemiti, in *Proc. of PVSEC-15, Shangai, China, 2005*, pp. 746−747
15. M. Moreno, M. Labrune, P. Roca i Cabarrocas, Sol. Energy. Mater. & Sol. Cells **94**, 402 (2010)
16. Y. Djeridane, A. Abramov, P. Roca i Cabarrocas, Thin Solid Films **515**, 7451 (2007)
17. A. Abramov, Y. Djeridane, R. Vanderhaghen, P. Roca i Cabarrocas, J. Non-Cryst. Solids **352**, 964 (2006)
18. E.A. Irene, Thin Solid Films **233**, 96 (1993)
19. P. Roca i Cabarrocas, S. Hamma, A. Hadjadj, J. Bertomeu, J. Andreu, Appl. Phys. Lett. **69**, 529 (1996)
20. G.E. Jellison Jr., V.I. Merkulov, A.A. Puretzky, D.B. Geohegan, G. Eres, D.H. Lowndes, J.B. Caughman, Thin Solid Films **377-378**, 68 (2000)
21. B. Kalache, A.I. Kosarev, R. Vanderhaghen, P. Roca i Cabarrocas, J. Appl. Phys. **93**, 1262 (2003)
22. N. Pham, Y. Djeridane, A. Abramov, A. Hadjadj, P. Roca i Cabarrocas, Mat. Sci. Eng. B **159-160**, 27 (2009)
23. N. Pham, A. Hadjadj, P. Roca i Cabarrocas, O. Jbara, F. Kail, Thin Solid Films **517**, 6225 (2009)
24. M. Labrune, M. Moreno, P. Roca i Cabarrocas, Thin Solid Films **518**, 2528 (2010)

2

An overview of molecular acceptors for organic solar cells

Piétrick Hudhomme[a]

L'UNAM Université, Université d'Angers, Laboratoire MOLTECH-Anjou, CNRS UMR 6200, 2 boulevard Lavoisier, 49045 Angers, France

Abstract Organic solar cells (OSCs) have gained serious attention during the last decade and are now considered as one of the future photovoltaic technologies for low-cost power production. The first dream of attaining 10% of power coefficient efficiency has now become a reality thanks to the development of new materials and an impressive work achieved to understand, control and optimize structure and morphology of the device. But most of the effort devoted to the development of new materials concerned the optimization of the donor material, with less attention for acceptors which to date remain dominated by fullerenes and their derivatives. This short review presents the progress in the use of non-fullerene small molecules and fullerene-based acceptors with the aim of evaluating the challenge for the next generation of acceptors in organic photovoltaics.

1 Introduction

The conversion of solar energy into electrical energy is probably one of the most exciting research challenges nowadays. The recent advances in the development of OSCs using carbon-based materials constitute an attractive alternative to silicon-based photovoltaics, essentially because of their potentially low cost and their ease of manufacture. In the last two decades, this tendency to move progressively from inorganic to organic-based materials for photovoltaic applications has emerged as the third generation of photovoltaic materials. These devices that contain organic materials in the active layer include dye-sensitized solar cells [1–4] and donor-acceptor heterojunction based solar cells [5–9]. This last topic has recently been the subject of intensive academic interest and has particularly captured a huge amount of industrial attention due to the rapid increase in power conversion efficiencies (PCE) of these solar cells and their potential for developing flexible, light-weight devices using low-cost production methods. A first significant advance was the contribution made in 1986 with the application of the p-n junction approach in organic photovoltaic cells [10]. A double-layer solar cell with the efficiency close to 1% was obtained by thermal evaporation of successive layers of copper phthalocyanine and a perylene-based small molecule playing the role of donor and acceptor, respectively. A major breakthrough arose with the discovery of fullerene C_{60} in 1985, followed by the demonstration of the photophysical, photochemical and photoinduced elec-

tron transfer properties of fullerene derivatives. In particular, it was demonstrated in 1992 that a π-conjugated polymer was able to transfer electrons efficiently to a C_{60} core giving rise to long-lived charge-separated states [11]. Considerable improvements in fundamental understanding, device construction, processing of the active layer, and optimization of new materials [12] have been made during these last few years and a certified 10.6% PCE was recently reported [13]. Extensive studies have been aimed at designing optimal conjugated polymers as electron donors, and this development has played a crucial role in the dramatic improvement of OSC performance in recent years. On the other hand, fullerene derivatives were rapidly considered as the most efficient acceptors and much less attention has been devoted to other classes of n-type acceptors. In this paper we present a short and non-exhaustive overview of non-fullerene small-molecules in a first part and fullerene-based acceptors in a second part which are developed so far. These two different families of n-type materials used in OSCs, with small-molecules non-fullerene materials [14,15] and fullerene based acceptors [16,17], have been recently and separately reviewed.

2 Principles of organic solar cells

Organic solar cells, based on heterojunctions using p-type donor (D) and n-type acceptor (A) semiconductors, are fabricated by sandwiching the p-type and n-type materials between two different electrodes, such as for examples Indium Tin Oxide (ITO) and aluminium layers for the anode and cathode, respectively. Donor and acceptor

[a] e-mail: `pietrick.hudhomme@univ-angers.fr`

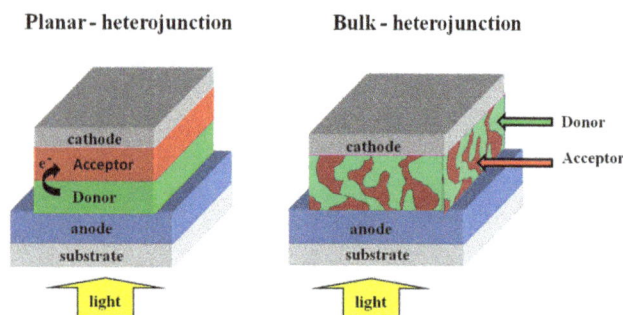

Fig. 1. Schematic representation of an OSC with a planar heterojunction (bilayer structure) or bulk-heterojunction (interpenetrating network of blended donor and acceptor materials).

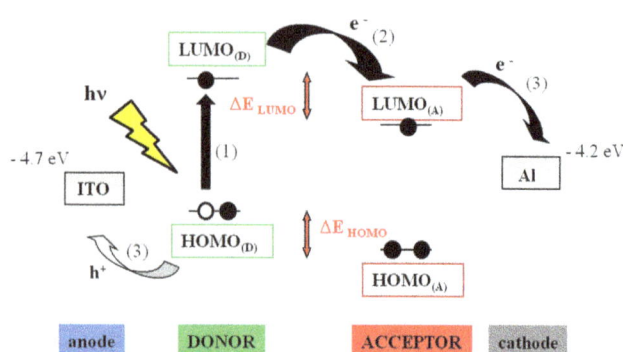

Fig. 2. HOMO-LUMO energy level diagram of donor and acceptor materials in an OSC.

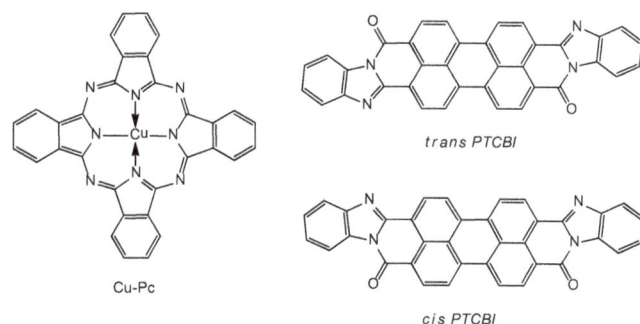

Fig. 3. Molecular structures of donor CuPc and acceptor PTCBI.

lowest unoccupied molecular orbital (LUMO) of the acceptor. If the off-sets of the energy levels of the donor and acceptor materials are higher than the exciton binding energy, photogenerated excitons in the donor side will dissociate by transferring the electron to the LUMO level of the acceptor, while those created in the other side will transfer the hole to the HOMO of the donor. Resulting free hole/electron charge carriers diffuse to the electrodes and are transported through semiconducting materials with the electron reaching the cathode (Al) and the hole reaching the anode (ITO).

3 Non-fullerenes acceptors for organic solar cells

The first heterojunction OSC was reported by Tang where a 3,4,9,10-perylenetetracarboxylic *bis*-benzimidazole (PTCBI) (50 nm) as the acceptor and copper phthalocyanine (CuPc) (30 nm) as the donor were successively evaporated between ITO-coated glass substrate and Ag electrode (Fig. 3). The resulting bilayer device provided a PCE of 0.95%, with a relatively high fill factor (FF) of 0.65 ($V_{oc} = 0.45$ V; $J_{sc} = 2.3$ mA cm^{-2}) under 75 mW cm^{-2} of AM2 illumination [10]. By modification of the device architecture with the same materials, efficiency exceeding 2% was attained [21,22]. It should be noted that the PTCBI acceptor is synthesized and used for devices as a mixture of *cis* and *trans* isomers. But the configuration *cis/trans* does not significantly interfere on the performance of the solar cell. Isomerically pure *cis* and *trans* PTCBI isomers were prepared by repeated fractionation in sulfuric acid. Bilayer devices consisting of CuPc and *trans-*, *cis-*, and mixed-PTCBI cells exhibited PCE of 1.1%, 0.93%, and 0.99%, respectively [23]. The higher efficiency for the *trans* isomer was attributed to a substantially better molecular ordering and a closer packing of *trans*-PTCBI molecules in corresponding films, mainly contributing to a higher exciton diffusion length.

Perylene-3,4:9,10-bis(dicarboximide) (PBI) derivatives are considered as the best n-type organic semiconductors available to date [24,25]. Because of their unique optical, redox, stability properties and charge carrier mobilities, PBI dyes provide significant prospects for investigations in photovoltaic devices [26]. Whereas solution processable perylene acceptors can be prepared by introducing

materials can be deposited as two distinct layers (planar p-n heterojunction) or blended in an homogeneous mixture throughout an interpenetrating network known as bulk-heterojunction (BHJ) which presents a larger interfacial area (Fig. 1). With the aim of covering more effectively the solar spectrum, another useful approach consists in stacking multiple photoactive layers with complementary absorption spectra in series to reach a tandem device [18]. From the technological point of view, the potential for low cost of these OSCs stems from the ability to produce these devices using high-throughput vacuum processing for small molecular weight based solar cells or roll-to-roll solution processing methods for polymer solar cells [19].

When incident photons are absorbed by the active layer, the photovoltaic effect is built following a precise sequence of events: (1) excitation of the donor upon absorption of light; (2) electron transfer from the donor to the acceptor component and formation of an electron-hole pair (exciton) with the positive charge (hole) on the donor and negative charge on the acceptor; and (3) diffusion then migration of the exciton at the donor/acceptor interface and collection of the charges at the corresponding electrodes [20].

From a theoretical point of view, the principal feature of the p-n heterojunction is the built-in potential at the interface between both materials presenting a difference of electronegativities. Photoexcitation of the absorbing material causes, through a photoinduced electron transfer process, the promotion of an electron from the highest occupied molecular orbital (HOMO) of the donor to the

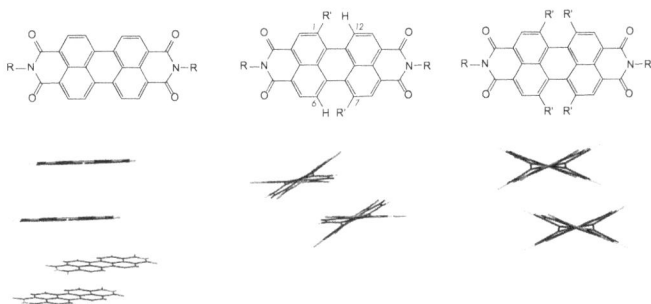

Fig. 4. Representation and X-ray crystallographic structures of planar non-bay substituted PBI (left) [30] and twisted *di*- and *tetra*bay-substituted PBI derivatives.

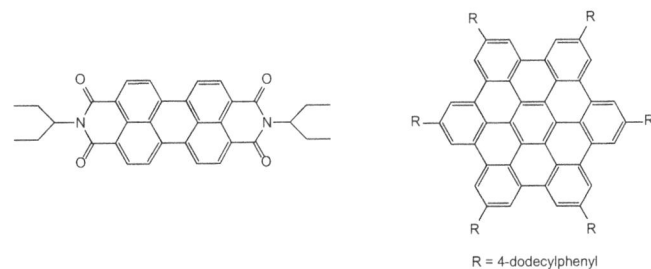

R = 4-dodecylphenyl

Fig. 5. Molecular structures of 2-ethylpropyl PBI and hexabenzocoronene (HBC).

solubilizing groups on the imide nitrogen atoms by keeping the planarity of the backbone, the synthesis of PBI derivatives bearing substituents in the 1,12- and 6,7-positions (the so-called bay region) has considerably promoted the development of emerging applications thanks to their increased solubility in organic solvents and the possibility to tune the electronic properties [24, 25, 27]. The introduction of substituents enforces a considerable twisting of the perylene skeleton as a result of electrostatic repulsion and steric effects among the substituents in positions 1,7 ($31.7°$ for 1,7-dibromoPBI) [28] or 1,6,7,12 positions ($37°$ for 1,6,7,12-tetrachloroPBI) [29] (Fig. 4).

An interesting work was carried out on ethylpropyl-substituted PBI at the imide periphery which was blended with a variety of donor polymers using spin coating techniques to prepare photovoltaic devices. It was shown that the tendency of such PBI acceptor to form crystalline domains was limiting the efficiencies of devices by creation of electron traps thus reducing the photocurrent [31,32]. On the contrary, their high crystallinity could be exploited when this ethylpropyl-substituted PBI was blended with discotic liquid crystal alkylated hexabenzocoronene (HBC) (Fig. 5) [26,33,34]. An external quantum efficiency (EQE) of 34% corresponding to about 2% PCE was observed for the PBI/HBC blend under illumination at 490 nm.

Interesting results were recently reported with the device using the 1,7-*bis*(4-*tert*-butylphenoxy)-bay substituted PBI derivative containing pyrene moieties at the imide positions. This Pyrene-PBI-Pyrene triad acceptor was blended with a phenylenevinylene based oligomer playing the role of the donor (Fig. 6). The corresponding device using the donor/acceptor couple in a

1:3.5 ratio yielded a PCE of 1.87% ($V_{oc} = 0.98$ V, $J_{sc} = 4.15$ mA cm^{-2}, $FF = 0.46$) which could reached 3.17% by insertion of a ZnO layer between the active organic layer and the cathode in order to enhance electron collection [35]. By introducing methylacenaphtho[1,2-b]pyrazine-8,9-dicarbonitrile at the PBI periphery and benzo[c][1,2,5]-selenadiazole in the donor system, the device (ITO/PEDOT-PSS/Donor: Acceptor/Al) provided an efficiency of 3.88% ($V_{oc} = 0.9$ V; $J_{sc} = 8.30$ mA cm^{-2}; $FF = 0.52$). This performance was largely attributed to the wide $400-800$ nm absorption range of the blend. This corresponds to the highest efficiency reported so far for a rylene-*bis*imide-based acceptor and also for a non-fullerene based organic BHJ solar cell [36].

Other classes of acceptors have been recently developed to be incorporated in OSCs. One of these corresponds to the diketopyrrolopyrrole (DPP) moiety. DPP-based polymers have recently received large attention for their excellent performance as donor materials in BHJ solar cells with an efficiency of 4.7% in combination with a fullerene derivative [37]. Interestingly the introduction of electron withdrawing groups such as fluoro or fluoroalkyl groups enhanced the DPP electron affinity to reach acceptor materials. For example, trifluoromethylphenyl groups have been introduced onto the DPP core and when blended in a BHJ device with the donor poly(3-hexylthiophene) (P3HT), a PCE of 1.0% ($V_{oc} = 0.81$ V, $J_{sc} = 2.36$ mA cm^{-2}, $FF = 0.52$) was achieved [38].

Similarly, metallophthalocyanines are widely known for their p-type conductivity but the introduction of fluorine atoms at the periphery of the aromatic rings increased their n-type character. A series of fluorinated boron subphthalocyanines used as acceptors in bilayer OSC devices were associated with phthalocyanine-based donors leading to PCE close to 1% [39]. Small molecular acceptors combining the highly electron deficient 2-vinyl-4,5-dicyanoimidazole moiety named Vinazene and benzothiadiazole were developed as processable oligomers for photovoltaics applications [40]. These oligomers displayed a wide range of optical and electronic properties thanks to the modification of the central aromatic moiety. Best results were described by blending this *bis*-vinazene – benzothiadiazole oligomer with an octylphenyl-substituted polythiophene giving a device for which a PCE of 1.4% ($V_{oc} = 0.62$ V, $J_{sc} = 5.5$ mA cm^{-2}, $FF = 0.4$) could be obtained.

A recent innovative approach introduced the concept of gain of aromaticity as a driving force to reach efficient small molecule acceptors represented by the use of a 9,9'-bifluorenylidene (99'BF) polycycle scaffold [41]. The addition of one electron onto the central double bond issued from the electron transfer from the donor should be highly favorable thanks to the formation of a less strained and twisted system which is characterized by a gain of aromaticity due to the generation of a 14-π electron structure. Preliminary BHJ solar cells (ITO/PEDOT-PSS/P3HT: 99'BF/Ba/Al) showed an impressive 1.7% PCE, and a relatively high V_{oc} of 1.10 V. Further promising results are expected in a near future

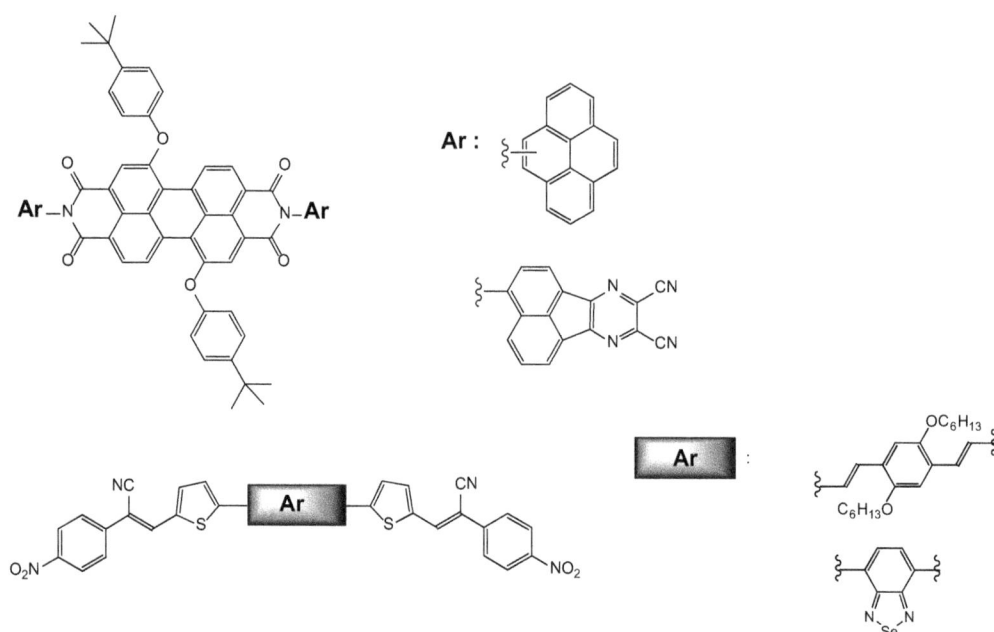

Fig. 6. Molecular structures of 1,7-*bis*(4-*tert*-butylphenoxy) bay-substituted PBI derivative acceptor and small-molecule based donor oligomer.

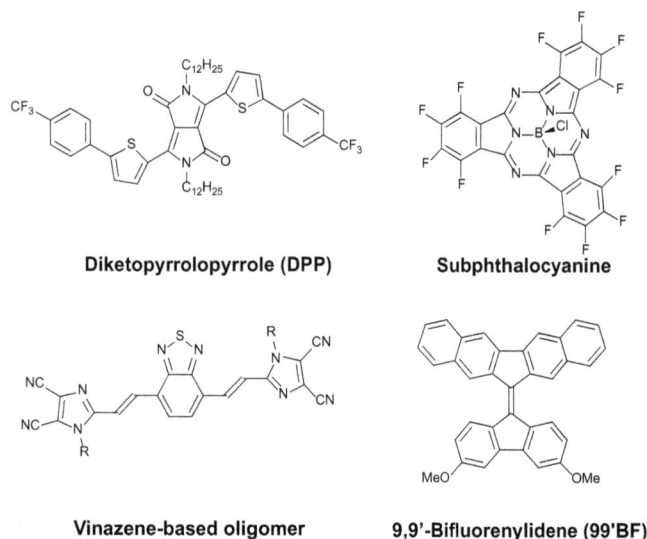

Fig. 7. Molecular structures of other families developed as acceptors.

using this versatile 9,9′-bifluorenylidene backbone considering that the possibility of functionalization on theoretically twelve positions constitutes advantages of particular interest for the development by organic chemists of this scaffold.

4 Fullerenes acceptors for organic solar cells

It is clear that at the present time and in the field of organic photovoltaics, fullerene derivatives play a dominant role as acceptor materials. Main reasons for the choice of fullerenes are their favorable LUMO energy and reversible

reduction leading to stable reduced species, their excellent electron transport properties, and their spherical shape allowing a three-dimensional charge transport, perfectly in agreement with this of the three-dimensional hole transport ability of conjugated polymers. Nevertheless, to be applied in the fabrication process of OSCs, the functional fullerene requires sufficient solubility in organic solvents, such as toluene, chlorobenzene, or o-dichlorobenzene. With the aim of designing new efficient fullerene acceptors, it is important to have in mind some critical factors known to affect the performance of BHJ solar cells. First, the open circuit voltage (V_{oc}) was found to correlate directly with the acceptor strength of the fullerene derivative [42]. Moreover, in order to reach highest efficiencies, practically around 10%, the difference between both LUMO levels of donor and acceptor needs to be around 0.3 eV [43] (Fig. 8). Consequently, the energy level of the fullerene derivative has to be engineered to maximize the ΔE LUMO(A) – HOMO(D) to achieve high V_{oc}, while maintaining proper LUMO offset [ΔE LUMO(D) – LUMO(A)] to facilitate exciton dissociation. In a first approach, the V_{oc} was proposed to be proportional to the difference between the HOMO level of the donor and the LUMO level of the acceptor [7]. Then it was more precisely evidenced that the maximum attainable and empirical limit for the V_{oc} is equal to the optical band gap energy minus ~0.60−0.66 V and that the V_{oc} correlates with the energy of the charge transfer states in donor-acceptor blend [44–46].

Soluble fullerene derivatives with the 6,6-phenyl-C_{61}-butyric acid methyl ester ([60]PCBM), historically synthesized by Hummelen et al. [47], and its C_{70} analogue ([70]PCBM) are the most widely used acceptors for BHJ solar cells. A major difference between C_{60}

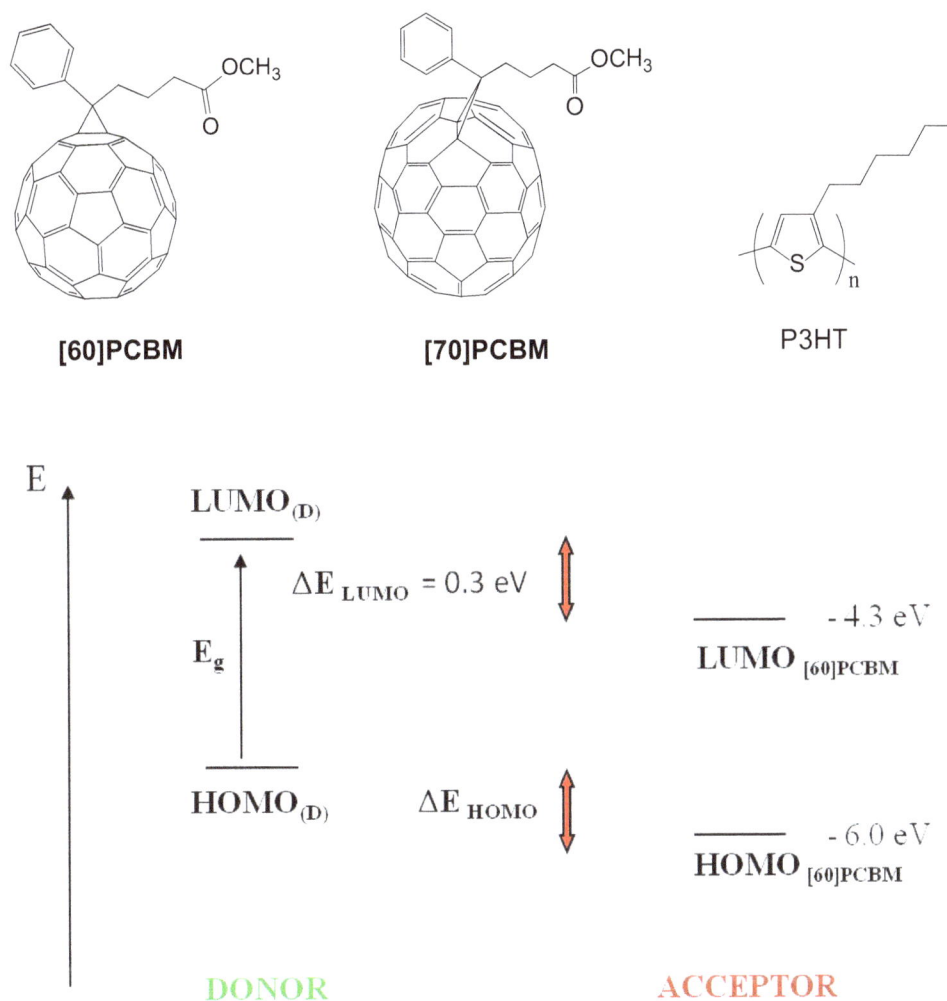

Fig. 8. Structures of PCBM molecules and schematic energy diagram with the bandgap energy (Eg) and the energy differences (ΔE) in a donor: [60]PCBM bulk-heterojunction.

and [60]PCBM concerns their relative acceptor strength, [60]PCBM appearing to be a less effective electron acceptor but more soluble in organic solvents than pristine C_{60}. The LUMO of [60]PCBM was reported to be between -3.7 eV [48] and -4.3 eV [43] because of the different measurement techniques, either in solution or solid form, or with different standards to estimate the values. Recently, the LUMO and HOMO levels of fullerene acceptors measured under the same conditions were correlated using the PCBM reference molecule with its widely accepted LUMO (-4.30 eV) and HOMO (-6.00 eV) energy levels [17]. Considering this LUMO value for [60]PCBM acceptor and if the bandgap of the polymer is in the range of $1.2-1.7$ eV, this would correspond to donor HOMO levels of -5.2 to -5.7 eV.

Since the advent of [60]PCBM, the blend which associated P3HT has been extensively investigated and exhibits a reproducible PCE around 5% [49]. One drawback of fullerenes is their relatively low light absorption in the visible region of the spectrum. Although [70]PCBM partially solved this problem [50], a major breakthrough towards efficient organic devices was obtained with the

development of low bandgap conjugated polymers able to harvest more light and to optimize corresponding energy levels to attain higher V_{oc} [51–53]. When blended with [60]PCBM or [70]PCBM, efficiencies have increased up to 7–8% [54] with a record which has recently reached over 10% [13].

Nevertheless, due to the different energetic HOMO and LUMO levels to be respected, it clearly appeared that PCBM was not the best acceptor for solar cells containing near infrared absorbing polymers. Consequently, alternative fullerene materials have to be developed and efforts are currently being made by organic chemists to obtain alternatives to [60]PCBM and [70]PCBM acceptors [12].

In the last decade, synthetic efforts were strengthened for chemical engineering on PCBM to change the properties of the fullerene π-system and to design novel fullerene acceptors with higher-lying LUMO levels compared with PCBM (Fig. 9). A systematic study of slight variations on the aryl or alkyl groups of PCBM can cause a significant change in its physical properties and in particular its solubility in organic solvents with strong effects on the morphology of the blend and the efficiency of solar cells.

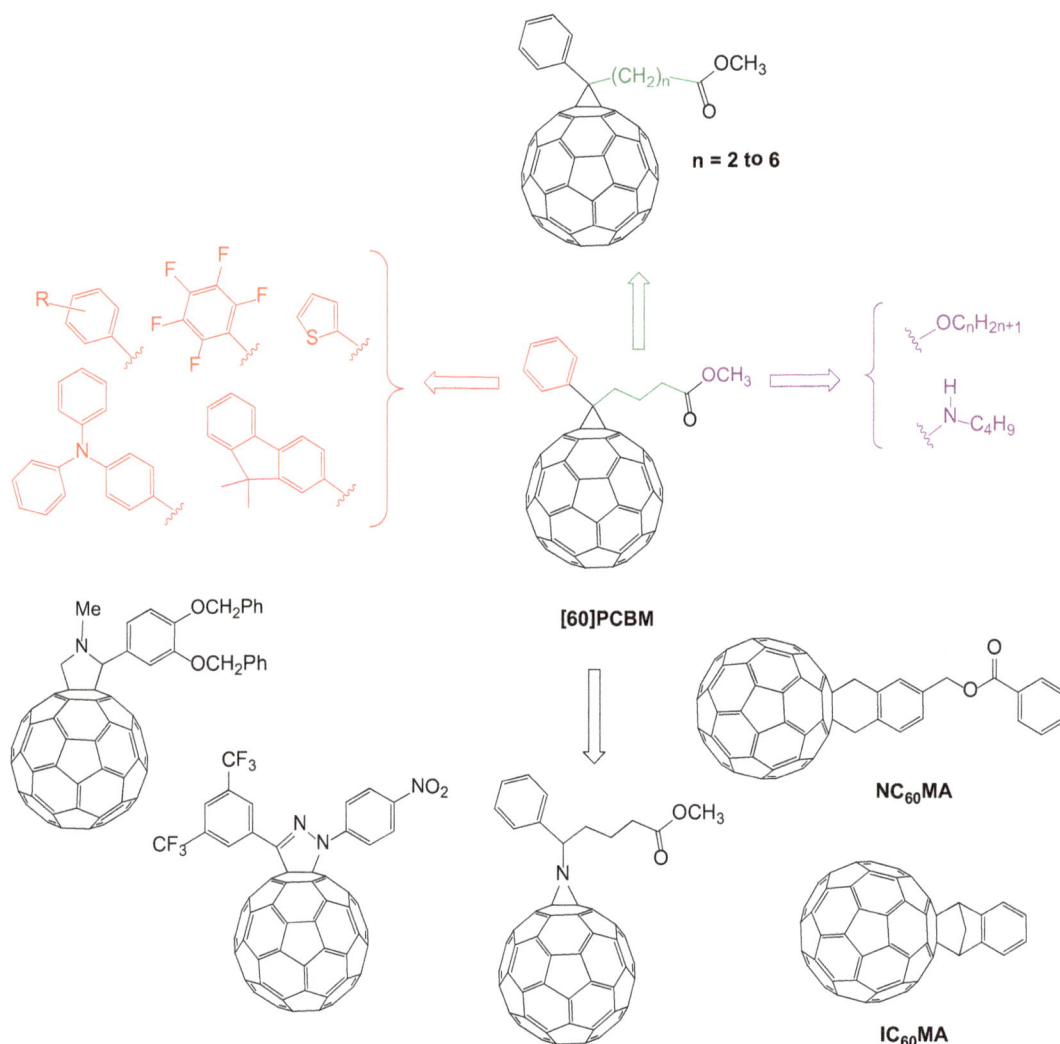

Fig. 9. An overview of modifications leading to C_{60} monoadducts.

It was concluded that the best combinations are those where donor and fullerene materials are of similar and sufficiently high solubility in the solvent used for the deposition of the active layer [55].

In order to evaluate the influence of electron-donating and -withdrawing groups on the electrochemical properties of [60]PCBM, the phenyl ring of [60]PCBM was substituted with different functional groups and fluorine atoms were shown to influence the most the first reduction potential [56]. On the other hand, the phenyl group of PCBM was replaced by a thienyl group [57] and the resulting BHJ solar cell based on P3HT: Th[60]CBM exhibited a PCE of 3.97% which was comparable to the reference P3HT: [60]PCBM device [58]. Another alternative consisted in replacing the phenyl ring with the bulky triphenylamine or 9,9-dimethylfluorene groups [59]. These acceptors showed electron mobility comparable to [60]PCBM, while solar cells using P3HT as the donor exhibited a high PCE of around 4% with a significant enhanced thermal stability thanks to the higher glass-transition temperature of these fullerene derivatives. Since

a slight structural modification of PCBM can influence the photovoltaic performance significantly, a series of [60]PCBM-like fullerene derivatives were synthesized in which the alkyl chain in PCBM was changed from 3 to 7 carbon atoms [60]. Devices based on P3HT as the donor were fabricated and PCE varied from 2.3 to 3.7%. These results indicated that the alkyl chain length significantly influenced the absorption intensity, the electron mobility, the morphology of the films, and the P3HT: PCBM analogue interface structure, so that it clearly influenced the photovoltaic performance of the solar cell. To enhance the solubility of [60]PCBM in organic solvents, several analogues were synthesized with different ester alkyl chains, varying from C_1 to C_{16} [61,62]. The length of the chain did not influence optical and electrochemical properties but it was noted that the morphology of the film was improved with length and the more soluble acceptors were able to produce a good quality film either alone or as a blend with donor components. The effect of the chain length on the terminal ester functionality was also established [63] as well as the replacement of the ester group by an acid

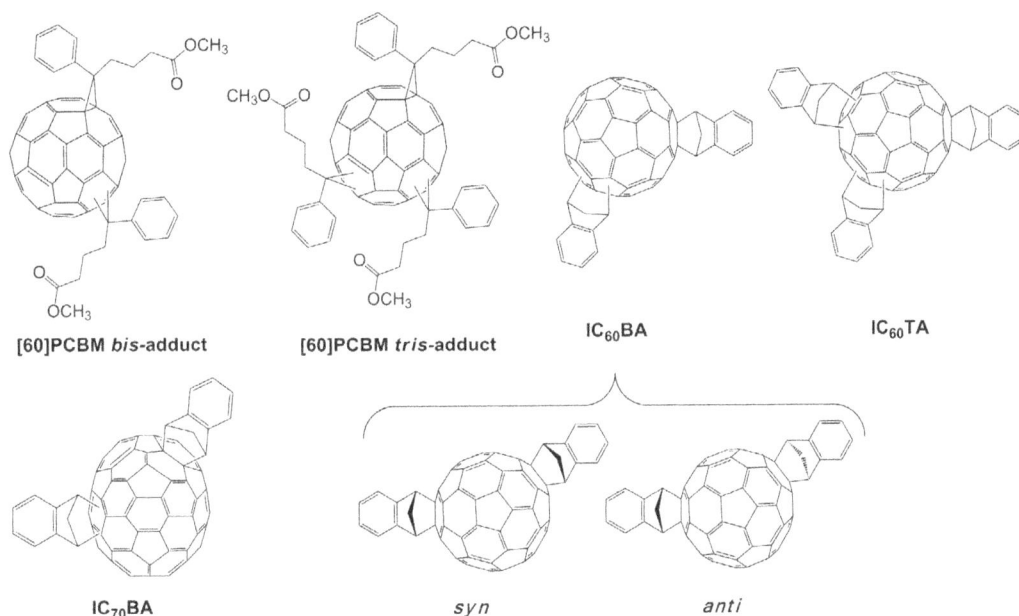

Fig. 10. Molecular structures of *bis-* and *tris-*adducts currently used in OSCs.

or amide functionality in order to evaluate the role of hydrogen bonds in the aggregation of fullerenes [64, 65]. Among the different strategies investigated for modification of the attachment mode on the fullerene moiety using cycloaddition reaction [66–70] (Fig. 9), new soluble fullerene derivatives were synthesized using the [4+2] Diels-Alder reaction onto C_{60} [71] to reach $IC_{60}MA$ [72] and $NC_{60}MA$ cycloadducts [73]. Blends P3HT: $NC_{60}MA$ were fabricated which exhibited a PCE up to 4.5% ($V_{oc} =$ 0.65 V, J_{sc} = 11.3 mA cm^{-2}, FF = 0.57, AM 1.5 G 80−100 mW cm^{-2}). This suggested that excellent alternatives to PCBM could be reasonably considered in the future.

The maximum V_{oc} delivered by the BHJ in polymer: fullerene solar cells being correlated to the difference between the energy of the HOMO of the donor and the LUMO of the acceptor, some research works have focused on the strategy to design fullerene C_{60} *bis-*adducts raising directly the LUMO energy level with an increase of the V_{oc} (Fig. 10). An increase of about 100 meV for the *bis*[60]PCBM LUMO level was reported [74], and the corresponding P3HT: *bis*[60]PCBM device exhibited a PCE of 4.5% (V_{oc} = 0.72 V, J_{sc} = 0.91 mA cm^{-2}, FF = 0.68, AM 1.5 G 100 mW cm^{-2}), this being 1.2 larger than that of similar P3HT: [60]PCBM solar cells. Consequently, the *bis-*adduct isomer mixture could be used without further separation of the theoretically eight regioisomers to yield enhanced cell performance compared to that of [60]PCBM.

A remarkable indene-C_{60} *bis-*adduct ($IC_{60}BA$) was more recently invented in 2006 by Laird et al. (Plextronics & NanoC patent) using a Diels-Alder cycloaddition of indene to C_{60} as a mixture of multiple isomers, considering that two *syn* and *anti* configurations could exist for each regioisomer [75]. The $IC_{60}BA$ acceptor is characterized by a LUMO energy level 0.17 V higher than

that of [60]PCBM [76]. The solar cell based on P3HT: $IC_{60}BA$ showed a high V_{oc} of 0.84 V and an impressive PCE of 5.1% (V_{oc} = 0.84 V, J_{sc} = 9.43 mA cm^{-2}, FF = 0.64, AM 1.5 G 100 mW cm^{-2}) [75] or 5.44% (V_{oc} = 0.84 V, J_{sc} = 9.7 mA cm^{-2}, FF = 0.67, AM 1.5 G 100 mW cm^{-2}) [76] compared to the 3.3 or 3.88% for a reference solar cell based on [60]PCBM, respectively. After optimization of the device morphology, a PCE of 6.48% was achieved (V_{oc} = 0.84 V, J_{sc} = 10.61 mA cm^{-2}, FF = 0.73, AM 1.5 G 100 mW cm^{-2}) as the highest values reported in the literature so far for P3HT-based solar cells [77]. Reproducibility on twenty P3HT: $IC_{60}BA$ devices from 6.06 to 6.76% indicated that the photovoltaic performance of ICBA as acceptor is greatly superior to that of the traditional [60]PCBM with a PCE of 3.73% obtained in the same experimental conditions. An efficient way to improve OSCs efficiency consists in using the tandem structure in order to cover a broad part of the emission solar spectrum. By combining P3HT: $IC_{60}BA$ in a front cell and PDTB-DFBT: [60]PCBM in a rear cell, with PDTB-DFBT acting as a benzothiadiazole based low bandgap polymer, a certified 10.6% PCE was very recently attained (V_{oc} = 1.53 V, J_{sc} = 10.1 mA cm^{-2}, FF = 0.68, AM 1.5 G 100 mW cm^{-2}) [13]. The indene-C_{70} *bis-*adduct ($IC_{70}BA$) analogue was synthesized and the LUMO energy level of $IC_{70}BA$ was shown to be 0.19 eV higher than that of [70]PCBM [78]. P3HT: $IC_{70}BA$ devices showed a higher V_{oc} of 0.84 V and higher PCE of 5.64%, while the devices based on P3HT: [60]PCBM (V_{oc} = 0.59 V) and P3HT: [70]PCBM (V_{oc} = 0.58 V) blends displayed PCE of 3.55% and 3.96%, respectively.

Nevertheless, a limiting factor to this polymer: C_{60} *bis-*adducts strategy was evidenced for some blends where a simultaneous dramatic loss in the J_{sc} and lower electron mobilities resulted in a reduced PCE for most polymers other than P3HT [79]. It was identified that the principal

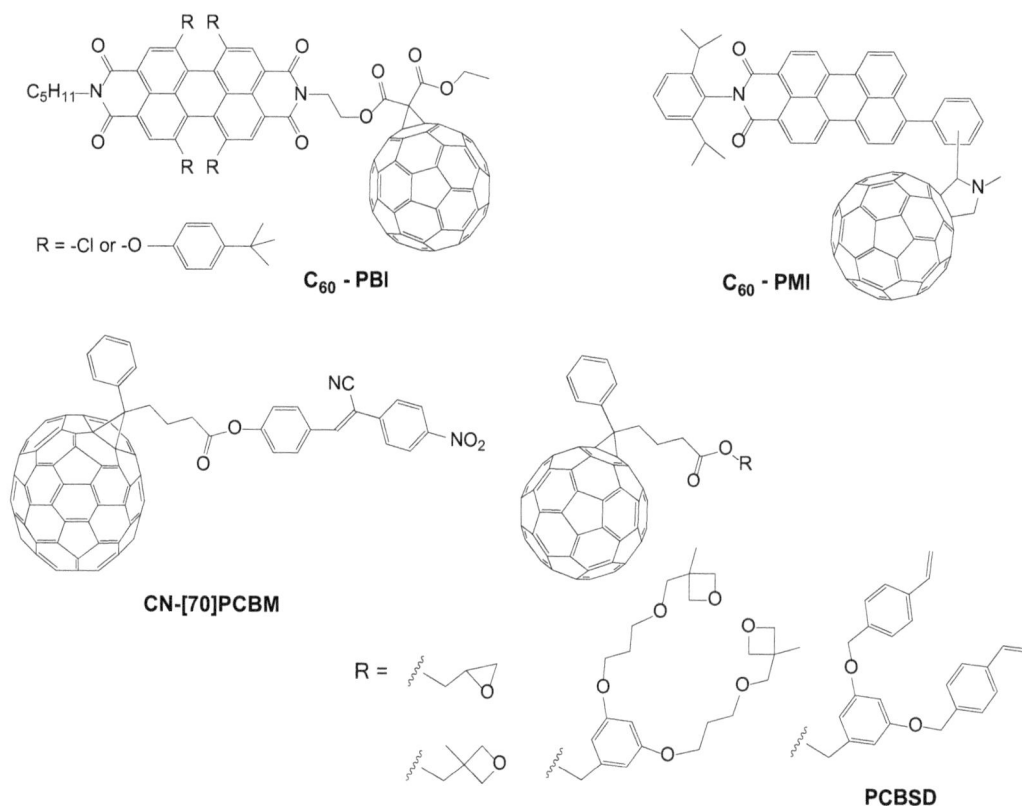

Fig. 11. Examples of highly absorbing and cross-linkable fullerenes for applications in OSCs.

reason was the correlation between J_{sc} and the free energy of the photoinduced charge generation ΔG_{CT}. An increase of this parameter ΔG_{CT} was shown to eventually inhibit the photoinduced electron transfer between the polymer and the fullerene *bis*-adduct [80].

In addition to the systematic decrease of the electron affinity of the acceptor with the number of additions onto the fullerene moiety, it was shown that the introduction of a third unit impacts all device parameters and performance. For example, the blend P3HT: [60]PCBM *tris*adduct, although having a high V_{oc} of 0.81 V, showed significantly reduced efficiency due to the deterioration of electron transport in the fullerene phase [81]. In the indene series, the V_{oc} values increase effectively with the degree of substitution for BHJ solar cells fabricated with P3HT (0.65, 0.83 and 0.92 V for $IC_{60}MA$, $IC_{60}BA$ and $IC_{60}TA$, respectively), but the P3HT: ICTA based device exhibited a much lower efficiency (1.56%) compared with similar device P3HT: ICBA (5.26%) [82]. It was suggested that an increase of the number of solubilizing groups added to the C_{60} decreased the electron mobility due to significant structural disorder and hindered electron transport within the fullerene phase resulting in lower current and fill factor [83].

Innovative approaches need to be applied for OSCs to cover the visible and near infrared region of the solar spectrum. One route, as an alternative strategy to the development of low bandgap polymers, consists in the improvement of the light-harvesting ability of the fullerene derivatives. The principle of associating a dye attached to

the fullerene was proposed to reach C_{60}-PBI or C_{60}-PMI dyads as super-absorbing fullerenes (Fig. 11) [84–87]. Even if the PCE were low using this dyad as an acceptor, the role of the perylene*bis*imide (PBI) acting as a light-harvesting antenna was evidenced by an efficient intramolecular energy transfer occurring from PBI onto C_{60} and the relationship between electrochemical properties of the dyad and the P3HT: C_{60}-PBI blend efficiency was evidenced [88]. Another efficient example was recently reported by modification of [70]PCBM with introduction of a cyanovinylene 4-nitrophenyl moiety. Resulting CN-[70]PCBM showed broader and stronger absorption in the visible region (350−550 nm) of the solar spectrum. The P3HT: CN-[70]PCBM blend showed a PCE of 4.88%, which is higher than that of devices based on [70]PCBM as the electron acceptor (3.23%). An increase in both the J_{sc} and the V_{oc} were noted thanks to the stronger light absorption of CN-[70]PCBM and a LUMO level higher by 0.15 eV compared to that of [70]PCBM [89].

Among aspects other than efficiency, improved processing, cost, and stability have been identified as important for the future development of OSCs [90]. It remains highly challenging to develop such devices that can provide a high PCE while maintaining good ambient stability, since prolonged exposure to air rapidly reduces the performance of unencapsulated conventional devices. The incorporation of an additional n-type C_{60} derivative as an interlayer or in the organic active layer might improve the device performance and the long-term stability thanks to the establishment of a thermal chemical cross-linking process. In this topic, main results involved the synthesis of cross-linkable fullerene materials for which epoxyde [91], oxetane [92], or styryl groups have been introduced (Fig. 11). In the latter case, the[60]PCB styryl dendron ester (PCBSD) was incorporated as an interlayer in P3HT: [60]PCBM [93] or P3HT: $IC_{60}BA$ [94] solar cells. Whereas the P3HT: $IC_{60}BA$ inverted device was leading to a remarkable PCE of 4.8%, the incorporation of PCBSD yielded to an exceptional record PCE of 6.2% then 7.3% using cross-linked-PCBSD nanorods [95].

5 Conclusion

Organic solar cells have shown a tremendous progress over the past decade with an important increase in the power coefficient efficiency. In parallel to the redaction of this mini-review, an accredited new record with 12% efficiency was achieved by Heliatek industry for organic photovoltaics[1]. Moreover, the estimation of the efficiency limits was very recently revisited and theoretical models suggested that $20-24\%$ could be attained in organic photovoltaics [96]. Research efforts have been dominated by the optimization of the materials by tuning their optical and electronic properties, the design and the control of nanoscale morphology of devices. Whereas the synthesis of p-type donor materials was particularly fruitful, n-type acceptors have been less regarded. Non-fullerene based acceptors are today gaining attention and the new challenge for them would be to attain in the near future the efficiency of fullerene derivatives for which PCBM and more recently ICBA are leaders. Whereas C_{60} was regarded two decades ago as an exotic material, costs have been dramatically reduced (20 $ per gram in 2013!) and their derivatives are now compatible for an industrial application in organic photovoltaics. In searching for the next generation of fullerene acceptors, some efforts should be directed to derivatives which possess a broader absorption coverage with high extinction coefficient, a controlled LUMO level to achieve the highest V_{oc}, a high electron mobility, and the development of derivatives which do not aggregate should also increase the stability of the device. Chemical and technological challenges will have to be overcome, but there is undoubtedly an exciting future for organic chemists to offer new acceptors for the development of organic photovoltaics.

The author acknowledges funding from the Agence Nationale de la Recherche (ANR) for financial support to the HABISOL 2010-003 (PROGELEC) program CEPHORCAS.

References

1. B. O'Regan, M. Grätzel, Nature **336**, 737 (1991)
2. M. Grätzel, Nature **414**, 338 (2001)
3. M. Grätzel, Acc. Chem. Res. **42**, 1788 (2009)

4. A. Hagfeldt, G. Boschloo, L. Sun, L. Kloo, H. Pettersson, Chem. Rev. **110**, 6595 (2010)
5. S. Günes, H. Neugebauer, N.S. Sariciftci, Chem. Rev. **107**, 1324 (2007)
6. B.C. Thompson, J.M.J. Fréchet, Angew. Chem. Int. Ed. **47**, 58 (2008)
7. G. Dennler, M.C. Scharber, C.J. Brabec, Adv. Mater. **21**, 1323 (2009)
8. R. Po, M. Maggini, N. Camaioni, J. Phys. Chem. C **114**, 695 (2010)
9. C.J. Brabec, S. Gowrisanker, J.J.M. Halls, D. Laird, S. Jia, S.P. Williams, Adv. Mater. **22**, 3839 (2010)
10. C.W. Tang, Appl. Phys. Lett. **48**, 183 (1986)
11. N.S. Sariciftci, L. Smilowitz, A.J. Heeger, F. Wudl, Science **258**, 1474 (1992)
12. F.G. Brunetti, R. Kumar, F. Wudl, J. Mater. Chem. **20**, 2934 (2010)
13. J. You, L. Dou, K. Yoshimura, T. Kato, K. Ohya, T. Moriarty, K. Emery, C.C. Chen, J. Gao, G. Li, Y. Yang, Nat. Commun. **4**, 1446 (2013)
14. P. Sonar, J.P.F. Lim, K.L. Chan, Energy Environ. Sci. **4**, 1558 (2011)
15. J.E. Anthony, Chem. Mater. **23**, 583 (2011)
16. P. Hudhomme, J. Cousseau, in *Fullerenes, principles and applications*, edited by F. Langa, J.-F. Nierengarten, 2nd edn. (RSC Publishing, 2011), Vol. 12, p. 416
17. C.-Z. Li, H.-L. Yip, A.K.-Y. Jen, J. Mater. Chem. **22**, 4161 (2012)
18. T. Ameri, G. Dennler, C. Lungenschmied, C.J. Brabec, Energy Environ. Sci. **2**, 347 (2009)
19. A. Mishra, P. Bäuerle, Angew. Chem. Int. Ed. **51**, 2020 (2012)
20. J.-L. Brédas, J.E. Norton, J. Cornil, V. Coropceanu, Acc. Chem. Res. **42**, 1691 (2009)
21. P. Peumans, V. Bulovic, S.R. Forrest, Appl. Phys. Lett. **76**, 2650 (2000)
22. A. Yakimov, S.R. Forrest, Appl. Phys. Lett. **80**, 1667 (2002)
23. S.B. Rim, R.F. Fink, J.C. Schöneboom, P. Erk, P. Peumans, Appl. Phys. Lett. **91**, 173504 (2007)
24. F. Würthner, Chem. Commun., 1564 (2004)
25. C. Huang, S. Barlow, S.R. Marder, J. Org. Chem. **76**, 2386 (2011)
26. L. Schmidt-Mende, A. Fechtenkotter, K. Müllen, E. Moons, R.H. Friend, J.D. MacKenzie, Science **293**, 1119 (2001)
27. L. Perrin, P. Hudhomme, Eur. J. Org. Chem. **28**, 5427 (2011)
28. P. Rajasingh, R. Cohen, E. Shirman, L.J.W. Shimon, B. Rybtchinski, J. Org. Chem. **72**, 5973 (2007)
29. S. Leroy-Lhez, J. Baffreau, L. Perrin, E. Levillain, M. Allain, M.-J. Blesa, P. Hudhomme, J. Org. Chem. **70**, 6313 (2005)
30. O. Guillermet, M. Mossoyan-Deneux, M. Giorgi, A. Glachant, J.C. Mossoyan, Thin Solid Films **514**, 25 (2006)
31. J.J. Dittmer, R. Lazzaroni, P. Leclère, P. Moretti, M. Granstrom, K. Petritsch, E.A. Marseglia, R.H. Friend, J.L. Brédas, H. Rost, A.B. Holmes, Sol. Energy Mater. Sol. Cells **61**, 53 (2000)
32. J.J. Dittmer, E.A. Marseglia, R.H. Friend, Adv. Mater. **12**, 1270 (2000)
33. J. Li, M. Kastler, W. Pisula, J.W.F. Robertson, D. Wasserfallen, A.C. Grimsdale, J. Wu, K. Müllen, Adv. Funct. Mater. **17**, 2528 (2007)

[1] http://www.heliatek.com/

34. G. De Luca, A. Liscio, M. Melucci, T. Schnitzler, W. Pisula, C.G. Clark, L.M. Scolaro, V. Palermo, K. Müllen, P. Samori, J. Mater. Chem. **20**, 71 (2010)
35. G.D. Sharma, P. Suresh, J.A. Mikroyannidis, M.M. Stylianakis, J. Mater. Chem. **20**, 561 (2010)
36. G.D. Sharma, P. Balraju, J.A. Mikroyannidis, M.M. Stylianakis, Synth. Met. **160**, 932 (2010)
37. J.C. Bijleveld, A.P. Zoombelt, S.G.J. Mathijssen, M.M. Wienk, M. Turbiez, D.M. de Leeuw, R.A.J. Janssen, J. Am. Chem. Soc. **131**, 16616 (2009)
38. P. Sonar, G.M. Ng, T.T. Lin, A. Dodabalapur, Z.K. Chen, J. Mater. Chem. **20**, 3626 (2010)
39. H. Gommans, T. Aernouts, B. Verreet, P. Heremans, A. Medina, C.G. Claessens, T. Torres, Adv. Funct. Mater. **19**, 3435 (2009)
40. C.H. Woo, T.W. Holcombe, D.A. Unruh, A. Sellinger, J.M.J. Fréchet, Chem. Mater. **22**, 1673 (2010)
41. F.G. Brunetti, X. Gong, M. Tong, A.J. Heeger, F. Wudl, Angew. Chem. Int. Ed. **49**, 532 (2010)
42. C.J. Brabec, A. Cravino, D. Meissner, N.S. Sariciftci, T. Fromherz, M.T. Rispens, L. Sanchez, J.C. Hummelen, Adv. Funct. Mater. **11**, 374 (2001)
43. M.C. Scharber, D. Mühlbacher, M. Koppe, P. Denk, C. Waldauf, A.J. Heeger, C.J. Brabec, Adv. Mater. **18**, 789 (2006)
44. D. Veldman, S.C.J. Meskers, R.A.J. Janssen, Adv. Funct. Mater. **19**, 1939 (2009)
45. K. Vandewal, K. Tvingstedt, A. Gadisa, O. Inganäs, J.V. Manca, Nat. Mater. **8**, 904 (2009)
46. M.A. Faist, T. Kirchartz, W. Gong, R.S. Ashraf, I. McCulloch, J.C. de Mello, N.J. Ekins-Daukes, D.D.C. Bradley, J. Nelson, J. Am. Chem. Soc. **134**, 685 (2012)
47. J.C. Hummelen, B.W. Knight, F. Lepeq, F. Wudl, J. Yao, C.L. Wilkins, J. Org. Chem. **60**, 532 (1995)
48. J.J. Benson-Smith, L. Goris, K. Vandewal, K. Haenen, J.V. Manca, D. Vanderzande, D.D.C. Bradley, J. Nelson, Adv. Funct. Mater. **17**, 451 (2007)
49. M.T. Dang, L. Hirsch, G. Wantz, Adv. Mater. **23**, 3597 (2011)
50. J.M. Kroon, M.M. Wienk, W.J.H. Verhees, J. Knol, J.C. Hummelen, P.A. van Hal, R.A.J. Janssen, Angew. Chem. Int. Ed. **42**, 3371 (2003)
51. Y.-J. Cheng, S.-H.Yang, C.-S. Hsu, Chem. Rev. **109**, 5868 (2009)
52. J. Chen, Y. Cao, Acc. Chem. Res. **42**, 1709 (2009)
53. H. Zhou, L. Yang, W. You, Macromolecules **45**, 607 (2012)
54. Y. Liang, Z. Xu, J. Xia, S.-T. Tsai, Y. Wu, G. Li, C. Ray, L. Yu, Adv. Mater. **22**, E135 (2010)
55. P.A. Troshin, H. Hoppe, J. Renz, M. Egginger, J.Y. Mayorova, A.E. Goryachev, A.S. Peregudov, R.N. Lyubovskaya, G. Gobsch, N.S. Sariciftci, V.F. Razumov, Adv. Funct. Mater. **19**, 779 (2009)
56. F.B. Kooistra, J. Knol, F. Kastenberg, L.M. Popescu, W.J.H. Verhees, J.M. Kroon, J.C. Hummelen, Org. Lett. **9**, 551 (2007)
57. L.M. Popescu, P. van't Hof, A.B. Sieval, H.T. Jonkman, J.C. Hummelen, Appl. Phys. Lett. **89**, 213507 (2006)
58. J.H. Choi, K-I. Son, T. Kim, K. Kim, K. Ohkubo, S. Fukuzumi, J. Mater. Chem. **20**, 475 (2010)
59. Y. Zhang, H.-L. Yip, O. Acton, S.K. Hau, F. Huang, A.K.-Y. Jen, Chem. Mater. **21**, 2598 (2009)
60. G. Zhao, Y. He, Z. Xu, J. Hou, M. Zhang, J. Min, H.-Y. Chen, M. Ye, Z. Hong, Y. Yang, Y. Li, Adv. Funct. Mater. **20**, 1480 (2010)
61. L. Zheng, Q. Zhou, X. Deng, M. Yuan, G. Yu, Y. Cao, J. Phys. Chem. B **108**, 11921 (2004)
62. C.-H. Yang, J.-Y. Chang, P.-H. Yeh, T.-F. Guo, Carbon **45**, 2951 (2007)
63. L. Zheng, Q. Zhou, X. Deng, W. Fei, N. Bin, Z.-X. Guo, G. Yu, Y. Cao, Thin Solid Films **489**, 251 (2005)
64. C. Yang, J.Y. Kim, S. Cho, J.K. Lee, A.J. Heeger, F. Wudl, J. Am. Chem. Soc. **130**, 6444 (2008)
65. C. Liu, Y. Li, C. Li, W. Li, C. Zhou, H. Liu, Z. Bo, Y. Li, J. Phys. Chem. C **113**, 21970 (2009)
66. N. Camaioni, L. Garlaschelli, A. Geri, M. Maggini, G. Possamai, G. Ridolfi, J. Mater. Chem. **12**, 2065 (2002)
67. N. Camaioni, G. Ridolfi, G. Casalbore-Miceli, G. Possamai, L. Garlaschelli, M. Maggini, Sol. Energy Mater. Sol. Cells **76**, 107 (2003)
68. X. Wang, E. Perzon, J.L. Delgado, P. de la Cruz, F. Zhang, F. Langa, M. Andersson, O. Inganäs, Appl. Phys. Lett. **85**, 5081 (2004)
69. X. Wang, E. Perzon, F. Oswald, F. Langa, S. Admassie, M.R. Andersson, O. Inganäs, Adv. Funct. Mater. **15**, 1665 (2005)
70. S.K. Pal, T. Kesti, M. Maiti, F. Zhang, O. Inganäs, S. Hellström, M.R. Andersson, F. Oswald, F. Langa, T. Österman, T. Pascher, A. Yartsev, V. Sundström, J. Am. Chem. Soc. **132**, 12440 (2010)
71. P. Hudhomme, C.R. Chim. **9**, 881 (2006)
72. A. Puplovskis, J. Kacens, O. Neilands, Tetrahedron Lett. **38**, 285 (1997)
73. S.A. Backer, K. Sivula, D.F. Kavulak, J.M.J. Fréchet, Chem. Mater. **19**, 2927 (2007)
74. M. Lenes, G.A.H. Wetzelaer, F.B. Kooistra, S.C. Veenstra, J.C. Hummelen, P.W.M. Blom, Adv. Mater. **20**, 2116 (2008)
75. D.W. Laird, R. Stegamat, H. Richter, V. Vejins, L. Scott, T.A. Lada, Patent US **8**, 217, 260
76. Y. He, H.-Y. Chen, J. Hou, Y. Li, J. Am. Chem. Soc. **132**, 1377 (2010)
77. G. Zhao, Y. He, Y. Li, Adv. Mater. **22**, 4355 (2010)
78. Y. He, G. Zhao, B. Peng, Y. Li, Adv. Funct. Mater. **20**, 3383 (2010)
79. M.A. Faist, S. Shoaee, S. Tuladhar, G.F.A. Dibb, S. Foster, W. Gong, T. Kirchartz, D.D.C. Bradley, J.R. Durrant, J. Nelson, Adv. Energy Mater. (2013), DOI: `10.1002/aenm.201200673`
80. D.Di Nuzzo, G.-J.A.H. Wetzelaer, R.K.M. Bouwer, V.S. Gevaerts, S.C.J. Meskers, J.C. Hummelen, P.W.M. Blom, R.A.J. Janssen, Adv. Energy Mater. **3**, 85 (2013)
81. M. Lenes, S.W. Shelton, A.B. Sieval, D.F. Kronholm, J.C. Hummelen, P.W.M. Blom, Adv. Funct. Mater. **19**, 3002 (2009)
82. H. Kang, C.-H. Cho, H.-H. Cho, T.E. Kang, H.J. Kim, K.-H. Kim, S.C. Yoon, B.J. Kim, Appl. Mater. Interfaces **4**, 110 (2012)
83. A.M. Nardes, A.J. Ferguson, J.B. Whitaker, B.W. Larson, R.E. Larsen, K. Maturová, P.A. Graf, O.V. Boltalina, S.H. Strauss, N. Kopidakis, Adv. Funct. Mater. **22**, 4115 (2012)
84. J. Baffreau, L. Perrin, S. Leroy-Lhez, P. Hudhomme, Tetrahedron Lett. **46**, 4599 (2005)
85. J. Baffreau, S. Leroy-Lhez, P. Hudhomme, M.M. Groeneveld, I.H.M. van Stokkum, R.M. Williams, J. Phys. Chem. A **110**, 13123 (2006)
86. J. Baffreau, S. Leroy-Lhez, N. Vân Anh, R.M. Williams, P. Hudhomme, Chem. Eur. J. **14**, 4974 (2008)

87. J. Baffreau, L. Ordronneau, S. Leroy-Lhez, P. Hudhomme, J. Org. Chem. **73**, 6142 (2008)

88. J. Baffreau, S. Leroy-Lhez, H. Derbal, A.R. Inigo, J.-M. Nunzi, M.M. Groeneveld, R.M. Williams, P. Hudhomme, Eur. Phys. J. Appl. Phys. **36**, 301 (2006)

89. S.P. Singh, Ch.P. Kumar, G.D. Sharma, R. Kurchania, M.S. Roy, Adv. Funct. Mater. **22**, 4087 (2012)

90. M. Jørgensen, K. Norrman, F.C. Krebs, Sol. Energy Mater. Sol. Cells **92**, 686 (2008)

91. M. Drees, H. Hoppe, C. Winder, H. Neugebauer, N.S. Sariciftci, W. Schwinger, F. Schäffler, C. Topf, M.C. Scharber, Z. Zhu, R. Gaudiana, J. Mater. Chem. **15**, 5158 (2005)

92. Y.-J. Cheng, F.-Y. Cao, W.-C. Lin, C.-H. Chen, C.-H. Hsieh, Chem. Mater. **23**, 1512 (2011)

93. C.-H. Hsieh, Y.-J. Cheng, P.-J. Li, C.-H. Chen, M. Dubosc, R.-M. Liang, C.-S. Hsu, J. Am. Chem. Soc. **132**, 4887 (2010)

94. Y.-J. Cheng, C.-H. Hsieh, Y. He, C.-S. Hsu, Y. Li, J. Am. Chem. Soc. **132**, 17381 (2010)

95. C.-Y. Chang, C.-E. Wu, S.-Y. Chen, C. Cui, Y.-J. Cheng, C.-S. Hsu, Y.-L. Wang, Y. Li, Angew. Chem. Int. Ed. **50**, 9386 (2011)

96. R.A.J. Janssen, J. Nelson, Adv. Mater. **25**, 1847 (2013)

3

Modelling on c-Si/a-Si:H wire solar cells: some key parameters to optimize the photovoltaic performance

I. Ngo, M.E. Gueunier-Farret[a], J. Alvarez, and J.P. Kleider

LGEP, UMR 8507 CNRS, SUPELEC, UPMC, Université Paris-Sud 11, 11 rue Joliot-Curie, Plateau de Moulon, 91192 Gif-sur-Yvette Cedex, France

Abstract Solar cells based on silicon nano- or micro-wires have attracted much attention as a promising path for low cost photovoltaic technology. The key point of this structure is the decoupling of the light absorption from the carriers collection. In order to predict and optimize the performance potential of p- (or n-) doped c-Si/ n-(or p-) doped a-Si:H nanowire-based solar cells, we have used the Silvaco-Atlas software to model a single-wire device. In particular, we have noticed a drastic decrease of the open-circuit voltage (V_{oc}) when increasing the doping density of the silicon core beyond an optimum value. We present here a detailed study of the parameters that can alter the V_{oc} of c-Si(p)/a-Si:H (n) wires according to the doping density in c-Si. A comparison with simulation results obtained on planar c-Si/a-Si:H heterojunctions shows that the drop in V_{oc}, linked to an increase of the dark current in both structures, is more pronounced for radial junctions due to geometric criteria. These numerical modelling results have lead to a better understanding of transport phenomena within the wire.

1 Introduction

Solar cells based on Si micro- or nanowire arrays with radial p-n junction are of great interest as potentially low-cost solutions in solar cell production and have been extensively studied in the last years [1–6]. Unlike planar structures, photon absorption and minority carrier collection are decoupled as shown in Figure 1, and are not anymore in competition when optimizing the cell dimensions. In fact, the wires can be grown long enough to optimize light absorption while being small enough in radius to collect photo-generated carriers. This last property enables the use of low purity Si substrates with a short minority carrier diffusion length.

The vapour-liquid-solid (VLS) technique [7] assisted by gold catalyst is one of the most current methods to grow micro- and nano-wires for photovoltaic applications. Despite many advantages in using Au as VLS catalyst (non toxicity, chemical stability, availability), Au is known to diffuse in the nanowires [8] creating deep-level defects in Si and leading to an increase of the carrier recombination rate. Among the large number of catalyst materials that have been tested to replace Au [9], aluminium is a good candidate since it does not create deep-level defects. Nevertheless, Al is a p-type dopant and the wires can be strongly p-doped. It is not obvious to know whether the performances of the solar device are strongly influenced by such a p-type doping level. In that case,

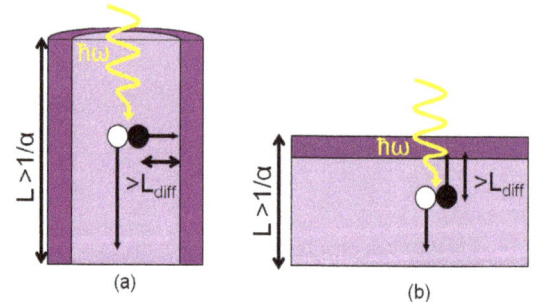

Fig. 1. Structures of solar cells based on (a) a radial p-n junction, (b) a planar p-n junction. Light penetration into the cell is characterized by the parameter $1/\alpha$, α being the wavelength-dependent absorption coefficient. The diffusion length of the generated minority carriers is given by L_{diff}. In the case of the radial junction, light absorption and carrier collection are decoupled.

numerical modelling can be of great help. In this modelling work, we have studied one single-wire Si solar cell which consists in a p-type crystalline silicon (c-Si) core surrounded by a n-type hydrogenated amorphous silicon (a-Si:H) thin layer. Cell efficiency, open-circuit voltage (V_{oc}), short-circuit current (I_{sc}) and fill factor (FF) under AM 1.5 spectrum have been calculated as a function of the p-type doping density (N_a) in the core. A drastic decrease in V_{oc} has been observed when N_a reaches values higher than 1×10^{17} cm^{-3}. An insight into the transport properties of the wire in the dark can help to understand this behavior.

[a] e-mail: farret@lgep.supelec.fr

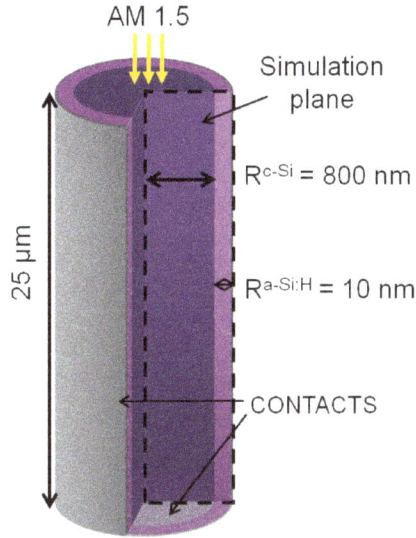

Fig. 2. Single-wire geometry of the model.

2 Silicon nanowire modelling: geometry and physical parameters

Simulations of the electrical characteristics of a single-wire c-Si/a-Si:H radial heterojunction solar cell were performed using ATLAS Silvaco software [10]. The geometry of the radial p-n junction studied here is given in Figure 2. The design was defined as a simulation plane in 2-dimensional cylindrical coordinates which models the 3D cylindrical structure. The wire dimensions were chosen according to experimental data of fabricated wire array cells, with a wire length of 25 μm and a radius of 810 nm. The p-type c-Si core (radius of 800 nm) is surrounded by a thin layer (10 nm) of n-type doped a-Si:H. Ohmic contacts were used, one at the bottom of the structure on the crystalline base and another one around the wire on the amorphous layer. At the bottom contact, the electron surface recombination velocity has been taken equal to 50 cm s^{-1}. For the surrounded contact, flat band conditions were imposed.

For the a-Si:H layer, the distribution of the density of states (DOS) is made of two exponential band tails (valence and conduction band tails) and two Gaussian distributions of deep defect states, one donor-like and one acceptor-like. The DOS and doping concentration were adjusted such that the Fermi level at room temperature was set at 0.2 eV below the conduction band edge E_c. The electron and hole mobilities were taken equal to $\mu_n = 20$ cm^2 V^{-1} s^{-1} and $\mu_p = 5$ cm^2 V^{-1} s^{-1}, respectively. In the c-Si core, we fixed a value of 1 ms for the carrier lifetime which corresponds to a diffusion length much bigger than the wire radius. The doping density N_a was varied between 1×10^{15} cm^{-3} and 1×10^{19} cm^{-3} such that the Fermi level at room temperature moves from 0.25 eV above the valence band edge E_v to a position very close to E_v, respectively.

The simulation is based on the resolution of Poisson's equation and electron and hole continuity equations. The

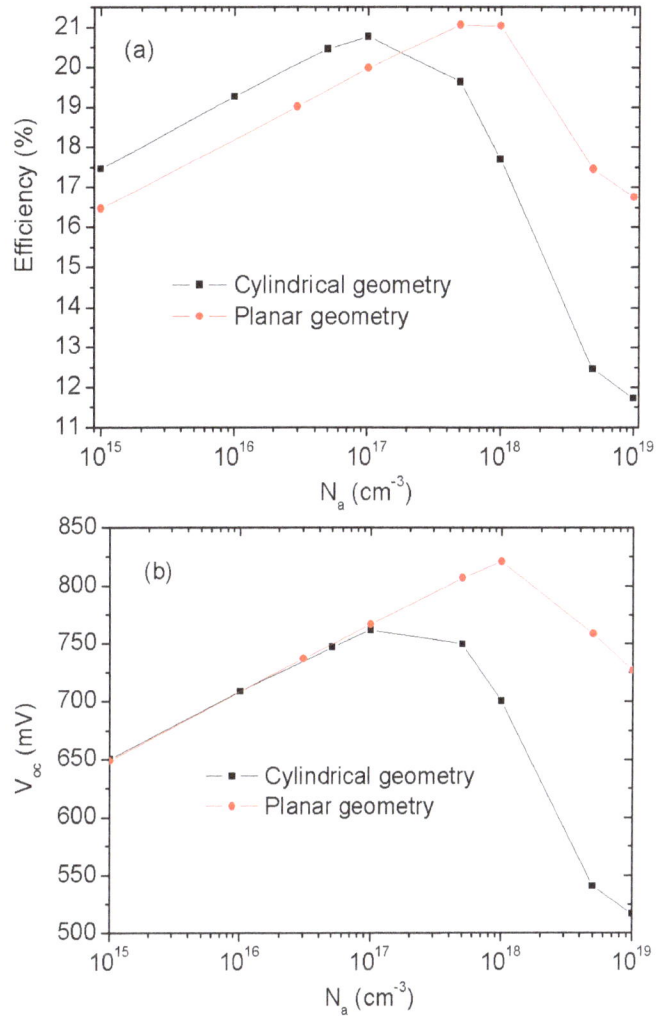

Fig. 3. Variations of the efficiency (a) and of the open-circuit voltage V_{oc} (b) of the modeled single-wire cell as a function of the p-type doping density Na in the c-Si core. These results are compared to those calculated for an equivalent c-Si(p)/a-Si:H(n) planar structure.

Boltzmann statistics is used with the drift-diffusion model in ATLAS, and Shockley-Read-Hall recombination was considered. The AM1.5 solar spectrum was used for the optical generation, the wire being under vertical illumination (see Fig. 2). Current-voltage characteristics ($I(V)$) were calculated in the dark and under standard one-sun illumination conditions. The open-circuit voltage, the short-circuit current, and the efficiency of the cell were deduced from the $I(V)$ curves under illumination.

3 Simulation results and discussion

The variations of the efficiency and of the open-circuit voltage of the single-wire device is given as a function of the doping density N_a in Figures 3a and 3b, respectively. These variations are compared to those of an equivalent planar solar cell. The planar structure was defined as a thin a-Si:H(n) layer (10 nm) on a 25 μm-thick c-Si(p)

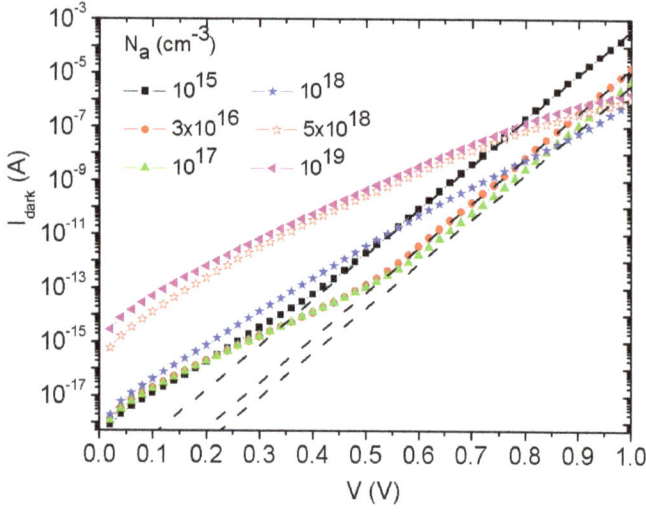

Fig. 4. Forward dark current characteristics $I_{dark}(V)$ for different doping density N_a in the c-Si(p) core of the wire. For $N_a \leqslant 1 \times 10^{17}$ cm^{-3}, the fits of the curves with the exponential law $I_0 \exp(qV/kT)$ corresponding to an ideal diode behavior is also given (dash lines).

Fig. 5. Forward dark current characteristics $I_{dark}(V)$ for various values of the defect density N_{defect} in the a-Si:H layer and a given doping density N_a in the c-Si(p) core equal to 1×10^{19} cm^{-3}.

wafer. The material properties and the illuminated surface are strictly the same in both structures. Moreover, whatever the doping density in the studied range, the short-circuit current density was found constant and equal to 32.1 mA/cm^2 for both cylindrical and planar geometries.

It can be observed on Figures 3a and 3b that the efficiency and V_{oc} increase with $\log(N_a)$ when N_a increases until an optimum value N_{aopt} which differs in both structures. In the case of the wire cell, N_{aopt} is found equal to 1×10^{17} cm^{-3}. Beyond N_{aopt}, there is a strong decrease of the efficiency which is mainly due to a heavy drop in the V_{oc}, and this behavior is more important in the silicon wire than in the planar device.

The V_{oc} is given by the relationship: $V_{oc} = nkT/q$ $\ln(I_{sc}/I_o)$, where n is the diode ideality factor, k the Boltzmann's constant, T the temperature, q the electronic charge and I_o the dark reverse saturation current. Since I_{sc} remains constant with N_a, the variation of the V_{oc} is related to the dark current. Thus, for the single-wire structure, the $I(V)$ characteristics in the dark (I_{dark}) were calculated for different values of N_a between 1×10^{15} cm^{-3} and 1×10^{19} cm^{-3} and the results are presented in Figure 4.

At low doping concentration ($N_a \leqslant 1 \times 10^{17}$ cm^{-3}), the trend of the curves remains the same and I_{dark} decreases when N_a increases, which is a classical behavior in a p-n junction. A change in the slope of the curves can be observed, which indicates two current regimes: at high forward bias ($V \geqslant 0.5$ V), the $I_{dark}(V)$ curves can be fitted by the expression $I_{dark} = I_0 \exp(qV/kT)$ as shown in Figure 4, which corresponds to an ideal diode behavior with a diffusion current. The values of I_0 deduced from these fits are found inversely proportional to N_a, which explains the variations of the V_{oc} previously described in this range of N_a. At low forward bias, the dark current seems to be dominated by a recombination/generation current in the

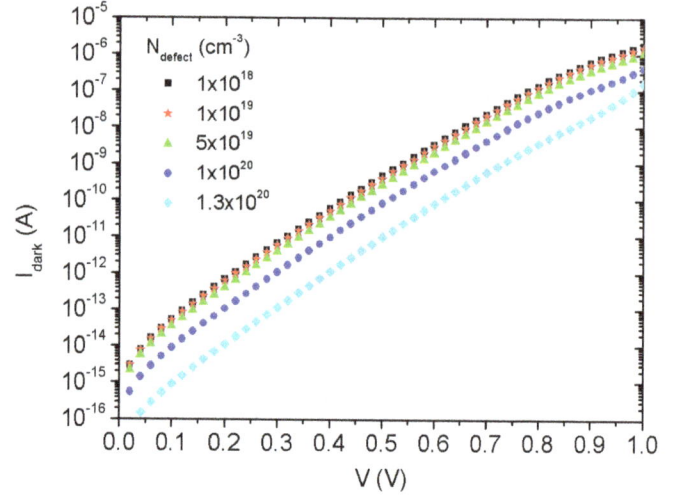

a-Si:H layer. At high doping density ($N_a > 1 \times 10^{17}$ cm^{-3}), no part of the $I_{dark}(V)$ characteristic at forward bias can be fitted by the previous exponential law and the recombination current in the a-Si:H shell becomes predominant all over the forward bias range. This is illustrated in Figure 5 where $I_{dark}(V)$ characteristics are calculated for various defect density N_{defect} in the a-Si:H layer and a given value of N_a equal to 1×10^{19} cm^{-3}. It can be observed that the dark current in the wire decreases when N_{defect} increases.

The heavy drop in V_{oc} at high doping density of the c-Si(p) core is directly related to a strong increase of the dark current which, at such N_a values, mainly becomes a recombination current in the defect-states rich a-Si:H layer. However, according to Figure 3b, this behavior is much more pronounced for the radial junction than for the planar one (it has to be reminded that this comparison is presented for the same illuminated area in both structures). A mapping of the dark current components near the c-Si(p)/a-Si:H(n) heterointerface of the single wire is presented in Figure 6 for $N_a = 1 \times 10^{17}$ cm^{-3} and $N_a = 1 \times 10^{19}$ cm^{-3}. It can be observed that for the lowest N_a value, most of the current is concentrated at the bottom interface of the structure whereas it is homogenously spread all over the junction for $N_a = 1 \times 10^{19}$ cm^{-3}. Simulations have shown that this difference in the current distribution appears around $N_a = 1 \times 10^{17}$ cm^{-3}. Consequently, the active junction surface that really contributes to the dark current becomes wider when N_a increases and it is expected that the ratio between the radius R of the wire (related to the illuminated area) and the wirelength L (related to the p-n junction) will affect the photovoltaic performance of the wire solar cell as soon as high doping density in the core is reached. This is also illustrated in Figure 7 where the dark current characteristics are plotted for two wirelength values $L = 25$ μm and $L = 250$ μm. For $N_a = 1 \times 10^{17}$ cm^{-3}, the high forward bias dark current does not depend on the wirelength

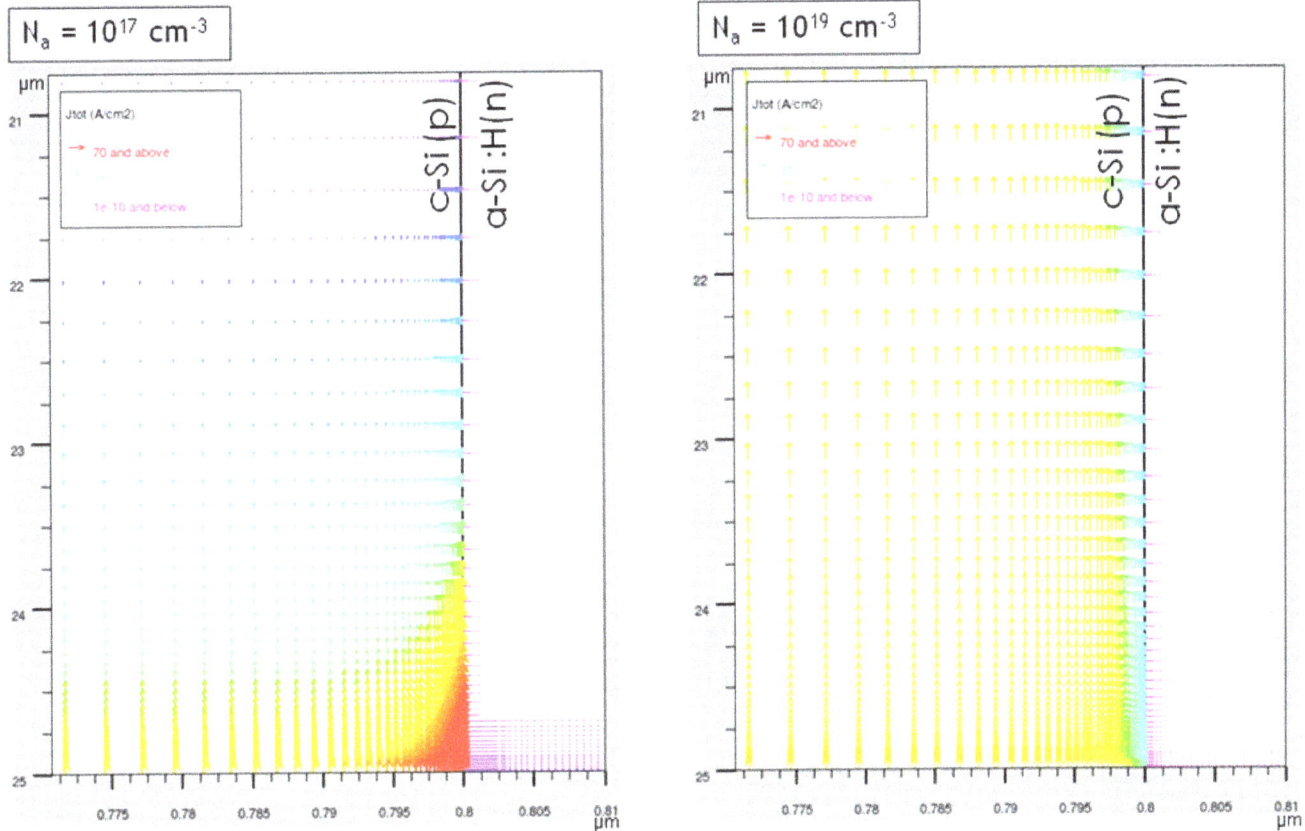

Fig. 6. Mapping of the dark current density distribution near the c-Si(p)/a-Si:H(n) heterointerface of the wire for $N_a = 1 \times 10^{17}$ cm^{-3} and $N_a = 1 \times 10^{19}$ cm^{-3}. Most of the current is concentrated at the bottom interface for the lowest value of N_a whereas it is homogeneously spread all over the p-n junction for the highest one.

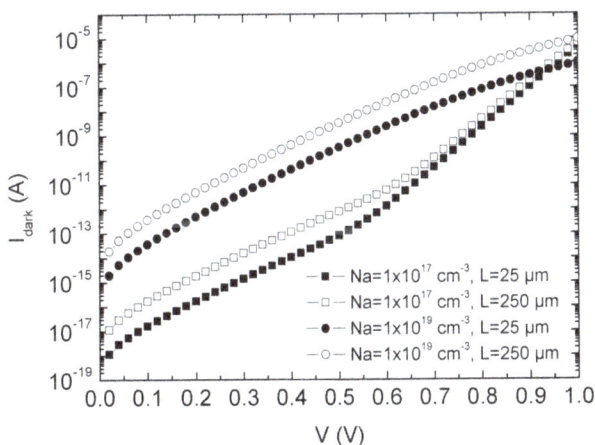

Fig. 7. Forward dark current characteristics $I_{dark}(V)$ for two wirelength values ($L = 25$ μm and $L = 250$ μm) at $N_a = 1 \times 10^{17}$ cm^{-3} and $N_a = 1 \times 10^{19}$ cm^{-3}.

whereas it increases with L for $N_a = 1 \times 10^{19}$ cm^{-3} due to an homogeneous current distribution all over the wirelength. Thus, at high values of N_a, we should obtain a comparable decrease in V_{oc} between a planar and a radial junction for an aspect ratio (ratio between the illuminated

area and the junction area) of the wire equal to 1. In the wire structures we have studied, the junction area is bigger than the illuminated area (ratio R/L much smaller than 1) such that the drop in V_{oc} with an increasing doping density is much more pronounced for the single-wire cell than for the planar one.

At low doping density, the current is dominated by the minority carriers diffusion and it mainly runs through a small active surface of the junction at the bottom of the structure as shown in Figure 6. This specific behavior seems to be related to the presence of a highly conductive electron interface layer at the c-Si(p) surface. Indeed, the band offsets between a-Si:H and c-Si can provide a strong band bending which results in a strong inversion layer at the c-Si surface [11–13]. For c-Si(p)/a-Si:H(n) structures, it has been shown from numerical calculations that the inversion layer occurs even at low values of the conduction band offset ΔE_c and that the electron concentration in the interface region of c-Si strongly increases with ΔE_c as soon as $\Delta E_c \geqslant 0.1$ eV [12]. In the case of c-Si(n)/a-Si:H(p) structures, a lower limit of the valence band offset ΔE_v has been found equal to 0.25 eV [13] such that ΔE_v can be varied in a wider range than ΔE_c to observe or not an interface inversion layer. Thus, in

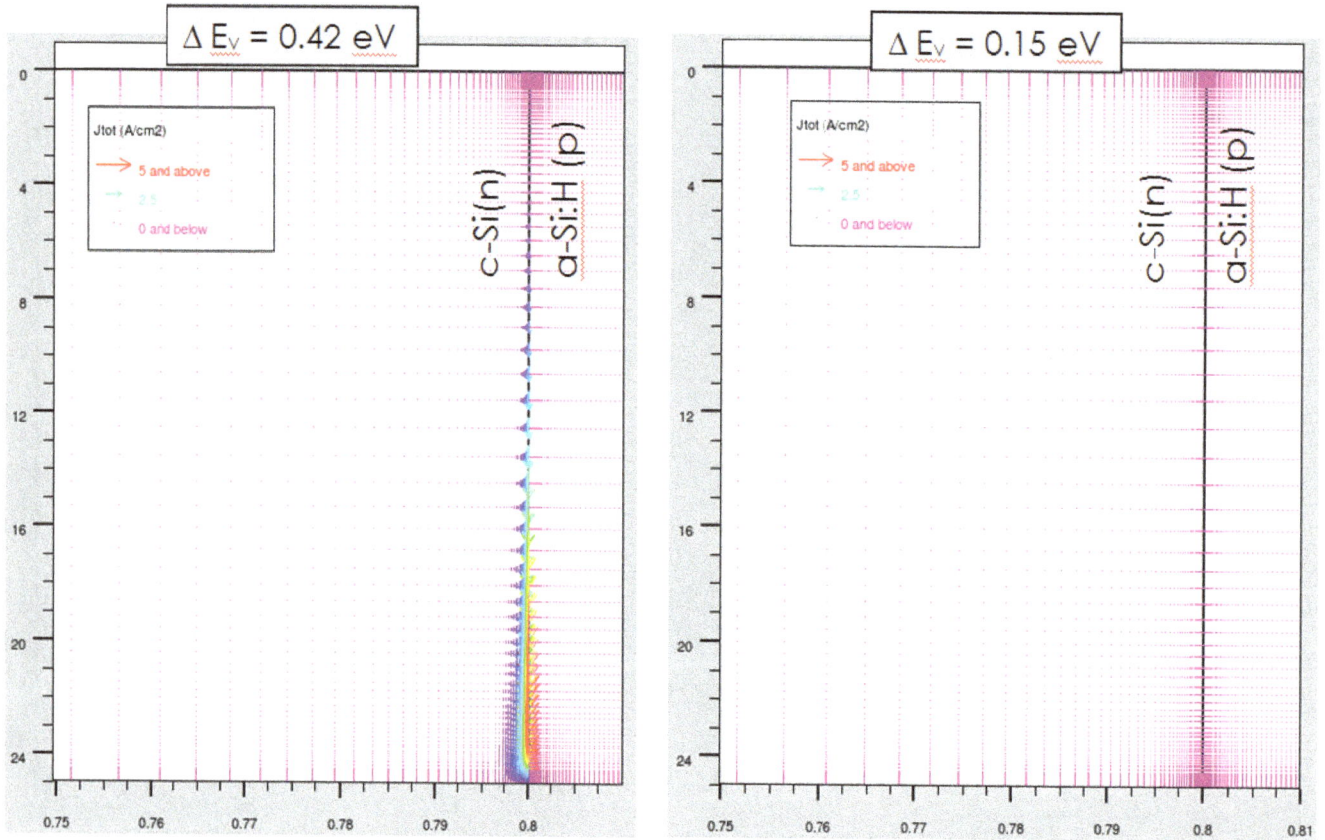

Fig. 8. Mapping of the dark current density distribution near the c-Si(n)/a-Si:H(p) heterointerface of the wire with $\Delta E_v =$ 0.42 eV and $\Delta E_v = 0.15$ eV. For the highest value of ΔE_v, most of the current is concentrated at the bottom interface in the crystalline core. For $\Delta E_v = 0.15$ eV, an homogeneous current distribution is obtained all along the heterointerface.

order to establish a link between the current distribution at the wire heterointerface and the presence-or not of an inversion surface layer, we have made some calculations on a c-Si(n)/ a-Si:H(p) radial cell for different values of the valence band offset. A mapping of the dark current components near the c-Si(n)/a-Si:H(p) interface is given in Figure 8 for $\Delta Ev = 0.42$ eV. and $\Delta Ev = 0.15$ eV. It can be observed that the current distribution is homogeneous along the wire for $\Delta Ev = 0.15$ eV whereas it is concentrated at the bottom of the cell for $\Delta Ev = 0.42$ eV which can be explained by a charge accumulation in the inversion layer at the interface. Furthermore, calculations performed on a single wire crystalline homojunction also show an homogeneous current distribution all along the wire interface whatever the doping density in the p-type core as shown in Figure 9. The $I_{dark}(V)$ characteristics follow the exponential law given above with a dependence on the wirelength (not shown here). To come back to our c-Si(p)/a-Si:H(n) wire structures, high values of the doping density in the crystalline core lead to a modification of the band bending such that no strong inversion layer occurs. An homogeneous current distribution is thus observed along the wire for $N_a \geqslant 1 \times 10^{17}$ cm^{-3}. Further studies have to be done to explain more precisely the location of the "hot spot" at the bottom of the interface

in the case of a non homogeneous current distribution. In particular, the geometry and position of the contacts can play a important role in that feature.

4 Conclusion

We have studied the potential performance of a single c-Si(p)/a-Si:H(n) wire based solar cell through 2D numerical modelling. We have noticed in particular a drastic drop of the open-circuit voltage when the p-type doping density N_a of the wire core increases beyond an optimum value equal to 1×10^{17} cm^{-3}. This loss in V_{oc} is linked to an increase of the dark current with N_a, the transport in the forward bias region being dominated by recombination current in the a-Si:H layer at high values of N_a. Moreover, the decrease in V_{oc} is less pronounced for an equivalent planar c-Si(p)/a-Si:H(n) heterojunction, the illuminated area of the planar and the radial modelled structures being the same. This difference can be explained by a dependence of the dark current in the wire with the wirelength. Thus, for $N_a \geqslant 1 \times 10^{17}$ cm^{-3} the wire based solar cell performance strongly depend on the ratio between the radius, which is linked to the illuminated area, and the wirelength, which is related to the p-n junction area. In conclusion, this study shows that the doping density N_a of the c-Si core should

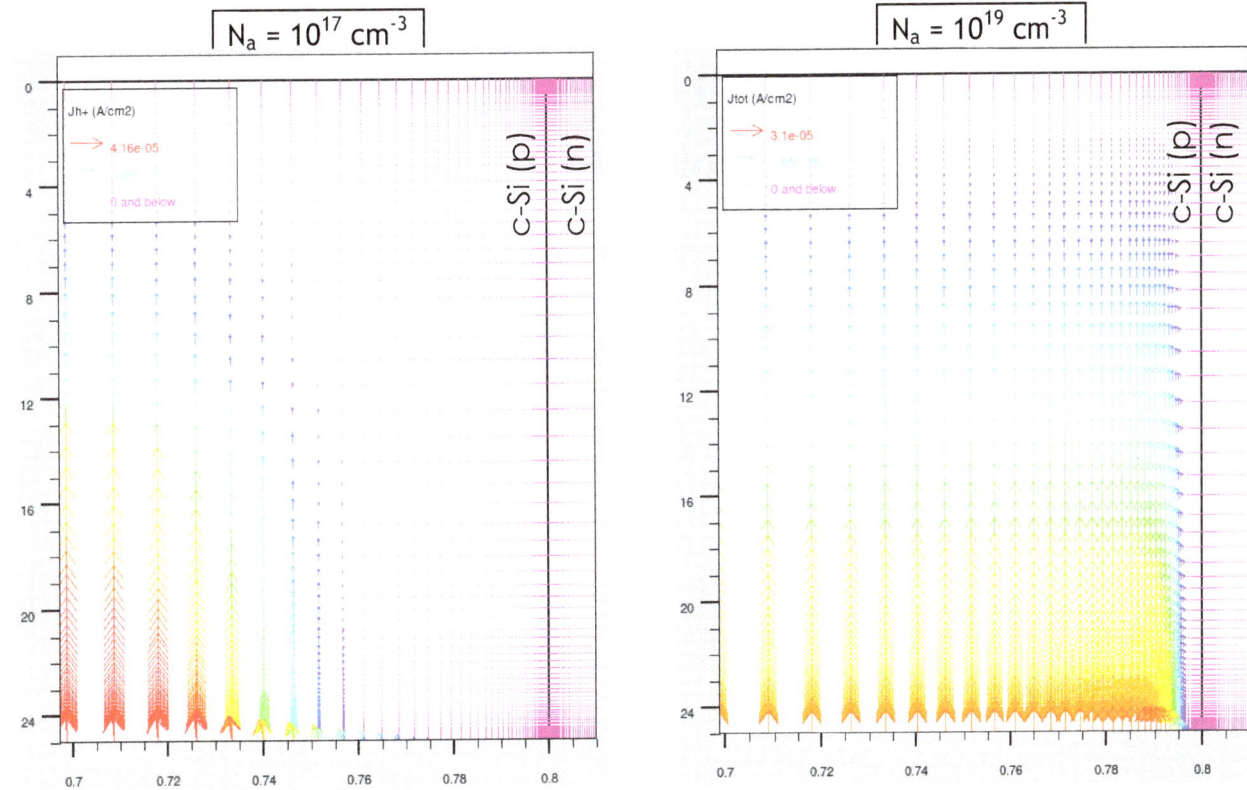

Fig. 9. Mapping of the dark current density distribution near the interface of a silicon crystalline homojunction wire with a doping density of the core $N_a = 1 \times 10^{17}$ cm^{-3} and $N_a = 1 \times 10^{19}$ cm^{-3}.

be kept below an optimum value of 1×10^{17} cm^{-3} and that any VLS catalyst which introduces a higher p-type doping density in the core should be avoided.

This work has been supported by French Research National Agency (ANR) through Habitat intelligent et solaire photo-voltaïque program (Projet Siflex No. ANR-08-HABISOL-010).

References

1. M. Law, J. Goldberger, P. Yang, Annu. Rev. Mater. Res. **34**, 83 (2004)
2. B.M. Kayes, H.A. Atwater, N.S. Lewis, J. Appl. Phys. **97**, 114302 (2005)
3. L. Tsakalakos, J. Balch, J. Fronheiser, B.A. Korevaar, O. Sulima, J. Rand, Appl. Phys. Lett. **91**, 233117 (2007)
4. B. Tian, X. Zheng, T.J. Kempa, Y. Fang, N. Yu, G. Yu, J. Huang, C.M. Lieber, Nature **449**, 885 (2007)
5. T. Stelzner, M. Pietsch, G. Andrä, F. Falk, E. Ose, S. Christiansen, Nanotechnology **19**, 295203 (2008)
6. M. Kelzenberg, S. Boettcher, J. Petykiewicz, D.B. Turner-Evans, M.C. Putman, E.L. Warren, J.M. Spurgeon, R.M. Briggs, N.S. Lewis, H.A. Atwater, Nat. Mater. **9**, 239 (2010)
7. R.S. Wagner, C.W. Ellis, Appl. Phys. Lett. **4**, 89 (1964)
8. D.E. Perea, J.E. Allen, S.J. May, B.W. Wessels, D.N. Seidman, L.J. Lauhon, Nano Lett. **6**, 181 (2006)
9. V. Schmidt, J.V. Wittemann, U. Gösele, Chem. Rev. **110**, 361 (2010)
10. *ATLAS Users' Manual* (Silvaco International, 2010)
11. J.P. Kleider, A.S. Gudovskikh, P. Roca i Cabarrocas, Appl. Phys. Lett. **92**, 162101 (2008)
12. J.P. Kleider, Y.M. Soro, R. Chouffot, A.S. Gudovskikh, P. Roca i Cabarrocas, J. Damon-Lacoste, D. Eon, P.J. Ribeyron, J. Non-Cryst. Solids **354**, 2641 (2008)
13. O.A. Maslova, J. Alvarez, E.V. Gushina, W. Favre, M.E. Gueunier-Farret, A.S. Gudovskikh, A.V. Ankudinov, E.I. Terukov, J.P. Kleider, Appl. Phys. Lett. **97**, 252110 (2010)

Laser annealing of thin film polycrystalline silicon solar cell

A. Chowdhury[1,a], A. Bahouka[2], S. Steffens[3], J. Schneider[4], J. Dore[4], F. Mermet[2], and A. Slaoui[1]

[1] ICube-University of Strasbourg and CNRS, Strasbourg, France
[2] IREPA Laser, Strasbourg, France
[3] HZB, Berlin, Germany
[4] SUNTECH, Thalheim, Germany
[5] SUNTECH, Botany, Australia

Abstract Performances of thin film polycrystalline silicon solar cell grown on glass substrate, using solid phase crystallization of amorphous silicon can be limited by low dopant activation and high density of defects. Here, we investigate line shaped laser induced thermal annealing to passivate some of these defects in the sub-melt regime. Effect of laser power and scan speed on the open circuit voltage of the polysilicon solar cells is reported. The processing temperature was measured by thermal imaging camera. Enhancement of the open circuit voltage as high as 210% is achieved using this method. The results are discussed.

1 Introduction

The past few years have seen an extensive use of rapid thermal processing (RTP) by means of halogen lamps in semiconductor technology for dopants activation, surface oxidation and for defects annealing of polycrystalline silicon on glass. On one hand, the RTP process is widely used for the crystallization of amorphous silicon on glass for solar cells and thin film transistors (TFTs) applications [1–5]. On the other hand, RTP on polycrystalline silicon materials can also be used to reduce the point defects present in the poly-Si layers and to activate the dopant elements [1, 2, 6–8]. However this method fails to achieve temperatures above 950 °C for long time (>1 min) while keeping the borosilicate glass substrate intact. Using laser thermal annealing (LTA), it is possible to achieve very high temperature very close to the melting point of crystalline silicon for few seconds with minimal or negligible deformation of borosilicate glass substrate. In this work we study the laser treatment of polysilicon solar cells in the solid phase regime and the consequence on their photovoltaic parameters.

Structures composed of Glass/SiON/SiN/N+/P-/P+ were routinely made by CSG Solar Pvt. Ltd. of Sydney, Australia [9]. The deposited stack on textured glass substrate employing plasma enhanced chemical vapour deposition (PECVD) is followed by solid phase crystallization at 600 °C. The defect passivation and dopant activation of un-activated dopants were done by rapid thermal

annealing at around 940 °C using halogen lamps. A plasma hydrogenation tool is used in the end to passivate the defects completely. Although 10% efficiency is quite good, simulations show an efficiency potential of $12-13\%$ with the present structure if the defects are more efficiently passivated. As the rapid thermal annealing process is already at its extreme before softening the glass, laser induced annealing may provide an alternative solution. The aim of this work is to anneal the polycrystalline silicon cell structure [10] composed of Glass/SiON/SiN/N+/P-/P+ at the highest temperature possible but below the melting point of silicon, in order to reduce the defect and to activate more dopants without damaging the stacked structure. For the study the polycrystalline silicon structures were made on planer glass substrates instead of textured glass substrates.

2 Experimental details

The samples used for laser annealing are deposited on planar borosilicate glass substrate. The structure of the samples is Glass/SiON/SiN/N$^+$/P$^-$/P$^+$ where the different layers were deposited by the Plasma Enhanced Chemical Vapor Deposition method. The annealed final structure is shown in Figure 1. The amorphous silicon N$^+$/P$^-$/P$^+$ stack was solid phase crystallized (SPC) prior to laser annealing. The laser source used for annealing is manufactured by LASERLINE Inc. and it is composed of a dual high power laser system simultaneously emitting at wavelengths of 808 nm and 940 nm. The maximum total power in the source is 3 kW.

---
[a] e-mail: amartyappb@yahoo.com

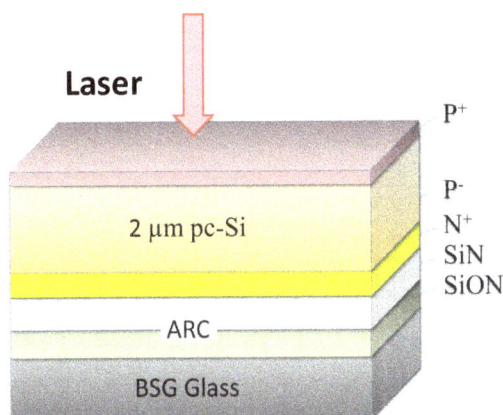

Fig. 1. Schematic diagram of the sample.

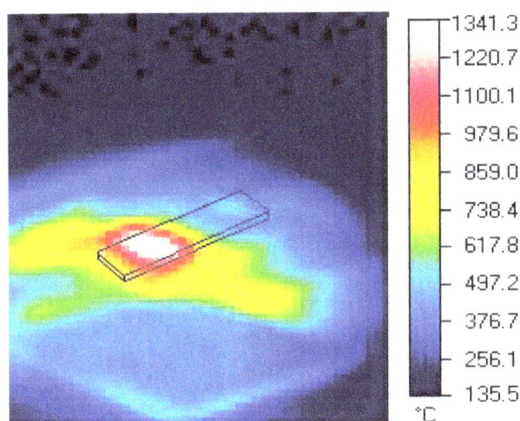

Fig. 2. Infrared imaging of temperature variation during laser annealing. The sample is schematically indicated as a rectangular object.

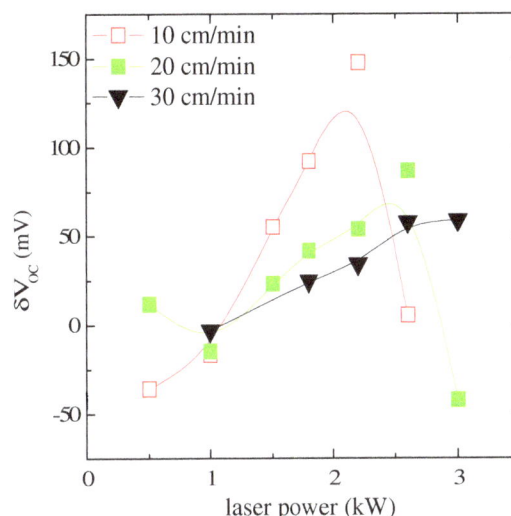

Fig. 3. Variation of open circuit voltage of laser annealed polysilicon structures as a function of laser power and at different scan speeds.

The open circuit voltage of the cell structure, before and after laser processing, was measured using the Suns-V_{OC} setup manufactured by Sinton Pvt. Ltd. [10]. Prior the measurements, the sample were etched at the corner to remove the P^+ and P layers, and to have access to the N^+ region. Electron Back Scattering Detraction (EBSD) experiment was also done to observe the presence of grain distribution with different laser annealing conditions.

3 Results and discussions

The samples were annealed by line shaped laser spot at different scan speeds and power densities. Figure 3 shows the variation of the open circuit voltage (V_{oc}) versus the operational parameters. Before laser annealing the open circuit voltage of the polysilicon structure was 130 ± 5 mV measured. The general observed trend is an increase of V_{oc} as the laser power increases and then a decrease above a certain power threshold. It is found that this threshold depends on the applied scan speed. The higher the scan speed, the higher is the power density needed reach this threshold. Later it will be seen that this threshold indicates the melting point of crystalline silicon. As a result, V_{oc} goes through an optimal value for each scan speed. A maximum V_{oc} value of 280 mV was obtained for a polysilicon structure annealed at 2.2 kW laser power and a scan speed of 10 cm/min. This indicates an enhancement of the V_{oc} of about 150 mV due to the laser annealing. Another noticeable result is that at very low power densities of 0.5 kW and scan speed of 10 cm/min, V_{oc} of the samples is lower than its original value of 130 ± 5 mV. This might be due to some defects (dislocations) induced by laser processing. These results indicate that benefit of laser applied to such structures is strongly dependent on the thermal budget assigned to the sample.

A thermal imaging camera was used to monitor the laser annealing process at different power and scan

Using a special setup the laser was made to incident on the sample in a shape of line with dimension of ~5 cm × 0.4 cm. The relative speed of the sample perpendicular to the laser line was varied from 10 cm/min to 30 cm/min. The laser power at the source was varied from 0.5 to 3 kW, but it should be noticed that there is always a ~33% optical laser power loss before reaching the sample surface in addition to loss due to reflection from sample surface. In all cases a substrate heater was used to elevate the sample temperature around 430 °C to enhance the light absorption in silicon and to reduce the thermal expansion coefficient mismatch between the substrate and the silicon layers.

During laser processing a thermal imaging camera named M7600PRO from Mikron Infrared Inc. was used extensively to monitor the thermal profile in each point of the sample. Figure 2 shows a captured image from a video sequence. The sample is schematically indicated as a rectangular object. The temperature estimation depend on the emissivity constant of the surface of the P^+ Si top layer and it is very tricky to find the exact values at $1000-1500$ °C due to a possible phase change at some parts of the silicon layer. Using suitable emissivity factors, the maximum processing temperature was estimated with ± 30 °C error in the $1000-1500$ °C range.

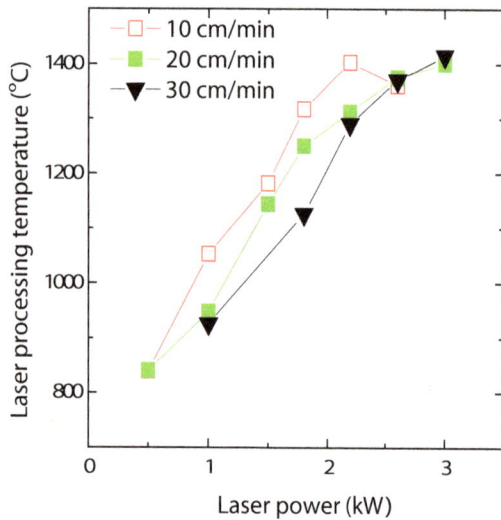

Fig. 4. Variation of processing temperature with respect to variation of laser power at different scan speed.

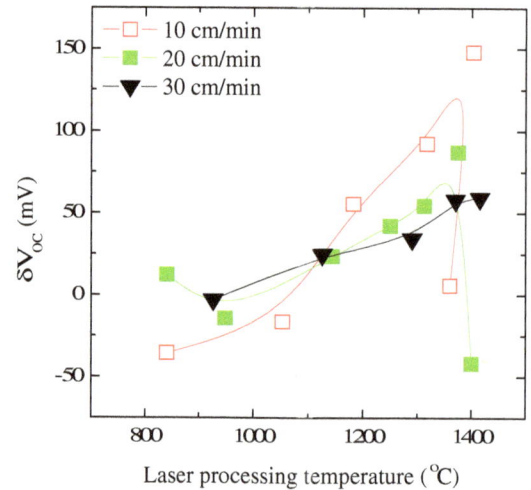

Fig. 5. Open circuit voltage of laser annealed polysilicon structures as a function of processing temperature.

speed. The corresponding temperatures were estimated. From Figure 4 it is observed that the sample temperature is monotonically increasing with laser power during annealing. A maximum temperature of $\sim$1410 °C, that is the melting point of polycrystalline silicon, was reached for a 2.2 kW laser power and 10 cm/min laser scan speed. As a substrate heater was used to set the sample temperature to 430 °C prior to laser processing, a processing temperature around 840 °C was obtained for only 500 W of laser power. Due to optical loss in the laser processing system and further losses with reflection from the sample itself, only around 200 W of laser power is incident on $\sim$5 cm $\times$ 0.4 cm area of sample. This indicates the effectiveness of the substrate heater in terms of direct elevation of sample temperature. Pre-heating the substrate increases the optical absorption coefficient which in turns helps the poly-silicon layers to absorb more efficiently the infrared laser power. The substrate temperature around 430 °C also helps to reduce the severe stress in glass-silicon interface by lowering the mismatch of thermal expansion coefficient of poly-silicon and glass substrate during high temperature laser processing.

Figure 5 plots the variation of V_{oc} as a function of processing temperature. The trend is similar for samples processed at different scan speed, first a steady increase in V_{oc} and then an abrupt decrease for processing temperatures around 1410 °C which is close to melting point of crystalline silicon. It is also observed here that for a similar processing temperature over 1200 °C, δV_{oc} is inversely proportional to the scan speed. This is indicating that higher duration of elevated temperature is beneficial for defect annealing and or dopant activation. It may also occur that a shorter duration of annealing is hampering the polysilicon layers to attain a stable structure before the cooling down.

Figure 6 shows the EBSD images of samples before and after laser processing. The non annealed polysilicon structure (Fig. 6a) has grains in the range 1–5 microns in

diameter. From Figure 6b it is obvious that laser thermal annealing did not affect the grain sizes or the distribution, while the open circuit voltage enhanced by 150 mV as reported above. Such improvement can therefore be only possible to a reduction in defects and/or to an electrical activation of dopants in the absorbing p layer. At 10 cm/min scan speed, further increase in laser power to 2.6 kW changes the grain structure completely and shows clear pattern of recrystallization (Fig. 6c). This recrystallization is also responsible for the sudden drop in open circuit voltage observed in Figure 3.

Some samples, on which EBSD study was performed, were also undergone SIMS experiment to monitor the boron and phosphorus distributions of the $p^+p^-n^+$ structure. Figures 7a and 7b plots respectively the boron and phosphorus profiles of reference and laser annealed sample at 2.2 KW. The B and P profiles of the non annealed sample are sharp and the junctions are well defined. After annealing, the boron and phosphorus profiles are graded and the elements are penetrating the p^- region from both sides. The B and P surface concentrations are lower than those of the reference sample and the corresponding junction depths are deeper. These results indicate that the B and P diffuse towards the volume probably in a sub-melting regime. Indeed despite the short time of the process (the scan speed is 10 cm/s) the dopants diffusion occurs thanks to the very high temperature process. The process can be compared to the annealing by flash lamps [11] here we offer temperatures above 1200 °C. yet the reconstruction of the layers after annealing is very good as witnessed by the EBSD images and by a higher open circuit voltage value obtained for this sample as compared to that of the reference. Increasing the laser power up 2.6 kW resulted in a strong distribution of the dopants (not shown here) within the $p^+p^-n^+$ structure and therefore to less defined junctions. As a consequence, a lower open circuit voltage is measured on this sample.

To complete our study, we have prepared 16 solar cell $p^+p^-n^+$ structures and post-annealed them using halogen

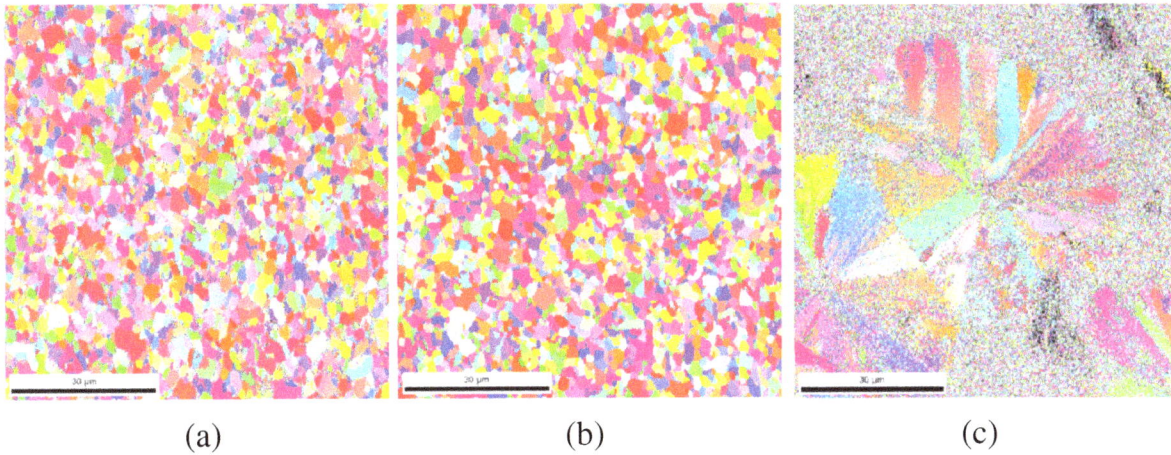

(a)   (b)   (c)

Fig. 6. EBSD images of samples (a) before laser processing, (b) 2.2 kW, 10 cm/s, (c) 2.6 kW, 10 cm/s.

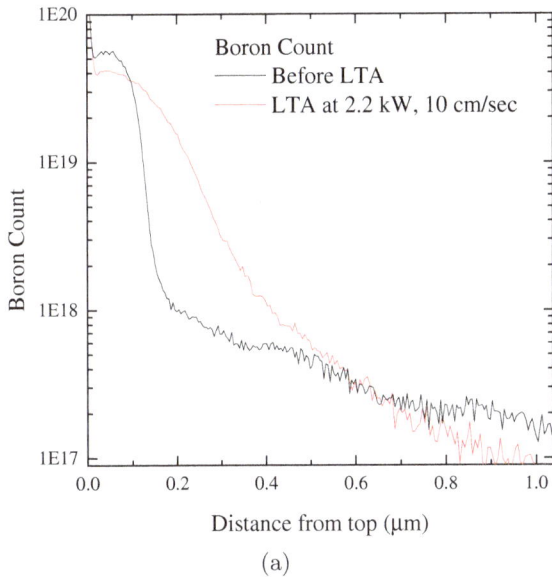

(a)

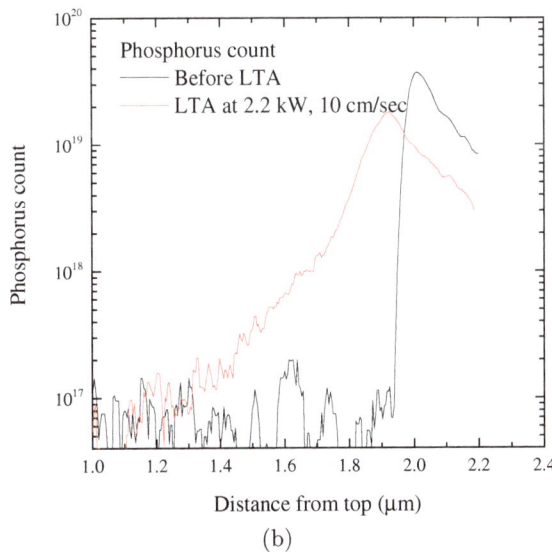

(b)

Fig. 7. SIMS study on the samples before and after LTA: (a) boron count, (b) phosphorus count.

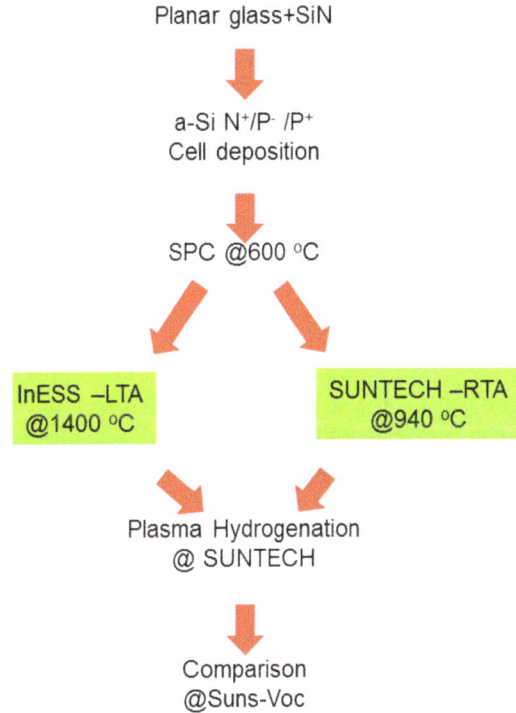

Fig. 8. Schematic diagram of experiment for LTA and RTP comparison.

Table 1. Laser annealing and rapid thermal annealing comparison.

	$V_{oc}(1)$	p_{FF}	p_{Eff}
LTA	480	76.6	7.3
RTA	463	75.5	7.0
Improvement by LTA	+3.6%	+1.4%	+4.3%

lamps (RTA) at Suntech Pty, Australia or CW IR laser (LTA) at ICube lab before to make contacts. The same hydrogenation step was applied to all cells. The open circuit voltage of these cells was measured using the SunsVoc system and the results are displayed in Table 1. An average open circuit voltage of 480 mV was measured after LTA, which is 17 mV higher than for RTA processed cells. Assuming a current density of the cells at 20 mA/cm^2, we found that the open circuit voltage, the pseudo fill factor and the pseudo efficiency are enhanced by 3.6%, 1.4% and 4.3% as compared to the values of the RTA annealed cells. While the hydrogenation step was previously thoroughly optimized for the routinely made RTA annealed structures [12], it is not the case for the LTA cells. Thus higher cells performances are expected for the late samples. Overall the LTA process shows its effectiveness in defect passivation when compared to the conventional RTA.

4 Summary

Polysilicon solar cells structures were CW laser irradiated in the sub-melting regime. We have demonstrated a 210% improvement in V_{oc} by using a fast laser scanning. A further enhancement in V_{oc} is possible by a suitable plasma hydrogenation process. Using LTA process and plasma hydrogenation a high open circuit voltage of 480 mV obtained, which is 17 mV higher when LTA is replaced by conventional RTA using halogen lamps. Similarly, in addition to improvement of pseudo fill factor, the pseudo cell efficiency also increases by 3.6% when the LTA process is applied to the cells. The increase of open circuit voltage after laser thermal annealing is probably due to passivation of defects and/or activation of the dopants. For annealing at power below 2.2 KW, the process is in sub-melt region of polycrystalline silicon and seemly the grain structures are not affected despite the high temperature processing beyond 1200 °C for a very short time. Above the melting point of polycrystalline silicon which occurs for high laser power, both boron and phosphorus dopants diffuse in the layers resulting in non defined junctions and therefore in low open circuit voltage values.

At processing temperature below 1000 °C, the open circuit voltage of the cells is lowered due to laser induced defects created by sudden change in temperature. These results show that the improvement of the open circuit voltage during laser thermal annealing is a result of a tread-off between thermal defect annealing and defects formation.

This work was funded by the European commission under project entitled POLYSIMODE.

References

1. D. Mathiot, A. Lachiq, A. Slaoui, S. Noël, J.C. Muller, C. Dubois, Mater. Sci. Semicond. Process. **1**, 231 (1998)
2. M.L. Terry, A. Straub, D. Inns, D. Song, A.G. Aberle, Appl. Phys. Lett. **86**, 1 (2005)
3. J.F. Michaud, R. Rogel, T. Mohammed-Brahim, M. Sarret, J. Non-Cryst. Solids **352**, 998 (2006)
4. G. Andra, J. Bergmann, F. Falk, Thin Solid Films **487**, 77 (2005)
5. T. Noguchi, Jpn J. Appl. Phys. **47**, 1858 (2008)
6. R. Gunawan, M.Y.L. Jung, E.G. Seebauer, R.D. Braatz, J. Process Control **14**, 423 (2004)
7. P. Doshi, A. Rohatgi, M. Ropp, Z. Chen, D. Ruby, D.L. Meier, Sol. Energy Mater. Sol. Cells **41**, 31 (1996)
8. M.L. Terry, D. Inns, A.G. Aberle, Adv. Optoelectron. 83657 (2007)
9. M.J. Keevers, T.L. Young, U. Schubert, M.A. Green, in *22nd EPVSEC, Milan, Italy, 2007*
10. M. Wolf, H. Rauschenbach, Adv. Energy Convers. **3**, 455 (1963)
11. W.S. Yoo, K. Kang, Nucl. Instrum. Methods Phys. Res. B **237**, 12 (2005)
12. M.J. Keevers, A. Turner, U. Schubert, P.A. Basore, M.A. Green, in *Proceedings of the 20th European Photovoltaic Solar Energy Conference, Barcelona, Spain, 2005*, p. 1305

5

Precise microstructuring of indium-tin oxide thin films on glass by selective femtosecond laser ablation

S. Krause[1], P.T. Miclea[1,2], F. Steudel[1], S. Schweizer[1,3], and G. Seifert[1,4,a]

[1] Fraunhofer Center for Silicon Photovoltaics CSP, Walter-Hülse-Str. 1, 06120 Halle (Saale), Germany
[2] Institute of Physics, Martin Luther University of Halle-Wittenberg, Heinrich-Damerow-Str. 4, 06120 Halle (Saale), Germany
[3] Department of Electrical Engineering, South Westphalia University of Applied Sciences, Lübecker Ring 2, 59494 Soest, Germany
[4] Centre for Innovation Competence SiLi-nano®, Martin Luther University of Halle-Wittenberg, Karl-Freiherr-von-Fritsch-Str. 3, 06120 Halle (Saale), Germany

Abstract Transparent conductive oxide (TCO) thin films were removed from glass substrates using femtosecond laser pulses. Irradiating through the glass, the threshold for complete TCO ablation was much lower than for front-side irradiation. Additionally, the former method created almost rectangular cross-sectional groove profiles despite the Gaussian laser beam. This indicates a non-thermal ultrafast ablation mechanism via critical carrier concentration achieved by the femtosecond pulse in the TCO at the interface. Very narrow scribes of only 5 μm width provided very good electrical separation, making this technique very attractive for micro-structuring applications like scribing of thin-film solar cells.

1 Introduction

Transparent electrodes consisting of conductive oxide thin films on flat glass are nowadays widely used as basis of technical applications like flat panel displays or solar cells. Traditionally, patterning of such transparent conductive oxide (TCO) films has been done by lithography. However, laser ablation techniques are becoming more and more attractive because they offer – instead of several processing steps including chemical wet etching – a maskless, one-step, dry process which can be conducted under ambient conditions. To comply with the need of precise microstructures like, for instance, groove widths below 10 μm providing perfect electrical isolation, ultrashort pulses have to be used in order to avoid any thermal damage or contamination in the ablated region. Previous empirical studies using pico- or femtosecond pulses at several wavelengths from the near infrared to ultraviolet spectral range showed that a fairly "clean" ablation of TCOs or photovoltaic absorber thin films is possible by irradiating successive pulses with considerable spatial overlap from the TCO side [1–7].

These investigations, however, did not consider in detail the physical mechanism of the TCO ablation. Very recent results reported by Rublack et al. [8,9] suggest an interesting approach for fs pulses: it was shown that dielectric coatings can be removed selectively from silicon wafers by fs laser ablation in such a way that the opened surface is perfectly smooth and the Si material below the ablated area is still perfectly crystalline. The key point here is free carrier absorption in the semiconductor created by the rising edge of the fs pulse itself, which leads to non-thermal decomposition of a very thin silicon layer of a few nanometers thickness only, the expansion of which blasts off the capping dielectric layer. Transferring this idea to the reverse case of TCO on glass (conductive layer on dielectric substrate), a similar ablation mechanism should be possible for irradiating fs pulses through the glass, thereby creating free carrier absorption in the TCO at the interface. So far, ablation experiments using rear-side irradiation with fs pulses have only been reported for metals like chromium [10], but not for TCO.

In this work, it is demonstrated that this approach really enables a damage-free (with respect to substrate and remaining film) ablation of TCO films. For this purpose, the ablation of indium-tin oxide (ITO) films on glass by 300 fs pulses at 1030 nm was studied; the laser irradiation was carried out from the glass as well as from the ITO side.

2 Experimental

The samples studied in this work were approximately 600 nm thick ITO layers on soda-lime float glass sheets

---

[a] e-mail: gerhard.seifert@physik.uni-halle.de

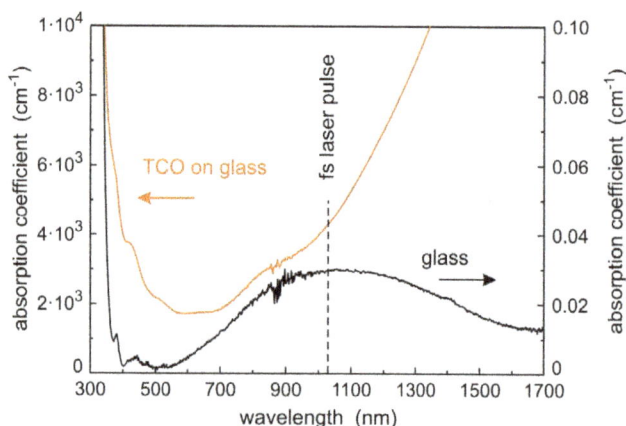

Fig. 1. Absorbance of an ITO/glass sample (orange curve) and a pure glass substrate (black curve); the laser wavelength is indicated by a vertical dashed line.

of approximately 3 mm thickness. Typically, such glasses are the base material for the production of inorganic thin-film solar cells. For the laser structuring of the ITO layer a femtosecond Yb:KGW laser system (PHAROS, Light-Conversion) was used with a wavelength of 1030 nm and pulse duration of 300 fs, average power of up to 6 W and maximum repetition rate of 350 kHz. These laser pulses were focused to the sample surface with an f-theta lens having a focal length of about 55 mm (yielding a nearly Gaussian beam of 13.0 ± 0.5 μm diameter at the beam waist), in connection with an xyz-scanner for positioning and motion control.

In Figure 1, the absorbance spectra (obtained from total transmittance and reflectance measured with an integrating sphere) of the ITO/glass samples and the bare glass substrate are compared. Considering the vastly different thickness of thin film and substrate, these data indicate a fairly large absorption coefficient of $\alpha \approx 4000$ cm^{-1} for the ITO layer at the laser wavelength of 1030 nm, but a comparably negligible one for the glass substrate ($\alpha \approx 0.03$ cm^{-1}). Due to this choice, experiments could be performed irradiating the samples from the ITO side as well as from the rear side, i.e. with the laser beam travelling through the glass before interacting with the ITO film. For comparison, also pure glass samples were irradiated in the same way. By varying the scribing speed, single ablation spots as well as lines consisting of successive pulses with well-defined overlap were produced. To be able to vary the pulse energies in a precise and reproducible way, an attenuator consisting of a half-wave plate and a thin-film polarizer were used. The actual pulse energies were then determined by measuring the average power in front of the samples with a thermopile detector, considering the chosen repetition frequency. Ablation thresholds were obtained in analogy to Liu's method [11] by evaluating the size of the ablated areas as a function of laser pulse energy (fluence), as described in detail previously [8].

After laser processing, the ablation spots or lines were characterized using optical microscopy (images were taken with a Leica DM RXE-650H microscope) and profilometry

(cross-sectional profiles were taken with an Ambios Technology profilometer XP-2). Scanning electron microscopy (SEM) images were acquired with a JSM-7401F (JEOL) electron microscope.

3 Results and discussion

In a first series of experiments, various "scribes" were made at different laser fluences, with a pulse-to-pulse distance of 7 μm (0.35 m/s at 50 kHz pulse repetition frequency); irradiation was carried out from the top (ITO surface) and from the rear side (through the glass substrate).

Figure 2 shows a selection of results in form of microscope images (Figs. 2a and 2b) and the corresponding cross-sectional profiles obtained at positions close to the spot centers of an arbitrary laser shot within the scribe. Figures 2a and 2c refer to irradiation from the TCO side, Figures 2b and 2d to the laser beam coming from the glass side. Discussing first the results in Figures 2a and 2c in the sequence of increasing fluence, one recognizes as main trend the creation of ablated regions of increasing maximum depth, d, and width, b, showing a cross-sectional profile which is more or less similar to the Gaussian laser beam profile. Starting with $b \approx 5$ μm and $d \approx 250$ nm at 1.5 J/cm^2, a value of $d \approx 600$ nm corresponding to the total ITO layer thickness is only achieved at 12.3 J/cm^2. At 18.4 J/cm^2, the groove is $b \approx 13$ μm wide and $d \approx 650$ nm deep; apparently, as indicated by the highest fluence result, the ablation does not stop at the former ITO/glass interface. Looking in more detail, the intermediate fluences exhibit a comparably steep hole in the crater centers, while the second lowest one even exhibits a small bulge there. Additionally, for the lowest fluence, the main deep hole seems to be surrounded by a shallow one of only a few nanometers of ITO material having been removed.

The results of irradiation through the glass exhibit a quite different behavior: first, ablation starts at considerably lower laser energy density; second, even at the lowest fluence the ablation depth $d \approx 600$ nm indicates complete removal of the ITO layer; third, the ablation cross-sectional profile looks almost rectangular in any case, with steep edges and almost flat bottom. Looking in detail, a 20–30 nm deep dimple towards the center of ablation can be observed in particular for the higher laser energies.

The rear-side irradiation, i.e. irradiation from the glass side, is similar to recent results described by Rublack et al. [8,9], who identified the fs laser induced non-thermal evaporation of an ultrathin ($\leqslant$5 nm) Si layer as the key process for selective ablation of SiO$_2$ or Si$_x$N$_y$ layers from silicon wafers without detectable damage to the substrate. If a similar, nonlinear absorption process occurs in ITO, already fluences slightly below the actual ablation threshold should cause a separation of the ITO layer from the substrate and a corresponding bulge. In fact, already at 0.6 J/cm^2, a shallow wall of 30–70 nm height is found instead of a 600 nm deep groove. A cross section through such a bulge is shown in the scanning electron microscopy (SEM) image of Figure 3, where the $\approx$600 nm thick ITO

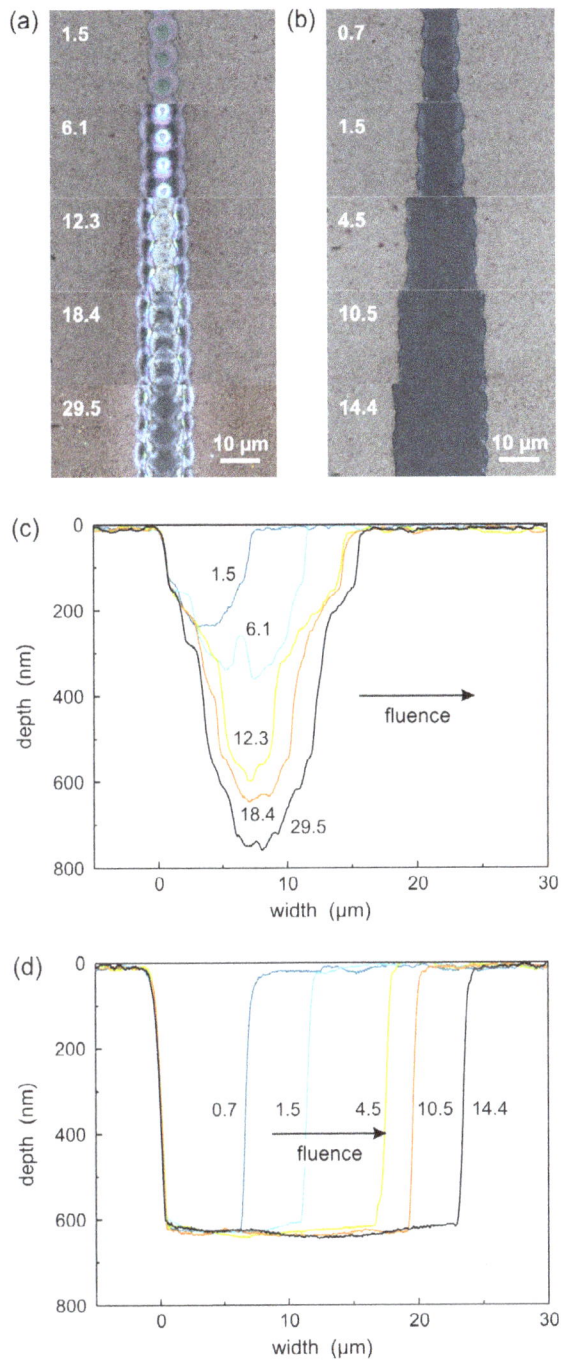

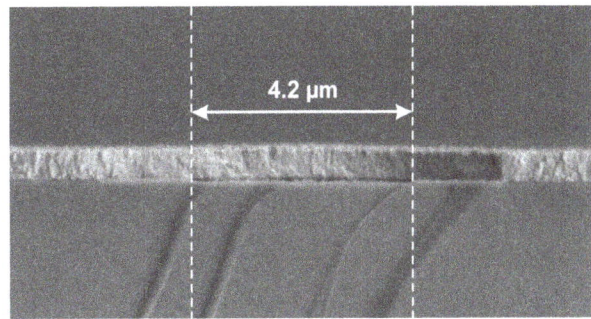

Fig. 3. SEM image of cross section through bulge after rear-side irradiation slightly below ablation threshold.

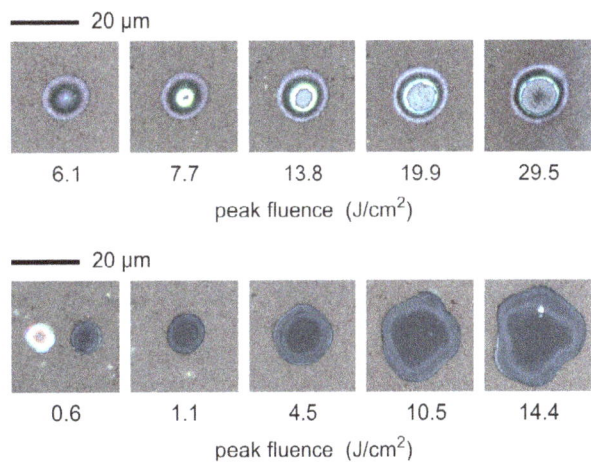

Fig. 4. Microscope images of single ablation spots. Upper row: irradiation from the TCO side; lower row: irradiation from the glass side.

Fig. 2. Microscope images (a, b) and cross-sectional profiles (c, d) of ablation lines; obtained after irradiation from the TCO side (a, c) or from the glass side (b, d). All numbers labeling the microscope images or depth profiles are fluence values in J/cm^2.

layer (light horizontal stripe in the center of the image) has clearly been delaminated from the glass substrate (lower region of the image), showing a dark gap of $b \approx 4 \ \mu m$ width and maximum height of ≈ 60 nm in the middle of the irradiated area. Apparently, this is the region where the intensity of the Gaussian beam was above the threshold for fs laser-induced non-thermal evaporation, but the

resulting pressure at the interface was still not high enough to break and lift off the TCO layer. The layer thickness of evaporated material can only be estimated to be of the order of 10 nm. Nevertheless this is a clear indication that the proposed mechanism has occurred in the ITO layer at the interface.

For a quantitative evaluation of the ablation thresholds another series of experiments was conducted with considerably increased spot-to-spot distance in order to exclude any possible effects due to interaction of two successive pulses. A selection of microscope images from this series obtained at different peak laser fluence from either side of the sample is collected in Figure 4. The images in the upper row of Figure 4 represent results of irradiation from the ITO side, while those in the bottom row have been obtained by sending the laser beam through the glass substrate first; the fluence increases from left to right in both cases. At the lowest fluence shown in the bottom row, some spots were not ablated, but only show a blister; here the fluence was apparently very close to ablation threshold, so that even the small laser fluctuations were sufficient to statistically lead to ITO ablation or only the above discussed blister formation. Therefore both effects were included in the bottom left image.

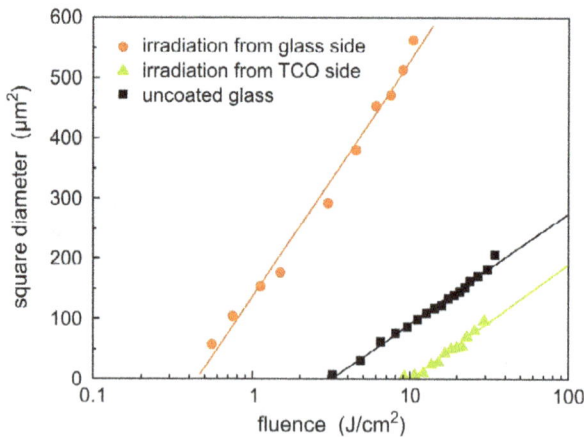

Fig. 5. Liu plot of squared diameters of ablation areas obtained by irradiation of TCO/glass samples from TCO (triangles) or glass side (circles), and of a pure glass substrate (squares) for comparison.

Apart from that, the images shown in Figure 4 indicate fairly well the same behavior as observed for considerable pulse overlap. Focusing on the difference between front-side and rear-side irradiation, it is noted that the rear-side illumination (lower panel in Fig. 4) creates in all cases craters with steep edges and a depth corresponding to the ITO thickness; the area of the opened glass surface increases monotonously with fluence. In contrast, irradiation from the layer side produces a trough of increasing depth first, before at higher fluence within this trough a smaller bulge with a narrow hole in the center occurs (situation at 6.1 J/cm^2, top left image). With further increase of fluence, the diameter of this central hole increases, then the crater bottom in the center reaches the glass (image at 13.8 J/cm^2); finally, upon further increase of laser pulse fluence, the opened glass surface area increases on expense of the surrounding rings, whereas the outer rim of the color change remains almost constant.

To determine the ablation thresholds for the complete series of ablation spots, the relevant areas (typically the squared diameter in case of a circular spot) have to be plotted versus laser pulse fluence on a logarithmic scale; then the ablation thresholds Φ_A can be derived by extrapolating to zero ablation area. The relevant diameters are clearly the sharp crater edges for the rear-side irradiation as well as for another series of ablation experiments on the pure glass substrate which has been performed for comparison. In case of the ITO-side irradiation, the relatively steep walls of the inner crater (the narrow hole occurring around 6 J/cm^2) were selected, because this is apparently the process which at higher fluence leads to complete ablation of the ITO layer. The results of this analysis are plotted in Figure 5. It is obvious that Φ_A for removing ITO is more than an order of magnitude lower for irradiation from the rear (glass) side (≈ 0.5 J/cm^2 and ≈ 9 J/cm^2, respectively). Apparently, a peak pulse fluence which does not cause visible changes when the ITO surface is irradiated causes already ablation of a small disk of the whole ITO layer when the same laser pulse enters through the

glass substrate. Removal of the whole layer by irradiation from the ITO side, on the other hand, consumes so much laser energy for evaporation of the material that the respective Φ_A is even higher than the ablation threshold for pure glass (≈ 3 J/cm^2).

All these observations confirm the above proposed mechanism for ITO removal upon femtosecond irradiation through the substrate glass: apparently, the fs laser pulse creates via the existing free carrier absorption a critical carrier density [12,13] in an ultrathin ITO layer adjacent to the ITO/glass interface, which leads to (non-thermal) evaporation of this layer, and the resulting high pressure breaks the ITO layer and leads to blasting off a disk, the outer rim of which is defined by the ablation threshold being reached within the Gaussian beam.

Irradiation from the ITO layer side will of course initiate the same mechanism. However, this means that comparable fluence can only ablate the ultra-thin layer having very high carrier density, because the resulting metal-like absorption will shield the deeper regions of the ITO layer from the highest pulse intensity. Therefore, higher fluence can only successively lead to deeper ablation, and since the pulse energy must be sufficient to evaporate the whole ITO material to be removed, the ablation threshold to achieve $d \approx 600$ nm is more than one order of magnitude above that for rear-side processing. This is clearly a disadvantage in view of technological applications, as is the rather irregular, non-rectangular profile of the grooves. In contrast, the almost rectangular profile and the low ablation threshold found for rear-side illumination makes this approach very attractive for micro-structuring applications of ITO and other transparent conductive oxide films on glass substrate.

The only open point to discuss here is the observation that the groove bottoms are not perfectly flat, but turn out to be 20–30 nm deeper in the groove center in all cases (see Fig. 2); in the microscope images these central areas appear dark grey, while the outer regions of grooves or craters exhibit a bluish ring in most cases. This can be taken as hint that the center really represents a glass surface, while the outer ring might indicate a thin residue of a layer of different composition. A possible explanation for such a rest comes from the fact that the glass substrate has several constituents with high diffusion coefficients (like Na^+ or K^+): during deposition of the ITO layer, inter-diffusion at the interface might create a transition zone from isolating glass to semiconducting ITO. Other than in the case of dielectric layers on silicon with an almost atomically sharp interface, this would cause a gradient of free carrier concentration and absorption over the transition zone. Under such circumstances, the critical carrier density required for the non-thermal ablation of the capping ITO layer may be achieved closer to the interface (original glass surface) at the highest pulse intensity in the center of the Gaussian beam profile, but in slightly larger distance in the outer regions. This could easily explain the shallow indentation observed in the laser trenches.

For technological applications, however, the indentation in the grooves is not a constraint, irrespective of its origin: preliminary conductivity measurements across the laser scribes showed that perfect electrical isolation (sheet resistance $> 10^6$ Ω/sq) is achieved even for the narrowest trenches (width ≈ 5 μm; see Fig. 2b) created by rear-side processing. More details on this point as well as further results for other TCO materials will be discussed in an extended study to be published separately.

4 Summary

In conclusion, single pulse femtosecond laser ablation of TCO (ITO) thin films from soda-lime glass substrates has been studied for irradiation both from the layer side and from the glass side. The rear-side illumination turned out to be a very effective tool to remove the ITO layer selectively and without noticeable thermal damage, enabling grooves with almost rectangular cross section, i.e. very steep edges and a flat bottom located at the initial position of the glass/TCO interface. The physical mechanism causing this fs laser lift-off process is apparently ultrafast non-thermal evaporation of a very thin TCO layer via laser-induced generation of a critical carrier density, in analogy to very recent results obtained on silicon wafers coated with thin dielectric layers. Technologically, the method studied in this work has a large potential for precise micro-structuring of TCO layers for various purposes. For thin film photovoltaics, the results presented here allow minimizing the area needed for the so-called P1 scribe (to produce electrically isolated TCO stripes), which is the first step towards the serial electrical connection of all the parallel stripes in a typical thin film solar cell. However, the applicability of this selective ablation technique is not limited to TCO, but can also be transferred to scribing of further layers of thin film solar cells, e.g. the P2 scribe to structure the absorber layer.

This work was supported by the FhG Internal Programs under Grant No. Attract 692 034. In addition, the authors would like to thank the German Federal Ministry for Education and Research ("Bundesministerium für Bildung und Forschung") for the financial support within the Centre for Innovation Competence SiLi-nano® (Förderkennzeichen: 03Z2HN11).

References

1. D. Ashkenasi, G. Müller, A. Rosenfeld, R. Stoian, I.V. Hertel, N.M. Bulgakova, E.E.B. Campbell, Appl. Phys. A **77**, 223 (2003)
2. J. Hermann, M. Benfarah, G. Coustillier, S. Bruneau, E. Axente, J.-F. Guillemoles, M. Sentis, P. Alloncle, T. Itina, Appl. Surf. Sci. **252**, 4814 (2006)
3. G. Raciukaitis, M. Brikas, M. Gedvilas, T. Rakickas, Appl. Surf. Sci. **253**, 6570 (2007)
4. S. Zoppel, H. Huber, G. Reider, Appl. Phys. A **89**, 161 (2007)
5. P. Gecys, G. Raciukaitis, M. Gedvilas, A. Selskis, Eur. Phys. J. Appl. Phys. **46**, 12508 (2009)
6. A. Schoonderbeek, V. Schütz, O. Haupt, U. Stute, J. Laser Micro Nanoeng. **5**, 248 (2010)
7. P. Gecys, G. Raciukaitis, M. Ehrhardt, K. Zimmer, M. Gedvilas, Appl. Phys. A **101**, 373 (2010)
8. T. Rublack, S. Hartnauer, P. Kappe, C. Swiatkowski, G. Seifert, Appl. Phys. A **103**, 43 (2011)
9. T. Rublack, M. Schade, M. Muchow, H.S. Leipner, G. Seifert, J. Appl. Phys. **112**, 023521 (2012)
10. W. Wang, G. Jiang, X. Mei, K. Wang, J. Shao, C. Yang, Appl. Surf. Sci. **256**, 3612 (2010)
11. J.M. Liu, Opt. Lett. **7**, 196 (1982)
12. A. Cavalleri, K. Sokolowski-Tinten, J. Bialkowski, M. Schreiner, D. von der Linde, J. Appl. Phys. **85**, 3301 (1999)
13. D. Arnold, E. Cartier, Phys. Rev. B **46**, 15102 (1992)

6

Factors limiting the open-circuit voltage in microcrystalline silicon solar cells

M. Nath[1,a], S. Chakraborty[1], E.V. Johnson[2], A. Abramov[2], P. Roca i Cabarrocas[2], and P. Chatterjee[1]

[1] Energy Research Unit, Indian Association for the Cultivation of Science, 700 032 Kolkata, India
[2] Laboratoire de Physique des Interfaces et des Couches Minces, École Polytechnique, CNRS, 91128 Palaiseau, France

Abstract In studying photovoltaic devices made with silicon thin films and considering them according to their grain size, it is curious that as the crystalline fraction increases, the open-circuit voltage (V_{oc}) – rather than approaching that of the single-crystal case – shows a decline. To gain an insight into this behavior, observed in hydrogenated microcrystalline silicon (μc-Si:H) solar cells prepared under a variety of deposition conditions, we have used a detailed electrical-optical computer modeling program, ASDMP. Two typical μc-Si:H cells with low ($\sim$79%) and higher ($\sim$93%) crystalline volume fractions (F_c), deposited in our laboratory and showing this general trend, were modeled. From the parameters extracted by simulation of their experimental current density – voltage and quantum efficiency characteristics, it was inferred that the higher F_c cell has both a higher band gap defect density as well as a lower band gap energy. Our calculations reveal that the proximity of the quasi-Fermi levels to the energy bands in cells based on highly crystallized μc-Si:H (assumed to have a lower band gap), results in both higher free and trapped carrier densities. The trapped hole population, that is particularly high near the P/I interface, results in a strong interface field, a collapse of the field in the volume, and hence a lower open-circuit voltage. Interestingly enough, we were able to fabricate fluorinated μc-Si:H:F cells having 100% crystalline fraction as well as very large grains, that violate the general trend and show a higher V_{oc}. Modeling indicates that this is possible for the latter case, as also for a crystalline silicon PN cell, in spite of a sharply reduced band gap, because the lower effective density of states at the band edges and a sharply reduced gap defect density overcome the effect of the lower band gap.

1 Introduction

When considering the full spectrum of thin film Si-based photovoltaic technologies and ranking them according to their crystalline volume fraction (F_c), it is remarked that the open-circuit voltage (V_{oc}) decreases with increasing F_c. This has been observed by various groups working with cells fabricated from hydrogenated microcrystalline silicon (μc-Si:H) films deposited by a variety of deposition techniques [1–4] (Fig. 1). This fact hinders the further development of thin-film silicon photovoltaics as the loss in V_{oc} offsets any gain in the short-circuit current density (J_{sc}), prompting Mai et al. [1, 2] to remark that the optimum μc-Si:H solar cells are always obtained with intermediate crystallinity. The physical reason behind this loss in V_{oc} is not immediately clear, as the maximum impact of the most obvious factor (the smaller band gap) is diminished by the example of single-crystal silicon (c-Si), which possesses the smallest band gap but a higher V_{oc}

than cells with smaller grains (Fig. 1). Since the observation is a decrease of V_{oc} with increasing F_c of the μc-Si:H absorber from 60% to over 90%, while a mono-crystalline silicon PN cell has an appreciably higher V_{oc}, it is expected that a missing link maybe found – a limiting case of a very well-crystallized and large grained μc-Si:H material – which when used as the absorber layer in a solar cell, should yield an open-circuit voltage intermediate between that of a cell based on a highly-crystallized μc-Si:H absorber and the considerably higher V_{oc} of a mono-c-Si PN cell. The parameters characterizing such a cell, as deduced from modeling, should then help us understand how the open-circuit voltage of μc-Si:H solar cells can be made to approach the higher V_{oc} that characterizes single crystal silicon cells.

As a matter of fact we have found that solar cells based on fluorinated μc-Si:H [5] (μc-Si:H:F), with a very high F_c as well as a significant fraction of large grains (F_{lg}), appear to violate the general rule (Fig. 1 – in fact this figure shows, besides the typical V_{oc} of classical diffused junction c-Si solar cells [6], also the V_{oc}'s of "heterojunction

---

[a] e-mail: madhumitanath_21@yahoo.co.in

Table 1. Comparison between the measured and simulated solar cell output parameters of the PIN devices having low and intermediate crystalline volume fraction (F_c) μc-Si:H as the intrinsic layer. F_{lg} gives the fraction of large grains in the I-layer. Also compared are the measured and modeled output parameters of the highly crystallized, large grained μc-Si:H:F cell.

Sample	F_c (%)	F_{lg} (%)		J_{sc} (mA/cm^2)	V_{oc} (V)	FF	Efficiency (%)
μc-Si:H (A)	79	0	Expt	16.76	0.54	0.687	6.22
			Model	16.76	0.54	0.685	6.20
μc-Si:H (B)	93	27	Expt	19.94	0.46	0.631	5.79
			Model	19.85	0.45	0.643	5.73
μc-Si:H:F	$\sim$100	50	Expt	23.20	0.523	0.680	8.30
			Model	22.93	0.526	0.704	8.49

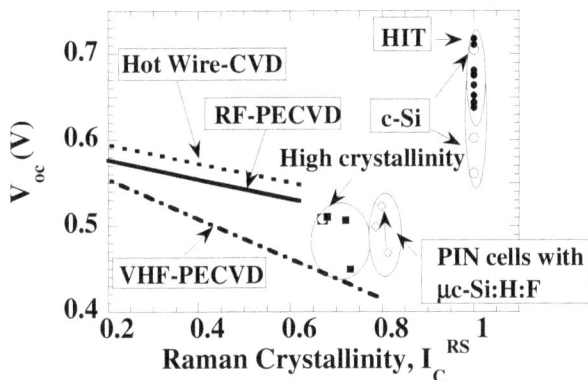

Fig. 1. Results taken from various literature sources [1–4] indicate the general trend of a decrease in the open-circuit voltage of μc-Si:H thin film PIN cells, with increase of the Raman crystallinity of the films. We also show an exception where the V_{oc} actually increases [5] in a highly crystallized μc-Si:H:F cell, specially after interface treatment (indicated by the arrow). Also shown in the figure are typical V_{oc}'s of classical diffused junction c-Si solar cells (open circles – 5, Green's ref), of a "heterojunction with intrinsic thin layer (HIT)" cell on a P-type c-Si substrate [7] and of a HIT cell on N-type c-Si [8] (closed circles).

with intrinsic thin layer (HIT)" cells on P-type [7] and N-type [8] c-Si substrates, for completeness). A material containing large grains is a dense material with low oxygen content, which can be obtained when using SiF$_4$ based plasma processes [5]. We then chose two of a series of typical (that follow the general rule of decreasing V_{oc} with increasing F_c – Fig. 1) μc-Si:H cells, as well as a cell based on μc-Si:H:F, for modeling using the detailed electrical-optical computer code ASDMP [9, 10]. Using parameters extracted by simulating the output experimental characteristics of the μc-Si:H thin film solar cells with absorber layers of different degrees of crystallinity, we identify in this article, the parameters responsible for the general decline of V_{oc} in such cells with increasing crystalline volume fraction (Fig. 1). Moreover by simulating the output characteristics of the μc-Si:H:F solar cells, we could identify the critical parameters which prevent μc-Si:H solar cells in achieving the higher V_{oc} attainable in c-Si PN cells.

2 Experimental details

Microcrystalline PIN solar cells having the structure textured ZnO/P-μc-Si:H/I-μc-Si:H/N-a-Si:H/Aluminum have been deposited in a multiplasma monochamber radio-frequency plasma-enhanced chemical vapor deposition (RF-PECVD) reactor [11]. Two sets of μc-Si:H solar cells have been deposited with $\sim$60 mW/cm^2 of RF-power at 175 °C. We employed different ratios of silane to hydrogen flow rates during the intrinsic layer deposition for the two sets – SiH$_4$:H$_2$ = 6:200 (cell A) and SiH$_4$:H$_2$ = 4:200 (cell B) [4], which results in different total crystalline volume fraction (F_c) and large grain fraction (F_{lg}) in the samples (Tab. 1). A third set has a fluorinated μc-Si:H intrinsic (I)-layer (Tab. 1), where the I-layer was deposited from a SiF$_4$, H$_2$ and Ar gas mixture [5]. The respective flow rates of the above gases were 3, 3 and 70 sccm respectively. The deposition pressure was 2.7 torr, the RF power 440 mW cm^{-2} and the substrate temperature 200 °C. The μc-Si:H P-layer and the a-Si:H N-layer were deposited in a second multi-chamber reactor, using TMB and PH$_3$ as dopant gases, and thus necessitating a vacuum break between the doped and intrinsic layers. A hydrogen plasma treatment (after both air breaks) was applied towards passivation of the P/I and I/N defects. The P- and N-layers were deposited using similar conditions for all cells. It maybe pointed out that a series of cells having similar output characteristics were deposited during each run and that the values cited in Table 1 are representative of each series.

The composition of the films was obtained from the Bruggeman effective medium approximation modeling [12] of the pseudo-dielectric function of the films, deduced from spectroscopic ellipsometry measurements. This approach has been shown to be well adapted to the modeling of μc-Si films [13]. In this approach we used as components for the model the dielectric functions of (i) amorphous silicon, (ii) voids to take into account the porosity of the films; (iii) small grain μc-Si and (iv) large grain μc-Si material produced by Chemical vapor deposition (CVD) at $\sim$650 °C [14]. In Table 1 we report the values of the total crystalline fraction F_c (which is the sum of the small grain and large grain fractions) and the large grain fraction

F_{lg}, for the two sets of μc-Si:H samples and those of μc-Si:H:F. Indeed, achieving a high value of F_{lg} in the case of μc-Si:H:F, indicates that the films obtained by PECVD at 175 °C have a grain size similar to that of polycrystalline layers produced by CVD at ~650 °C [14]. The greatly improved spectral response at long wavelengths also confirms the very high crystalline volume fraction of μc-Si:H:F [5].

3 Simulation model

The one-dimensional electrical-optical model ASDMP [9, 10] (amorphous semiconductor device modeling program) used in this study solves the Poisson's equation and the two carrier continuity equations under steady state conditions for the given device structure, and yields the resulting J-V characteristics and the quantum efficiency. The electrical part of the modeling program is described in references [15, 16]. The expressions for the free and trapped charges, the recombination term, the boundary conditions and the solution technique in this program are similar to the AMPS computer program developed by McElheny et al. [17].

The gap state model used in these calculations consists of the tail states, as well as a donor-like and an acceptor-like set of Gaussian distribution functions to simulate the deep dangling bond states. The contact barrier heights for a cell with the P-layer in contact with the ZnO at $x = 0$ and the N-layer in contact with aluminum at $x = L$, are taken to be 1.11 eV and 0.2 eV, respectively. Since the activation energies of our P-μc-Si:H and N-a-Si:H layers are 0.09 eV (P-μc-Si:H band gap is 1.2 eV) and 0.2 eV respectively, this implies that no effective surface band bending has been assumed at the ZnO/P-μc-Si:H contact and the N-a-Si:H/Al contact is ohmic. The assumption of no band bending at the ZnO/P-μc-Si:H interface originates from the fact that we had to assume an extremely high P-layer doping density (3×10^{19} cm^{-3}), therefore also a very high P-layer defect density, to simulate these cells. This means that when the TCO and the P-layer are put in contact, the trapped electrons on the P-μc-Si:H side are confined to a very thin layer on the P-side, resulting in a very high surface band bending that however does not appreciably extend into the bulk of the P-layer. For the purpose of calculating the built-in potential (V_{bi}) and V_{oc}, the bulk activation energy of the P-layer is then already achieved almost at the TCO/P interface, and the band-bending does not extend to any appreciable thickness of the P-layer.

The generation term in the continuity equations has been calculated using a semi-empirical model [18], that has been integrated into the modeling program [9, 10]. Both specular interference effects and diffused reflectances and transmittances due to interface roughness are taken into account. It is now well-recognized that TCO texture is a key issue in increasing cell efficiency, as it reduces optical reflection loss and greatly increases light diffusion. Weakly absorbed radiation, when scattered, can be partly absorbed in a high refractive index layer, such as

amorphous or micro-crystalline Si ($n \approx 3.7$), due to total internal reflection at the interfaces giving rise to optical confinement. However because of the broad distribution of grain size and shape, the interaction between the multi-layer device and incident light is very complex and so, rather than a rigorous three-dimensional electromagnetic treatment of the diffused radiation, a rather sophisticated semi-empirical model [18], was integrated [9, 10] with the electrical model [15, 16]. Here diffused reflectances and transmittances, are derived from angular resolved photometric measurements, and used as input parameters. In the model, the electromagnetic field's specular reflection and transmission are assumed proportional to the Fresnel coefficients, the proportionality factor depending on the amount of total diffused light. In the specular part light coherence is kept, resulting in interference effects when the TCO is more or less flat. However in the diffused part, light coherence with the incident light is assumed lost, so, the point where light is diffused, is considered as a new source emitting in several directions in the stack. The latter light in each direction is assumed to be a plane wave, and each wave, when it meets the next rough interface, is again divided into specular and diffused components. Instead of calculating and successively adding each of these components, the total electromagnetic field is directly derived by the matrix method of Abeles [19–21]. In the model it is possible to consider up to two rough interfaces. These are taken to be the TCO/P and N/metal interfaces in the present case. The complex refractive indices for each layer of the structure are also required as input, and have been measured in-house by spectroscopic ellipsometry. These are presented in Figure 2.

4 Experimental results and analysis

Since the aim of this article is to understand the general trend in μc-Si:H solar cells, which is that the open-circuit voltage (and the fill factor (FF) to a certain extent) decreases with increasing crystalline volume fraction, as also to understand why the most ordered c-Si PN cell has a higher V_{oc} (Fig. 1), we need to model a variety of output characteristics of μc-Si:H cells in order to extract parameters that characterize a representative cell of each series. The first 2 cells of Table 1 follow the general trend, as is obvious from the appreciably lower V_{oc} of the cell with the intermediate crystalline volume fraction. However the fluorinated μc-Si:H cell with a very high value of the crystalline volume fraction (F_c) violates this general trend (Tab. 1) and exhibits both higher V_{oc} and J_{sc}. It may be noted that this cell has a particularly high large grain fraction, which may have a bearing to its exceptional behavior. The interest in modeling the latter cell is to gain an insight into the possible reasons why the most ordered mono-c-Si PN cell has a higher V_{oc} than μc-Si:H cells.

The experimental solar cell output parameters for the "low F_c" cell (device A, $F_c \sim 79\%$, no large grain fraction detected), the "intermediate F_c" cell (device B, $F_c \sim 93\%$, with large grain fraction ~27%), and the fluorinated μc-Si:H cell ($F_c \sim 100\%$, $F_{lg} \sim 50\%$) are compared to the

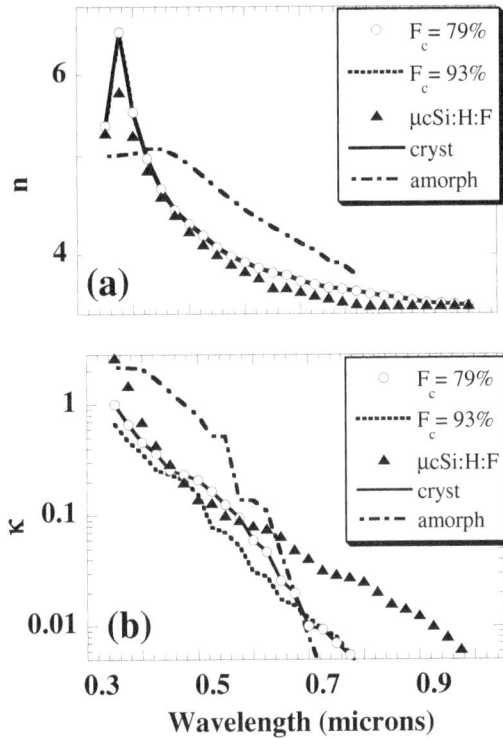

Fig. 2. Values of the complex refractive indices (a) real part, n and (b) imaginary part, κ for low (79%), intermediate (93%) crystalline volume fraction (F_c) μc-Si:H, and for μc-Si:H:F, $F_c \sim 100\%$, compared to the respective values of a-Si:H (amorph) and c-Si (crys).

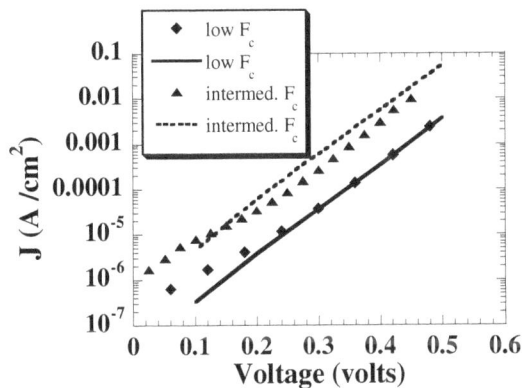

Fig. 4. Calculated values (open symbols) of (a) the reverse saturation current density (J_0) and (b) the diode ideality factor (n) for low and intermediate F_c μc-Si:H solar cells, compared to experiments (closed symbols) at temperatures from 10 °C to 50 °C. The lines are guides to the eye.

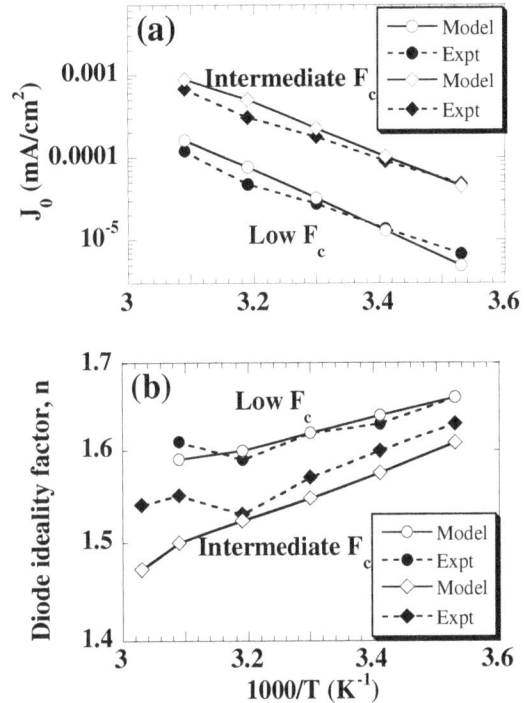

Fig. 3. Calculated dark J-V characteristics of the intermediate F_c and low F_c μc-Si:H cells at 30 °C, compared to experimental results. The lines are our modeling results, while symbols represent experimental measurements.

the lower F_c cell, A. This comparison has been done for temperatures from 10 to 50 °C, and the modeled and experimental dark saturation current density (J_0) and diode ideality factor (n) of these cells are presented in Figures 4a and 4b. The external quantum efficiency (EQE) curves of all three types of solar cells are compared to their experimental counterparts in Figures 5a–5c. The match with experiments appears to be satisfactory.

All experimental results of a particular type of cell – μc-Si:H cell A, μc-Si:H cell B and of the fluorinated μc-Si:H cell – have been simulated with the same set of input parameters, which are given in Table 2. It may be mentioned that the electron and hole mobilities shown in Table 2 and used as input to the modeling program, are the band microscopic mobilities. The drift mobilities measured in actual experiments are the band microscopic mobilities reduced by trapping and de-trapping of carriers, and are therefore one to two orders of magnitude lower. Moreover, we have assumed in general that the carrier mobilities are higher as the material becomes more crystallized. To accurately model all aspects of the experimentally measured solar cell performance, we had to assume that the more crystallized μc-Si:H cell B (with however a large grain fraction that is considerably lower than in the highly crystallized fluorinated μc-Si:H cell) has a lower band gap, higher carrier mobilities, a higher mid-gap defect density, and broader band tails (Tab. 2) relative to the μc-Si:H cell A. We further infer from modeling that the μc-Si:H:F cell ($F_c \sim 100\%$) has an even

modeling results in Table 1. The dark current density vs. voltage (J-V) characteristics of the first two types of cells at 30 °C have been both measured and simulated, and the good agreement between these results can be noted from Figure 3. Brammer and Stiebig [22,23] have also observed that the dark forward current at low forward voltages is a strong function of the silane concentration (SC) and is lower for higher values of SC, in other words, lower for

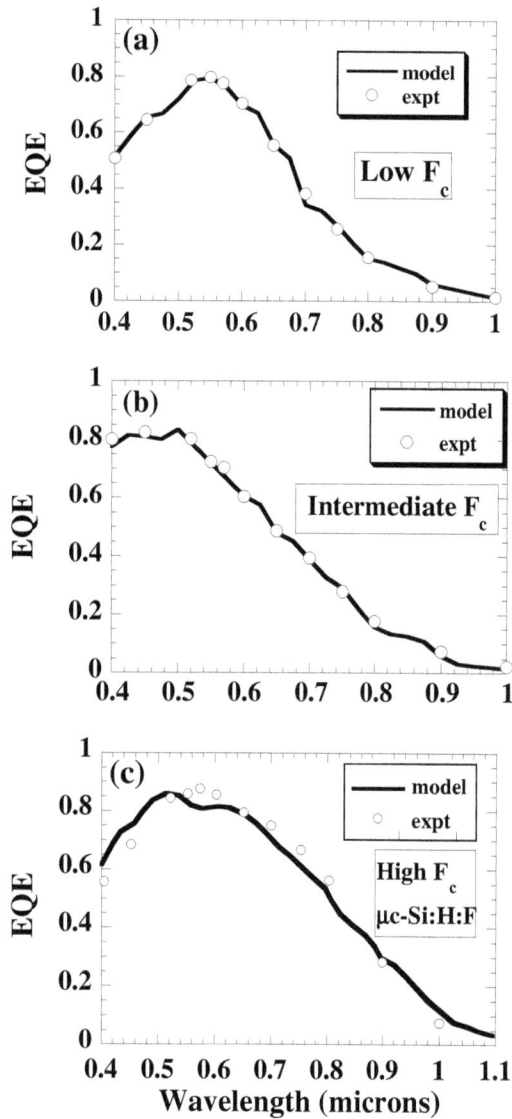

Fig. 5. Calculated external quantum efficiency (EQE) curves under AM1.5 bias light and short-circuit conditions for the (a) low F_c μc-Si:H cell A, (b) intermediate F_c μc-Si:H cell B and (c) high F_c, large grained μc-Si:H:F cell, compared to experimental results.

lower band gap and higher carrier mobilities. This dense material with low oxygen content and very high large grain fraction [5], should have fewer grain boundaries and modeling indicates that it has a sharply reduced dangling bond defect density and effective density of states at the band edges, the latter similar to c-Si. (Tab. 2). Only its valence and conduction band tails appear to be fairly broad (the characteristic energies are 40 meV and 20 meV respectively for the valence and conduction band tails, as in the case of the intermediate F_c μc-Si:H cell B). This implies that this highly crystallized material nevertheless has a strained lattice. Justification of the parameters inferred from modeling that characterize the different types of cells (Tab. 2) will be presented in Section 6.

However, it maybe relevant at this point to compare the parameters extracted by the present modeling (Tab. 2) to some other modeling results in the literature. Strengers et al. [24] and Sturiale et al. [25] have modeled the dark and illuminated J-V characteristics of μc-Si solar cells, where the intrinsic layer is deposited by the hot wire CVD (HWCVD) technique. They have observed [24] that in these samples the optical absorption in the red region is much higher than in amorphous silicon. Since the optical absorption is related to the imaginary part of the complex refractive index (κ) and our Figure 2b indicates that this is the case for our μc-Si:H:F samples, it is reasonable to compare the parameters used in their modeling of HWCVD deposited μc-Si cells to those of our highest F_c μc-Si:H:F cell. The comparison is shown in Table 2 in brackets with asterisks. We note that the values of references [24, 25] are quite close to what we have assumed for our μc-Si:H:F I-layer. In particular the effective DOS in the valence and conduction bands; as well as the tail prefactors G_{D0}, G_{A0} are similar to ours and an order of magnitude lower than those normally assumed for hydrogenated amorphous silicon and also assumed for the low (A) and intermediate (B) F_c μc-Si:H I-layers here, for reasons to be justified in Section 6. The mid gap defect density is $\sim 10^{15}$ cm^{-3}, also like our value for μc-Si:H:F and the band gap of 1.25 eV is close to ours, and higher than that of c-Si (1.12 eV). Only the capture cross-sections of the dangling bond states are more than two orders of magnitude higher than our case. Probably this had to be assumed for the HWCVD μc-Si I-layers [24, 25] since the current density from these devices ($13-16$ mA cm$^{-2}-22$) is much lower than ours (23.20 mA cm^{-2}) for comparable values of the absorption coefficients. On the other hand, our average mid gap defect density is the same and their capture cross-sections in the highest F_c μc-Si:H I-layer are much closer to the values extracted from modeling the dark J-V characteristics of PECVD μc-Si I layers in reference [22].

We had also attempted to model the experimental characteristics of the "intermediate F_c" cell B, without decreasing the mobility gap, and thus by increasing the gap state defect density alone. This is because, Yan et al. [26] have mentioned without employing detailed modeling to support their statement, that this decrease of V_{oc} is due to distorted bonds in the grain boundary regions that lead to increased band gap defects. Modeling however reveals that a Gaussian defect density of some $\sim 6 \times 10^{17}$ cm^{-3} must be assumed to match the low V_{oc} in this case, which also causes a sharp fall in the short-circuit current density (J_{sc}) and fill factor (FF), resulting in our determination that all aspects of the solar cell output characteristics of cell B cannot be matched by increasing the band gap defect density alone.

Modeling of the EQE curves in Figure 5a and 5b suggests that the P-layer in the intermediate F_c cell B is thinner than that in the low F_c cell, although the P-layers in all types of cell were deposited under the same experimental conditions. This means that some etching of this layer occurred during the subsequent I-layer deposition for

Table 2. Parameters that characterize intrinsic μc-Si:H of different degree of crystallinity (as extracted by modeling). The quantities in brackets marked with asterisks in the column of parameters of μc-Si:H:F correspond to the values extracted by modeling similar μc-Si:H samples (refs TSF, JAP of Rubinelli).

Parameters	μc-Si:H (A) ($F_c = 79\%$)	μc Si:H (B) ($F_c = 93\%$)	μc-Si:H:F ($F_c = 100\%$)
Thickness (μm)	0.77	0.84	1.50
Mob. gap (E_μ) (eV)	1.4	1.33	1.2 (1.25–1.32)*
Elec. affinity (eV)	4.0	4.0	4.2
Eff. DOS in bands N_c	2×10^{20}	2×10^{20}	2.8×10^{19} (3×10^{19})*
(cm^{-3}) N_v	2×10^{20}	2×10^{20}	1.04×10^{19} (3×10^{19})*
Charac. En. E_D (E_A) (meV)	20 (10)	40 (20)	40 (20) (35 (23))*
G_{D0}, G_{A0} (cm^{-3}eV^{-1})	4.0×10^{21}	4.0×10^{21}	3.0×10^{20} (2×10^{20})*
Elec. (hole) mobility	32 (8)	120 (30)	200 (50)
(cm^2/V s)			(120–150 (40–75))*
Gaussian defect den. (cm^{-3})	4.0×10^{16}	8.0×10^{16}	1.0×10^{15} (2.5–3.75×10^{15})*
Gaussian σ charged	10^{-15}	10^{-15}	10^{-16} (6×10^{-14})*
Neutral	10^{-16}	10^{-16}	10^{-17} (6×10^{-15})*
Tail σ charged	10^{-15}	10^{-15}	10^{-15} (10^{-15})*
Neutral	10^{-16}	10^{-16}	10^{-16} (10^{-17})*

the intermediate F_c case. This is not surprising, since to obtain the more crystallized μc-Si:H layer a higher hydrogen dilution was employed. Also, the P/I interface layer is both thinner and demonstrates lower capture cross-sections than in the case of the low F_c μc-Si:H cell. These factors combine to yield the very high blue response in the case of the intermediate F_c μc-Si:H solar cell (Fig. 5b). An interface layer is expected when the I-layer is deposited on top of the P-layer, as occurs in a PIN device deposition process. Hence, physically one might expect a thinner interfacial layer in the intermediate F_c I-layer case, since the band gap mismatch between its I-layer ($E_g = 1.33$ eV) and the highly crystallized P-layer ($E_g = 1.2$ eV, deposition parameters same for both cells) is smaller than for the low F_c μc-Si:H solar cell (I-layer band gap 1.4 eV).

5 Discussion

As already stated in the previous section the intermediate F_c μc-Si:H cell B has higher carrier mobilities, a higher dangling bond defect density, broader band tails and lower band gap than μc-Si:H cell type A, having the lowest crystalline volume fraction (Tabs. 1, 2). It is a combination of these factors that results in cell B having a higher current density but lower V_{oc}, FF and conversion efficiency relative to cell A. In the following we will study how the higher gap defect density and lower band gap affect cell performance. In studying the sensitivity of μc-Si:H cell performance to each of the above-mentioned parameters, all other parameters are held constant at the values of the μc-Si:H cell, type A having the lowest F_c, which we may call our reference case.

5.1 Effect of changes in the I-layer band gap defects on the photovoltaic response

5.1.1 Sensitivity to the characteristic energy of the band tails

The μc-Si:H cell A (Tab. 2) presents values for the characteristic energy of the valence and conduction band tails of $E_D = 20$ meV and $E_A = 10$ meV, respectively. To model the intermediate F_c cell B, one of the changes that we had to assume was broader band tails, as characterized by $E_D = 40$ meV and $E_A = 20$ meV. Figure 6a depicts the illuminated J-V characteristics for two cases differing only in the characteristic energy of the band tails and indicating a fall of V_{oc} by $\sim$0.03 V when the band tails broaden. Figure 6b plots the electric field in the two devices, with the high field at the P/I interface shown in a different scale in the inset. We find that higher photo-generated hole trapping in the broader valence band tail near the P/I interface, where the quasi-Fermi level lies close to the valence band, results in a stronger electric field near this interface, and a consequent fall of the field in the volume of the absorber layer. This fact is known [15], to bring down the open-circuit voltage.

5.1.2 Sensitivity to the Gaussian defect density

Again, using as reference the case of the lowest F_c μc-Si:H cell type A with parameters as given in Table 2, we have studied the impact of an increase in only the Gaussian defect density starting from a value of 4×10^{16} cm^{-3}, characteristic of cell A, to 8×10^{16} cm^{-3}, inferred from modeling of the μc-Si:H cell B. This leads to

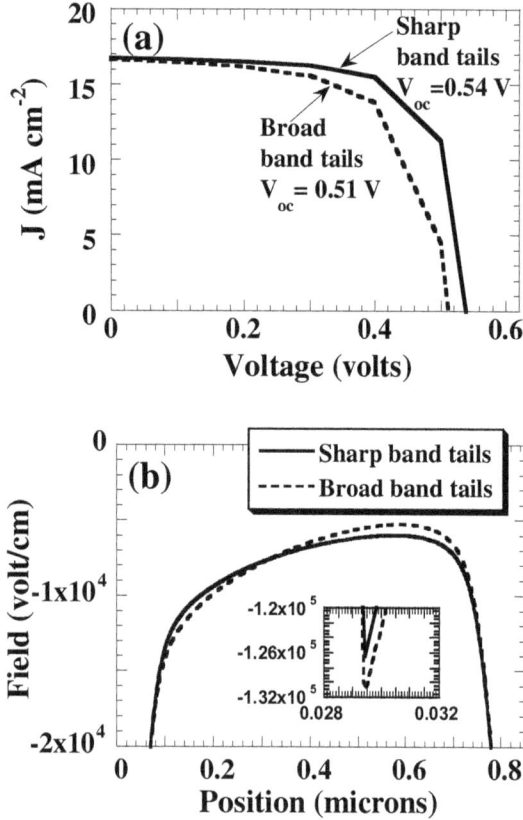

Fig. 6. (a) Sensitivity of the illuminated J-V characteristic to the valence and conduction band tail characteristic energies and (b) the electric field in the device when the band tails are sharp (characteristics energies of band tails E_D, E_A = 20 meV, 10 meV respectively, characteristic of μc-Si:H cell type (A) and when they are broader (E_D, E_A = 40 meV, 20 meV respectively, characteristic of cell type (B). The electric field over the P/I interface region is plotted on a different scale in the inset.

a decrease in V_{oc} from 0.54 V down to 0.52 V and an increase in the dark saturation current J_0 from 1.15×10^{-5} mA/cm^2 up to 8.35×10^{-5} mA/cm^2. We again find that a higher field near the P/I interface due to higher photo-generated hole trapping for the case having higher mid gap DOS is responsible for the collapse in the bulk electric field and the fall in V_{oc}.

We may confirm our inferences above (obtained by detailed modeling using ASDMP) regarding a lower V_{oc} in the cell having a higher defect density by considering the approximate analytical formula which ignores the shunt and series resistances:-

$$V_{oc} = (nkT/q) \ln [(J_{sc}/J_0) + 1]. \qquad (1)$$

Here n is the diode ideality factor, J_0 the dark saturation current density, q the electronic charge and T the absolute temperature. It indicates that as the dark reverse saturation current density J_0 increases, V_{oc} decreases. We indeed find that J_0 is increased by a higher mid gap defect density, and that this corresponds to a lower V_{oc}, in agreement with equation (1).

5.2 Effect of the band gap on the open-circuit voltage

As has already been stated, in order to model all aspects of the dark and illuminated characteristics of the intermediate F_c μc-Si:H cell B (Tab. 1), one of our assumptions was a lower mobility band gap in this material (Tab. 2) relative to case A having a lower crystalline volume fraction. In Table 4 we compare the J-V parameters of two μc-Si:H solar cells having exactly the same parameters as the lowest F_c μc-Si:H cell A (Tab. 2), except that in one case (named "high E_g") the I-layer band gap is 1.4 eV (the value used to model the lowest F_c cell A, Tab. 2), while in the other case (named "low E_g") the I- μc-Si:H band gap is 1.33 eV, the value required to simulate the intermediate F_c cell B (Tab. 2). As a result we observe a drop in V_{oc} from 0.54 V down to 0.50 V, accompanied by a factor of 10 increase in J_o, while the J_{sc} and FF are practically not affected.

As mentioned already, all parameters for the low and high "E_g" cells, as also the light absorbed in every segment of the cells, have been assumed the same for the two cases that differ only in the band gap. Nevertheless, modeling reveals that to accommodate the difference in the band gap between the two cells, both the quasi-Fermi level separation and the distance of the Fermi levels from the band edges are less for the case of the cell with a lower intrinsic layer band gap. This latter fact results in a higher free carrier density in the bands as quantified by the following expressions of the free carrier densities:

$$n = N_C \exp -[E_C - E_{F_n}]/kT \qquad (2)$$
$$p = N_V \exp -[E_{F_p} - E_V]/kT. \qquad (3)$$

The trapped carrier density at any point in the device depends on the corresponding free carrier density, the defect density at that location as well as the relative values of the charged and neutral capture cross-sections of these defect states. Hence all else remaining the same, an increase of the free carrier density due to a lowering of the band gap, results in increased carrier trapping in the defect states. As the quasi-Fermi level for holes is closer to the valence band of the intrinsic layer at the P/I interface, the higher free hole density (from Eq. (3)) in the "low E_g, intermediate F_c" cell, results in particularly high hole trapping and hence electric field over the P/I interface region, that in a manner similar to Figure 6b for the band tail case, leads to a fall of the electric field in the volume of the absorber layer. The latter fact is known [15] to bring down the open-circuit voltage, and explains the lower V_{oc} of the cell having μc-Si:H of lower band gap, higher F_c.

In Sections 5.1 and 5.2 we have analyzed the reasons why a solar cell having a higher crystalline volume fraction μc-Si:H intrinsic layer may have a lower V_{oc} by examining specific sets of models. However, as seen in the experimental results (Tab. 1), a μc-Si:H cell with a higher F_c will show a higher J_{sc}, primarily because of higher free carrier mobilities (Tab. 2). Nevertheless, as Table 1 indicates, the energy conversion efficiency for the higher F_c μc-Si:H cell B is ultimately lower than that of cell A, due to the accompanying reduction in V_{oc} and FF. In other words,

Table 3. The solar cell output of the fluorinated μc-Si:H solar cell compared to the output of a hypothetical cell having higher effective DOS at the band edges and higher exponential tail pre-factors.

Case	N_c, N_v (cm^{-3})	G_{D0}, G_{A0} (cm^{-3} eV^{-1})	J_{sc} (mA cm^{-2})	V_{oc} (V)	FF	Efficiency (%)
μc-Si:H:F cell	2.80×10^{19}, 1.04×10^{19}	3.0×10^{20}	22.93	0.526	0.704	8.49
Hypothetical cell, D	2.0×10^{20}	4.0×10^{21}	22.96	0.412	0.653	6.18

Table 4. The solar cell output parameters of two μc-Si:H cells having exactly the same parameters as the lowest F_c μc-Si:H cell A , except the band gap of the intrinsic material.

E_g (eV)	J_{sc} (mA cm^{-2})	V_{oc} (V)	FF	Efficiency (%)	J_0 (mA cm^{-2})
1.4	16.76	0.54	0.69	6.20	1.15×10^{-5}
1.33	16.97	0.50	0.69	5.85	1.02×10^{-4}

due to three interacting effects – higher mid gap and tail defect density and lower band gap – the intermediate F_c μc-Si:H cell B (Tab. 1) will show a fall in V_{oc} significant enough to cancel the advantage of a higher J_{sc} due to higher carrier mobilities.

We now examine how it is possible for the fluorinated μc-Si:H cell having an even higher F_c, as well as a very high large-grain fraction, to have a higher V_{oc} than the intermediate F_c cell B, in spite of a further reduction of its band gap, as predicted by modeling.

5.3 V_{oc} in the highly crystallized large grained μc-Si:H:F cell

So far we have compared the parameters as deduced from modeling (Tab. 2) that characterize the μc-Si:H solar cells A ($F_c = 79\%$, no large grains detected) and the more crystallized cell B ($F_c = 93\%$, $F_{lg} = 27\%$), and explained in Sections 5.1 and 5.2, why the latter shows a lower open-circuit voltage and fill factor than the former. This is normal behavior for μc-Si:H cells, as evidenced from numerous experimental observations (Fig. 1) and is the reason why the best μc-Si:H solar cells have so far been produced close to the a-Si:H/μc-Si:H transition. However experiments indicate an all-round improvement in the output characteristics of the highly crystallized large grained fluorinated μc-Si:H ($F_c \sim 100\%$, $F_{lg} \sim 50\%$) cell relative to type B μc-Si:H cells, indicating that the output properties of this cell violates the general trend of decreasing V_{oc} with increasing crystalline volume fraction (Fig. 1). By modeling its output characteristics (Tab. 1 and Fig. 5c), we have extracted the parameters that characterize this material (Tab. 2). The salient features of the μc-Si:H:F cell parameters are: (a) a reduced band gap compared to μc-Si:H cells A and B, (b) higher carrier mobilities, (c) sharply reduced dangling bond defect density, (d) reduced effective DOS at the band edges, that match those of mono-crystalline silicon and (e) higher absorption over a large portion of the longer visible wavelengths compared to low and intermediate F_c μc-Si:H (Fig. 2b, the absorption coefficient is

proportional to the imaginary part of the complex refractive index κ). The characteristic energies of its band tails however are similar to those of cell B, indicating that this highly crystallized material nevertheless has a strained lattice. The higher absorption and, to a smaller extent the larger carrier mobilities, are responsible for the high current density in this cell (Tab. 1). Assumption of the lowest band gap for this material, having the highest crystalline volume fraction, follows the general trend of the parameters inferred by modeling the output characteristics of the three types of μc-Si:H cells (Tab. 2). What is surprising however is that this fact does not lead to a further fall in V_{oc} following the general rule in μc-Si:H solar cells (Fig. 1). One reason for this is the sharp fall in the dangling bond defect density in this case that, as described in Section 5.1.2, leads to an improvement of the electric field over the intrinsic layer and increased V_{oc}. This fact partially cancels the negative effect of a lower band gap on V_{oc}. In the following sub-section we examine the effect of reduced effective DOS at the band edges for the case of the fluorinated μc-Si:H I-layer (Tab. 2) and address how this can also partly explain the observed improvement in V_{oc} for this type of cells.

5.3.1 Sensitivity of V_{oc} to the effective DOS at the band edges

Table 3 compares the solar cell output parameters of the fluorinated μc-Si:H cell with that of a hypothetical cell D, having identical parameters as the former, except that the effective DOS at the band edges in cell D are like those in hydrogenated amorphous silicon (a-Si:H) or the other μc-Si:H cells A and B. In other words N_c, N_v for cell D are 2×10^{20} cm^{-3}, while they are 2.8×10^{19} cm^{-3} and 1.04×10^{19} cm^{-3} for N_c, N_v respectively for fluorinated μc-Si:H (Tab. 2). Note that the exponential band tail pre-factors are related to the N_c, N_v via the relations:

$$G_{A0} \sim N_c/kT; \quad G_{D0} \sim N_v/kT, \qquad (4)$$

so that a fall in N_c, N_v automatically reduces G_{A0} and G_{D0}. This impacts on the tail defect density according to the relations:

$$g_A = G_{A0} \exp(-E/E_A) \qquad (5)$$

$$g_D = G_{D0} \exp(-E'/E_D), \qquad (6)$$

where g_A, g_D are the tail defect densities (cm^{-3} eV^{-1}) at energy locations E and E' respectively from the conduction and valence band edges; and E_A and E_D are the characteristic energies of the respective band tails. Thus reduced $N_{c(v)}$ lead to reduced band tail defect density, even for the same values of the characteristic energies of the band tails. We thus note that although the fluorinated μc-Si:H cell has the same values of E_D and E_A as the μc-Si:H cell B (Tab. 2), the band tail defect density is smaller for the former. This then is one reason for the higher open-circuit voltage for this case relative to cell B (Tab. 1), according to the arguments presented in Section 5.1.1.

The position of the Fermi-level is determined by the relaxation, trapping, and recombination dynamics of the photo-generated carriers, and thus for a given density of free-carriers, a greater quasi-Fermi level separation can be achieved with a lower $N_{c(v)}$ (from Eqs. (2), (3)) and a higher V_{oc} will result. Also lower $N_{c(v)}$ means lower free – and therefore trapped carrier densities – for a given quasi-Fermi level separation (Eqs. (2), (3)). Lower values of trapped carrier densities lead to lower P/I interface field and hence more field penetration into the bulk of the device (as explained in Sect. 5.2); therefore [15] an improved V_{oc}. Table 3 indicates large improvements in V_{oc} and FF possible as a consequence of the fall in N_c, N_v. We thus conclude that the fluorinated μc-Si:H cell has a higher V_{oc} and FF relative to the μc-Si:H cell B, because the combined effect of a lower dangling bond DOS, a lower effective DOS at the band edges and lower exponential tail pre-factors, overcome the negative influence of its reduced band gap.

5.4 V_{oc} in crystalline silicon PN solar cells

The original aim of this article was to investigate the general trend in μc-Si:H solar cells, which is that their open-circuit voltage decreases with increasing crystalline volume fraction (Fig. 1) and our modeling has indicated that one of the principal reasons for this is the lower energy band gap (Sect. 5.2) in more crystallized material. Crystalline silicon (c-Si) has a band gap of only 1.12 eV, so all else being equal, it should produce cells with an even lower V_{oc}. However, the typical V_{oc} of c-Si solar cells is *higher* than in those of good quality μc-Si:H and typically lies between 0.55 V and 0.6 V [6], while world record c-Si cells possess a V_{oc} above 0.7 V [27]. Fortunately we have been able to produce in our laboratory a series of μc-Si:H solar cells – the fluorinated μc-Si:H series of cells – that violate the observed general trend in μc-Si:H cells (Fig. 1) and exhibit a *higher* V_{oc}, in spite of having a lower energy band gap. This series has therefore provided the necessary insight to explain why the limiting case of c-Si solar cells can

possess higher V_{oc} in spite of a strongly reduced band gap. In the previous sub-section we have shown for the case of μc-Si:H:F, that the higher V_{oc} in spite of a reduced band gap was made possible by sharply reduced dangling bond (DB) and tail defects, as well as reduced effective DOS at the band edges. We have assumed the effective DOS at the band edges in μc-Si:H:F to be similar to the low values of c-Si. Moreover, c-Si has a DB DOS that is three orders of magnitude lower than even the relatively low mid-gap DOS of μc-Si:H:F (Tab. 2), and the tail states are absent. Therefore it is now only to be expected that c-Si solar cells will have higher V_{oc} and FF than those of the μc-Si:H:F cells, which are indeed far superior to those observed in highly crystalline μc-Si:H cells (example cell B in Tabs. 1 and 2), obeying the general trend of V_{oc} as a function of the crystalline volume fraction (Fig. 1).

6 Justification of the parameters deduced by modeling in the three types of μc-Si:H solar cells studied

Table 2 shows the parameters that characterize the intrinsic layers of the low F_c μc-Si:H series of cells A, the intermediate F_c cell series B and the large grained high F_c fluorinated μc-Si:H solar cells as inferred by modeling their output characteristics. These indicate that the intermediate F_c μc-Si:H cell B, has a lower band gap, higher carrier mobilities, and both higher mid-gap defect density and broader band tails as compared to the cell having low F_c μc-Si:H A. In order to simulate the improved cell performance (Tab. 1) of the μc-Si:H:F cell, we had to assume even higher carrier mobilities, lower mid-gap defects and lower effective DOS at the band edges (similar to crystalline silicon), while its band gap was assumed to be smaller than that of cells of type B. We have throughout assumed higher carrier mobilities for more crystallized materials. The presence of a significant fraction of large grains in a material produced by PECVD at 175 °C has been shown to correlate with improved transport properties of the films [28], and in particular we have shown elsewhere that the electron mobility as measured by time resolved microwave conductivity , increases with the fraction of large grains [28].

We have assumed a decreasing band gap with increasing crystalline volume fraction. This assumption is well-supported by reports from the literature. For example, Delley and Steigmeier [29] – who computed the band gap of μc-Si:H as a function of cluster diameter using the density functional approach for finite structures – have shown that the band gap of μc-Si:H increases as the cluster size decreases. A higher band gap for less-crystallized μc-Si:H was also previously measured by Hamma and Roca i Cabarrocas [30] using in situ Kelvin probe analysis and the "Flat Band Heterojunction" technique. Merdzhanova et al. [31] have studied the photoluminescence (PL) in thin film μc-Si:H PIN solar cells deposited by the Hot wire chemical vapor deposition technique [32] and have observed that the PL band shifts to higher energy with decreasing crystalline volume fraction.

Another property that had to be assumed in order to model the experimental characteristics are lower band tail characteristic energies in the μc-Si:H film of the lowest F_c, the absorber layer in cell A (Tab. 2). This assumption is also supported by experimental evidence, such as the previously cited work of Merdzhanova et al. [31]. Their assumption that PL originates from transitions between localized band tail states indicates that less-crystallized μc-Si:H has sharper band tails. Additionally, μc-Si:H films with lower F_c are expected to have a larger fraction of hydrogenated amorphous silicon (a-Si:H), which encourages structural relaxation of the μc-Si:H network [31], thus giving rise to less strained films with sharper band tails. Since a-Si:H is also expected to passivate grain boundary defects, our additional assumption of a lower Gaussian defect density for the least crystallized μc-Si:H in cell A (Tab. 2), compared to the more crystallized I-layer in cell B, also appears to be justified.

However it may be noted that while the intermediate F_c μc-Si:H I-layer B is predicted to have a higher DB defect density, compared to the less crystallized I-layer A, modeling the highly crystallized fluorinated solar cell requires a sharp decrease of this defect density. To understand this, we first note that although the intermediate F_c ($F_c = 93\%$) I-layer B (Tab. 2) has a high crystalline volume fraction, its large grain fraction F_{lg} is considerably lower than in the case of μc-Si:H:F. High F_c with low F_{lg} implies the presence of a large number of tiny crystallites and hence many grain boundaries, with a higher probability of defects. Additionally this material B, with a high F_c, possesses a low amorphous fraction (F_a). The latter is known to passivate grain boundary defects. Thus a large number of grain boundaries, together with a low F_a would necessarily lead to a high number of DB defects. Also more crystallized μc-Si:H (case B, Tab. 2) is known to have a fairly large oxygen content [33]. One possibility is that this oxygen occupies substitutional sites in the μc-Si:H lattice and produces N-type doping. However any appreciable doping of an intrinsic layer of a solar cell containing band gap defects where carriers can be trapped, would lead to sharp interface fields, weak penetration of field and flat bands in the volume of the intrinsic layer resulting in high recombination that should strongly reduce the current. This is not observed in the μc-Si:H cells studied (Tab. 1), thus excluding the possibility of oxygen producing any appreciable doping in the μc-Si:H cell B. These oxygen atoms, which are probably located at grain boundaries, therefore produce defects in this material leading to a higher DB DOS in the more crystallized I-layer B (Tab. 2). On the other hand fluorinated μc-Si:H has not only a high crystalline volume fraction but also a high fraction of large grains resulting in fewer grain boundaries and therefore less grain boundary defects. Also, the use of SiF_4 as gas precursors allows to reduce the concentration of oxygen leading to a dense large grained material with low defects [28] and justifies the sharply reduced defect density in this material as inferred from modeling

(Tab. 2), that is consistent with the enhanced electronic properties of this material [28].

6.1 Band-edge effective DOS (N_c, N_v) in amorphous and microcrystalline solids

We have underlined in Sections 5.3.1 and 5.4 that a key factor for improving the open-circuit voltage in the large grained fluorinated μc-Si:H and c-Si PN cells is the reduced band edge effective DOS ($N_{c(v)}$), lower by nearly an order of magnitude compared to amorphous and disordered μc-Si:H (I-layers A and B, Tab. 2). However, the reason behind the higher $N_{c(v)}$ in disordered silicon has not yet been addressed. To do so, it must first be noted that the effective DOS at the valence band edge (N_v) is calculated from the relationship satisfying:

$$\int_{-\infty}^{E_V} DOS(E)\, e^{-(E_F-E)/kT} dE \approx N_v\, e^{-(E_F-E_V)/kT} \quad (7)$$

and similarly for N_c, assuming that the Fermi level E_F is sufficiently far away from the valence band to approximate the Fermi distribution with the exponential relationship shown. Because N_v is calculated using the product of the true DOS with the Fermi distribution of holes, the states closer to the band-edge influence the value of N_v more strongly than those far from the edge. For this reason, the shifting of states towards the band-edge will increase N_v, though the absolute number of states may not change.

Theoretical evidence strongly suggests that such a shifting of states closer to the band edges occurs when a tetrahedrally bonded crystalline solid is amorphized. Solving the appropriate Bethe lattice for amorphous silicon, Joannopoulos [34, 35] showed a significant shifting of the DOS towards the band-edge. The tight-binding Hamiltonian used showed that the shift was particularly great for the P-like states at the valence band edge [34, 35] and also accounted for the steepening of the valence band edge density-of-states with disorder as observed in X-ray photoemission experiments [36]. Using a Continuous Random Network (CRN) model, Singh [37] has also shown that dihedral-angle and topological disorders lead to an increase of the DOS at the valence and conduction band edge respectively. The effect is equally seen when the effect of hydrogen is included. The strain in stretched Si-Si bonds may partially be released by hydrogen incorporation (resulting in hydrogenated amorphous or micro- crystalline silicon, the materials of interest in PIN solar cells), and hence removed from the tails [38]. The addition of hydrogen thus widens the band gap, moving the peak in the valence band DOS closer to the band edge, and increasing N_v at the band edge. Monte Carlo calculations examining mixed phase nanocrystalline-amorphous silicon showed that the location of the peak in the valence band DOS is little affected by the addition of a considerable volume fraction of crystalline material [39], although the mid-gap and lower energy DOS were significantly modified. This result indicates that the amorphized fraction has a dominant effect on the location of this peak, and that the assumption of higher N_c and N_v at the band edges, is justified even for microcrystalline silicon with a considerable crystalline volume fraction (in other words for the μc-Si:H I-layers A and B, Tab. 2). It should be noted that in going from c-Si to a-Si:H, the peak in the valence band DOS (for example) is not shifted up in absolute terms, but only relative to the relevant mobility/energy gap. It is this relative shift that is important in determining $N_{c(v)}$.

We thus justify the assumption of a higher (by $\sim$1 order of magnitude) $N_{c(v)}$ for hydrogenated amorphous and μc-Si:H I-layers A and B, relative to $N_{c(v)}$ of c-Si. For the case of the highly crystalline, large-grained μc-Si:H:F I-layer, showing both high J_{sc} and V_{oc}, transport measurements [28] revealed an improved electron mobility that we correlate to a reduced defect density, justifying our assumption of a reduced dangling DOS of 10^{15} cm^{-3} for this material (Tab. 2). However our detailed modeling revealed that with a band gap of only $\sim$1.2 eV (Tab. 2), the high values of V_{oc} measured in this case could not be achieved by a DOS of 10^{15} cm^{-3} alone, and since this dense large-grained μc-Si:H:F, in its properties appears to be very close to mono-c-Si, we assumed $N_{c(v)}$ for this case to be like the latter and were thus able to reproduce all the measured solar cell output (Tab. 2, Fig. 5c). As already discussed similar values of $N_{c(v)}$ have been assumed by other workers in this field [24, 25].

7 Conclusions

In this article we have simulated the dark and illuminated J-V and quantum efficiency characteristics of typical solar cells having low (case A, Tabs. 1 and 2) and intermediate crystalline volume fraction μc-Si:H (case B, Tabs. 1 and 2). The lower F_c material A has been assumed to have a higher energy band gap than the intermediate F_c material B. Both experimental and modeling results indicate a higher J_{sc} but lower V_{oc}, FF and conversion efficiency for the solar cell based on intermediate F_c μc-Si:H B (Tab. 1). In fact the general trend in μc-Si:H solar cells is that V_{oc} decreases with increasing F_c (Fig. 1). We have analyzed the reasons for this and have found that this can be explained by broader band tails and higher Gaussian defect density in the intermediate F_c material B, since structural relaxation of the μc-Si:H network and passivation of grain boundary defects cannot properly take place due to the low amorphous silicon content in this well-crystallized material, as well as due to high oxygen content. Another very important factor is the lower band gap of the intermediate F_c μc-Si:H B, resulting in higher free carrier density in the bands, due to the proximity of the band edges to the quasi-Fermi levels. This leads to higher photo-generated hole trapping, especially near the P/I interface, which in turn leads to a collapse of the electric field over the volume and a lower V_{oc}.

We have also shown that high F_c, large grained μc-Si:H:F, having low oxygen content, is an exception to this general rule (Fig. 1), since its sharply reduced band gap defect density and lower effective DOS at the band edges

overcome the negative influence of the lower band gap to produce a higher V_{oc}. In fact fluorinated μc-Si:H solar cells serve as a link to explain why c-Si PN solar cells, in spite of having a sharply reduced band gap, can have open-circuit voltages considerably higher than μc-Si:H solar cells.

The work at CNRS-LPICM has been partly supported by the European Project "SE Powerfoil" (Project number 038885 SES6). The computer modeling program was developed by P. Chatterjee during the course of a project funded by MNRE and DST, Government of India, and partly during her tenure as Marie Curie fellow at the Laboratoire de Physique des Interfaces et des Couches Minces, Ecole Polytechniuque, Palaiseau, France. E.V. Johnson acknowledges the support of NSERC.

References

1. Y. Mai, S. Klein, R. Carius, H. Steibig, X. Geng, F. Finger, Appl. Phys. Lett. **87**, 073503 (2005)

2. S. Klein, F. Finger, R. Carius, M. Stutzmann, J. Appl. Phys. **98**, 024905 (2005)

3. C. Droz, E. Vallat-Sauvain, J. Bailat, L. Feitknecht, J. Meier, A. Shah, Solar Energy Mater. Solar Cells **81**, 61 (2004)

4. M. Nath, P. Roca i Cabarrocas, E.V. Jonson, A. Abramos, P. Chatterjee, Thin Solid Films **516**, 6974 (2008)

5. M. Moreno, R. Boubekri, P. Roca i Cabarrocas, Solar Energy Mater. Solar Cells, in press (2011).

6. S.M. Sze, *Physics of Semiconductor Devices* (John Wiley, New York, 1981), p. 807

7. A. Datta, J. Damon-Lacoste, P. Roca i Cabarrocas, P. Chatterjee, Solar Energy Mater. Solar Cells **92**, 1500 (2008)

8. E. Maruyama, A. Terakawa, M. Taguchi, Y. Yoshimine, D. Ide, T. Baba, M. Shima, H. Sakata, M. Tanaka, *Proceedings 4th World Conf. on Photovoltaic Solar Energy Conversion* (Hawaii, USA, IEEE, 2006), pp. 1455–1460

9. P. Chatterjee, M. Favre, F. Leblanc, J. Perrin, Mater. Res. Soc. Symp. Proc. **426**, 593 (1996)

10. N. Palit, P. Chatterjee, Solar Energy Mater. Solar Cells **53**, 235 (1998)

11. P. Roca i Cabarrocas, J.B. Chévrier, J. Huc, A. Lioret, J.Y. Parey, J.P.M. Schmitt, J. Vac. Sci. Technol. A **9**, 2331 (1991)

12. D.A.G. Bruggeman, Ann. Phys. (Leipzig) **24**, 636 (1935)

13. P. Roca i Cabarrocas, S. Hamma, A. Hadjadj, J. Bertomeu, J. Andreu, Appl. Phys. Lett. **69**, 529 (1996)

14. G.E. Jellison Jr., M.F. Chisholm, S.M. Gorbatkin, Appl. Phys. Lett. **62**, 3348 (1993)

15. P. Chatterjee, J. Appl. Phys. **76**, 1301 (1994)

16. P. Chatterjee, J. Appl. Phys. **79**, 7339 (1996)

17. P.J. McElheny, J.K. Arch, H.-S. Lin, S.J. Fonash, J. Appl. Phys. **64**, 1254 (1988)

18. F. Leblanc, J. Perrin, J. Schmitt, J. Appl. Phys. **75**, 1074 (1994)

19. H.A. Macleod, *Thin Film Optical Filters* (Hilger, Bristol, 1986)

20. F. Abeles, Ann. Phys. Paris **5**, 596 (1950)

21. F. Abeles, Ann. Phys. Paris **5**, 706 (1950)
22. T. Brammer, H. Stiebig, J. Appl. Phys. **94**, 1035 (2003)
23. T. Brammer, H. Stiebig, Mater. Res. Soc. Symp. Proc. **715**, 641 (2002)
24. J.J.H. Strengers, F.A. Rubinelli, J.K. Rath, R.E.I. Schropp, Thin Solid Films **501**, 291 (2006)
25. A. Sturiale, Hongbo T. Li, J.K. Rath, R.E.I. Schropp, F.A. Rubinelli, J. Appl. Phys. **106**, 014502 (2009)
26. B. Yan, G. Yue, J. Yang, A. Banerjee, S. Guha, Mater. Res. Soc. Symp. Proc. **762**, 369 (2003)
27. J. Zhao, A. Wang, M.A. Green, Solar Energy Mater. Solar Cells **65**, 423 (2001)
28. Y. Djeridane, A. Abramov, P. Roca i Cabarrocas, Thin Solid Films **515**, 7451 (2007)
29. B. Delley, E.F. Steigmeier, Phys. Rev. B **47**, 1397 (1993)
30. S. Hamma, P. Roca i Cabarrocas, Appl. Phys. Lett. **74**, 3218 (1999)
31. T. Merdzhanova, R. Carius, S. Klein, F. Finger, D. Dimova-Malinovska, Thin Solid Films **451-452**, 285 (2004)
32. R. Carius, T. Merdzhanova, F. Finger, S. Klein, O. Vettrl, J. Mater. Sci. Mater. Electron. **14**, 625 (2003)
33. A. Abramov, P. Roca i Cabarrocas. Phys. Stat. Sol. C **7**, 529 (2010)
34. J.D. Joannopoulos, Phys. Rev. B **16**, 2764 (1977)
35. J.D. Joannopoulos, J. Non-Cryst. Solids **35-36**, 781 (1980)
36. L. Ley, S. Kowalczyk, R. Pollak, D.A. Shirley, Phys. Rev. Lett. **29**, 1088 (1972)
37. J. Singh, Phys. Rev. B **23**, 4156 (1981)
38. M.E. Eberhart, K.H. Johnson, D. Adler, Phys. Rev. B **26**, 3138 (1982)
39. S. Nomura, X. Zhao, Y. Aoyagi, T. Sugano, Phys. Rev. B **54**, 13974 (1996)

7

In-situ determination of the effective absorbance of thin μc-Si:H layers growing on rough ZnO:Al

Matthias Meier[a], Karsten Bittkau, Ulrich W. Paetzold, Jürgen Hüpkes, Stefan Muthmann, Ralf Schmitz, Andreas Mück, and Aad Gordijn

IEK5-Photovoltaik, Forschungszentrum Jülich GmbH, 52425 Jülich, Germany

Abstract In this study optical transmission measurements were performed in-situ during the growth of microcrystalline silicon (μc-Si:H) layers by plasma enhanced chemical vapor deposition (PECVD). The stable plasma emission was used as light source. The effective absorption coefficient of the thin μc-Si:H layers which were deposited on rough transparent conductive oxide (TCO) surfaces was calculated from the transient transmission signal. It was observed that by increasing the surface roughness of the TCO, the effective absorption coefficient increases which can be correlated to the increased light scattering effect and thus the enhanced light paths inside the silicon. A correlation between the in-situ determined effective absorbance of the μc-Si:H absorber layer and the short-circuit current density of μc-Si:H thin-film silicon solar cells was found. Hence, an attractive technique is demonstrated to study, on the one hand, the absorbance and the light trapping in thin films depending on the roughness of the substrate and, on the other hand, to estimate the short-circuit current density of thin-film solar cells in-situ, which makes the method interesting as a process control tool.

1 Introduction

Light trapping in thin-film silicon solar cell devices plays a major role at today's research activities to increase the efficiency of solar modules. Rough surfaces are commonly used to scatter the light in the thin absorber layer of the solar cell. Therefore, random textures are used which can be realized by e.g. the transparent conductive oxide (TCO) window and front contact layer applied in thin-film silicon solar cells [1–5]. By light scattering and diffraction, the path of the light traversing through the absorber layer is enhanced and the light absorption is increased [6]. It was shown that different types of TCO with different roughnesses are able to scatter and diffract light in a different way such that the scattering property can be varied over a broad range [6–8]. This is important to adapt the interface texture to different types of solar cells, e.g. single junction or multi junction concepts, whereby always the optimal texture for different absorber layer thicknesses is used.

In the present study, we show a way to determine the effective absorbance of silicon thin-films, which depends on the light path enhancement and thus on the substrate texture. The transient signal of transmission measurements, which were performed in-situ during the deposition of microcrystalline silicon, was used. In our earlier work, we have demonstrated that these transmission measurements can be used to determine the thickness and the crystallinity of the growing layers [9]. Additionally, we showed that especially by in-situ controlling the thickness of the deposited absorber layers, it was possible to fabricate tandem solar cells in which the top and bottom cell generate the identical short-circuit current density which is referred to as "current matched" [10]. In these two studies, it was demonstrated that in-situ transmission measurements are very interesting to use them as process control in the industrial production line. In the present study, we focus on the interpretation of the transient transmission signal which can be correlated to the effective absorbance of the silicon absorber layers and can be used to estimate short-circuit current density of solar cell devices.

2 Experimental setup

Figure 1 shows the setup of the in-situ transmission measurement. Plasma enhanced chemical vapor deposition (PECVD) processes were used to fabricate silicon thin films integrated in solar cell devices [11]. A parallel plate

[a] e-mail: ma.meier@fz-juelich.de

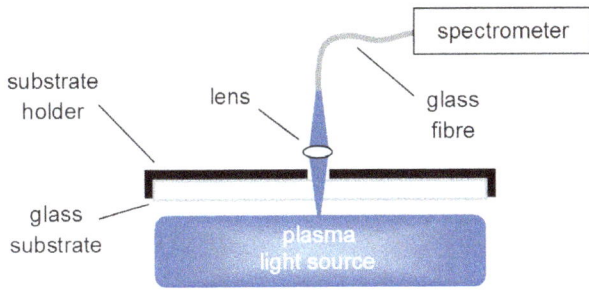

Fig. 1. Experimental setup. PECVD process including an optical port on the backside of the substrate carrier for transmission measurements. The plasma emission is used as light source.

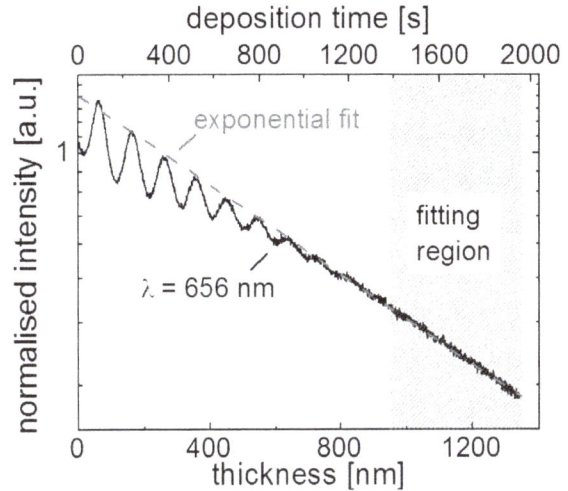

Fig. 2. Transient transmission of the plasma emission line at 656 nm through a growing μc-Si:H film. An exponential fit function (dashed curve) was applied at the last 400 nm of the μc-Si:H layer deposition (see gray fitting region).

reactor with a showerhead electrode was used to deposit the intrinsic microcrystalline silicon (μc-Si:H) absorber layers at which the transmission measurements were performed. The plasma emission was used as light source for this purpose. The plasma emission spectrum contains dominant peaks at wavelengths of 656 nm, 760 nm and 890 nm which were used in this study. An example of such emission spectrum can be found in reference [10]. A hole in the backing plate of the substrate holder allows the light collection from the plasma via a lens through the substrate and the growing silicon thin films. The lens focusses the light into a glass fiber which is connected to a spectrometer.

As substrate for the solar cell deposition, 10×10 cm^2 glasses (Corning Eagle X) were used on which ZnO:Al layers were sputtered. The ZnO:Al layers were etched in liquid HCl solution for texturing [4]. Thus, the textured ZnO:Al films serve as front electrode of the solar cell and also as scattering layer to increase the light absorbance in the solar cell. By varying the etching time between 5 s and 40 s, various random surfaces textures were fabricated to study the influence of the surface texture on the effective absorbance of the silicon thin films.

Microcrystalline thin-film silicon solar cells in p-i-n configuration were fabricated to investigate the influence of the in-situ determined effective absorbance of the intrinsic layers on the short-circuit current density of the device. In the PECVD processes, a gas mixture of silane and hydrogen was used. The excitation frequency was 13.56 MHz at a power of 1 W/cm^2. The deposition pressure was 10 Torr at a silane concentration ($SC_i = [SiH_4]/([SiH_4]+[H_2])$) of 0.58% resulting in a Raman-crystallinity of about 60% for the intrinsic μc-Si:H layers. The p-type and n-type layers of the solar cells were deposited in separate chambers with cross flow configuration using trimethyl-boron and phosphine as doping gas, respectively. A ZnO:Al/Ag layer stack was deposited through a shadow mask as back reflector and electrical contact resulting in active solar cell areas of 1×1 cm^2.

The ZnO:Al surface textures, on which the solar cells were deposited, were characterized by atomic force microscopy (AFM). The Raman-crystallinity of the solar cell material was measured after the deposition of the whole layer stack through the n-layer using a laser wavelength of 647 nm [12]. The microcrystalline silicon solar cells were electrically characterized under AM1.5 illumination.

3 Results

Figure 2 shows an example of the transient optical transmission signal at a wavelength of 656 nm which was recorded during the deposition of μc-Si:H on an etched ZnO:Al substrate with a root mean square roughness of 110 nm. The shape of the transient transmission curve is known from our earlier studies [9, 10]. We show it here again to briefly summarize the signal shape which helps to follow our further evaluation and the discussion. The lower axis of abscissae shows the film thickness of the μc-Si:H layer and at the upper axis of abscissae the actual deposition time of the μc-Si:H layer is provided. Using the maxima and minima of the curve, the deposition time can be transferred to film thickness as described in reference [9]. The axis of ordinates shows the (normalized) intensity of the plasma emission at a wavelength of 656 nm. The signal is normalized to the first minimum of the curve. The fringes of the transient transmission signal result from interference effects due to multiple reflections inside the growing silicon layer. The exponential decrease of the transmission signal intensity is due to the absorbance of the silicon which increases with increasing film thickness.

By using an exponential function, the transient transmission signal, which is plotted as function of film thickness, was fitted for the last 400 nm of the silicon deposition (see fit in Fig. 2 – a detailed discussion of the fit will follow in the discussion section). With the fit function the effective absorption coefficient (α_{eff}) of the growing layer was determined using the Lambert-Beers law:

$$I = I_0 e^{-\alpha_{\text{eff}} d} = I_0 e^{-w\alpha d}. \tag{1}$$

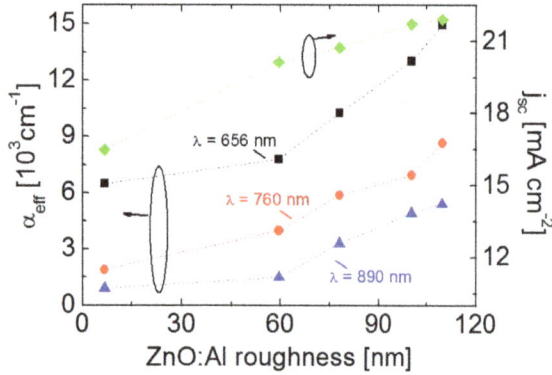

Fig. 3. Effective absorption coefficient α_{eff} and short-circuit solar cell current density j_{sc} as a function of RMS roughness of the different ZnO:Al layers. The thickness of the cells is ~ 1.4 μm.

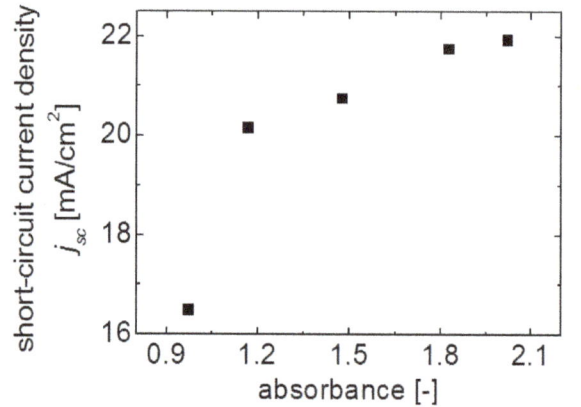

Fig. 4. Short-circuit current density j_{sc} as a function of effective absorbance $= \alpha_{\mathrm{eff}}d$. The short-circuit current density of the cells was measured under standard AM1.5 illumination (ex-situ). The effective absorption coefficient was determined using exponential fits of the transient transmission signal at a wavelength of 656 nm (in-situ). The final thickness d of the solar cells was measured after deposition using a surface profiler (Dektak).

In equation (1) I is the signal intensity, I_0 is the initial signal intensity, d is the film thickness and α_{eff} is the effective absorption coefficient which is the product of the absorption coefficient α of the growing silicon (at a distinct wavelength of 656 nm) and the light path enhancement factor w which derives from the scattering of the light by the textured ZnO:Al layer. The light path enhancement factor w is known from reference [6] and will be further discussed later.

In classical reflection-transmission (RT) measurements, equation (1) is transformed into:

$$\frac{T}{1-R} = e^{-w\alpha d}. \qquad (2)$$

For layer thicknesses d large enough to prevent effects of light coherence, a constant reflection R can be assumed. Therefore, the light path enhancement factor w can be determined with the transmission signal T only.

In the experiment five different ZnO:Al layers on glass substrates with different root mean square roughnesses (6.7 nm, 69.5 nm, 78.0 nm, 100.4 nm and 109.8 nm) were investigated. On the different ZnO:Al layers, μc-Si:H solar cells were fabricated. During the deposition of the intrinsic absorber layer, in-situ transmission measurements were performed. After fabrication, the solar cells were characterized using a class A sun simulator.

Figure 3 shows the effective absorption coefficient α_{eff} of the intrinsic absorber layer as a function of ZnO:Al roughness for three different wavelengths. Additionally, the short-circuit current density j_{sc} of the solar cells as a function of ZnO:Al roughness is plotted. (Please note, that the roughness is used in this study to distinguish the different ZnO:Al films by a simple parameter, though the roughness is known to be insufficient to describe the surface texture completely [4].)

By increasing the ZnO:Al roughness, the value of the effective absorption coefficient α_{eff} increases for all wavelengths. This is caused by the enhanced light scattering effect of rougher interfaces and, thus, an enhanced light path represented by w in equation (1). Also, the short-circuit current density j_{sc} of the solar cells is increased by

increasing the roughness of the ZnO:Al which is in agreement with earlier studies [4,6].

With this experiment, it is demonstrated that an increasing absorbance of the i-layer of the solar cell with increasing substrate roughness can be detected in-situ and that it is possible to correlate the increasing absorbance to an increasing short-circuit current density of the solar cell device. Figure 4 shows the correlation plot between these two parameters. The absorbance (here) is the product of the effective absorption coefficient determined at a wavelength of 656 nm and the thickness of the solar cell which was measured using a surface profiler (dektak). The correlation shown in Figure 4 points out the attractiveness of the in-situ method for a j_{sc} prediction during the deposition of solar cells. It is interesting for an active process control in which the thickness of the solar cell absorber layer can be controlled to generate a distinct current which is important, e.g. for the fabrication of multi-junction solar cells [10].

4 Discussion

The first part of the discussion covers the light path enhancement factor w which is derived from the fitting equation of the transient transmission signal (see Fig. 2 and Eq. (1)). The wavelength dependent absorption coefficient α is an intrinsic material property. In transmission and reflection measurements for example, α is derived from well defined, flat sample geometries. The present experiment on textured surfaces has shown a dependence of the effective absorption coefficient α_{eff} on the roughness of the substrate surface. Since rough surfaces scatter the light into the silicon, the optical path length of incident light in the silicon is enhanced. In the past, it was shown that the light path enhancement factor w can be derived

Table 1. Light path enhancement factors as function of ZnO:Al roughness. The light path enhancement factors of three different wavelengths (656 nm, 760 nm, 890 nm) were calculated from in-situ transmission measurements ($w_{\text{in-situ}}$) and quantum efficiency measurements of the solar cells ($w_{\text{ex-situ}}$).

	656 nm		760 nm		890 nm	
ZnO:Al roughness [nm]	$w_{\text{in-situ}}$	$w_{\text{ex-situ}}$	$w_{\text{in-situ}}$	$w_{\text{ex-situ}}$	$w_{\text{in-situ}}$	$w_{\text{ex-situ}}$
6.7	1.3	1.36	1.6	2.31	3.1	3.09
59.5	1.56	1.61	3.33	3.96	5.09	5.81
78.0	2.05	1.77	4.92	4.49	11.08	6.61
100.4	2.61	1.92	5.80	5.45	16.51	8.49
109.8	2.99	1.97	7.22	5.97	18.13	9.16

from e.g. quantum efficiency (QE) measurements [6] according to:

$$w(\lambda) = -\frac{\ln(1 - QE(\lambda))}{\alpha(\lambda)d}. \qquad (3)$$

It was demonstrated that for certain textures, the light path enhancement factor increases with increasing surface roughness of the ZnO:Al, which is in agreement with the observations in this study. But here, the light path enhancement factor is derived from in-situ measurements during the deposition of the intrinsic absorber layer of µc-Si:H single junction solar cells (see Eq. (1)). Thus, the absorbance of only the absorber layer is measured using the in-situ method in contrast to QE measurements which are performed after the accomplishment of the complete solar cell device. Hence, the influence of the front contact layer, the back reflector and the doped layers of the solar cell device is not included using the in-situ transmission measurements and thus, e.g. parasitic absorption has no contribution. Furthermore, because only the change in the in-situ transmission signal during the deposition of the intrinsic absorber layer is evaluated, the initial reflectance of the solar cell device has no influence on the signal in contrast to the QE measurements were the initial reflectance, which is in the range of 10%, has an significant impact on the measurement signal. Another difference in the two experiments is given by the layer stack and the direction of light propagation. In the QE measurements, a complete solar cell with back reflector is specular illuminated from the glass side. This means that, for the rough samples, light scattering at the front contact occurs at an interface between TCO and silicon which morphology is only defined by the texture-etched TCO. In contrast to the QE measurement setup, at the in-situ transmission measurement setup the sample is illuminated from the silicon layer side by a diffuse (plasma) light source (see Fig. 1). Furthermore a layer stack without back reflector is investigated in-situ. Here, light scattering at the in-coupling to silicon occurs at an interface between air and silicon which morphology is not only defined by the texture-etched TCO, but also by the growth of the silicon layer [12,13]. Light scattering properties are mainly defined by the interface morphology and the difference in refractive indices. Both are different in the two experiments, in particular for samples with large roughness. Furthermore, the in-situ

measurement detects light which is scattered within the acceptance angle of the lens (see Fig. 1). Assuming a change in the scattering properties with increasing silicon layer thickness (e.g. by changing surface morphology), the portion of transmitted light, which is detected by the setup, modifies. This issue is also addressed below in the second part of the discussion.

Therefore, the light path enhancement factor $w_{\text{in-situ}}$ which is derived from the in-situ transmission measurements at single absorber layers is not the same as the light path enhancement factor $w_{\text{ex-situ}}$ which is derived from QE measurements at solar cell devices. Table 1 shows examples of values for $w_{\text{in-situ}}$ calculated from α_{eff} in comparison to values for $w_{\text{ex-situ}}$ calculated from QE measurements of the considered cells according to equation (3). The material specific absorption coefficient α was taken from photo thermal deflection spectroscopy measurements at µc-Si:H reference samples. The thickness d of the silicon was measured using a dektak surface profiler after the solar cell fabrication.

It is found that for the samples with a small roughness, the values of $w_{\text{in-situ}}$ agree well with the light path enhancement factors $w_{\text{ex-situ}}$, showing that for samples with low light scattering properties, any differences in the two experimental techniques are well compensated by the transient approach of the in-situ measurement. With increasing roughness, and therefore increasing light scattering, the light path enhancement factors determined by in-situ transmission measurement are higher than those determined by QE measurement. The higher difference in refractive index at the front interface in the in-situ experiment leads to an improved light scattering efficiency compared to the configuration in the QE measurement [14]. This effect is the strongest pronounced for the largest wavelength (890 nm), since light scattering has the strongest impact on absorption. Beside this effect, discrepancies in the light path enhancement factor at a wavelength of 890 nm for larger roughness can also be explained by an increased parasitic absorption in the front TCO. Although the effective absorbance of the intrinsic silicon layer is strongly increased, parts of this improvement are compensated in the QE measurement by the reduced light intensity that reaches the silicon due to the absorbance in the TCO which is increased as well with the roughness. Since the in-situ measurement is directly sensitive to the effective absorbance in the intrinsic layer, higher values are measured. In summary, different optical effects occur which are hard to fully take into account but which are reasonable to explain the discrepancies in the light path enhancement factor determined by the two experiments. Since a quantitative comparison of the light path enhancement factors is difficult, the main focus will be on the correlation of the effective absorbance, determined by in-situ measurements, and the short-circuit current density of the solar cells.

The second point of the discussion covers the fitting procedure of the transient transmission signal which leads to the determination of the effective absorption coefficient. The estimation of α_{eff} was found to be reasonable for the deposition of the final 400 nm of the silicon where equation (1) was applied and α_{eff} can be correlated to j_{sc} of the solar cells. In Figure 2, the region wherein the exponential fit was performed is highlighted. In addition, it can be observed in Figure 2 that the slope of the transient transmission curve does not follow a single exponential function over the whole thickness range and a single exponential term is not sufficient to describe the whole curve. We have to mention that only for the example shown in Figure 2 the exponential fitting curve strikes the maxima of the fringes in the beginning. For other roughnesses this trend is different. The different slopes of the transmission signal can be explained by the modification of the growing surface of the μc-Si:H layer on the rough ZnO:Al.

First, in the beginning of the deposition, only a very thin silicon film is on top of the ZnO:Al. In that case the two interfaces, ZnO/Si and Si/air, are coplanar to a large extend and most light coherently can interfere upon reflection at both interfaces, whereas after deposition of a certain silicon thickness, these two interfaces are more and more separated and the light rays do not necessarily hit a parallel interface after passing through the thick absorber layer. Thus, after a certain silicon thickness, the coherence is destroyed, which is represented in the transient transmission signal by vanished interference fringes and a changed signal slope (see Fig. 2).

Second, the texture of the silicon surface which is deposited on the textured ZnO:Al gradually changes with increasing the layer thickness of the silicon [12, 13]. At the beginning, when only a very thin silicon layer is deposited, the ZnO:Al texture is present also on the silicon surface due to the conformity of the thin silicon on the rough ZnO:Al. At the end of the deposition, the texture on the silicon surface changed in comparison to the original ZnO:Al texture due to the non-conformal growth of thick silicon layers. Thus, the light scattering texture on the silicon surface is different at the beginning and at the end of the absorber layer deposition.

Both effects change the scattering behavior events at the silicon surface and, consequently, the light path enhancement during the silicon deposition. At the same time, the spatial coherence of the light is destroyed and interference fringes vanish with thickness of the silicon layer, leading to a reflection at the air-silicon interface which is independent on the layer thickness. For the usually quite thick cells, the diminished interference fringes and the steeper decay of transmission is already present, thus, it is reasonable, in accordance to equation (2), to fit only the last part of the transient transmission signal for the calculation of the effective absorption coefficient and the light path enhancement factor and this actual optical system can be correlated to the photovoltaic parameters of the accomplished device.

In summary, it is demonstrated in this study that the transient transmission signal of the in-situ measurements can be used for light scattering and light trapping studies in silicon thin-film solar cells. The effective light absorbance of only the absorber layer is investigated what makes this method unique and useful. Additionally, the correlation between the in-situ measured effective absorbance of the solar cell absorber layer and the resulting short-circuit current density of the solar cell device makes this method attractive as an active process control in which for example the deposition time of the absorber layer can be adapted to produce solar cells generating a restricted current.

5 Conclusion

It is shown that in-situ transmission measurements are feasible for the determination of the effective absorbance of the deposited intrinsic μc-Si:H absorber layers. The measured effective absorption coefficient α_{eff} can be correlated to the short-circuit current density of solar cell devices. This demonstrates, on the one hand, a tool to study optical scattering and light trapping effects at growing thin-film silicon absorber layers and, on the other hand, the possibility of an active process control.

The authors thank U. Rau and R. Carius, O. Gabriel from Helmholtz Zentrum Berlin and G. Dingemans, M. Creatore and A. Bronneberg from Technical University Eindhoven for fruitful discussions and T. Guo and W. Appenzeller for technical assistance. The present work was financially supported by the Federal Ministry of Education and Research (BMBF) and the state government of Berlin (SENBWF) in the framework of the program "Spitzenforschung und Innovation in den Neuen Ländern" (Grant No. 03IS2151).

References

1. W. Beyer, J. Hüpkes, H. Stiebig, Thin Solid Films **516**, 147 (2007)
2. S. Faÿ, J. Steinhauser, N. Oliveira, E. Vallat-Sauvain, C. Ballif, Thin Solid Films **515**, 8558 (2007)
3. M. Kambe, M. Fukawa, N. Taneda, K. Sato, Sol. Energy Mater. Sol. Cells **90**, 3014 (2006)
4. O. Kluth, G. Schöpe, J. Hüpkes, C. Agashe, J. Müller, B. Rech, Thin Solid Films **442**, 80 (2003)
5. J. Müller, B. Rech, J. Springer, M. Vanecek, Sol. Energy **77**, 917 (2004)
6. M. Berginski, B. Rech, J. Hüpkes, H Stiebig, M. Wuttig, Proc. SPIE **6197**, 61970Y (2006)

7. J.I. Owen, J. Hüpkes, H. Zhu, E. Bunte, S.E. Pust, Phys. Stat. Sol. A **208**, 109 (2011)

8. M. Boccard, P. Cuony, C. Battaglia, S. Hänni, S. Nicolay, L. Ding, M. Benkhaira, M. Bonnet-Eymard, G. Bugnon, M. Charrière, K. Söderström, J. Escarre-Palou, M. Despeisse, C. Ballif, From single to multi-scale texturing for high efficiency of micromorph thin film silicon solar cells, in *Proc. 37th IEEE Photovoltaic Specialists Conference, Seattle, USA, 2011*

9. M. Meier, S. Muthmann, A.J. Flikweert, G. Dingemans, M.C.M. van de Sanden, A. Gordijn, Sol. Energy Mater. Sol. Cells **95**, 3328 (2011)

10. M. Meier, T. Merdzhanova, U.W. Paetzold, S. Muthmann, A. Mück, R. Schmitz, A. Gordijn, IEEE J. Photovolt. **2**, 77 (2012)

11. T. Roschek, T. Repmann, J. Muller, B. Rech, H. Wagner, J. Vac. Sci. Technol. **20**, 492 (2002)

12. L. Houben, M. Luysberg, P. Hapke, R. Carius, F. Finger, H. Wagner, Philos. Mag. A **77**, 1447 (1998)

13. V. Jovanov, U. Planchoke, P. Magnus, H. Stiebig, D. Knipp, Sol. Energy Mater. Sol. Cells **110**, 49 (2013)

14. C. Rockstuhl, S. Fahr, K. Bittkau, T. Beckers, R. Carius, F.-J. Haug, T. Söderström, C. Ballif, F. Lederer, Opt. Express **18**, A335 (2010)

8

Ultrafast laser direct hard-mask writing for high efficiency c-Si texture designs

Kitty Kumar[1], Kenneth K.C. Lee[2], Jun Nogami[1], Peter R. Herman[2], and Nazir P. Kherani[1,2,a]

[1] Department of Materials Science and Engineering, 184 College Street, Toronto, Ontario, M5S 3E4, Canada
[2] Department of Electrical and Computer Engineering, 10 King's College Rd. Toronto, Ontario, M5S 3G4, Canada

Abstract This study reports a high-resolution hard-mask laser writing technique to facilitate the selective etching of crystalline silicon (c-Si) into an inverted-pyramidal texture with feature size and periodicity on the order of the wavelength which, thus, provides for both anti-reflection and effective light-trapping of infrared and visible light. The process also enables engineered positional placement of the inverted-pyramid thereby providing another parameter for optimal design of an optically efficient pattern. The proposed technique, a non-cleanroom process, is scalable for large area micro-fabrication of high-efficiency thin c-Si photovoltaics. Optical wave simulations suggest the fabricated textured surface with 1.3 μm inverted-pyramids and a single anti-reflective coating increases the relative energy conversion efficiency by 11% compared to the PERL-cell texture with 9 μm inverted pyramids on a 400 μm thick wafer. This efficiency gain is anticipated to improve further for thinner wafers due to enhanced diffractive light trapping effects.

1 Introduction

A major trend driving the development of low-cost high-efficiency c-Si based photovoltaics is a reduction in material cost through the use of thinner wafers [1,2]. However, thin Si wafers do not readily lend themselves to high-efficiency photovoltaic devices owing to large penetration depth of infrared wavelengths. Therefore, along with surface reflection reduction, an effective light-trapping scheme is imperative for the use of thinner wafers in photovoltaic devices.

In present commercial solar cells, 160–200 μm thick c-Si wafers are chemically etched in hot potassium hydroxide (KOH) solution to form random pyramids of few to ten microns in size on the wafer surface [3] that reduces the surface reflection by promoting multiple bounces of the incident light. However, such texture with large-scale pyramids is not appropriate for thin wafers due to the high Si consumption during etching and little to no optical diffraction for efficient light-trapping. In contrast, optical modeling has shown reduced surface reflection and strong diffraction for effective light-trapping in thin wafers only when grating textures are reduced to feature size comparable with the wavelength [4]. Although various textures consisting of rods [5,6], cones [6], inverted-pyramids [7], holes (honey-comb structure) [8], grooves [9], etc. have

been applied in different lattice configurations and with different pitches to decrease reflectivity. Inverted-pyramid texture and honey-comb texture are the most preferred in practice due to the simple, cost-effective wet chemical processes that are widely available for Si wafers in industry [10]. Further, assessment by Zhao et al. [11] favours an inverted-pyramidal texture over a honeycomb pattern of holes. In addition, the exposed (111) planes in the inverted-pyramidal texture can be readily passivated, a critical requirement for efficient collection of photogenerated carriers. Photolithography has produced inverted-pyramidal texture with 9 μm feature size that demonstrates record energy conversion efficiency in solar cells [12], while further improvements are anticipated from nano-imprint and colloidal lithography [7] that pushes the inverted-pyramid size to sub-micron scale for thin wafers. These methods have been researched extensively and have yet to be adopted for mass production.

In this paper, we present an alternative approach of patterning silicon with inverted pyramids using a non-cleanroom laser direct hard-mask writing technique. The process avails femtosecond laser interaction for blistering (Fig. 1a) and catapulting (Fig. 1b) of thin-film dielectric coatings on c-Si (100) to form a high-resolution hard-mask on c-Si which in turn facilitates chemical etching of Si into inverted-pyramidal structure (Fig. 1c). The technique enables inverted-pyramidal texturing of Si with flexible pattern designs, thus exploiting the low reflection, high

---

[a] e-mail: `kherani@ecf.utoronto.ca`

diffractive light-trapping capability and passivation benefits of such texture. While femtosecond lasers have been used to texturize c-Si with chemical assistance [13], the produced 'black' silicon has significant structural damage that prevents passivation and decreases carrier lifetime. Alternatively, femtosecond lasers have been applied to SiO_x thin films on c-Si to form micro-blisters [13,14] and induce catapulting which we extend to SiN_x and further exploit here to form a high resolution hard mask for KOH etching.

2 Optimization of the texturing process

Various thicknesses of SiO_x and SiN_x films were grown to serve as hard masks (100 nm to 300 nm for SiO_x; 20 nm to 266 nm for SiN_x) for alkaline etching. The SiN_x layer was grown by plasma enhanced chemical vapor deposition (PECVD) on single-side polished p-type (100) c-Si wafers of 400 μm thickness in a PlasmaLab 100 PECVD system (Oxford Instruments) at 300 °C and 650 mT chamber pressure using a gas mixture of 5% silane in nitrogen (400 sccm), ammonia (20 sccm) and pure nitrogen (600 sccm). The deposition was carried out at the rate of 14 nm/minute by using alternating combinations of high frequency (13.56 MHz) plasma for 13 s and low frequency (100 KHz) plasma for 7 s successively. The RF power was set to 50 W and 40 W for high and low frequencies, respectively. For SiO_x, the PECVD procedure was modified to 1000 mT chamber pressure and a gas mixture of 5% silane in nitrogen (170 sccm) and nitrous oxide (710 sccm). The deposition was carried out at the rate of 55 nm/minute by using 30W RF power. Owing to the high etching rate of 18 nm/min of SiO_x in KOH compared with 1.2 nm/min for SiN_x, a minimum of 100 nm thick SiO_x and 20 nm thick SiN_x is required for hard-masking Si during KOH etching. Thus, a systematic study on laser-induced blistering and catapulting was carried out on 100 nm to 300 nm thick SiO_x and 20 nm to 266 nm thick SiN_x film to establish fluence thresholds and exposure windows that produced the smallest ablation craters in hard mask with high reproducibility and minimal collateral damage to expose Si for alkaline etching. The femtosecond fiber laser (FCPA μJewel D-400-VR, IMRA) output was frequency doubled to 522 nm via second harmonic generation, and applied at 100 kHz pulse repetition rate to avoid cumulative heating effects by such rapidly arriving laser pulses. The beam was linearly polarized and of 170 fs pulse duration. By monitoring the back reflection on a CCD camera, a plano-convex lens of 8mm focal length (5724-H-A New Focus) focused the Gaussian-shaped laser beam to a diffraction-limited spot size of 1.25 μm diameter $(1/e^2)$ precisely onto the sample surface that was mounted on a XY motorized stages (ABL1000, Aerotech). A linear polarizer and waveplate power attenuator varied the laser pulse energy between 0.2 and 1.25 nJ to drive different levels of surface modification in hard masks of various thicknesses. Additional laser processing details are provided in reference [15]. Further, the smallest craters in all the films were tested for effective anisotropic etching in 30 wt% aqueous

solution of KOH maintained at 60 °C using a two-step procedure. First, the sample was etched for 30 s, and then cleaned and washed in DI water followed by nitrogen drying. Second, the sample was etched again over a variable time to yield a high-fidelity inverted-pyramid structure. The first step served to mainly remove ablation debris.

The fluence threshold for blistering and catapulting of SiN_x was found to increase strongly from 0.29 J/cm^2 and 0.45 J/cm^2, respectively, for a 20 nm thick film to 0.67 J/cm^2 and 1.02 J/cm^2, respectively for a 266 nm thick film. However, the values modulate with varying Fabry-Pérot interference effects as the optical film thickens. Over this film thickness range, the minimum ablation crater diameter was found to increase monotonically from 0.6 μm to ~2 μm. Generally, higher fluence was necessary to generate larger internal ablation pressure to delaminate and lift thicker films, but at the cost of creating larger diameter craters in the c-Si substrate. The observed fluence thresholds and the ablation crater diameters were nearly identical in SiO_x films in comparison with SiN_x films of the same thickness, and are in accord with the results reported for 147 nm SiO_x films [13]. During KOH etching, the smallest ablation craters in $\geqslant$100 nm thick SiN_x/SiO_x films did not etch to form inverted-pyramids. This is attributed to the high fluence ($\geqslant$0.96 J/cm^2) used to delaminate $\geqslant$100 nm thick films that crystallographically damages the underlying Si and thus rendering it unetchable in KOH [16]. The results show that the thinnest possible dielectric film is favoured to yield the smallest ablation crater diameter with minimum Si damage. Hence, a 20 nm SiN_x film was deemed optimal for creating the smallest possible mask aperture with sufficient thickness to resist KOH etching.

The blister dynamics observed in 20 nm SiN_x with increasing laser fluence was similar to that reported for a 100 nm SiO_x film [17]. The blister grows in diameter until a threshold for the perforation of the blister is reached at which point a nano-hole is formed, followed by mechanical ejection of the blistered SiN_x film at higher fluences, leaving behind approximately a 50 nm deep crater in the underlying Si substrate. Figure 2a shows scanning electron micrographs (SEMs) of this sequence beginning with the threshold for blister formation (0.29 J/cm^2), and following with the threshold for perforating the blister with a nanohole (0.31 J/cm^2), the collapse of a blister with a nanohole (0.41 J/cm^2), the catapulting threshold fluence for mechanically ejecting the blister (0.45 J/cm^2), and the catapulted blister at a fluence above the threshold (0.49 J/cm^2). The corresponding atomic force micrographs and line profiles of this morphology are shown in Figures 2b and 2c, respectively, while Figure 2d shows SEMs of the corresponding features after KOH etching for 2.5 min (step1: 30 s, step2: 2 min). The blistered SiN_x layer (intact, perforated and collapsed) was found to protect the underlying Si from KOH etching whereas the unprotected Si crater formed due to a blister catapulting ($\geqslant$0.45 J/cm^2) resulted in typical inverted-pyramid structure after KOH treatment. Further, the ablation crater diameter and the effect of KOH etching was, comprehensively, examined over a broad range

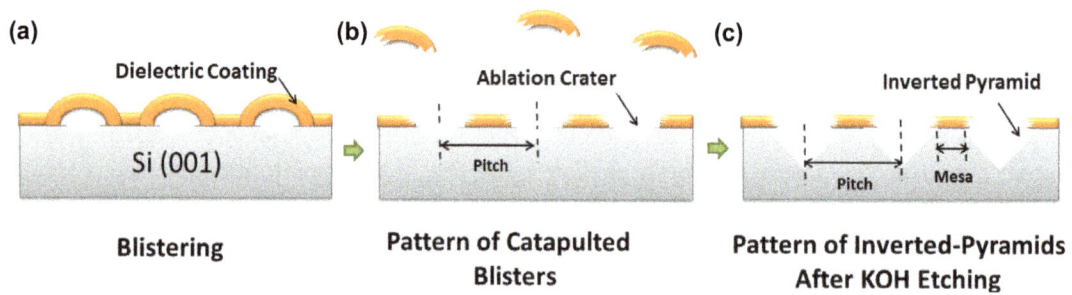

Fig. 1. A schematic illustrating (a) blistering of a dielectric coating upon interaction with femtosecond pulses, (b) catapulting of the blisters at higher laser fluence to form a pattern of shallow ablation craters in c-Si, and (c) the resulting inverted-pyramid structure following KOH etching.

of laser exposure conditions to determine the smallest possible inverted-pyramid structure that could be reproducibly formed with the tightest packing density and the results are summarized in Figure 2e. The graph shows the crater diameter and the resulting inverted-pyramid width after etching in KOH for 2.5 min to increase from 0.6 μm and 1.13 μm, respectively, at 0.43 J/cm^2 to 1.25 μm and 1.31 μm, respectively, at 0.53 J/cm^2 fluence. The degree of undercutting increases dramatically as the fluence decreases towards the catapulting threshold, below which only blistering is observed. This is attributed to the decrease in laser-induced Si damage at low fluence. The undercutting was essential to etch beyond the laser damage zone in the c-Si surface to expose damage free (111) planes. The increasing damage apparent in the SEM images (Fig. 2e inset) with increasing laser fluence required longer etching times to compensate for damage but with the trade-off of forming larger sized inverted-pyramid structures. The smallest (1.13 μm) clean inverted-pyramid was reproducibly formed at 0.45 J/cm^2 whereas a 1.07 μm wide inverted-pyramid was formed at 0.43 J/cm^2 with only 33% reproducibility.

A high density grid of inverted-pyramids was next investigated using the optimized 0.45 J/cm^2 fluence exposure. At 100 kHz repetition rate, the pitch (Λ) of craters along the scanning direction (x) was examined over a range of $\Lambda_x = 1.2$ to 1.5 μm by varying the scan speed between 120 to 150 mm/s, while line-to-line offsets (y-direction) of $\Lambda_y = 1.2$ to 2 μm were tested to create tightly packed arrays with minimum collateral damage. The densest packing was found for 1.5 μm spacing in both directions, yielding the grid of craters shown in Figure 3a that etched into a high-fidelity array of inverted-pyramids seen in top and cross-sectional views (inset) in Figure 3b. In the array of ablation craters produced at optimized fluence, the inverted-pyramid size can be varied from 1.13 μm to 1.3 μm by increasing the KOH etching time from 2.5 to 5 min, leaving a 370 nm to 200 nm wide flat mesa, respectively, between the inverted-pyramids. A further increase in the inverted-pyramid size would lead to occasional over-etching that manifests in the fusion of neighboring pyramids. Larger area SEM observation of the inverted-pyramidal texture (Fig. 3c) did not reveal any pyramid defects over our whole sample set ($\sim$14,000 holes

viewed) in spite of a non-cleanroom processing environment, suggesting a high reproducibility of the devised technique with less than 1 defect per 10^4 holes.

3 Flexibility offered by the texturing process

The method offers the advantage of independently varying pattern pitches in x- and y-directions with the help of computer-controlled motion stages, as illustrated in Figure 3d. Also, the laser exposure can be modulated during wafer scanning with the help of an acoustic optical modulator (AOM) driven by position-synchronized output (PSO) of the motion stages to create patterns of inverted-pyramids; this is illustrated through the creation of the University of Toronto crest which is shown in SEM and optical images (Figs. 3e and 3f, respectively). A micro-pattern of the inverted-pyramid structure is shown in Figure 3e (inset). In the optical image, the bare silicon appears bright, whereas the texturized regions appear dark, demonstrating their strong optical anti-reflection property.

4 Optical performance

After optimization, 2 cm × 2 cm textures with inverted pyramids of 1.13 μm and 1.3 μm sizes arrayed at our highest pitch density ($\Lambda_x = \Lambda_y = 1.5$ μm) were fabricated to measure the anti-reflection efficacy and the effect of mesa width on the anti-reflectance efficacy of the textures. Figures 4a and 4b show total reflectance at normal incidence and specular reflectance at different angles of incidence, respectively, together with the total reflectance spectrum measured for a bare Si wafer. With 1.13 μm wide inverted-pyramids, the surface reflectance was reduced to an average of 18.6% compared with untextured silicon which reflects 34.7% of incident light in the wavelength range 400 nm–1000 nm. For 1.3 μm pyramids, the reflectance was reduced to 13.6% due to the 42% decrease in the width of mesas. For the same sample, the total reflectance was further reduced to 4% with the addition of an incompletely optimized 70 nm thick SiN$_x$ anti-reflective coating (ARC). Calculated optical reflectivity curves are also plotted for the cases of 1.3 μm inverted-pyramids (with 200 nm

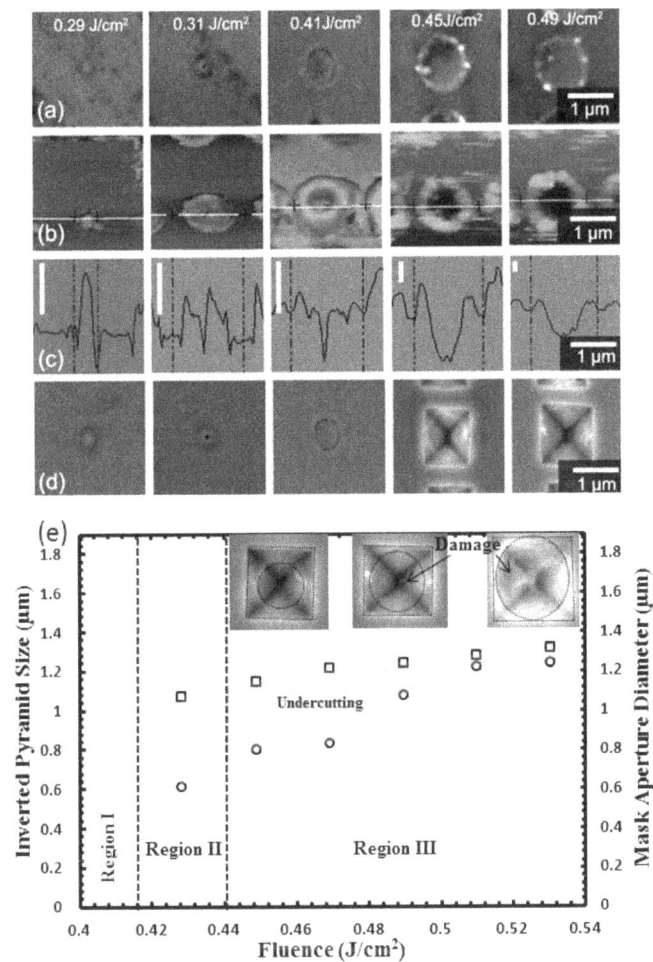

Fig. 2. (a) SEM images of surface modifications in a 20 nm SiN$_x$ film on c-Si as a function of laser fluence. Corresponding atomic force micrographs and line profiles of the features are shown in (b) and (c), respectively. The vertical scale bar is 20 nm in each panel of (c). (d) Shows post-etching SEM micrographs of the morphology shown in (a), when etched in KOH for 2.5 min. The ablation crater diameter (circular marker (o)) and the inverted-pyramid size (square marker (□)) observed after 2.5 min KOH etching are shown in (e) together with select SEM images as a function of fluence. The mask aperture and the inverted pyramid are outlined by a dashed circle and a square, respectively, in each of the SEM images. Fluence zones for blistering (Region I) and low (Region II) and high (Region III) reproducible ejection are identified.

Fig. 3. (a) Top view of a Si wafer with a grid of catapulted blisters at 1.5 μm spacing. KOH etching results in a clean array of 1.3 μm inverted-pyramids seen in top and cross-sectional views (inset) (b) and in a large area oblique-view (c). The defect-free texture indicates the high reproducibility offered by the devised technique. (d) Shows an array of 1.27 μm sized inverted-pyramids placed at a $\Lambda_x = 1.5$ μm and $\Lambda_y = 1.7$ μm. (e) SEM image of a University of Toronto crest micro-patterned with inverted-pyramid structure shown in (e) (inset) demonstrates the capability of the technique to selectively texture areas, leaving untextured planar areas for front contacts. The patterned areas appear bright in this image. (f) The optical image of the crest demonstrates the strong anti-reflective effect in the textured areas.

mesas) and the untextured Si surface. Optical wave analysis based on the scattering matrix method [18] was used to simulate the total optical reflectance at normal incidence over the 280 nm–1000 nm spectrum. Reflectivity of the inverted pyramid texture on a 400 μm thick polished silicon wafer was calculated for a set pitch of 1.5 μm and inverted pyramid size of 1.13 μm and 1.3 μm. For 70 nm thick SiN$_x$ PECVD antireflective coating, the optical constants were experimentally obtained by ellipsometry. The simulated and measured curves match in the case of untextured Si, whereas a slight incongruity is seen for the 1.3 μm inverted-pyramid texture; this deviation

is possibly due to our theoretical assumption of a square grid arrangement of inverted pyramids, whereas the laboratory sample had skewed alignment of adjacent rows due to the limited control available in the motion stages. The average specular reflectance over 280 nm–1000 nm for the AR coated (70 nm thick SiN$_x$ deposited by PECVD) texture with 1.13 μm inverted-pyramids lies below 2% for incidence angles of 8° to 40° and rapidly increases to 3%, 4.6% and 7.9% for incident angles of 48°, 56° and 64°, respectively, indicating better optical performance when compared with the grid-less PERL-cell [19] that is also plotted in Figure 4b for comparison.

5 Efficiency calculations

Given the strong anti-reflective characteristics observed in the present inverted-pyramid structures, it is instructive to simulate the cell efficiency as a function of

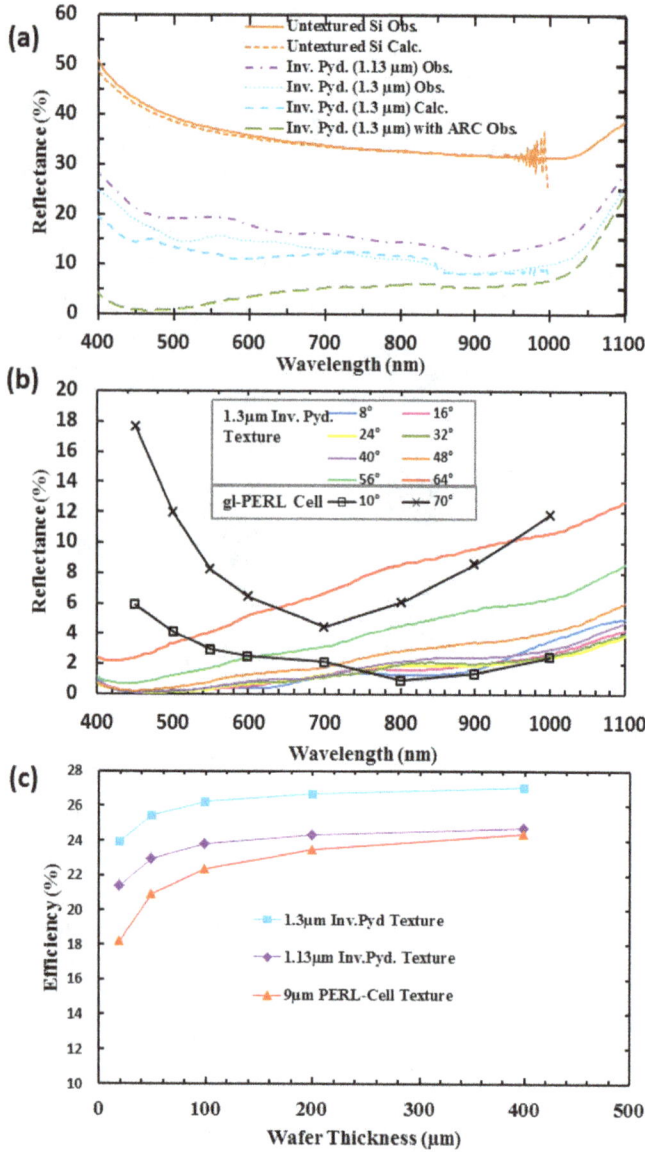

both cases. For efficiency calculations, the total number of photons absorbed in the solar cell N_{ph} was calculated by integrating absorption of AM1.5 solar radiation over the wavelength range of 280 nm–1000 nm. Consistent with Feng et al. [20], a collection efficiency of 85% was assumed. The short circuit current was given by $J_{sc} = 0.85 \times eN_{ph}$, where e is the electronic charge. The open circuit voltage was obtained from $V_{oc} = (kT/e)\ln(J_{sc}/J_o + 1)$, where k is the Boltzmann constant, $T = 300$ K, and J_o is the reverse bias saturation current. Given that the smallest J_o value of about 10^{-15} A cm^{-2} [21] for Si at 300 K, here we use $J_o = 1.5 \times 10^{-15}$ A cm^{-2}. A fill factor of 80% was assumed, representing an achievable value for a well-designed photovoltaic device operating at the maximum power point [21]. Finally, the solar cell efficiency was expressed as $\eta = 0.8 \times J_{sc} \times V_{oc}/P_{\text{AM1.5}}$, where $P_{\text{AM1.5}} = 0.1$ W cm^{-2} is the incident power of AM1.5 solar radiation.

The simulation results (Fig. 4c) show a moderate efficiency gain of 2.7% and 0.4% for the 1.3 μm and 1.13 μm inverted-pyramids, respectively, in relation to the 24.4% for the PERL-cell texture in the case of a 400 μm thick wafer. Smaller mesas are clearly favoured as expected, while the smaller pyramids on the size scale of optical wavelengths benefit from reduced reflection due to the graded refraction index effect. However, much stronger enhancement is found in thinner wafers. Efficiency gain of 5.7% is obtained in the case of 1.3 μm sized inverted pyramids on 20 μm thin wafers in relation to the 18.2% PERL-cell texture, due to the enhanced light trapping from the larger diffraction angles possible in the smaller periodic surface structure that directs the weakly absorbing infrared light laterally and thus resulting in a marked increase in performance in such thin wafers. Further, the 1.3 μm texture effectively etches less silicon for a given wafer thickness compared to the PERL-cell texture, and hence it provides more silicon for light absorption.

Fig. 4. (a) Experimental (solid lines) and simulated (dashed lines) total reflectance spectra at normal incidence from fabricated samples (1.13 μm and 1.3 μm inverted-pyramids) and bare silicon as a function of wavelength. (b) Measured specular reflectance spectra from 1.3 μm inverted pyramid texture coated with a 70 nm thick SiN$_x$ ARC and from a grid-less PERL-cell with dual layer ARC [19] for the various incidence angles shown. (c) Predicted efficiency of a solar cell with the presently fabricated (1.13 and 1.3 μm inverted-pyramid) and the PERL-cell textures at various wafer thicknesses. A 70 nm thick PECVD deposited SiN$_x$ARC is considered in efficiency calculations.

6 Results and discussion

We have demonstrated a simple and versatile hard-mask writing technique for inverted-pyramid texturing of c-Si with texture feature size and pitch on the order of wavelength. The technique uses individual laser pulses to define the pattern of catapulting blisters in a 20 nm-thin SiN$_x$ layer to expose the underlying c-Si with minimal laser damage, which is then effectively etched with KOH to form a high fidelity inverted-pyramidal texture. The technique offers control over the periodicity through the use of computer controlled motion stages whereas feature size can be modulated by varying laser fluence or alternatively by chemical etching time. It also offers precise control of the patterned areas. Further, the proposed patterning technique along with the feature size of the order of wavelength leads to minimal removal of silicon ($\sim$1 μm) and consequently has beneficial implications for high-efficiency ultra-thin ($\sim$20 μm) silicon PV where excess Si for etching

wafer thickness for the textures fabricated in this study and the 9 μm inverted-pyramidal texture with $\Lambda_x = \Lambda_y = 10$ μm that was used in the grid-less PERL-cell. Inverted pyramids were placed on the light facing side of the wafer, etched in to a polished silicon wafer of a given thickness. A 70 nm thick SiN$_x$ ARC and a perfect back reflector were assumed for the front and rear surfaces, respectively, in is not available. A relative Si material savings of $\sim 80\%$ is anticipated when using our inverted pyramids of 1.3 μm size in comparison with the 9 μm PERL-cell inverted pyramids.

Detailed optical wave calculations on a 400 μm thick wafer suggest relative efficiency enhancement of 11% and 1.4% for the 1.3 μm and 1.13 μm inverted-pyramids, respectively, when compared with the PERL-cell texture of 9 μm inverted-pyramids. Light trapping is not imperative in thick wafers and hence, the predicted enhancement in 400 μm thick wafer is mainly due to the graded refractive index effectively formed by the inverted-pyramids of the order of wavelength that couples more light into the Si substrate thereby increasing the cell efficiency. Smaller mesas result in stronger enhancement as expected. However, in thin wafers the amount of light trapping significantly affects the cell efficiency. Specifically, for a 20 μm thick wafer a much stronger enhancement of 31.6% and 17.7% is calculated in the case of 1.3 μm and 1.13 μm inverted-pyramids, respectively, relative to the PERL-cell texture. This enhancement in efficiency is due to two factors. First, due to the diffraction of light from the fabricated textures that directs the weakly absorbing infrared light laterally which is effectively absorbed in such thin wafers. Second, 1.3 μm inverted pyramid texture etches $\sim 80\%$ less Si and hence offers more material for light absorption relative to the PERL-cell texture.

In the present study, catapulted blisters were formed at the rate of 10^5 s^{-1} with 0.55 mW laser power. The current development of $\geqslant 100$ W femtosecond lasers together with multi-lens focusing arrays suggests a $\sim 2 \times 10^5$ fold increase in the processing rate, i.e., 2×10^{10} blisters per second, equivalent to 200 cm^2 area production per second that can meet the requirements of current c-Si solar cell manufacturing processes. With further advances in lasers and beam delivery methods the proposed approach has the potential of becoming a practical texturing technique for high-efficiency thin c-Si photovoltaics.

This work was supported by the Natural Sciences and Engineering Research Council of Canada, the Ontario Research Fund – Research Excellence program, and the University of Toronto.

References

1. K.A. Munzer, K.T. Holdermann, R.E. Schlosser, S. Sterk, IEEE Trans. Electron Devices **46**, 2055 (1999)
2. C.T.M. Group, International Technology Roadmap for Photovoltaics (ITRPEV.net) Results 2010 2, (2011)
3. D. Munoz, P. Carreras, J. Escarre, D. Ibarz, S. Martin de Nicolas, C. Voz, J.M. Asensi, J. Bertomeu, Thin Solid Films **517**, 3578 (2009)
4. H. Sai, Y. Kanamori, K. Arafune, Y. Ohshita, M. Yamaguchi, Prog. Photovolt. Res. Appl. **15**, 415 (2007)
5. C.H. Sun, W.-L. Min, N.C. Linn, P. Jiang, B. Jiang, J. Vac. Sci. Technol. B **27**, 1043 (2009)
6. C.M. Hsu, S.T. Connor, M.X. Tang, Y. Cui, Appl. Phys. Lett. **93**, 133109 (2008)
7. C.H. Sun, W.L. Min, N.C. Linn, P. Jiang, B. Jiang, Appl. Phys. Lett. **91**, 231105 (2007)
8. G. Kumaravelu, M.M. Alkaisi, A. Bittar, in *Proc. of the 29th IEEE Photovoltaic Specialists Conference, New Orleans, 2002*, pp. 258–261
9. Y. Kanamori, M. Sasaki, K. Hane, Opt. Lett. **24**, 1422 (1999)
10. M. Moynihan, CircuiTree **22**, 16 (2009)
11. J. Zhao, A. Wang, Martin A. Green, F. Ferrazza, Appl. Phys. Lett. **73**, 1991 (1998)
12. M.A. Green, K. Emery, Y. Hishikawa, W. Warta, E.D. Dunlop, Prog. Photovolt. Res. Appl. **19**, 565 (2011)
13. B.R. Tull, J.E. Carey, E. Mazur, J.P. McDonald, S.M. Yalisove, MRS Bull. **31**, 626 (2006)
14. J.P. McDonald, A.A. McClelland, Y.N. Picard, S.M. Yalisove, Appl. Phys. Lett. **86**, 264103 (2005)
15. K. Kumar, K.K.C. Lee, P.R. Herman, J. Nogami, N.P. Kherani, Appl. Phys. Lett. **101**, 222106 (2012)
16. T. Rublack, M. Schade, M. Muchow, H.S. Leipner, G. Seifert, J. Appl. Phys. **112**, 023521 (2012)
17. T. Rublack, S. Hartnauer, P. Kappe, C. Swiatkowski, G. Seifert, Appl. Phys. A **103**, 43 (2011)
18. A. Chutinan, N.P. Kherani, S. Zukotynski, Opt. Express **17**, 8871 (2009)
19. A. Parretta, A. Sarno, P. Tortora, H. Yakubu, P. Maddalena, J. Zhao, A. Wang, Opt. Commun. **172**, 139 (1999)
20. N.-N. Feng, J. Michel, J. Michel, L. Zeng, J. Liu, C.-Y. Hong, L.C. Kimerling, X. Duan, IEEE Trans. Electron Devices **54**, 1926 (2007)
21. S.M. Sze, *Physics of semiconductor devices*, 2nd edn. (Wiley, New York, 1981)

9

Microcrystalline silicon absorber layers prepared at high deposition rates for thin-film tandem solar cells

S. Michard, V. Balmes, M. Meier[a], A. Lambertz, T. Merdzhanova, and F. Finger

Institute of Energy and Climate Research 5 – Photovoltaik, Forschungszentrum Jülich, 52425 Jülich, Germany

Abstract We have investigated high deposition rate processes for the fabrication of thin-film silicon tandem solar cells. Microcrystalline silicon absorber layers were prepared under high pressure depletion conditions at an excitation frequency of 81.36 MHz. The deposition rate was varied in the range of 0.2 nm/s to 3.2 nm/s by varying the deposition pressure and deposition power for given electrode spacings. The silane-to-hydrogen process gas mixture was adjusted in each case to prepare optimum phase mixture material. The performance of these tandem solar cells was investigated by external quantum efficiency and current-voltage measurements under AM1.5 illumination before and after 1000 h of light degradation. Up to deposition rates of 0.8 nm/s for the microcrystalline silicon absorber layer high quality tandem solar cells with an initial efficiency of 10.9% were obtained (9.9% stabilized efficiency after 1000 h of light degradation).

1 Introduction

Microcrystalline silicon is widely used as absorber layer of the bottom solar cell in the amorphous/microcrystalline thin-film solar cell concept. This solar cell concept was pioneered by research groups in Neuchâtel, Jülich and at Kaneka Corporation [1–3]. The use of stacked solar cells with different optical bandgaps facilitates the efficient utilization of the solar spectrum [4,5]. Due to the indirect semiconductor bandgap of microcrystalline silicon a thickness of more than 1 μm is favorable to have enough light absorption, particularly in the long wavelength region. Increasing the thickness of the bottom solar cell absorber layer is also one approach to achieve current matching between top and bottom solar cell.

Increasing the efficiency or reducing the manufacturing cost of thin-film solar cells are two approaches to increase the cost competitiveness of the thin-film silicon technology. We investigate processes which target high material growth rates to reduce the fabrication costs per unit and to increase the throughput of the deposition system. Studies on how to reach world record solar cell efficiencies were given by Hänni et al. [6] for μc-Si:H single junction solar cells and by Kim et al. [7] and Guha et al. [8] for multijunction solar cells. Further details about the recent developments in the field of thin-film solar cells from research

scale to industrial scale can be found in a comprehensive overview published by Shah et al. [9].

In this study we investigated based on the results of our earlier studies [10–12] the effect of an increased deposition rate for the μc-Si:H absorber layer on the quality of state of the art thin-film tandem solar cells. Here, we also investigate the effect of high deposition rate used for fabrication of μc-Si:H bottom cell on the stability of the tandem solar cells under light induced degradation. To achieve high deposition rates while maintaining the material quality two approaches were combined. These were: deposition processes in the high pressure depletion (HPD) regime [13,14], and the use of high excitation frequencies in the VHF band ($\nu_{ex} > 30$ MHz) [15–17].

2 Experimental details

For the device fabrication commercially available glass covered with SnO2:F (Asahi type VU) served as substrate. Silicon layers were deposited by plasma enhanced chemical vapor deposition (PECVD) on 10 cm × 10 cm large substrates using a clustertool deposition system which is described in detail in reference [12]. For the a-Si:H top solar cell and the μc-Si:H bottom solar cell the p-i-n deposition sequence was used in the superstrate configuration. The top solar cell is formed by a layer stack consisting of p-type a-SiC:H, intrinsic a-Si:H, and n-type μc-Si:H layers. The bottom solar cell is formed by a layer stack

---
[a] e-mail: `ma.meier@fz-juelich.de`

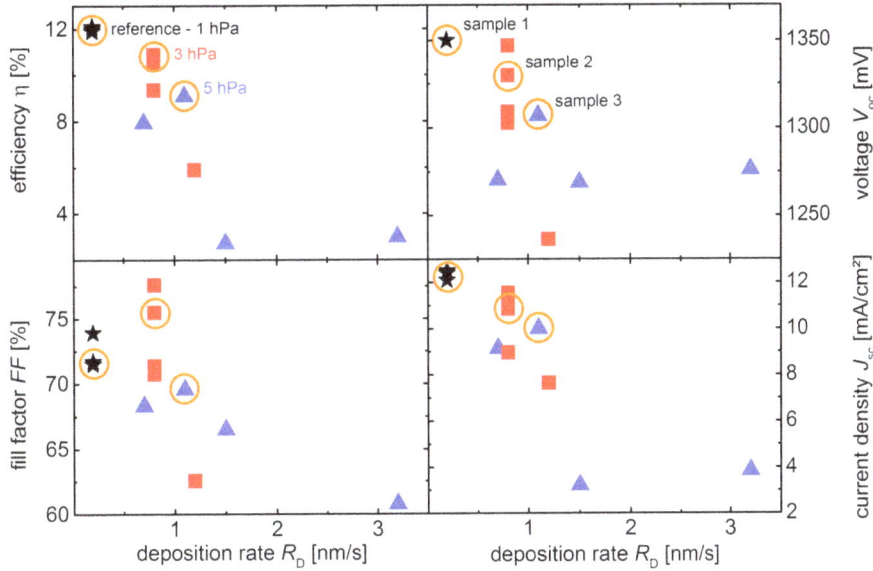

Fig. 1. Photovoltaic-parameters – efficiency η, fill factor FF, open-circuit voltage V_{oc}, and short-circuit current density J_{sc} – as a function of the deposition rate R_{D}. The circles indicate samples 1, 2, and 3, which were further investigated by light induced degradation measurements.

of p-type μc-Si:H, intrinsic μc-Si:H, and n-type a-Si:H. A layer sequence of sputtered ZnO:Al/Ag/ZnO:Al served as the back contact of the final device. The active device area of 1 cm^2 was defined by a laser scribing step. Apart from the deposition conditions of the μc-Si:H absorber layer of the bottom solar cell, all deposition parameters remained unchanged during the experiments. The deposition rate of the absorber layer of the bottom solar cell was varied by adjusting the deposition power, the silane concentration, and the deposition pressure. To ensure a similar total thickness of the terminated devices the deposition time of the bottom solar cells absorber layer was adjusted according to the deposition rate. The deposition power was varied between 20 W and 400 W, the silane concentration was varied between 2.4% and 7%, and the deposition pressure was varied between 1 hPa and 5 hPa. The silane concentration SC is defined by the ratio of the silane $Q[\mathrm{SiH_4}]$ and the hydrogen $Q[\mathrm{H_2}]$ mass flow into the deposition chamber:

$$SC = \frac{Q[\mathrm{SiH_4}]}{Q[\mathrm{SiH_4}] + Q[\mathrm{H_2}]}.$$

The absorber layers of the solar cells used as reference devices in this study were processed at a deposition power of 20 W, an electrode distance of 11 mm, a deposition pressure of 1 hPa, and an excitation frequency of 81.36 MHz.

The photovoltaic parameters of the solar cell – the conversion efficiency η, fill factor FF, open-circuit voltage V_{oc}, and short-circuit current density J_{sc} – were measured using a double beam WACOM-WXS-140S-Super (Class A) sun simulator. External quantum efficiency measurements were performed at 25 °C without bias voltage under short-circuit conditions. The J_{QE} current density for each sub-cell of the tandem solar cells is calculated

by the spectral integration of the EQE data and the AM 1.5 spectrum.

The photovoltaic-parameters and the external quantum efficiency (EQE) were measured on selected samples before and after 1000 h of light induced degradation (LID). During the light-soaking procedure the illumination was constant at 1000 W/m^2 and the temperature was held constant at 50 °C. The following results present the data for the best solar cell out of the 18 solar cells fabricated for each sample set. The solar cell active area is 1 cm^2.

3 Results

Figure 1 shows the photovoltaic-parameters as a function of the deposition rate of the absorber layer of the μc-Si:H bottom solar cell prepared at various deposition pressures. The best device performance with an efficiency of 12.1% was observed for the reference process at a deposition rate of 0.2 nm/s. With the exception of the reference solar cells all μc-S:H i-layers were processed at 3 hPa, and 5 hPa. The reference solar cell μc-S:H i-layer was processed at 1 hPa. For the solar cells where the μc-Si:H i-layer was processed at 3 hPa the deposition power was varied between 50 W and 100 W with varying SC between 3.4% and 3.9%. For the solar cells where the μc-Si:H i-layer was processed at 5 hPa the deposition power was varied between 50 W and 400 W while the SC was varied from 2.4% to 7.0%. The simultaneous variation of the deposition pressure, the deposition power, and the silane concentration was necessary to achieve device grade μc-Si:H i-layers at elevated deposition rates [12] and thus to guarantee the optimal processing conditions for the fabrication of high quality thin-film solar cells. Processes leading to device grade microcrystalline silicon were presented

Table 1. Deposition and photovoltaic-parameters for sample 1 (reference process) and samples 2 and 3. Samples 2 and 3 are the samples which show the best efficiency for the sample set where the absorber layer of the bottom solar cell was processed at 3 hPa and 5 hPa, respectively. The deposition parameters are abbreviated as follows: deposition pressure p, electrode distance d, deposition power P, silane concentration SC, and deposition rate R_D. The photovoltaic parameters describe the efficiency η, the fill factor FF, the open-circuit voltage V_oc, and the short-circuit current density J_sc.

	p [hPa]	d [mm]	P [W]	SC [%]	R_D [nm/s]	η [%]	FF [%]	J_sc [mA/cm^2]	V_oc [V]
Sample 1	1	11	20	5.2	0.2	12.1	73.9	12.1	1.35
Sample 2	3	10	50	3.8	0.8	10.9	75.5	10.8	1.33
Sample 3	5	10	100	3.8	1.1	9.1	69.6	10	1.31

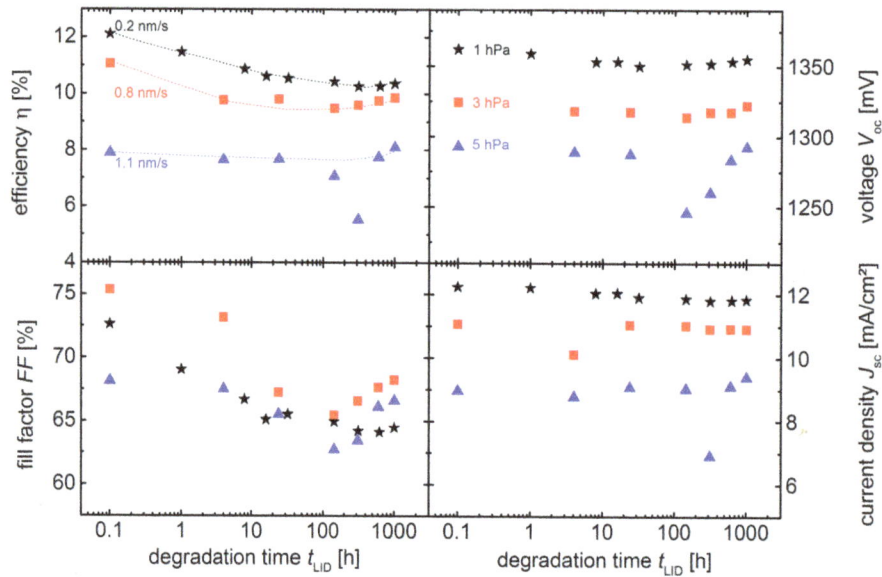

Fig. 2. Photovoltaic-parameters – efficiency η, fill factor FF, open-circuit voltage V_oc, and short-circuit current density J_sc – as a function of time for light induced degradation (LID). The LID was performed with an illumination of 1000 W/m^2, and a controlled temperature of approximately 50 °C.

in an earlier parameter study [12]. The criteria to classify silicon layers as device grade material have been taken from [2,18]. With increasing deposition rate, the efficiency of the solar cells decreases around 3% (absolute scale) until a deposition rate of 1.1 nm/s. Increasing the deposition rate above 1.2 nm/s resulted in a significant decrease of the device performance. The open-circuit voltage for those samples is below 1.3 V while the short-circuit current density is below 4 mA/cm^2. The best solar cells where the μc-Si:H i-layers were processed at elevated deposition pressures of 3 hPa and 5 hPa show efficiencies of 10.9% and 9.1% and deposition rates of 0.8 nm/s and 1.1 nm/s, respectively. The photovoltaic-parameters and the depositions conditions of these solar cells together with the reference solar cell are summarized in Table 1. In the following these solar cells will be referred to as sample 1, sample 2, and sample 3, according to Table 1. Comparing sample 1 and sample 3 an increase of the deposition rate from 0.2 nm/s to 1.1 nm/s resulted in a decrease of 45 mV in the open-circuit voltage, and a decrease in the short-circuit current density of approximately 2 mA/cm^2. When compared to sample 1, sample 2 also showed reduced open-circuit voltage and reduced short-circuit current density with increasing deposition rate, although the

effect is not as strong as for sample 3. The open-circuit voltage decreases by 20 mV while the short-circuit current density decreases by 1.3 mA/cm^2, with an increase of the deposition rate from 0.2 nm/s to 0.8 nm/s. Samples 1 to 3 were further investigated by light-induced degradation (LID) and EQE measurements.

Figure 2 shows the photovoltaic-parameters of samples 1 to 3 as function of the degradation time. The efficiency of sample 1 decreases from 12.1% to 10.4% after 1000 h of LID. For sample 2 the efficiency of the solar cell decreases from 10.9% to 9.9% after the light soaking procedure. For sample 3 a reduction in the solar cells efficiency from 9.1% to 7.9% within the first LID cycle is observed. Afterwards the efficiency stabilizes around 8% and doesn't show any further degradation with proceeding LID. There is a strong decrease in J_sc and V_oc of sample 3 at degradation times of 150 h and 300 h. This is attributed to contacting issues during the measurement of the IV characteristic. Within the uncertainty of the measurement setup, there is no dependency of the J_sc or V_oc on the degradation time, for any sample.

For the timeframe of 0–100 h of LID, the fill factors of all samples decreased with increasing degradation time. After 100 h of LID the fill factors for all three samples

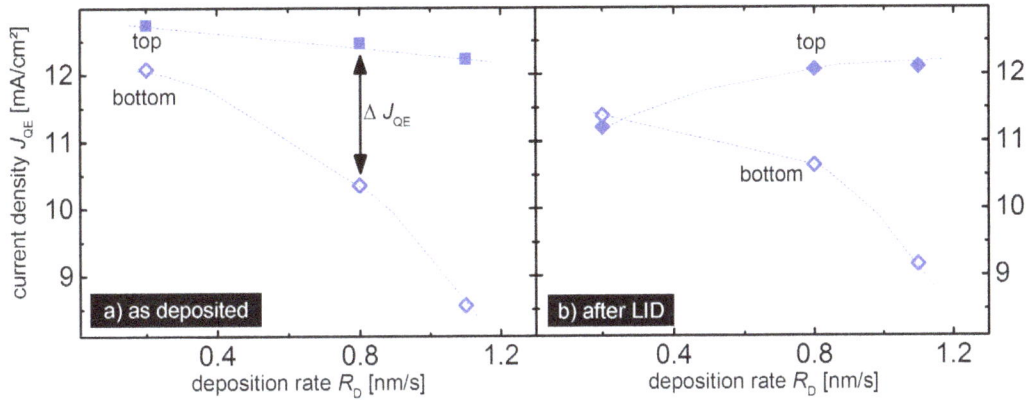

Fig. 3. Current density J_{QE} of the top and the bottom solar cells, derived from EQE measurements, as a function of deposition rate R_D before (a) and after (b) 1000 h of light induced degradation.

increases slightly with increasing degradation time. However, the interpretation of the fill factor should be taken with care. For example a high value of FF can be caused by a mismatch between the currents generated by the top and the bottom solar cell. Thus, it is seen that the current matching between a-Si:H top solar cell and μc-Si:H bottom solar cell changes with proceeding degradation time.

Figure 3 shows the current density J_{QE} derived from EQE measurements for top and bottom solar cell as a function of deposition rate R_D before (Fig. 3a) and after (Fig. 3b) 1000 h of LID. Before LID the J_{QE} generated by the top and the bottom solar cell decreases with increasing deposition rate.

After LID the J_{QE} generated by the top solar cell increases slightly with increasing deposition rate, whereas the J_{QE} generated by the bottom solar cell decreases with increasing deposition rate. The values for the J_{QE} for the top solar cells are reduced after LID compared to the initial values. Prior to LID all samples are limited by the current of the bottom solar cell in the as-deposited state. After the LID samples 2 and 3 are still limited by the current of the bottom solar cell whereas sample 1 shows good current matching conditions between the a-Si:H top and μc-Si:H bottom solar cell. The quality of the matching between the top and the bottom solar cell current is visually expressed by the difference (ΔJ_{QE}) between the J_{QE} values for top and bottom solar cell for each sample. After LID, the difference ΔJ_{QE} decreases for sample 1 from 0.66 mA/cm^2 to 0.17 mA/cm^2, for sample 2 from 2.1 mA/cm^2 to 1.4 mA/cm^2, and for sample 3 from 3.7 mA/cm^2 to 3 mA/cm^2.

Figures 4a–4c show the results of EQE measurements for samples 1, 2, and 3, respectively. The current density values J_{QE} for the top and bottom solar cells before and after LID are displayed in each figure. The initial values are displayed in bold face, and the EQE curves for the initial case are displayed as solid lines. The EQE curves for the stabilized cases are displayed by dashed lines. The EQE curve of the top solar cell as well as the bottom solar cell of the reference sample are both reduced by the LID (see Fig. 4a). The EQE of the top solar cell is reduced in the wavelength region between 380 nm and 580 nm,

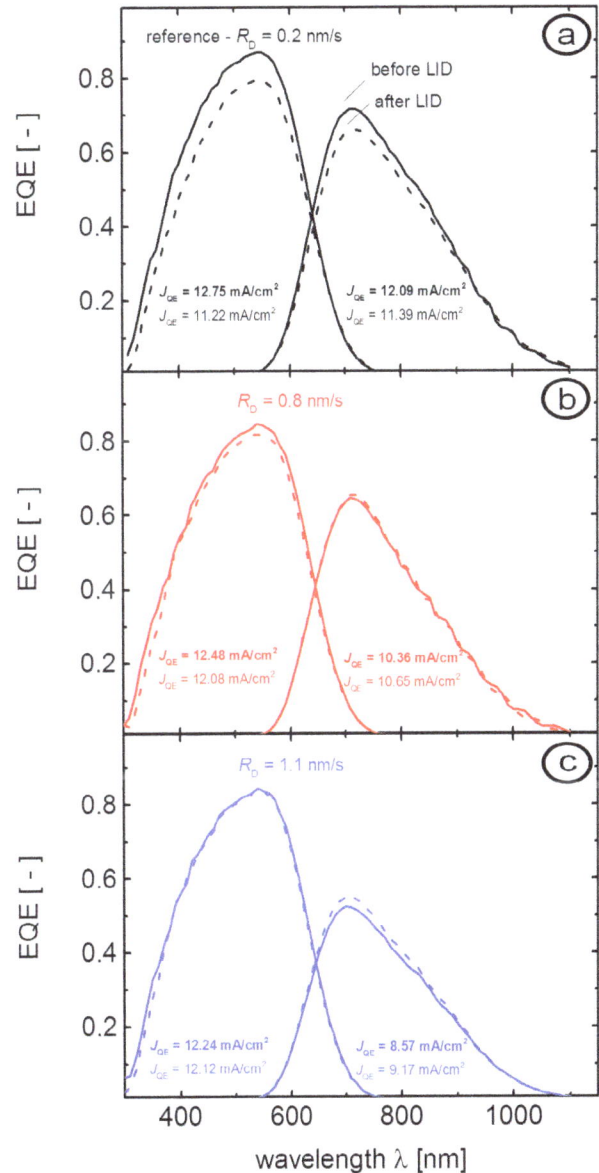

Fig. 4. EQE measurements for the top and bottom solar cells of sample 1 (a), sample 2 (b), and sample 3 (c).

whereas the EQE of the bottom solar cell is affected in the wavelength region between 650 nm and 870 nm. Both samples deposited at elevated deposition rates show a reduced EQE for the bottom solar cell compared to sample 1. Furthermore the reduction is more pronounced for the sample 3 compared to sample 2. The top and bottom solar cells of sample 2 and 3 are less affected by the LID compared to the reference sample 1.

4 Discussion

The discussion part is organized in two parts. In the first part we discuss the effect of a varied deposition rate on the device performance. In the second part we present a model calculation about potential cost benefits when applying elevated deposition rates on an industrial scale.

To achieve high deposition rates for device grade microcrystalline silicon it is necessary to adjust the deposition parameters over a broad range. Adapting the deposition parameters may strongly influence the reactions in the glow discharge. Varying the deposition power or the deposition pressure for example leads to a modification of the ion bombardment of the growing film which in turn influences the growth properties of microcrystalline silicon [12]. The fact that the quantum efficiency of the top solar cell is not strongly affected by the application of the processes leading to high deposition rates for the absorbing layer of the bottom solar cell shows the energies of the bombarding ions are not sufficient to damage the i-n interface of the top solar cell. The decrease of the open-circuit voltage and the short-circuit current density can be attributed to a deterioration of the p-i interface of the bottom solar cell.

Samples 2 and 3 show less degradation of the bottom solar cells compared to sample 1. This can be attributed to the effect that microcrystalline layers processed at elevated deposition rates grow with a pronounced Raman intensity ratio[1]. This assumption is supported by the reduced open-circuit voltage seen in the as-deposited state of the solar cells with increased deposition rate.

A slight increase of the efficiency of sample 2 and 3 is observed for degradation times beyond 100 h (see Fig. 2). Additionally, an increase of the fill factor is observed for all samples for degradation times exceeding 100 h. Both effects might be attributed to a bottom limited current device and the improved current matching conditions caused by light induced degradation (see Fig. 3).

Together with industrial partners we are investigation within the "Fast Track" project (see acknowledgements) the possibility to scale processes which yield high deposition rates from laboratory scale to industry scale. To deduce a value for the cost benefits that can be used to compare both processes, a model calculation is presented. We assume that there are no further losses in the device performance during the up-scaling of the processes to large areas. Furthermore we do not include interconnection losses which may be introduced during the module fabrication

process. Considering the efficiency of sample 1 of 12.1% and an illumination intensity of 1000 W/m^2 together with the processing time of 180 min we can calculate the output for the chamber to be 40.33 W/(m^2 h).

$$12.1\% \xrightarrow{1000\frac{W}{m^2}} 121 \text{ W/m}^2 \xrightarrow{/180 \text{ min}} 40.3 \text{ W/(m}^2 \text{ h)}.$$

For sample 2, with an efficiency of 10.9% and a processing time of 70 min for the absorber layer of the bottom solar cell, we can calculate in an analogous way an output value of 93.4 W/(m^2 h). Thus by the introduction of processes leading to elevated deposition rates it is possible to increase the output of the i-chamber by a factor of 2.3.

5 Conclusion

The effect of the deposition rate of μc-Si:H absorber layers was investigated in the range of 0.2 nm/s and 3.2 nm/s on the performance of thin-film silicon tandem solar cells. For a deposition rate of 0.8 nm/s, which corresponds four times the deposition rate of the reference process, an initial efficiency of 10.9% with a stabilized efficiency of 9.9% was observed. Despite the reduction in efficiency of 1.2% for the sample processed at a deposition rate of 0.8 nm/s compared to the reference solar cell, calculations show that the output of the deposition system in terms of Watts per hour and area can be increased by a factor of 2.3 due to reduced processing times.

The authors thank A. Schmalen and J. Wolff for technical support and L. Nießen, C. Grates, C. Zahren for support with characterization setups. Assistance provided by A. Bauer, H. Siekmann, and U. Gerhards for the fabrication of ZnO/Ag back contacts is greatly appreciated. The authors thank U. Rau for constant support, encouragement and for rewarding discussions. We acknowledge the "Bundesministerium für Umwelt, Naturschutz und Reaktorsicherheit" (Project "Quick μc-Si", Contract No. 03225260C). A part of this work was carried out in the framework of the FP 7 project "Fast Track", funded by the European Union under grant agreement No. 283501.

References

1. H. Keppner, J. Meier, P. Torres, D. Fischer, A. Shah, Appl. Phys. A: Mater. Sci. Proc. **69**, 169 (1999)
2. O. Vetterl, F. Finger, R. Carius, P. Hapke, L. Houben, O. Kluth, A. Lambertz, A. Mück, B. Rech, H. Wagner, Sol. Energy Mater. Sol. Cells **62**, 97 (2000)
3. K. Yamamoto, T. Suzuki, M. Yoshimi, A. Nakajima, Jpn J. Appl. Phys. **36**, 569 (1997)
4. G. Nakamura, K. Sato, Y. Yukimoto, in *16th IEEE Photovolt. Specialist Conf., New York, 1982*, pp. 1331–1337
5. J. Yang, A. Banerjee, S. Guha, Appl. Phys. Lett. **70**, 2975 (1997)
6. S. Hänni, G. Bugnon, G. Parascandolo, M. Boccard, J. Escarré, M. Despeisse, F. Meillaud, C. Ballif, Prog. Photovolt.: Res. Appl. **21**, 821 (2013)

---

[1] S. Michard, to be published (n.d.).

7. S. Kim, J. Chung, H. Lee, J. Park, Y. Heo, H. Lee, Sol. Energy Mater. Sol. Cells **119**, 26 (2013)

8. S. Guha, J. Yang, B. Yan, Sol. Energy Mater. Sol. Cells **119**, 1 (2013)

9. A. Shah, E. Moulin, C. Ballif, Sol. Energy Mater. Sol. Cells **119**, 311 (2013)

10. Y. Mai, S. Klein, R. Carius, J. Wolff, A. Lambertz, F. Finger, X. Geng, J. Appl. Phys. **97**, 114913 (2005)

11. A. Gordijn, A. Pollet-Villard, F. Finger, Appl. Phys. Lett. **98**, 211501 (2011)

12. S. Michard, M. Meier, B. Grootoonk, O. Astakhov, A. Gordijn, F. Finger, Mater. Sci. Eng. B **178**, 691 (2012)

13. L. Guo, M. Kondo, M. Fukawa, K. Saitoh, A. Matsuda, Jpn J. Appl. Phys. **37**, 1116 (1998)

14. A. Matsuda, J. Non-Cryst. Solids **338–340**, 1 (2004)

15. F. Finger, P. Hapke, M. Luysberg, R. Carius, H. Wagner, M. Scheib, Appl. Phys. Lett. **65**, 2588 (1994)

16. J. Meier, E. Vallat-Sauvain, S. Dubail, U. Kroll, J. Dubail, S. Golay, L. Feitknecht, P. Torres, S. Fay, D. Fischer, A. Shah, Sol. Energy Mater. Sol. Cells **66**, 73 (2001)

17. U. Graf, J. Meier, U. Kroll, J. Bailat, C. Droz, E. Vallat-Sauvain, A. Shah, Sol. Energy Mater. Sol. Cells **427**, 37 (2003)

18. O. Vetterl, A. Gross, T. Jana, S. Ray, A. Lambertz, R. Carius, F. Finger, J. Non-Cryst. Solids **299–302**, 772 (2002)

10

Kësterite thin films for photovoltaics

S. Delbos[1,2,3,a]

[1] EDF R&D, Institut de Recherche et Développement sur l'Énergie Photovoltaïque (IRDEP), 6 quai Watier, 78401 Chatou, France
[2] CNRS, UMR 7174, 78401 Chatou, France
[3] Chimie ParisTech, 75005 Paris, France

Abstract In the years to come, electricity production is bound to increase, and $Cu_2ZnSn(S,Se)_4$ (CZTS) compounds, due to their suitability to thin-film solar cells, could be a means to fulfill the demand. After explaining the reasons of the sudden interest of the CIGS scientific community for CZTS solar cells, this paper reviews recent papers published on the subject of kësterites-based solar cells. After a description of crystallographic and optoelectronic properties, including CZTS crystalline structure, defect formation and metal composition, this review paper focuses on CZTS synthesis processes and device properties. Synthesis strategies, including one- or two-step processes, deposition temperature, binary formation control via atmosphere control and their effect on device properties are discussed.

1 Sustainability of photovoltaic industry

In the years to come, electricity production is bound to increase (from 17 PWh in 2009 to 28–34 PWh in 2035 [1]) and as a consequence the increased demand on fossil fuels as well as possible regulations on CO_2 emissions are bound to increase the wholesale electricity costs. On the other hand, the price-experience factor of 22% that was observed for the last 4 decades should stay at the same level or slightly below [2, 3], leading to a decrease of photovoltaic (PV) electricity costs [2, 4]. A first change of paradigm is now happening, because in some places like South Italy and other sunny/high electricity price regions [5], the price of retail electricity is higher than the cost of PV electricity (grid parity). A second parity, the "fuel parity", will happen when the cost of PV electricity is lower than the marginal costs of operating fossil fuel based power plants [6]. According to deployment scenarios, this fuel parity could happen between 2020 and 2040.

The massive deployment of PV foreseen in years 2020−2040 could lead to 1.5 to 2.5 TWp of worldwide installed PV capacity in 2030, with a demand of 20−150 GWp/y and around 30% of the market taken by thin films [2, 4].

The question of the sustainability of the production of 20−150 GWp/y of PV was raised by a few publications, and in particular the worldwide mining and refining capacity of a few critical elements could be the main problem of the photovoltaic industry in the years to come. Each

PV technology has one or more limiting elements, such as silver for crystalline silicon, indium and gallium for CIGS, Tellurium for CdTe, Ruthenium for dye-sensitized solar cells, silver and indium for thin-film silicon, germanium, gold, indium for III-V PV [7, 8]. The chalcogenide technologies are particularly vulnerable to In, Ga, Te supply, which has been noted critical or near critical by the US Department of Energy [9] or the European Commission [10].

According to various authors, the production of PV modules could be limited below the demand reported earlier in this paper, as summarized in Table 1. Crystalline silicon PV seems suitable for supplying the demand, but one should keep in mind that the basic assumption used for the third and fourth column of this table is that all worldwide production of silver is used for crystalline PV. The fifth column shows the forecast 2030 demand for Ag, Te, In with respect to the 2008 production. It appears that the demand will grow enormously. If we focus on In, in 2030 the total demand (including PV and other uses) could be 3 to 7 times the 2008 production [10,11], whereas, as In is a byproduct of Zn extraction, the elasticity of the production with respect to demand is low. One author suggests that CIGS and CdTe technologies will not be impeded by resources problems, but he does not take the increase of other sources of demand for In and Te into account [12].

The sustainability of PV production is therefore a real question, and the development of a new PV technology based on abundant and preferably non-toxic elements would alleviate the pressure on all PV technologies in term of resources.

---
[a] e-mail: `delbos.sebastien@gmail.com`

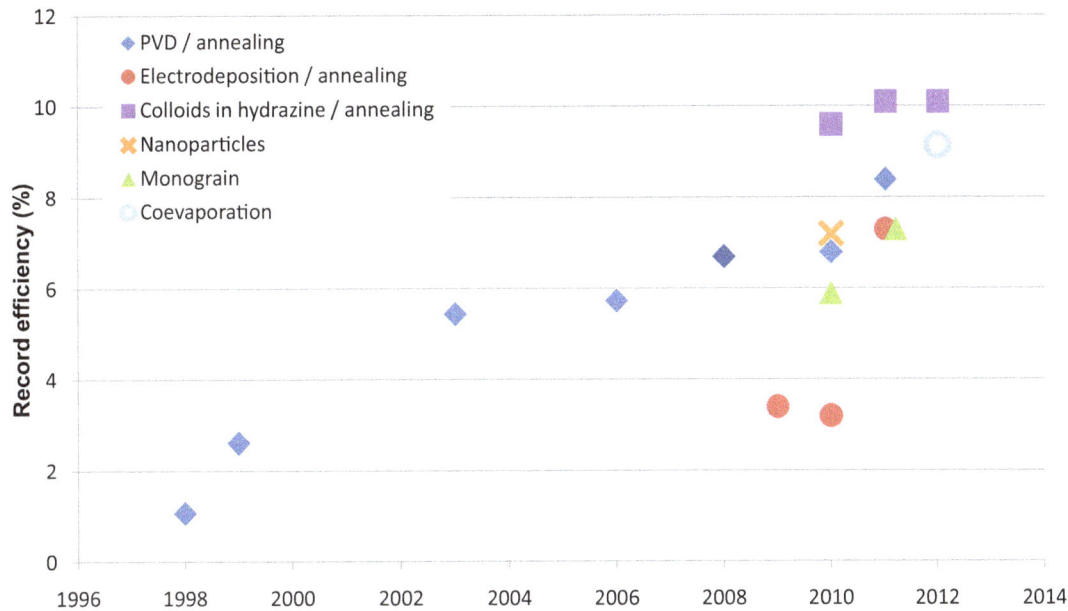

Fig. 1. Evolution of the record efficiency for CZTS solar cells. See Table 2 for the references.

Table 1. Production limitations of the PV technologies [7,8,10–12,129].

PV technology	Limiting element	Max installed capacity based on the limiting element reserves (1.5–2.5 TWp demand)	Max yearly production based on annual production of limiting material (20–150 GWp demand)	Ratio (demand 2030/2008 production) of limiting element
c-Si	Ag	5–15 TWp	305 GWp	20
CdTe	Te	0.1–0.8 TWp	2.3–40 GWp	50
CIGS	In	0.1–0.6 TWp	13–26 GWp	3
Dye-sensitized solar cells	Ru, In (TCO)	0.06–0.9 TWp	2.2 GWp	n.a.

2 Advantages of CZTS

Recently the CIGS scientific community started to work on $Cu_2ZnSn(S,Se)_4$, (CZTS), and the enthusiasm was kindled by the fast increase of record efficiency of CZTS-based solar cells, from the 2009 record of 6.7% [13] to 2011 record of 10.1% [14], as shown in Figure 1. CZTS has numerous advantages that could lead to its massive use as an abundant, non toxic, low cost absorber for thin-film photovoltaic solar cells:

– it is a compound whose intrinsic point defects lead to p-type semiconductor behavior;

– it has a direct bandgap and an absorption coefficient $>10^4$ cm^{-1}, which is suitable for thin film photovoltaics applications [15–17];

– its band gap has been predicted [18] to be 1.0 eV for $Cu_2ZnSnSe_4$ and 1.5 eV for Cu_2ZnSnS_4 and evidences of the bandgap tunability have been found via the variations of V_{OC} [19], as well as direct measurements of the variation of the bandgap between 1.0 eV for $Cu_2ZnSnSe_4$ and 1.5 eV for Cu_2ZnSnS_4 [20–24]. The band gap of CZTS can also be tuned by incorporation of Ge, Ge-containing materials having smaller bandgap than their Ge-free counterparts [25]. This tunability is of particular interest for the manufacturing of absorbers with a band gap between 1.1 and 1.5 eV, which allow theoretical efficiencies higher than 30% [26];

– its crystallographic structure can accept some shifts from the stoichiometric composition [27];

– it includes Zn and Sn which are respectively produced in quantity 20 000 and 500 times bigger than In [28];

– it is possible to make CZTS solar cells just by replacing CIGS by CZTS in CIGS solar cells, and such solar cells yielded efficiencies up to 10.1% [14,29]. The knowledge gathered on back contact, buffer layers and window layer by CIGS scientists can therefore be used and adapted for CZTS solar cells. Nevertheless, for high efficiency solar cells the processes used for CIGS solar cells will need to be tuned. The buffer layer, in particular, will have to be tailored in order to adjust the lattice-matching, the valence and band offsets with CZTS;

– the grain boundaries (GB) seem to have the same beneficial properties for CZTS as for CIGS, such as enhanced minority carrier collection taking place at the GB [30].

3 Crystallographic and optoelectronic properties

This paper is focused on synthesis processes of CZTS, and thus will only briefly address the crystallographic and optoelectronic properties. For more information on crystal and band structure and defects in kësterites, the reader can find a recent review dedicated to these topics here [31].

3.1 Crystalline structure

In the literature Cu_2ZnSnS_4 and $Cu_2ZnSnSe_4$ (as well as $Cu_2ZnGe(S,Se)_4$ [32]) are described by the structural models of two natural minerals: stannite (space group $I\text{-}42m$) [33–35] and kësterite (space group $I\text{-}4$) [33]. These crystal structures are very close, in both structures the cations are located on tetrahedral sites but their distributions on planes perpendicular to the c-axis are not the same. In addition, the position of the chalcogen atom is slightly different in these structures [33–35].

– kësterite: (Cu + Sn) / (Cu + Zn) / (Cu + Sn), chalcogen in position (x, y, z), Figure 2;
– stannite: Cu / (Zn + Sn) / Cu, chalcogen in position (x, x, z).

In the rest of the paper, "kësterite" will be used for the crystallographic structure, and "kësterites" will be used for $Cu_2ZnSn(S,Se)_4$ compounds that are crystallized in kësterite type structure. X will be used for the chalcogens S and Se when an effect or properties can apply to both of them (for example $Cu_2(S,Se)$ will be noted Cu_2X).

Due to their structural similarities, kësterite and stannite are very difficult to distinguish by X-Ray Diffraction and Raman spectroscopy, and it is necessary to use neutron diffraction [36, 37] to tell them apart. According to ab initio calculi, the most stable crystalline structure is kësterite [38–42] and one study by ab initio calculation even suggests that all observations of stannite structure for CZTS compounds were due to partial disorder in the I-II (001) layer of the kësterite phase [32]. Recent neutron diffraction [37] and anomalous diffusion [43] characterizations show that CZTS crystallizes in the kësterite structure, and synchrotron X-Ray experiments show that this structure is dominant for temperature <876 °C (transition to a sphalerite cubic structure) [37].

The structural difference between Cu_2ZnSnS_4 and $Cu_2ZnSnSe_4$ is not clear. It seems that the Se compound has larger lattice parameters as well as higher electric conductivity than the S compound [20]. Concerning mechanical and thermo-physical properties of CZTS, no experimental data are available, but ab initio simulations were performed [44, 45].

Sulfur kësterites have XRD peaks very close to those of ZnS and Cu_2SnS_3 [36], and Raman spectroscopy is necessary to tell them apart [46]. Nevertheless, it is not possible to have a quantitative determination of the secondary phases by Raman. For such purpose, XENAS has been used [47].

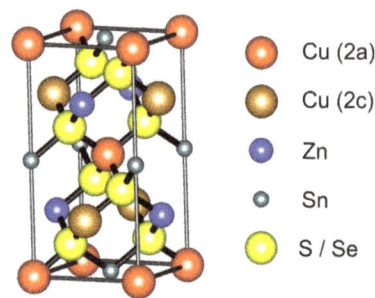

Fig. 2. $Cu_2ZnSn(S,Se)_4$ compound in kësterite structure, according to [33].

3.2 Intrinsic point defect formation

The kësterite structure allows for several types of point defects which make discrete energy levels appear, allowing p-type doping. As in the CIGS materials, the doping is obtained by stoichiometry variations and not by extrinsic doping. Nevertheless, according to ab initio calculi, the kësterite structure is less tolerant to these variations than the chalcopyrite structure [39, 48–51]. According to point-defect measurements, the kësterites are susceptible to higher carrier recombination than chalcopyrite [52]. Measurements of the deep defects of CZTS are reported here [53].

Generally, the Sn-Se bond is much stronger than the Zn-Se and Cu-Se bond, leading to a high formation energy of V_{Sn} [51]. Contrary to CIGS, the accepting defect the most easily formed is not V_{Cu} but Cu_{Zn} [39,49,54,55]. Nevertheless some authors suggest the contrary [52,56,57]. Zn_{Cu} substitutions occur at the 2c site and Cu_{Zn} at the 2d site [36,58] (see Fig. 2). The accepting defects are easily formed, leading to difficult n-type doping [54]. The easy formation of Cu_{Zn} accepting defects also leads to low charge separation, a drawback that could be overcome thanks to the type-II band alignment of the CZTS/CdS interface [49]. The most active recombination centers are expected to be Cu_{Sn} (In_{Cu} for CIGS), and the lack of defect complex ($2V_{Cu} + In_{Cu}$ for CIGS) is thought to lead to the higher defect concentration observed [59].

3.3 Metal composition

The single phase composition region is much narrower for CZTS than for CIGS. According to the Tallinn team, single phase CZTS monograin powders can be synthesized only from a precursor mixture comprising metal ratios of $Cu/(Zn + Sn) = 0.92-0.95$ and $Zn/Sn = 1.0-1.03$ [60]. Nevertheless, a Rietveld refinement of anomalous dispersion measurements revealed a $Cu/(Zn + Sn) = 0.97$, and $Zn/Sn = 1.42$ composition for kësterites crystals [43], suggesting that the single phase composition region is not as narrow as the Tallinn team suggests.

Until now, only Zn-rich and Cu-poor materials gave high efficiency (see Fig. 3), as was already noted by the Nagaoka team [61]. In Zn-rich growth conditions, the kësterite structure is tolerant to stoichiometry deviations

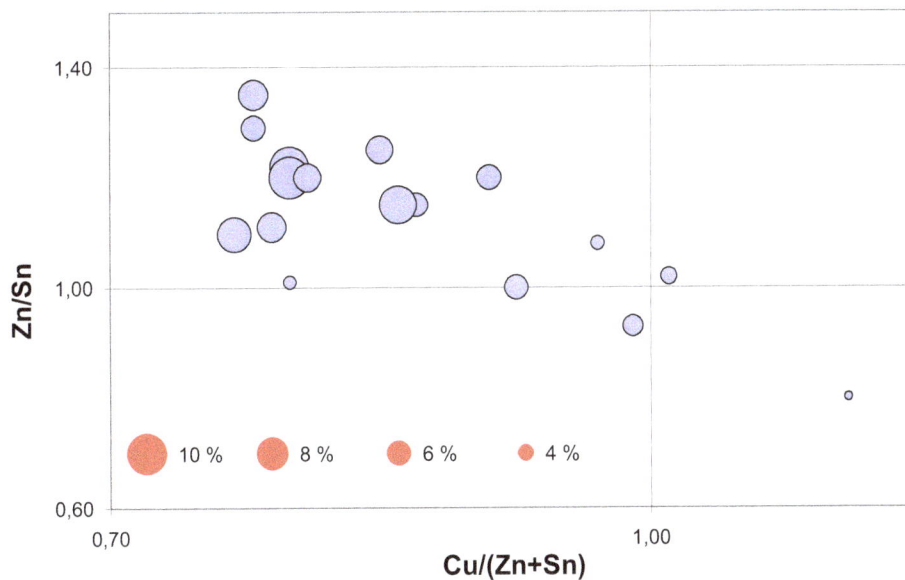

Fig. 3. Composition map of the best CZTS solar cells. The diameter of the circles is proportional to the conversion efficiency. See Table 2 for the references.

by zinc excess and copper lacking [61–64], in the form of Zn_{Cu} and V_{Cu} vacancies [51,65], but high Zn content also leads to structural disorder [66].

3.4 Bandgap measurement

Absorbance measurements are not suitable for extracting the bandgap values, because the absorption coefficient derived from spectrophotometric data on thin films is determined by defect absorption and by measurement accuracy limitations [66]. In particular, ZnSe [67] and Zn-caused disorders [66] lead to incorrect bandgap determination when it is determined from absorbance measurements. PL measurements now confirm that the bandgap of $Cu_2ZnSnSe_4$ is indeed around 1.0 eV at room temperature [68], whereas before that publication the measured band gap was around 1.5 eV. Suitable methods for measurement of CZTS bandgap are PL, TR-PL and EQE measurements.

4 Synthesis strategies

4.1 One-step or two step processes for thin films

For CIGS, the deposition techniques used to be classified as vacuum and non-vacuum techniques. For CZTS, the same approach was also used in a review [69]. Nevertheless, for CZTS, this classification is not accurate, because recent works showed that low-cost deposition does not mean low-cost solar modules [70], and because of the problematic of binaries control during CZTS synthesis (see next paragraph). We therefore decided to classify the processes into one-step or two-step processes:

– Two-step processes, where the needed elements are first incorporated during an ambient temperature process, followed by an annealing step. The chalcogen can be incorporated into the precursor or during the annealing step. These processes allow the use of fast and low-cost techniques for precursor deposition. Until now, and contrary to the CIGS, two-steps processes yielded the best efficiencies (more than 10%) [14]. IBM works on three such processes, the hybrid solution-particle technique [14], electrodeposition [71] and co-evaporation [72]. Other teams investigated numerous other deposition techniques, such as codeposition by evaporation, sputtering, OA-CVD (Open Atmosphere Chemical Vapor Deposition), nanoparticles, sol-gel (see Tab. 2).

– One-step processes, where all the elements are incorporated simultaneously. This type of processes yields the better results for CIGS, but until now it was not the case for CZTS. The best efficiency until now is 9.15% by coevaporation [73]. Only a few groups published results for such processes, and very few had significant results in terms of solar cell efficiency: NREL [73], Tallinn University [74], HZB [75]. Other groups tried reactive sputtering and pulsed laser deposition (PLD) (see Tab. 2). The method of Tallinn University also introduces a huge difference in cell architecture, where 50 μm CZTS monograins are synthesized in molten KI, wrapped in the buffer layer and attached to the substrate by epoxy glue [76].

These processes are different, but the two types of processes (one-step and two-steps) reached high efficiencies (>9%). Until now, it does not seem that one or the other is intrinsically better, because, as we will see in the next paragraphs, the key point is temperature and atmosphere control during the one-step deposition or annealing step.

Table 2. Synthesis methods for CZTS solar cells.

Synthesis method	Chalcogen and method of insertion	η (%)	I_{sc} (mA/cm^2)	V_{oc} (mV)	FF (%)
Spin-coating of metal-chalcogen colloids in a hydrazine solution, 540 °C thermal treatment [14, 29, 64, 130, 131]	S, Se (in the colloids) S/(S+Se) $\sim$ 0.4 S/(S+Se) $\sim$ 0.03	10.1 [14] 10.1 [29]	31 39	517 423	64 62
Coevaporation of metals and Se, three-stage process [73]	Se during coevaporation, 40 Å/s	9.15 [73]	37.4	377	65
Coevaporation of metals and cracked S on low-temp (150 °C substrate) followed by 5 min anneal at 570 °C at atmospheric pressure [72, 120] Coevaporation of metals followed by 540 °C with S, 5 min [94, 103]	Pure sulfur, coevaporated. Mix of coevaporated S and Se (7.5%) [121]	8.4 [72]	19.5	661	66
Monograin solar cells with S/Se gradient[74]	CZTSe monograin ($\sim$50 μm), SnS_2 annealing causing a 0.1–2 μm S diffusion.	7.4 [74]	18.4	720	60
Monograin cells [19, 23, 76, 132]	$S_{0.75}, Se_{0.25}$ [76]	7.3 [23]	19.9	696	58
Sequential electrodeposition [71, 133–141]	585 °C in S/N_2 for 12 min	7.3 [71]	22	567	58
Blade-coating of CZTS nanoparticles in hexanethiol. Best efficiency (7.2%) after light-soaking. 6.8% with $Cu_2ZnSn_{0.3}Ge_{0.7}(S_{0.5},Se_{0.5})_4$	S in the colloids, Se thermal treatment (20 min 500 °C)	6.7 [142] 6.8 [25]	30.4 21.5	420 640	52 49
Cosputtering of ZnS, SnS, Cu [13, 65]	3 h at 580 °C with H_2S	6.7	17.9	610	62
Stacked deposition of ZnS (sputtering) / Cu (evaporation) / Sn (evaporation) [106]	30 min in Se atmosphere	6.0	31.5	390	49
Stacks of evaporated metals [17, 113], 6.1% with Cd buffer, 5.8 with Cd-free buffer. Variation with evaporation of ZnS, Sn, Cu [100]	Sulfurization	5.8	15.8	618	60
Open-Atmosphere Chemical Vapor Deposition (OA-CVD). Spray pyrolysis of metal-oxide precursors [143]	520–560 °C for three hours in 5 vol% H_2S balanced with N_2.	6.0	16.5	658	55
Coevaporation of Cu, Zn, Sn, Se [98]	Thermal treatment in the presence of Sn and S	5.4	20.0	497	54
Sol-gel: metals and sulfur in ethanol [144]	Annealing under N_2+H_2S (5%) atmosphere at 530 °C for 30 min	5.1	18.9	517	53

Table 2. Continued.

Synthesis method	Chalcogen and method of insertion	η (%)	I_{sc} (mA/cm²)	V_{oc} (mV)	FF (%)
Sputtering of Cu/ZnSn/Cu [108]	Annealing 30 min 560 °C	4.6	15.4	545	55
Metallic precursors in alcohol [99]	Thermal treatment in Se atmosphere	4.2	23.2	305	55
Metal chlorines in dimethyl sulfoxide [145]	Thiourea in solution and Se atmosphere, 500 °C	4.1	24.9	400	41
Fast coevaporation (~20 min) of Cu, Sn, ZnS [75]	Cracked sulfur during evaporation	4.1	13.0	541	60
Cosputtering of Cu, Sn, Zn [146]. Variations with Cu/Sn, Zn and Cu/Sn, ZnS [147,148] and with Se thermal treatment [149]	S in sealed glass tube	3.7 [146]	16.5	425	53
One-step electrodeposition of metals [16,62,85,104,150–153]. One-step electrodeposition (including sulfur) [154–157]	H_2S in Ar at 550 °C for the record. Other techniques includes S in Ar, H_2S in Ar, S in N_2, Se and H_2	3.4 [62]	14.8	563	41
Reactive cosputtering of Cu, Zn, Sn in H_2S/Ar [30]	50% H_2S / Ar mixture at 5 mtorr	3.4	12.4	428	64
Sequential sputtering of Cu, Zn, Sn [158]	Se vapors at 500 °C for 30 min	3.2	20.7	359	43
PLD from CZTS based on sintered Cu_2S, ZnS and SnS_2. N_2 thermal treatment at 500 °C during 1 h [159–162]	S in the CZTS pellets	3.14	8.8	651	55
Sol-gel solution (metals/methoxyethanol) [91,163].	H_2S 3% in N_2	2.23	10.2	529	42
Nanoparticles on flexible Al foil, ZnS as buffer layer [118]		1.94			
Successive ionic layer adsorption and reaction S (SILAR) [164,165] (measurements at 30 mW/cm°)	1.85	3.2	280	62	
Sequential e-beam evaporation of Cu, Zn, Sn [166]	S vapors for 2h at 560 °C	1.7	9.78	478	38
Reactive sputtering from Cu_2ZnSn target [167,168]	H_2S plasma	1.35	9.52	0.343	41
Chemical bath deposition of SnS, ZnS, Cu ion exchange [169]	H_2S at 500 °C for 2 h	0.16	2.4	210	31
Evaporation from a quaternary source [170]		0.36			

(a)   (b)

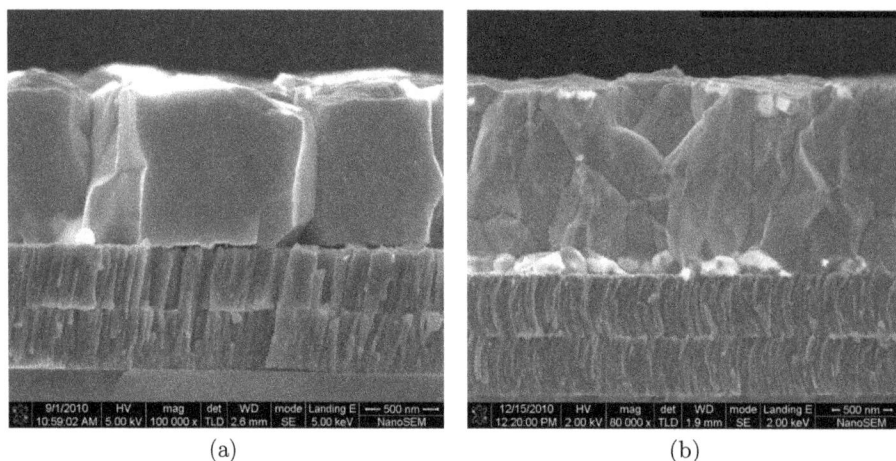

Fig. 4. SEM cross-section of CZTSe films grown at the same temperature and constant deposition fluxes, except Cu rates were adjusted to produce only (a) Cu-rich growth or (b) Cu-poor growth [59].

Other methods of synthesis that did not produce solar cells include: cosputtering of Cu, Sn, Zn followed by annealing in a hot tube in presence of H_2S [77] , dip-coating in a methanol solution followed by annealing at 200 °C in air [78], CZTS nanoparticles in oleylamine [79] , chemical bath deposition [80] , sputtering of a sintered mix of CuSe, $CuSe_2$, ZnS, and SnS_2 [81].

4.2 CZTS formation

Similarly to CIGS, CZTS needs temperature between 500 and 600 °C to be synthesized. Nevertheless, the chemical reactions are not yet completely understood, even if a reaction path was proposed [82]. There are successive reactions taking place in the bulk of the layer between the elements leading to binaries, ternaries and finally the quaternary.

At room temperature, only metal binaries can form, for example Cu_6Sn_5 [83], Cu_5Zn_8 [84] , Cu_3Sn and CuZn [85]. At higher temperature, between 200 and 450 °C, metal-chalcogenide binaries such as CuX, Cu_2X, and SnX form [86–88], and for temperature >450 °C, Cu_xX binaries then react with Sn to form Cu_xSn_yS [80]. At higher temperature (between 550 and 580 °C [71,85,86,89]), and for longer annealing time ($\sim$8 h) or one-step deposition at 500 °C [73, 83] , ZnX reacts with Cu_2SnX_3 to form Cu_2SnZnX_4 according to the equation proposed by [82] :

$$Cu_2SnX_3 + ZnX \rightarrow Cu_2ZnSnX_4. \qquad (1)$$

In order to control which reactions take place during the annealing step of a two-step process, it is necessary to closely monitor the temperature of the layer, because, for example, too fast an annealing can lead to liquid Sn bubbles formation, preventing the formation of large grains [90]. For processes using elemental sulfur or selenium, fast annealing (a few minutes) is possible, whereas for processes using H_2S, which is less reactive than elemental sulfur, a long annealing time (2−3 h) is necessary for

the binaries to react completely and to form large grains of CZTS [91, 92].

For one-step processes, Cu-rich growth conditions are needed at the beginning of the reaction in order to foster growth of large grains [73, 93], as shown in Figure 4. Nevertheless, for both processes, the finished layer must be Zn-rich (see part 4.3), and in particular Cu_2X-free [60] in order to get high photovoltaic efficiencies. If the layer is not Cu_2X-free at the end of the synthesis process, the subsequent cyanide etching leads to possible voids and defects [89]. If Zn-rich growth conditions are used, ZnX formation is promoted [48, 49].

Therefore, for one-step processes it is necessary to adapt the rate of deposition of each element with respect to the reaction path, as well as control deposition temperature, because higher temperatures can also increase grain size [59]. For two-step processes it is necessary to closely monitor growth conditions, and in particular prevent Cu_2X formation and increase ZnX reaction with Cu_2SnS_3.

4.3 Atmosphere control

Unfortunately, contrary to CIGS, the elements involved in CZTS synthesis are prone to evaporation and sublimation. Zn sublimates at 430 °C [94], SnSe at 350 °C [88], SnS at 370 °C and Sn evaporates at 460 °C [95, 96]. Moreover, high temperatures promote CZTS decomposition, according to the chemical equilibrium between CZTS and solid and gaseous binaries [97,98].

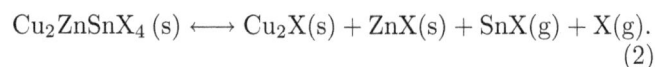

$$Cu_2ZnSnX_4\,(s) \longleftrightarrow Cu_2X(s) + ZnX(s) + SnX(g) + X(g). \qquad (2)$$

If this equilibrium is displaced towards the right member, ZnX, SnX and X formation and losses will occur, leading to a Cu_2X-rich layer, increasing layer resistance and preventing good photovoltaic efficiencies (see 5.6). It is therefore of the utmost importance to prevent CZTS decomposition as well as binaries losses. The key to that result is atmosphere control.

The equilibrium of equation (2) can be displaced towards CZTS by saturating the atmosphere in one of the right-side members of the equation [93,97]. It is possible to counteract SnX evaporation by introducing gaseous SnX during the thermal treatment [98], or by increasing the partial pressure of the chalcogen [99], by performing the chalcogenation in a small closed volume [100], or by using a cap on the precursor layer during annealing [101]. If the annealing takes place in a H_2 atmosphere, Zn loss can be prevented by using ZnS as a precursor [102]. Continuous evaporation of Zn towards the substrate prevents the decomposition of CZTS in a coevaporation process [73]. It also seems that the presence of MoX_2 at the back contact pushes the reaction towards the right member [103].

If atmosphere control is not good, annealing time is the next best lever on CZTS decomposition, because if CZTS is synthesized and cooled rapidly, only a small quantity of binaries will be lost. One paper compared slow and fast annealing [104]. It concluded that fast annealing is better than slow annealing because many different secondary phases were found with slow annealing, without more explanation. Nevertheless, in this case slow annealing could be linked to a badly controlled reaction atmosphere, leading to CZTS decomposition into binary and ternary phases.

Concerning the key point of atmosphere control, strategies of synthesis should be slightly different for one-step or two-step processes. For two-step processes, there seems to be two stages during the annealing step: one short (a few minutes) step of CZTS formation, and one longer (up to a few hours) step of grain growth. The control of the atmosphere should be very accurate in the second step, and the reaction atmosphere could also be completely different from the first step [93]. Fortunately, it is quite easy and inexpensive to design annealing equipments that provide good static atmosphere control (that is to say, equipments where nothing can either go in or go out). Nevertheless, vapor injection is more tricky, as it requires carrier gas, pressure measurements, leading to an equipment resembling an evaporator.

For one-step synthesis, atmosphere control is not so easy, because reaction chambers are designed for controlling the deposition rate and not the partial pressure of each element. In order to prevent CZTS decomposition and binaries losses, Sn and chalcogen deposition rate should remain high during the cooling phase, in order to counteract Sn loss [73].

4.4 Composition gradient

The composition gradient either created by a variation of the rate of deposition of each element in a codeposition or by stacked elements, has an influence on CZTS film properties, because of Cu diffusion, Sn loss and chemical mechanisms of formation.

Concerning Cu, its diffusion can create voids at the back contact. Two-step processes, especially when the precursor layer is a codeposited metal layer or stacked metal layers with Cu as first layer, are subject to these voids [89,103]. Secondary phases are observed, preventing the total formation of CZTS [105].

As explained in the previous paragraph (see Eq. (2)), Sn evaporates if the atmosphere is not well controlled. Nevertheless, the loss of Sn could also be prevented by having the Cu layer on top [63,93].

Concerning the reaction mechanism, for stacked metals, it is beneficial to have Cu and Sn in close contact for the formation of large CZTS grains [106]. If they are separated by Zn or Zn(S, Se), the reaction between copper chalcogenides and tin chalcogenides, which is necessary to form Cu_2SnX_3 [82], could be inhibited [107]. For stacked metals, Cu on top seems to foster formation of Cu_xS, which may result in larger grains and denser films [108]. For one-step processes, high Cu content at the beginning of the deposition promotes the formation of large grains [73].

Concerning the composition gradient,

- for two-steps processes, the Cu must not be deposited first, and Cu and Sn should be in close contact. The possible combinations for stacks are therefore Mo/Zn(X)/Cu/Sn or Mo/Zn(X)/Sn/Cu. The codeposition seems also possible, but with a Cu-poor composition near the back-contact;
- for one-step processes, the layer must be Cu rich at the beginning of the deposition in order to promote the growth of large grains.

4.5 Sodium

Sodium seems to have the same effects on CZTS as on CIGS: it promotes the growth of larger grains, enhances conductivity and has a significant effect on film morphology, as was demonstrated by SLG / borosilicate substrate comparison and dipping in Na_2S [109,110]. Nevertheless, contrary to CIGS, Na was not detected by XPS at the surface, meaning that the Na diffusion in CZTS is smaller than in CIGS [111].

4.6 Device properties

CZTS devices are similar to CIGS devices. The architecture is usually SLG/Mo/CZTS/CdS/i-ZnO/ZnO:Al, as presented in Figure 5. Calculation showed that CdS has a suitable band offset with CZTS [112] and until now, the most efficient solar cells were obtained with a CdS buffer layer. Oxidation of the CZTS absorber by dipping in deionized water [13] or O_2 annealing [73] before deposition of CdS seems to improve efficiencies. As for CIGS, reducing the time between absorber synthesis and CdS deposition to the minimum is of the utmost importance [106].

A few variations of the architecture are known, especially on the buffer layer. Pure ZnS does not seem to be a suitable buffer layer because of too high a band offset [112], but good results were obtained with Zn(S, O, OH) by Solar

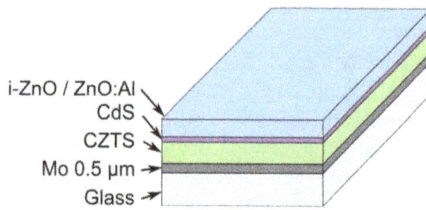

i-ZnO / ZnO:Al
CdS
CZTS
Mo 0.5 μm
Glass

Fig. 5. Classical architecture for CZTS solar cells.

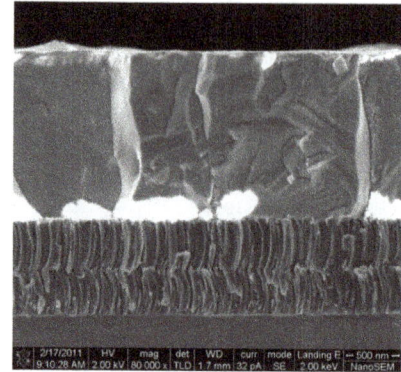

Fig. 6. CZTSe with ZnSe segregation at the back contact [59].

Frontier [113]. Nevertheless for such a buffer layer the optimal composition of the absorber layer seems to be less Zn-rich (Zn/Sn ∼ 1) [113]. Other architectures were proposed: SLG/Mo/CZTS/a-Si/ITO [114], a superstrate architecture (glass/FTO/CZTS) [115, 116], another superstrate (glass/FTO/TiO$_2$/In$_2$S$_3$/CZTS/Mo) [117], a flexible substrate (Al foil/Mo/CZTS/ZnS/i-ZnO/ITO/Al-Ni) [118].

The literature is still quite poor concerning device properties, because very few high-efficiency devices were synthesized yet. The IBM team started to investigate this field [14, 72, 119, 120] concluding that their devices were limited by the recombination at absorber/buffer interface, minority carrier lifetime and Schottky-type barrier in the back contact, which appears to be caused by secondary phases and/or MoS$_x$ interfacial layer between CZTS and Mo [121].

Concerning the absorber / buffer interface, it seems that the surface of the CZTS is Cu-poor, similarly to CIGS [111], or even Cu-free [122]. The KCN etching, which is also used in CIGS processes to remove binaries from the surface, seems to etch Cu preferentially and Sn in a lesser extent [93]. It seems also beneficial for removing Cu$_2$X that increase the resistivity of the layer [123]. This causes a widening of the bandgap without type inversion, which could be an easy way to tailor bandgap alignments [124,125]. It seems also that the electronic structure of a Sn-depleted surface is not favorable for the formation of well working p-n junction [74,93]. The Conduction Band Offset, about 0.4−0.5 eV, is somewhat above the optimal range of 0−0.4 eV and may contribute to lower J_{sc} and fill factor in the CZTSSe devices [126]. Better lattice matching between absorber and buffer could limit recombination at the interface, and using different buffer with smaller lattice such as In$_2$S$_3$ or Zn(S, O, OH) could improve the efficiency of the devices [127,128].

Concerning the back contact, ZnX was observed at the back contact [59, 106], as shown in Figure 6, and it seems that increasing the ZnX content strongly impairs efficiency [47]. It was nevertheless speculated that ZnS could be less detrimental than previously thought [72], acting as a simple "dark" material. Binaries losses that occur during annealing makes Cu$_2$X to diffuse towards the rear of the cell, could lead to a Cu$_2$X/MoX$_2$ mix, leading to high resistance at the back contact.

Concerning minority carrier lifetime, it might not be such a problem, because as in CIGS and CdTe, grain boundaries collect minority carrier and provide a current pathway for them to reach the n-type CdS and ZnO layers [30] and for coevaporated absorbers, high minority carrier diffusion length was measured (several hundreds

of nm [72]). Nevertheless, it seems that grain boundaries are Cu-rich, which could increase recombination rates and thus diminishing V_{OC} [29]. Cu-rich absorbers were found to have lower V_{OC} than Cu-poor ones [60], probably in relationship with Cu$_2$X formation. Longer minority carrier lifetimes were measured in low band gap materials, such as CZTSe, than in high band gap devices [29]. The difference between E_g in eV and V_{OC} in V, called V$_{OC}$ deficit, can therefore be decreased by diminishing the bang gap as well as tightly controlling the formation of Cu$_2$X or its removal by KCN etching.

For high efficiency, it is important to focus both on absorber quality and on interfaces between the absorber and the back contact and buffer. The Cu$_x$X binaries control seems to be a key point for both absorber and interface quality. Fortunately, the CIGS community has a strong experience on such problems, and hopefully it will be possible to optimize these interfaces much faster than for the CIGS.

5 Conclusion

The problematics encountered by research teams working on CZTS-based solar cells were presented and discussed. The field of CZTS solar cells is young, and until now it is difficult to point to certainties. Nevertheless, it seems that these assumptions start to be viewed as "common knowledge" in the community:

- CZTS has kësterite structure;
- the bandgap is tunable between 1.0 eV (Cu$_2$ZnSnSe$_4$) and 1.5 eV (Cu$_2$ZnSnS$_4$);
- the synthesis atmosphere must be controlled carefully;
- binary formation control is a key point for high efficiency devices;
- Zn-rich materials yield better results (when using a CdS buffer layer).

Nevertheless, there are numerous questions still, such as:

- what are the crystalline defects, how are they formed; what role do they play in the structure;
- what is the reaction path to CZTS, how to control binaries formation;
- what is the role of sodium during growth.

More and more scientists from the CIGS community work on CZTS solar cells. The problems encountered in CZTS solar cells are quite similar to the ones of CIGS solar cells, and similar methods can be used to solve them. There is no clear superiority of two-step processes over one-step processes, even if until now the best results were obtained with one-step processes. For both types of processes, a focus on binary control formation by atmosphere control and adaptation of interfaces to kësterites materials are the key points for increasing photovoltaic efficiencies for both types of processes.

The author is thankful for financial support of the French Agence Nationale de la Recherche (ANR) under grant NovACEZ (ANR-10-HABISOL-008).

References

1. International Energy Agency, *World Energy Outlook 2011* (2011)
2. Greenpeace and the European Photovoltaic Industry Association, *Solar Generation VI* (2011)
3. European Climate Foundation, *Roadmap 2050, a practical guide to a prosperous, low-carbon Europe* (2010)
4. International Energy Agency, *Technology Roadmap, Solar photovoltaic energy* (IEA, 2009)
5. A. Zaman, S. Lockman, *Solar industry growth: you ain't seen nothing yet.* The *grid parity decade* (Piper Jaffray Investment Research, 2011)
6. C. Breyer, M. Görig, J. Schmid, Fuel-parity: impact of photovoltaics on global fossil fuel fired power plant business, presented at the 26, *Symposium Photovoltaische Solarenergie, Bad Staffelstein* (2011)
7. C.S. Tao, J. Jiang, M. Tao, Solar Energy Mater. Solar Cells **95**, 3176 (2011)
8. A. Feltrin, A. Freundlich, Renew. Energy **33**, 180 (2008)
9. U.S. Department of Energy, *Critical Materials Strategy* (2010)
10. European Commission, *Critical Raw Materials for the EU* (2010)
11. A. Zuser, H. Rechberger, Resour. Conserv. Recycl. **56**, 56 (2011)
12. V. Fthenakis, Renew. Sust. Energy Rev. **13**, 2746 (2009)
13. H. Katagiri, K. Jimbo, W.S. Maw, K. Oishi, M. Yamazaki, H. Araki, A. Takeuchi, Thin Solid Films **517**, 2455 (2009)
14. D.A.R. Barkhouse, O. Gunawan, T. Gokmen, T.K. Todorov, D.B. Mitzi, Prog. Photovol. Res. Appl. **20**, 6 (2012)
15. K. Ito, T. Nakazawa, Jpn J Appl. Phys. **27**, 2094 (1988)
16. C.P. Chan, H. Lam, C. Surya, Solar Energy Mater. Solar Cells **94**, 207 (2010)
17. W. Xinkun, L. Wei, C. Shuying, L. Yunfeng, J. Hongjie, J. Semiconductors **33**, 022002 (2012)
18. S. Chen, X.G. Gong, A. Walsh, S.H. Wei, Appl. Phys. Lett. **94**, 041903 (2009)
19. K. Timmo, M. Altosaar, J. Raudoja, K. Muska, M. Pilvet, M. Kauk, T. Varema, M. Danilson, O. Volobujeva, E. Mellikov, Solar Energy Mater. Solar Cells **94** 1889 (2010)
20. J. He, L. Sun, S. Chen, Y. Chen, P. Yang, J. Chu, J. Alloys Compd. **511**, 129 (2012)
21. M. Moynihan, G. Zoppi, R. Miles, I. Forbes, Investigating synthesis of Cu2ZnSn(Se$_{1-x}$,S$_x$)4 for values of 0≤ x≤ 1 by S for Se substitution and direct sulphidisation of metallic precursors, *presented at the 26th European Photovoltaic Solar Energy Conference and Exhibition, Hamburg* (2011)
22. J. He, L. Sun, N. Ding, H. Kong, S. Zuo, S. Chen, Y. Chen, P. Yang, J. Chu, J. Alloys Compd. **529**, 34 (2012)
23. E. Mellikov, D. Meissner, M. Altosaar, M. Kauk, J. Krustok, A. Öpik, O. Volobujeva, J. Iljina, K. Timmo, I. Klavina, J. Raudoja, M. Grossberg, T. Varema, K. Muska, M. Ganchev, S. Bereznev, M. Danilson, AMR **222**, 8 (2011)
24. S. Levcenco, D. Dumcenco, Y.P. Wang, Y.S. Huang, C.H. Ho, E. Arushanov, V. Tezlevan, K.K. Tiong, Opt. Mater. **34**, 1362 (2012)
25. G.M. Ford, Q. Guo, R. Agrawal, H.W. Hillhouse, Chem. Mater. **23**, 2626 (2011)
26. W. Shockley, H.J. Queisser, J. Appl. Phys. **32**, 510 (1961)
27. L. Choubrac, A. Lafond, C. Guillot-Deudon, Y. Moëlo, S. Jobic, Inorg. Chem. **51**, 3346 (2012)
28. USGS, *Commodity Statistics and Information, USGS Minerals Information* (2010), http://minerals.usgs.gov/minerals/pubs/commodity/
29. S. Bag, O. Gunawan, T. Gokmen, Y. Zhu, T.K. Todorov, D.B. Mitzi, Energy Environ. Sci. **5**, 7060 (2012)
30. J.B. Li, V. Chawla, B.M. Clemens, Adv. Mat. **24**, 720 (2012)
31. S. Siebentritt, S. Schorr, Prog. Photovolt. Res. Appl. in press
32. S. Chen, X.G. Gong, A. Walsh, S.H. Wei, Phys. Rev. B **79**, 165211 (2009)
33. S.R. Hall, J.T. Szymanski, J.M. Stewart, Can. Mineral. **16**, 131 (1978)
34. L.O. Brockway, Z. Kristaloogr. **89**, 434 (1934)
35. P. Bonazzi, L. Bindi, G.P. Bernardini, S. Menchetti, Can. Mineral. **41**, 639 (2003)
36. S. Schorr, H.J. Hoebler, M. Tovar, Eur. J. Min. **19**, 65 (2007)
37. S. Schorr, Sol. Energy Mater. Solar Cells **95**, 1482 (2011)
38. J. Paier, R. Asahi, A. Nagoya, G. Kresse, Phys. Rev. B **79**, 115126 (2009)
39. A. Walsh, S. Chen, X.G. Gong, S.H. Wei (CZTS): Theoretical insights, http://web.mac.com/aronwalsh/publications/10/icps_czts_10.pdf
40. A. Walsh, S.H. Wei, S. Chen, X.G. Gong, Design of quaternary chalcogenide photovoltaic absorbers through cation mutation, in *34th IEEE Photovoltaic Specialists Conference* (2009)
41. A. Walsh, S. Chen, S. Wei, X. Gong, Kesterite Thin-Film Solar Cells: Advances in Materials Modelling of Cu2ZnSnS4, Adv. Energy Mater. **2**, 400 (2012)
42. Y. Zhang, X. Sun, P. Zhang, X. Yuan, F. Huang, W. Zhang, J. Appl. Phys. **111**, 063709 (2012)
43. H. Nozaki, T. Fukano, S. Ohta, Y. Seno, H. Katagiri, K. Jimbo, J. Alloys Compd. **524**, 22 (2012)
44. X. He, H. Shen, Physica B: Condensed Matter **406**, 4604 (2011)
45. X. He, H. Shen, Phys. Scr. **85**, 035302 (2012)

46. P.A. Fernandes, P.M.P. Salomé, A.F. da Cunha, Thin Solid Films, **517**, 2519 (2009)

47. J. Just, D. Lützenkirchen-Hecht, R. Frahm, S. Schorr, T. Unold, Appl. Phys. Lett. **99**, 262105 (2011)

48. A. Nagoya, R. Asahi, R. Wahl, G. Kresse, Phys. Rev. B **81**, 113202 (2010)

49. S. Chen, X.G. Gong, A. Walsh, S.H. Wei, Appl. Phys. Lett. **96**, 021902 (2010)

50. P.J. Dale, K. Hoenes, J. Scragg, S. Siebentritt, A review of the challenges facing kësterite based thin film solar cells, in *Photovoltaic Specialists Conference (PVSC), 2009 34th IEEE* (2010), pp. 002080–002085

51. T. Maeda, S. Nakamura, T. Wada, Thin Solid Films **519**, 7513 (2011)

52. M.J. Romero, H. Du, G. Teeter, Y. Yan, M.M. Al-Jassim, Phys. Rev. B **84**, 165324 (2011)

53. E. Kask, T. Raadik, M. Grossberg, R. Josepson, J. Krustok, Energy Procedia **10**, 261 (2011)

54. S. Chen, J.H. Yang, X.G. Gong, A. Walsh, S.H. Wei, Phys. Rev. B **81**, 245204 (2010)

55. T. Maeda, S. Nakamura, T. Wada, Jpn J. Appl. Phys. **50**, 04DP07 (2011)

56. A. Jeong, W. Jo, S. Jung, J. Gwak, J. Yun, Appl. Phys. Lett. **99**, 082103 (2011)

57. A. Nagaoka, K. Yoshino, H. Taniguchi, T. Taniyama, H. Miyake, J. Cryst. Growth **341**, 38 (2012)

58. T. Washio, H. Nozaki, T. Fukano, T. Motohiro, K. Jimbo, H. Katagiri, J. Appl. Phys. **110**, 074511 (2011)

59. I. Repins et al., Kesterites and Chalcopyrites: A Comparison of Close Cousins, in *MRS Proceedings* (2011), Vol. 1324

60. K. Muska, M. Kauk, M. Altosaar, M. Pilvet, M. Grossberg, O. Volobujeva, Energy Procedia **10**, 203 (2011)

61. H. Katagiri, K. Jimbo, M. Tahara, H. Araki, K. Oishi, The influence of the composition ratio on CZTS-based thin film solar cells, in *Materials Research Society Symposium Proceedings* (2009), p. 1165

62. A. Ennaoui, M. Lux-Steiner, A. Weber, D. Abou-Ras, I. Kötschau, H.-W. Schock, R. Schurr, A. Hölzing, S. Jost, R. Hock, T. Voß, J. Schulze, A. Kirbs, Thin Solid Films **517**, 2511 (2009)

63. P.A. Fernandes, P.M.P. Salomé, A.F. da Cunha, Semicond. Sci. Technol. **24**, 105013 (2009)

64. T.K. Todorov, K.B. Reuter, D.B. Mitzi, Adv. Mat. **22**, (2010)

65. K. Jimbo, R. Kimura, T. Kamimura, S. Yamada, W.S. Maw, H. Araki, K. Oishi, H. Katagiri, Thin Solid Films **515**, 5997 (2007)

66. M. Valentini, C. Malerba, F. Biccari, R. Chierchia, P. Mangiapane, E. Salza, A. Mittiga, Growth and characterization of Cu2ZnSnS4 thin films prepared by sulfurization of evaporated precursors, in *26th EUPVSEC* (Hamburg, 2011)

67. S.J. Ahn, S. Jung, J. Gwak, A. Cho, K. Shin, K. Yoon, D. Park, H. Cheong, J.H. Yun, Appl. Phys. Lett. 021905 (2010)

68. D. Park, D. Nam, S. Jung, S.J. An, J. Gwak, K. Yoon, J.H. Yun, H. Cheong, Thin Solid Films **519**, 7386 (2011)

69. H. Wang, Int. J. Photoenergy **2011** (2011)

70. M. Edoff, S. Schleussner, E. Wallin, O. Lundberg, Thin Solid Films **519**, 7530 (2011)

71. S. Ahmed, K.B. Reuter, O. Gunawan, L. Guo, L.T. Romankiw, H. Deligianni, Adv. Energy Mater. in press

72. B. Shin, O. Gunawan, Y. Zhu, N.A. Bojarczuk, S.J. Chey, S. Guha, Prog. Photovolt. Res. Appl. in press

73. I. Repins, C. Beall, N. Vora, C. DeHart, D. Kuciauskas, P. Dippo, B. To, J. Mann, W.-C. Hsu, A. Goodrich, R. Noufi, Solar Energy Mater. Solar Cells, in press

74. M. Kauk, K. Muska, M. Altosaar, J. Raudoja, M. Pilvet, T. Varema, K. Timmo, O. Volobujeva, Energy Procedia **10**, 197 (2011)

75. B.-A. Schubert, B. Marsen, S. Cinque, T. Unold, R. Klenk, S. Schorr, H.-W. Schock, Prog. Photovolt Res. Appl. **19**, 93 (2011)

76. E. Mellikov, D. Meissner, T. Varema, M. Altosaar, M. Kauk, O. Volobujeva, J. Raudoja, K. Timmo, M. Danilson, Energy Mater. Solar Cells **93**, 65 (2009)

77. X.Y. Li, D.C. Wang, Q.Y. Du, W.F. Liu, G.S. Jiang, C.F. Zhu, Adv. Mater. Res. **418−420**, 67 (2011)

78. T.K. Chaudhuri, D. Tiwari, Solar Energy Mater. Solar Cells **101**, 46 (2012)

79. T. Rath, W. Haas, A. Pein, R. Saf, E. Maier, B. Kunert, F. Hofer, R. Resel, G. Trimmel, Solar Energy Mater. Solar Cells **101**, 87 (2012)

80. N.M. Shinde, C.D. Lokhande, J.H. Kim, J.H. Moon, J. Photochem. Photobiol. A: Chem., in press

81. R.A. Wibowo, W.S. Kim, E.S. Lee, B. Munir, K.H. Kim, J. Phys. Chem. Solids **68**, 1908 (2007)

82. F. Hergert, R. Hock, Thin solid films **515**, 5953 (2007)

83. A.J. Cheng, M. Manno, A. Khare, C. Leighton, S. Campbell, E. Aydil, J. Vac. Sci. Technol. A Vac. Surf. Films **29**, 051203 (2011)

84. O. Volobujeva, J. Raudoja, E. Mellikov, M. Grossberg, S. Bereznev, R. Traksmaa, J. Phys. Chem. Solids **70**, 567 (2009)

85. R. Schurr et al., Thin Solid Films **517**, 2465 (2009)

86. M. Ganchev, J. Iljina, L. Kaupmees, T. Raadik, O. Volobujeva, A. Mere, M. Altosaar, J. Raudoja, E. Mellikov, Thin Solid Films **519**, 7394 (2011)

87. K. Maeda, K. Tanaka, Y. Nakano, H. Uchiki, Jpn J. Appl. Phys. **50**, 05FB08 (2011)

88. A. Redinger, S. Siebentritt, Appl. Phys. Lett. **97**, 092111 (2010)

89. H. Yoo, J. Kim, Thin Solid Films **518**, 6567 (2010)

90. J. Ge, Y. Wu, C. Zhang, S. Zuo, J. Jiang, J. Ma, P. Yang, J. Chu, Appl. Surf. Sci. **258**, 7250 (2012)

91. K. Maeda, K. Tanaka, Y. Fukui, H. Uchiki, Solar Energy Mater. Solar Cells **95**, 2855 (2011)

92. K. Maeda, K. Tanaka, Y. Nakano, Y. Fukui, H. Uchiki, Jpn J. Appl. Phys. **50**, 05FB09 (2011)

93. J.J. Scragg, *Studies of Cu2ZnSnS4 Films Prepared by Sulfurisation of Electrodeposited Precursors*, Ph.D. thesis, University of Bath, Department of Chemistry, Bath, 2010

94. G. Teeter, H. Du, J.E. Leisch, M. Young, F. Yan, S. Johnston, P. Dippo, D. Kuciauskas, M. Romero, P. Newhouse, S. Asher, D. Ginley, Combinatorial study of thin-film Cu2ZnSnS4 synthesis via metal precursor sulfurization, in *35th IEEE-PVSEC* (2010)

95. A. Weber, R. Mainz, H.W. Schock, J. Appl. Phys. **107**, 013516 (2010)

96. H. Yoo, J. Kim, L. Zhang, Curr. Appl. Phys. in press

97. A. Redinger, D.M. Berg, P.J. Dale, R. Djemour, L. Gutay, T. Eisenbarth, N. Valle, S. Siebentritt, IEEE J. Photovoltaics **1**, 200 (2011)

98. A. Redinger, D.M. Berg, P.J. Dale, S. Siebentritt, J. Am. Chem. Soc. 156 (2011)

99. C.M. Fella, G. Ilari, A.R. Uhl, A. Chirilă, Y.E. Romanyuk, A.N. Tiwari, Non-vacuum deposition of Cu2ZnSnSe4 absorber layers for thin film solar cells, in *26 EUPVSEC* (Hamburg, 2011)

100. F. Biccari, R. Chierchia, M. Valentini, P. Mangiapane, E. Salza, C. Malerba, C.L.A. Ricardo, L. Mannarino, P. Scardi, A. Mittiga, Energy Procedia **10**, 187 (2011)

101. S. Guha, D.B. Mitzi, T.K. Todorov, K. Wang, *Annealing Thin Films*, US Patent United States Patent Application 2012007093622, 2012

102. P. Salomé, J. Malaquias, P. Fernandes, M. Ferreira, J. Leitão, A. da Cunha, J. González, F. Matinaga, G. Ribeiro, E. Viana, Solar Energy Mater. Solar Cells **95**, 3482 (2011)

103. K. Wang, B. Shin, K.B. Reuter, T. Todorov, D.B. Mitzi, S. Guha, Appl. Phys. Lett. **98**, 051912 (2011)

104. R. Juškėnas, S. Kanapeckaitė, V. Karpavičienė, Z. Mockus, V. Pakštas, A. Selskienė, R. Giraitis, G. Niaura, Solar Energy Mater. Solar Cells **101**, 277 (2012)

105. S.W. Shin, S. Pawar, C.Y. Park, J.H. Yun, J.H. Moon, J.H. Kim, J.Y. Lee, Solar Energy Mater. Solar Cells **95**, 3202 (2011)

106. L. Grenet, S. Bernardi, D. Kohen, C. Lepoittevin, S. Noël, N. Karst, A. Brioude, S. Perraud, H. Mariette, Solar Energy Mater. Solar Cells **101**, 11 (2012)

107. H. Yoo, J. Kim, Solar Energy Mater. Solar Cells **95**, 239 (2011)

108. R. Chalapathy, G.S. Jung, B.T. Ahn, Solar Energy Mater. Solar Cells **95**, 3216 (2011)

109. W. Hlaing Oo, J. Johnson, A. Bhatia, E. Lund, M. Nowell, M. Scarpulla, Journal of Electronic Materials **40**, 2214

110. T. Prabhakar, N. Jampana, Solar Energy Mater. Solar Cells **95**, 1001 (2011)

111. M. Bär, B.A. Schubert, B. Marsen, S. Krause, S. Pookpanratana, T. Unold, L. Weinhardt, C. Heske, H.W. Schock, Appl. Phys. Lett. **99**, 112103 (2011)

112. A. Nagoya, R. Asahi, G. Kresse, J. Phys. Condens. Matter **23**, 404203 (2011)

113. N. Sakai, H. Hiroi, H. Sugimoto, Development of cd-free buffer layer for Cu2ZnSnS4 thin-film solar cells, in *37th IEEE PVSC Conference* (2011)

114. F. Jiang, H. Shen, W. Wang, L. Zhang, Appl. Phys. Expr. **4**, 074101 (2011)

115. P.K. Sarswat, M.L. Free, Phys. Status Solidi **208**, 2861 (2011)

116. P.K. Sarswat, M. Snure, M.L. Free, A. Tiwari, Thin Solid Films **520**, 1694 (2012)

117. Q.-M. Chen, Z.-Q. Li, Y. Ni, S.-Y. Cheng, X.-M. Dou, Chin. Phys. B **21**, 038401 (2012)

118. Q. Tian, X. Xu, L. Han, M. Tang, R. Zou, Z. Chen, M. Yu, J. Yang, J. Hu, Cryst. Eng. Commun., **14**, 3847 (2012)

119. O. Gunawan, T.K. Todorov, D.B. Mitzi, Appl. Phys. Lett. **97**, 233506 (2010)

120. K. Wang, O. Gunawan, T. Todorov, B. Shin, S.J. Chiy, N.A. Bojarczuk, D. Mitzi, S. Guha, Appl. Phys. Lett. **97**, 143508 (2010)

121. B. Shin, K. Wang, O. Gunawan, K.B. Reuter, S.J. Chey, N.A. Bojarczuk, T. Todorov, B. Mitzi, S. Guha, in *37th IEEE PVSC Conference* (Seattle, 2011), Vol. 1

122. M. Bär, B.-A. Schubert, B. Marsen, R.G. Wilks, M. Blum, S. Krause, S. Pookpanratana, Y. Zhang, T. Unold, W. Yang, L. Weinhardt, C. Heske, H.-W. Schock, J. Mater. Res. FirstView, 1

123. T. Tanaka, T. Sueishi, K. Saito, Q. Guo, M. Nishio, K.M. Yu, W. Walukiewicz, J. Appl. Phys. **111**, 053522 (2012)

124. M. Bär, B.-A. Schubert, B. Marsen, S. Krause, S. Pookpanratana, T. Unold, L. Weinhardt, C. Heske, H.-W. Schock, Appl. Phys. Lett. **99**, 152111 (2011)

125. M. Bär, B.-A. Schubert, B. Marsen, R.G. Wilks, S. Pookpanratana, M. Blum, S. Krause, T. Unold, W. Yang, L. Weinhardt, C. Heske, H.-W. Schock, Appl. Phys. Lett. **99**, 222105 (2011)

126. R. Haight, A. Barkhouse, O. Gunawan, B. Shin, M. Copel, M. Hopstaken, D.B. Mitzi, Appl. Phys. Lett. **98**, 253502 (2011)

127. J.M. Raulot, C. Domain, J.F. Guillemoles, J. Phys. Chem. Solids **66**, 2019 (2005)

128. A. Opanasyuk, D. Kurbatov, M. Ivashchenko, I.Y. Protsenko, H. Cheong, Journal of Nano- and Electronic Physics **4**, 01024 (2012)

129. C. Tao, J. Jiang, M. Tao, ECS Transactions **33**, 3 (2011)

130. D.B. Mitzi, T.K. Todorov, *Method of Forming Semiconductor Film and Photovoltaic Device Including the Film*, US Patent US2011094557 (A1)Apr-2011

131. D.B. Mitzi, T.K. Todorov, Aqueous-based method of forming semiconductor film and photovoltaic device including the film, US Patent US 2011097496 (A1), 2011

132. E. Mellikov, M. Altosaar, J. Raudoja, K. Timmo, O. Volobujeva, M. Kauk, J. Krustok, T. Varema, M. Grossberg, M. Danilson, K. Muska, K. Ernits, F. Lehner, D. Meissner, Materials Challenges in Alternative and Renewable Energy: Ceramic Transactions **224**, 137

133. J.J. Scragg, P.J. Dale, L.M. Peter, Electrochem. Commun. **10**, 639 (2008)

134. J.J. Scragg, D.M. Berg, P.J. Dale, J. Electroanal. Chem. **646**, 52 (2010)

135. H. Araki, Y. Kubo, A. Mikaduki, K. Jimbo, W.S. Maw, H. Katagiri, M. Yamazaki, K. Oishi, A. Takeuchi, Solar Energy Mater. Solar Cells **93**, 996 (2009)

136. J.J. Scragg, P.J. Dale, L.M. Peter, Thin Solid Films **517**, 2481 (2009)

137. J.J. Scragg, P.J. Dale, L.M. Peter, G. Zoppi, I. Forbes, Phys. Stat. Sol. B **245**, 1772 (2008)

138. M. Kurihara, D. Berg, J. Fischer, S. Siebentritt, P.J. Dale, Physica Status Solidi **6**, 1241 (2009)

139. X. Zhang, X. Shi, W. Ye, C. Ma, C. Wang, Appl. Phys. A **94**, 381 (2009)

140. H. Deligianni, L. Guo, R. Vaidyanathan, *Electrodeposition of Thin-Film Cells Containing Non-Toxic Elements*, US Patent United States Patent Application 2012004837803, 2012

141. S. Ahmed, H. Deligianni, L.T. Romankiw, K. Wang, *Structure and Method of Fabricating a CZTS Photovoltaic Device by Electrodeposition*, US Patent United States Patent Application 2012006179015, 2012

142. Q. Guo, G.M. Ford, W.C. Yang, B.C. Walker, E.A. Stach, H.W. Hillhouse, R. Agrawal, J. Am. Chem. Soc. **132**, 2844 (2010)

143. T. Washio, T. Shinji, S. Tajima, T. Fukano, T. Motohiro, K. Jimbo, H. Katagiri, J. Mater. Chem., **22**, 4021 (2012)

144. K. Woo, Y. Kim, J. Moon, Energy Environ. Sci. **5**, 5340 (2012)

145. W. Ki, H.W. Hillhouse, Adv. Energy Mater. **1**, 732 (2011)
146. N. Momose, M.T. Htay, T. Yudasaka, S. Igarashi, T. Seki, S. Iwano, Y. Hashimoto, K. Ito, Jpn. J. Appl. Phys. **50**, 01BG09 (2011)
147. H. Flammersberger, *Experimental study of Cu2ZnSnS4 thin films for solar cells*, Master thesis, Uppsala University, 2010
148. C. Platzer-Björkman, J. Scragg, H. Flammersberger, T. Kubart, M. Edoff, Solar Energy Mater. Solar Cells **98**, 110 (2012)
149. K.H. Kim, I. Amal, Electronic Materials Letters **7**, 225 (2011)
150. H. Araki, Y. Kubo, K. Jimbo, W.S. Maw, H. Katagiri, M. Yamazaki, K. Oishi, A. Takeuchi, Phys. Status Solidi **6**, 1266 (2009)
151. H. Kühnlein, J. Schulze, T. Voss, *Metal Plating Composition And Method For The Deposition Of Copper-zinc-tin Suitable For Manufacturing Thin Film Solar Cell*, US Patent US2009/0205714 (A1)20, 2009
152. H.H. Kühnlein, *Elektrochemische Legierungsabscheidung zur Herstellung von Cu2ZnSnS4 Dünnschichtsolarzellen*, Ph.D. thesis, Dresden, 2007
153. M.L. Free, P.K. Sarswat, A. Tiwari, M. Snure, *Modified copper-zinc-tin semiconductor films, uses thereof and related methods*, US Patent United States Patent Application 20110132462A106, 2011
154. B.S. Pawar, S.M. Pawar, S.W. Shin, D.S. Choi, C.J. Park, S.S. Kolekar, J.H. Kim, Appl. Surf. Sci. **257**, 1786 (2010)
155. S.M. Pawar, B.S. Pawar, A.V. Moholkar, D.S. Choi, J.H. Yun, J.H. Moon, S.S. Kolekar, J.H. Kim, Electrochimica Acta **55**, 4057 (2010)
156. M. Jeon, Y. Tanaka, T. Shimizu, S. Shingubara, Energy Procedia **10**, 255 (2011)
157. M. Jeon, T. Shimizu, S. Shingubara, Mater. Lett. **65**, 2364 (2011)
158. G. Zoppi, I. Forbes, R.W. Miles, P.J. Dale, J.J. Scragg, L.M. Peter, Progress in Photovoltaics: Research and Applications **17**, 315 (2009)
159. K. Moriya, K. Tanaka, H. Uchiki, Jpn J. Appl. Phys. **46**, 5780 (2007)
160. A.V. Moholkar, S.S. Shinde, A.R. Babar, K.-U. Sim, Y. Kwon, K.Y. Rajpure, P.S. Patil, C.H. Bhosale, J.H. Kim, Solar Energy **85**, 1354 (2011)
161. L. Sun, J. He, H. Kong, F. Yue, P. Yang, J. Chu, Solar Energy Mater. Solar Cells **95**, 2907 (2011)
162. A. Moholkar, S. Shinde, A. Babar, K. Sim, K. Hyun, K. Rajpure, P. Patil, C. Bhosale, J. Kim, J. Alloys Compd. **509**, 7439 (2011)
163. K. Tanaka, Y. Fukui, N. Moritake, H. Uchiki, Solar Energy Mater. Solar Cells **95**, 838 (2011)
164. C. Lokhande, N. Shinde, J. Kim, J. Moon, Invertis Journal of Renew. Energy **1**, 142 (2011)
165. S.S. Mali, B.M. Patil, C.A. Betty, P.N. Bhosale, Y.W. Oh, S.R. Jadkar, R.S. Devan, Y.-R. Ma, P.S. Patil, Electrochimica Acta **66**, 216 (2012)
166. H. Araki, A. Mikaduki, Y. Kubo, T. Sato, K. Jimbo, W.S. Maw, H. Katagiri, M. Yamazaki, K. Oishi, A. Takeuchi, Thin Solid Films **517**, 1457 (2008)
167. F. Liu, Y. Li, K. Zhang, B. Wang, C. Yan, Y. Lai, Z. Zhang, J. Li, Y. Liu, Solar Energy Mater. Solar Cells **94**, 2431 (2010)
168. V. Chawla, B. Clemens, Inexpensive, abundant, non-toxic thin films for solar cell applications grown by re-active sputtering, in *Photovoltaic Specialists Conference (PVSC), 2010 35th IEEE* (2010), pp. 001902–001905
169. A. Wangperawong, J.S. King, S.M. Herron, B.P. Tran, K. Pangan-Okimoto, S.F. Bent, A chemical bath process for depositing Cu2ZnSnS4 photovoltaic absorbers, in *35th IEEE-PVSEC* (Hawai, 2010)
170. C. Shi, G. Shi, Z. Chen, P. Yang, M. Yao, Mater. Lett. **73**, 89 (2012)

Upconversion of 1.54 μm radiation in Er^{3+} doped fluoride-based materials for c-Si solar cell with improved efficiency

F. Pellé[1,a], S. Ivanova[1,2], and J.-F. Guillemoles[3]

[1] Laboratoire de Chimie de la Matière Condensée de Paris LCMCP UMR7574 CNRS/UPMC/Chimie ParisTech, Chimie ParisTech, 11 rue Pierre et marie Curie, 75235 Paris, France
[2] Center for Information and Optics Technology, University of Information Technology, Mechanics and Optics, 199034, St. Petersburg, Russia
[3] IRDEP, UMR 7174CNRS/EDF/Chimie ParisTech, 6 quai Watier, 78401 Chatou, France

Abstract Upconverted emission from erbium ions in fluoride materials (glass and disordered crystal of the system CaF$_2$-YF$_3$) are observed in a wide spectral range (from the visible to the near infrared) under infrared excitation at 1.54 μm. In both cases, the upconverted emission in the near infrared ($\sim$1 μm) dominates the spectrum. Absolute UC efficiency defined as the ratio between the UC luminescence power and the absorbed pump power has been experimentally measured. The NIR ($\sim$1 μm) luminescence energy yield for the glass and the disordered crystal varies from 2.4 to 11.5% for the glass and from 7.7 to 16% for the crystal for an infrared excitation power density ranging from 2 W/cm^2 to 100 W/cm^2. This is of a particular interest for their use as upconverter to improve the c-Si cells quantum efficiency since the energy of the excitation lies below the c-Si absorption edge (1.12 eV at 300 K) and is well located compared to the AM1.5G solar spectrum, outside of the absorption lines due to different atmospheric gases. Furthermore, the most efficient upconverted emission recorded in the investigated materials occurs at an energy just above the gap. A current generated in a bifacial c- Si solar cell is observed when the Er^{3+} doped material (1.55 mA and 2.15 mA for the glass and the crystal respectively), placed at the rear face of the cell, is excited at 1.54 μm. The current dependence as a function of the sub-bandgap excitation power has been measured and modelled. Finally the EQE of the complete device is deduced from the measured short-circuit current and the incident photon flux on the cell. An increase of the cell quantum efficiency of 2.4% and 1.7% is obtained at 1.54 μm with adding the glass and the crystal respectively at the rear face of the c-Si cell. The results are compared to those already obtained with Er: NaYF$_4$ known as the most efficient upconverter.

1 Introduction

Researches on sustainable energy sources underwent a strong impulse these last years. Among different opportunities, solar energy offers the most abundant sustainable source but, up to now, its use still remains low compared to other sustainable energy sources such as hydraulic, biomass, wind. The main hindrance to a wide development of solar energy has several causes such as the cost of the kWh produced (20−40 cts/kWh), that is in part limited by the efficiency of commercially available solar cells (5−20%)[1]. Important technological breakthroughs only will allow overcoming these difficulties and investigations in this field are directed toward researches on third generation solar cells. Hot carriers, solar concentrators, nanostructured silicon, tandem cells represent the most popular orientations to achieve third generation solar cells.

In these cases, predicted efficiency is drastically improved and may lead to devices with a much higher cost. In this frame, a possibility based on down- (photon cutting) and up- conversion (photon addition) has been proposed. The efficiency increase is provided by adding materials able to convert photons of high energy or sub-bandgap photons into useful ones for the semiconductor [2,3]. Theoretically, the Shockley-Queisser efficiency limit [4] can be pushed from close to 30% up to 40.2% for a silicon solar cell with an upconverter illuminated by non concentrated light [5].

Upconversion processes represent a convenient way to convert sub-bandgap photons into visible ones, they are easily observed with rare earth (RE) doped materials as well as with organic materials [6]. Moreover, the particular energy level scheme of the rare earth ions offers many opportunities for different semiconductors (c-Si, a-Si, AsGa ...); several RE combinations and their upconversion ability can be adapted to improve the enhanced photon current generation by the semiconductor

[a] e-mail: `fabienne-pelle@chimie-paristech.fr`

depending on its energy gap. This approach exhibits several advantages. First of all, the introduction of an upconverter in the device does not require drastic modification of the fabrication process since the optical material and its reflector will be placed at the rear face of the cell. Moreover, each part of the complete device (semiconductor and upconverter) can be separately optimized, thus a low cost modification should be possible to improve the solar irradiance – current efficiency of solar cells with upconversion. Crystalline silicon is the most common material for solar cells today and in this case, about ~20% of solar energy contained in the air mass global AM1.5G terrestrial solar spectrum is lost due to the transmission, conversion of sub-band gap infrared into near infrared and/or visible emissions should provide improvement. In this aim, the optical conversion of infrared excitation (IR) at about 1.5 μm to near infrared (NIR) photons near 1 μm in Er^{3+} doped materials is of particular interest since the energy of excitation photons lies below the c-Si absorption edge (1.12 eV at 300 K) and the converted photons can be efficiently absorbed by the semiconductor leading to an efficient light – current conversion. Furthermore, the Er^{3+} absorption in this spectral range, corresponding to the $^4I_{15/2} \rightarrow ^4I_{13/2}$ transition, exhibits a quite important cross-section; it is usually broad due to the large number of populated Stark components of the ground state. Another interest to work with Er^{3+} ions is linked to the location of the fundamental absorption compared to the terrestrial AM 1.5G solar spectrum as shown on Figure 1. Up to now, the validity of the concept has been demonstrated using $NaYF_4$ as upconverter by studying the effect of upconverted emissions obtained by a sub-bandgap excitation of a composite material composed of Er^{3+}− doped $NaYF_4$ nanoparticles dispersed into a mixture of white oil and rubberizer on a photocurrent generation [7–9] or Er^{3+}: $NaYF_4$ compacted powder [10]. In [7–9], the unoptimized device has been demonstrated to respond effectively to wavelengths in the range of 1480−1580 nm with an external quantum efficiency (EQE) of 3.4% occurring at 1523 nm at an illumination intensity of 2.4 W/cm^2. A selection of some materials has been already reported although their absolute conversion efficiency is unknown [11].

In this paper, we are especially interested in upconversion, from Er^{3+} ions, observed under IR excitation at 1.54 μm and its conversion efficiency at about 1 μm. Disordered fluoride based materials, a ZBLAN glass and a disordered crystal of the YF_3-CaF_2 system, were selected as hosts for three reasons. The first one is obvious, it is now well known that upconversion processes are more efficient in materials with a low phonon cut-off frequency in which non-radiative de-excitation by multiphonon relaxation rates are minimized. Among them, halides satisfy this criterion, however only fluorides represent interest, others are instable with regard to moisture. The second one is related to the ability of Er^{3+} to absorb the widest possible spectral range, so we have chosen to investigate glasses and disordered crystals in which the spectral features exhibit inhomogeneous broadening of the spectral features induced by the local disorder.

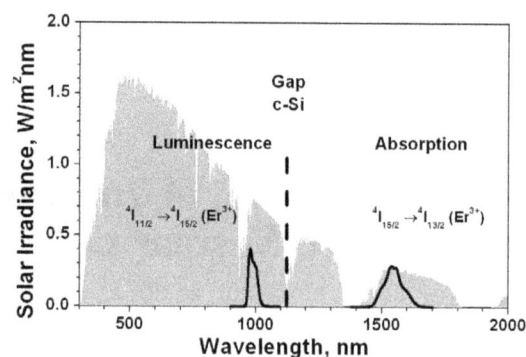

Fig. 1. Er^{3+} Absorption from the ground state to the first excited state ($^4I_{15/2} \rightarrow ^4I_{13/2}$) and upconverted emission in the near infrared (($^4I_{11/2} \rightarrow ^4I_{15/2}$) in comparison with the AM1.5G solar spectrum.

Since their discovery [12, 13], upconversion processes observed in RE doped materials are extensively studied. However, if a huge amount of papers on upconversion have been published in literature, very few of them report absolute conversion efficiency values and all of them concern the upconversion process between Yb^{3+}-RE^{3+} (RE = Er^{3+}, Tm^{3+} mainly) upon 980 nm excitation [14–17]. For the targeted application, the knowledge of the upconverted luminescence yield is the most important factor to evaluate the potentiality of a material as upconverter. Recently, we have reported the procedure to measure absolute upconversion conversion efficiencies [18]. In this paper, we are especially interested in upconversion from Er^{3+} ions under IR excitation at 1.54 μm and the conversion efficiency at about 1 μm. Only the results obtained with the the optimized Er^{3+} doped fluoride based glass and disordered crystal giving the highest upconversion efficiency will be presented. The absolute conversion efficiency measured for the IR → NIR (@ ~ 1 μm) upconverted luminescence depends as usually on the excitation power, from 2.4 to 11.5% for the glass and from 7.7 to 16% for the crystal with an infrared excitation power density ranging from 2 W/cm^2 to 100 W/cm^2. The measurements have been done between the available laser diode power ranging from 20 mW to 100 mW so in our experimental conditions, the lower excitation power density was limited to 2 W/cm^2. From extrapolation of the experimental data, an absolute conversion efficiency of 1% and 4% is expected at 1 W/cm^2 for the glass and the crystal respectively.

In a second part, we will concentrate on the current generated in a c-Si cell following the absorption of the upconverted emission obtained with an infrared excitation @ 1.54 μm of the investigated samples placed at rear face of the cell. The current dependence as a function of the sub-bandgap excitation power is observed and modelled. Finally the EQE of the complete device is deduced from the measured short-circuit current and the incident photon flux on the cell. An increase of the cell efficiency of 2.4% and 1.7% is obtained at 1.54 μm with adding the glass and the crystal respectively at the rear face of the c-Si cell.

2 Experiments

The composition of the investigated glass is the following 54.0 ZrF_4 − 26.6 BaF_2 − 2.8 LaF_3 − 4.8 AlF_3 − 6.6 NaF −0.5 InF_3 − 4.7 ErF_3 (mol%). This glass, noted Z18 in the following, was synthesized according to the particular procedure described in the reference [19]. The $Ca_{0.89}Y_{0.11}F_{2.11}$ disordered crystal (noted as CYF), synthesized using the Storbarger-Bridgman technique, crystallizes with the fluorite structure. The synthesis procedure and optical properties of Er^{3+}: CYF are described in [20, 21]. The crystal doped with 5% of Er^{3+} ions was investigated in this study.

Infrared excitation at 1.54 μm was provided by a CW pig-tailed Er fiber laser (ELT-100 IRE Polus Group) delivering power up to 100 mW via a single mode fiber with 8 μm core diameter. A splitter and a photodiode integrated into the device permitted to control the excitation power.

The experimental device and procedure to measure the absolute IR → NIR conversion efficiency have been already reported [18], the details are only briefly recalled here. Plates with polished parallel faces of samples under study were brought as close as possible to the pumping laser fiber termination. The ensemble was introduced inside an integrating sphere covered with Spectralon™, the luminescence was sent to a spectrometer (AVASpec-1024TEC) by an optical fiber (Fig. 2a). The whole setup (integrating sphere + spectrometer) is calibrated with the use of a halogen tungsten lamp (10 W tungsten halogen fan-cooled Avalight-HAL). The IR-visible luminescence spectra (1100−380 nm) are recorded in absolute energy units. The infrared luminescence of the samples in the range of 900−1700 nm was dispersed by a monochromator (SpectraPro 750, resolution 1.2 mm/nm), detected by a liquid nitrogen cooled PbS detector and corrected from the spectral response of the system. To scale the IR part (900−1700 nm) of the emission spectrum recorded in the same experimental conditions (excitation power, geometry of the experiment...), the luminescence intensity has been multiplied by a constant factor in order to obtain the same luminescence power of the band at ~0.98 μm as that in spectra recorded using the calibrated setup. Both setups allow to record spectra in energy per wavelength units.

The bifacial Si cell used in this set of experiments was prepared at the University of New South Wales – UNSW (Australia). The cell is an untextured monocrystalline cell with buried contacts and front and rear sides antireflection coatings. Measured Quantum Efficiency of this cell (using a double beam, calibrated setup) at 0.980 nm was found equal to 0.27 and the measured efficiency under 1000 W/m² AM1.5 spectrum was found to be 12%.

Photocurrent generated in the bifacial c-Si cell by upconverted photons has been collected through connectors welded to the opposite sides of the cell. The intensity of the current is measured by a 485/4853 Keithley picoammeter, the scheme of the experimental setup is represented on Figure 2b. The samples were attached to a reflector using a glue to ensure a contact between the samples

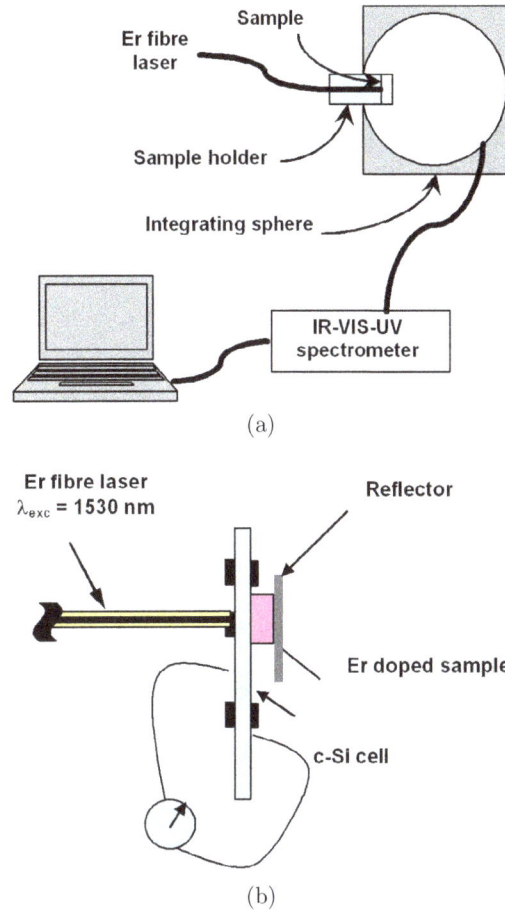

(a)

(b)

Fig. 2. Experimental setup (a) for the 1.5 μm → Visible, NIR absolute conversion efficiency of studied samples; (b) for the photocurrent measurement in the bifacial c-Si solar cell generated by sub bandgap excitation of an upconverter material.

Table 1. Relevant parameters used for short – circuit current calculations (the effective absorption cross-section is calculated by integrating σ_λ (abs) ($^4I_{15/2} \rightarrow {}^4I_{13/2}$) over in the spectral range corresponding to the FWHM of the excitation laser line).

	Z18	CYF
N (ions/cm³)	8.07×10^{20}	6.488×10^{20}
d (cm)	0.289	0.093
n	1.51	1.426
$\int \sigma_{abs}(\lambda)\, d\lambda$ (cm)	1.41×10^{-18}	2.05×10^{-18}
σ_{eff} (cm²)	1.6×10^{-20}	2.6×10^{-20}
$R_{reflector}$	0.90	0.96

and the reflector as good as possible; the glue was deposited only at the corners of the samples, out of the excitation/collection zones . The reflection coefficient of the reflector, tabulated in Table 1, has been checked by measuring, after removing the samples, the laser power before and after reflection at angle as small as possible to reproduce the experimental conditions.

The photocurrent measurement as a function of excitation power was done either changing the diode current (high power regime) or modifying the excitation density

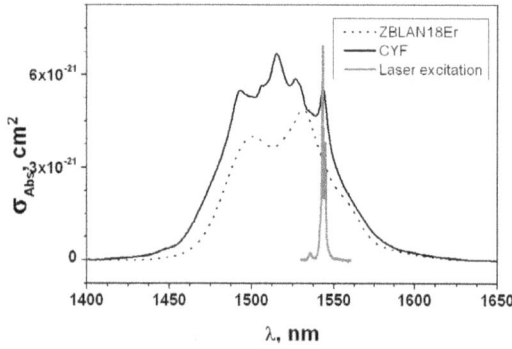

Fig. 3. Er^{3+} fundamental absorption ($^4I_{15/2} \rightarrow {}^4I_{13/2}$) for both studied materials and the laser excitation.

by varying the distance between the cell and the end fiber, keeping the diode current for the maximum output power (low power regime).

3 Experimental results

3.1 Absolute conversion efficiency

Absorption spectrum of Er^{3+} from the ground state to the first excited state ($^4I_{13/2}$) in Z18 is represented on Figure 1. A shown in Figure 1, the Er^{3+} transition from the ground state to the first excited state is beneficially located compared to the AM1.5G solar spectrum, out of the absorption due to O_2, H_2O and CO_2 in the atmosphere and below the c-Si gap absorption edge. For all investigated samples, this absorption band is quite large ($\Delta\lambda \approx 70$ nm) and exhibits a high oscillator strength that is required for the concerned application (Fig. 3). Furthermore, the excited states from which upconverted emissions are observed are well located well above the c-Si gap, especially the transition $^4I_{11/2} \rightarrow^4 I_{15/2}$ (Er^{3+}) as shown on Figure 1.

In order to measure the absolute conversion 1.5 μm $\rightarrow$ NIR-VIS efficiency of the samples under study, the IR excited luminescence was registered in power per wavelength units. The spectra, represented on Figure 4a−b, were recorded with an excitation power of 75 mW and 65 mW for Z18 and CYF repectively. The contribution of each observed transition to the total emitted power is calculated following:

$$\eta_i^{\exp} = \frac{\int\limits_{\lambda_0}^{\lambda_1} P_{meas}(\lambda)d\lambda}{P_{lum}^{int}} \qquad (1)$$

where, $P_{meas}(\lambda)$ represents the luminescence power displayed in energy per nanometer units for the transition i extending from λ_0 to λ_1, and $P_{lum}^{int} = \int\limits_{380\ nm}^{1700\ nm} P_{meas}(\lambda)d\lambda$ is the total emitted power. The fractions $\eta_i^{\exp}$ calculated from the luminescence spectrum (displayed in Figure 4a−b) are summarized in Table 2.

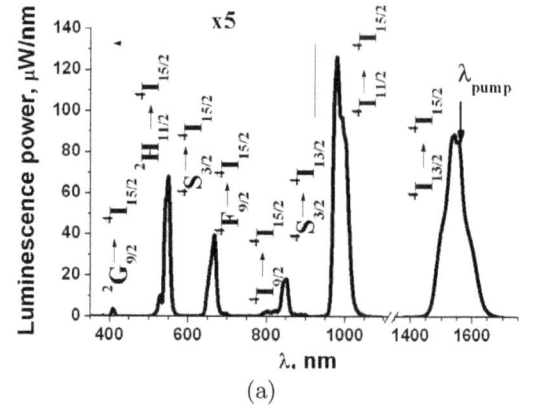

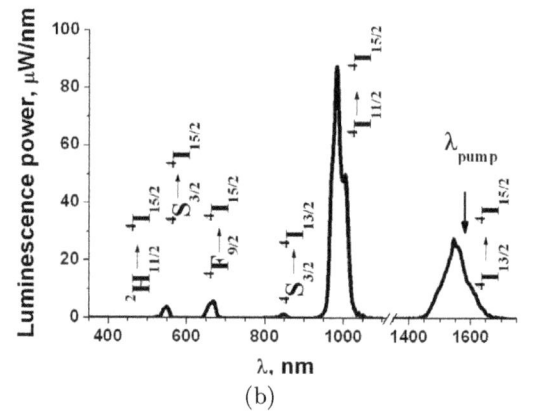

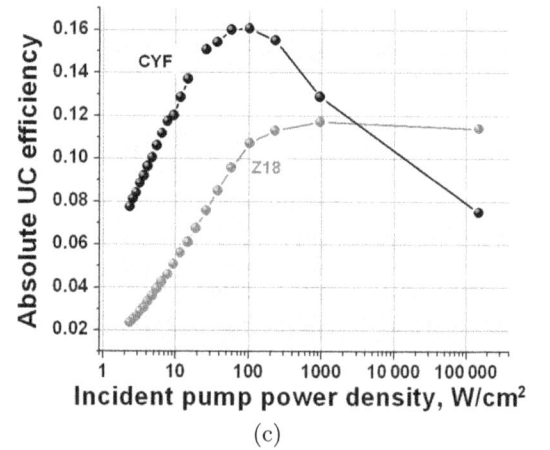

Fig. 4. Er^{3+} calibrated emission spectrum excited at 1.54 μm for CYF ($P_{exc} = 65$ mW) (a) and Z18 ($P_{exc} = 75$ mW) (b); Absolute conversion efficiency as a function of the 1.5 μm excitation power density (c).

For applications such as display or upconverter for PV, the most important criteria to select a material is the ratio of the luminescence power emitted in the whole spectral range P_{lum}^{int}, or in a particular range of interest, to the absorbed pump power. Thus, the absolute conversion efficiency is defined following:

$$\eta_{tot}^{\exp} = \frac{P_{lum}^{int}}{P_{abs}^{pump}}. \qquad (2)$$

Table 2. Er^{3+} Electronic transitions observed in emission, η_i^{exp} is the fraction of power emitted in a specific spectral domain of the transition "i", η_i, the absolute efficiency of the emission "i".

		Z18		CYF	
i	Electronic transition	η_i^{exp}	η_i	η_i^{exp}	η_i
1	$^4I_{13/2} \rightarrow {}^4I_{15/2}$	61.5%	20.3%	49.9%	16.8%
2	$^4I_{11/2} \rightarrow {}^4I_{15/2}$	34.8%	11.5%	47.6%	16.0%
3	$^4S_{3/2} \rightarrow {}^4I_{13/2}$	0.7%	0.2%	0.2%	0%
4	$^4F_{9/2} \rightarrow {}^4I_{15/2}$	1.2%	0.4%	1.6%	0.5%
5	$^4S_{3/2} \rightarrow {}^4I_{15/2}$	1.7%	0.6%	0.7%	0.2%
6	$^2G_{9/2} \rightarrow {}^4I_{15/2}$	0.1%	0%	0%	0%

The total absolute energy yield of the 1.5 μm $\rightarrow$ NIR – VIS upconversion is calculated by considering all upconverted emissions (from the NIR to the visible) $\sum_{i=2}^{6} \eta_i^{exp} \cdot \eta_{tot}^{exp}$ and for a particular emission "i" the absolute energy yield is obtained from the product $\eta_i^{exp} \cdot \eta_{tot}^{exp}$. Then, for the 1.5 μm $\rightarrow$ NIR conversion which is of a particular interest in our purpose, the absolute energy yield was found equal to 11.5% and 16.7% for Z18 and CYF respectively for an excitation power density of 100 W/cm². The excitation power dependence of the IR $\rightarrow$ NIR absolute conversion efficiency is represented in Figure 4c for both studied samples. Such a power dependence of the conversion efficiency is commonly observed and results from a competition between several processes which occur at high excitation power. A complete analysis of the power dependence of the upconverted luminescence is developed in [22]. From the values gathered in Table 2, it is clear that most part of upconversion emission energy is contained in the NIR part of the spectrum (at ~1 μm), which demonstrates that the studied samples are promising for application as active layer for solar cells with enhanced efficiency, since the NIR light has an absorption depth of only 100 μm and is thus strongly absorbed in a wafer-based silicon solar cell. For crystalline-silicon-based solar cells, approximately a half of the energy contained in the VIS part of the UC spectrum will be released as heat during thermalization of carriers created after absorption of VIS photons.

3.2 Generated current with a sub band gap excitation of a c-Si solar cell

The light-generated current in a c-Si cell was measured following two approaches. In both cases, the sample was pasted on a mirror in order to avoid excitation and emission losses and the sample was closely brought on the rear face of the cell. In a first step, the photocurrent was measured as a function of the excitation power, the power was varied between 75 mW and 20 mW (the lower limit of the fiber laser operation), the distance between the fiber and the front face was kept constant (equal to 1.3 mm). In a second experiment, the excitation power was set at 75 mW and the light-generated current measured as a function of the distance between the fiber and the entrance face of the cell allowing a larger range of the excitation density

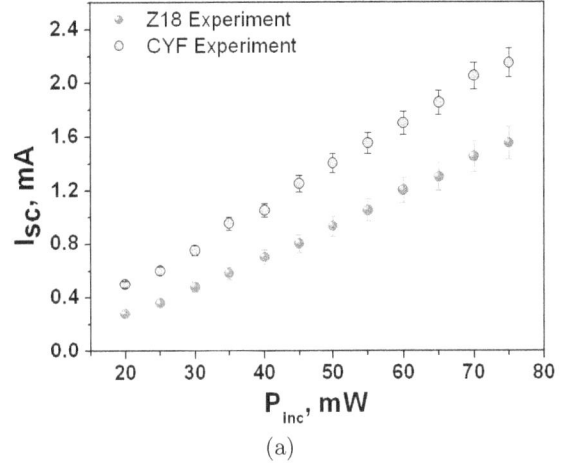

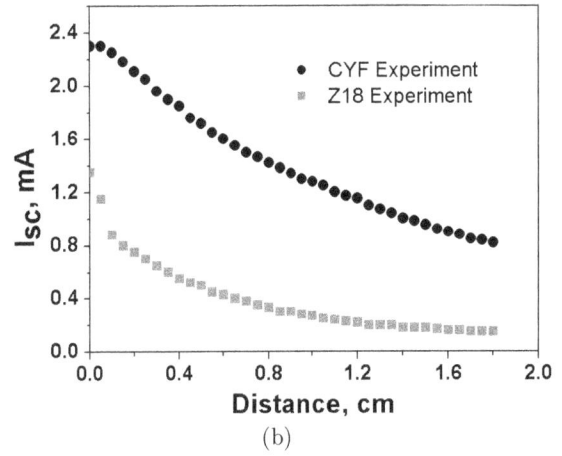

(a)

(b)

Fig. 5. Photocurrent generated in the bifacial c-Si solar cell measured by excitation at 1.54 μm as a function of the excitation power (a) and as a function of the distance between the sample and the end fiber, the laser power was constant and set equal to 75 mW (b).

than the previous experiment due to the divergence of the laser. Without the upconverting material, infrared excitation at 1.54 μm does not produce measurable current in the cell. The results of both experiments are represented on Figure 5.

4 Modelling the experimental results

A first model has been proposed by Byung-Chul Hong and Katsuyayasu Kawano, concerning the short circuit current generated by KMgF₃:Sm layer placed at a top of a CdS/CdTe solar cell [23]. Here the configuration is quite different since the light converter is placed on the rear face of the solar cell, the used formalism, very close to that proposed in [23], is developed in Appendix A. The calculation of the short circuit current generated by upconverted emission was performed in a first step using equations (A.1−A.9) and, in a second step, using equations (A.10−A.13) which include internal reflections in the upconverter. These results were checked using a transfer

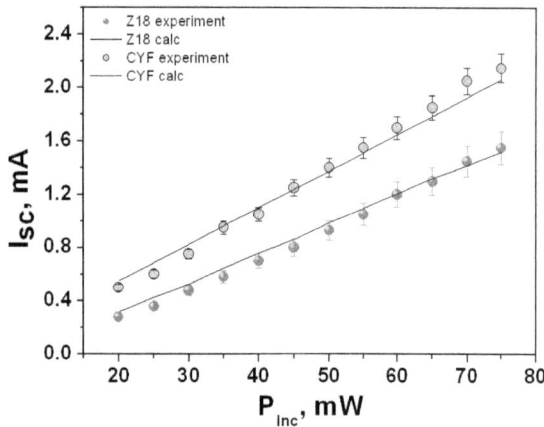

Fig. 6. Photocurrent as a function of the 1.54 μm excitation power of the upconverter: experimental results and calculated values using the simple model.

matrix formalism [24]. The emission flux is calculated by taking into account either only the NIR emission or all observed upconverted emissions.

Even if the laser peak is fixed at 1.54 μm, the laser excitation has a finite line width and an effective absorption cross section should be taken into account. Effective absorption cross-sections were determined by integrating the curves represented on Figure 3 over the FWHM of the laser line, all values considered in the following are gathered in Table 1 with the refractive index.

4.1 Dependence on the excitation power

Applying equation (A.2) with the emission flux given by equation (A.9) and using the set of data summarized in Table 1, the calculated photogenerated current is represented on Figure 6 in comparison with the experimental data obtained with both samples. Taking account for the geometry of the different samples (polished and parallel faces) internal reflections are expected to occur, the short circuit current has been calculated using equation (A.13) for the emission flux from the upconverter. As shown on Figure 7, addition of internal reflections does not introduce great modifications of the results obtained by equation (A.9). This is explained by the quite high absorption coefficient and the poor reflectivity of the mirror in the case of Z18. The correction due to the term $(1 - R_1 R_2 \exp(-2\alpha d))^{-1}$ is close to 1 (1.0006 for Z18 and 1.007 for CYF). Calculating step by step the contribution of the different possible reflections, the second reflection contributes for only 1% to the first step.

As it can be observed on Figure 7, the calculated I_{sc} values are slightly underestimated compared to experimental data for high excitation power. In the case of CYF, other upconverted emissions can contribute to the photovoltaic effect. Equation (A.9) has been applied by taking into account for the conversion efficiencies of the green and red emissions to the total short-circuit current. From the result of the calculation (Fig. 7b), these contributions become non negligible from 40 mW of incident power and

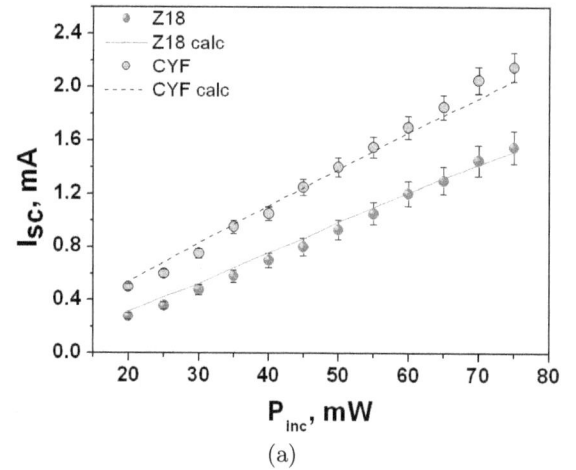

(a)

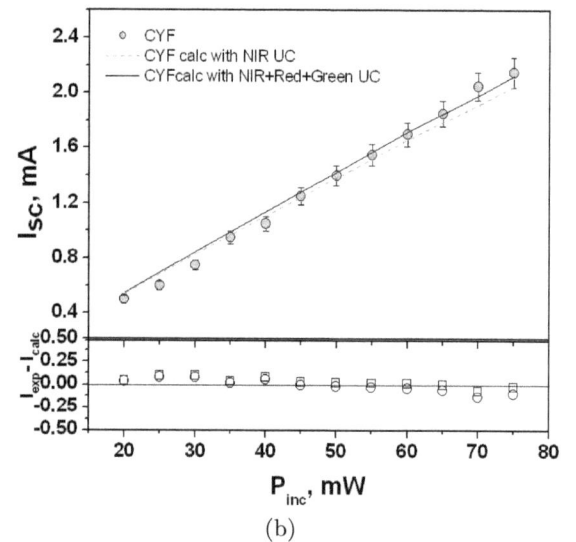

(b)

Fig. 7. Photocurrent as a function of the 1.54 μm excitation power of the upconverter: experimental results and calculated values considering multiple reflections in the upconverter (a) only the upconverted in the NIR is considered in the calculations; (b) all upconverted emissions are considered in the calculations.

the agreement between experimental and calculated values is better.

4.2 Dependence on excitation power density

The short circuit current registered as a function of the distance between the cell and the end fiber results, due to the divergence of the laser beam, in a decrease of the power density. According to the divergence and the distances involved in the experiment, the irradiated surface can be approximated to: $S = 0.01 \pi D^2$, here D is the distance between the upconverter surface and the fiber (in cm) and gives the same result as the exact formulae for $D > 0.05$ cm. This experiment was performed keeping the laser power constant and equal to 75 mW, the excitation density varies from 141 to 0.65 W/cm^2. In this range, the

(a)

(b)

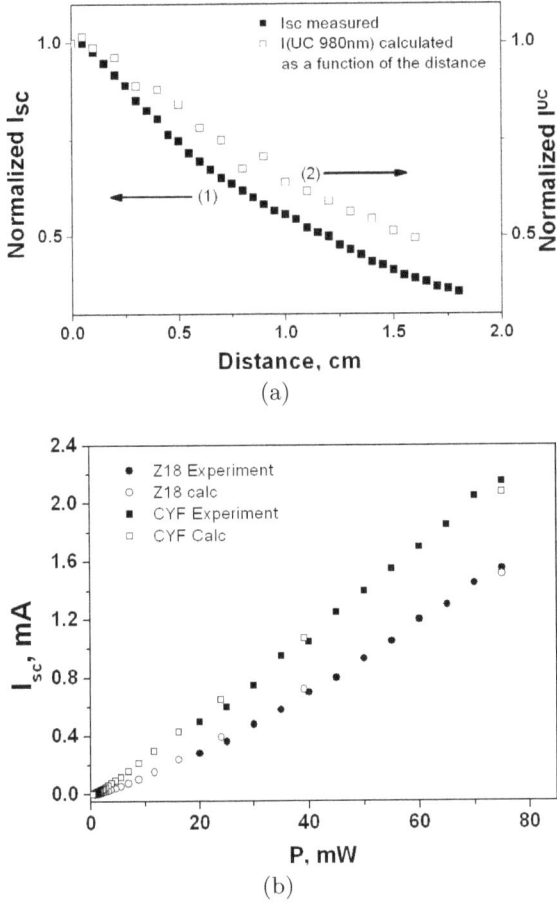

Fig. 8. (a) Comparison between the photocurrent and the intensity of the NIR upconverted emission as a function of the distance sample-end fiber laser; (b) comparison between the calculated values of the photocurrent (I_{sc}) using equation (A.9) and the experimental data including both experiences (I_{sc} measured as a function of the excitation power, I_{sc} mesured as a function of the sample-end fiber distance).

Fig. 9. Excitation power dependence of EQE and the IR → NIR absolute conversion efficiency.

varying distance. Then, applying equation (A.9), the I_{sc} calculated is in perfect agreement with experimental data recorded in the first experiment (Fig. 8b).

The simple model, we propose, reproduces quite well the experimental data and can help to predict roughly the expected short circuit current if all spectroscopic parameters are carefully determined as well as the absolute energy conversion efficiency and its excitation power density dependence. Visible upconverted emissions with noticeable conversion efficiency can contribute also for a small part to the photocurrent even if the most part is due, in the case of c-Si cell, to the NIR upconverted emission.

5 External quantum efficiency (EQE)

The performance of a photovoltaic cell is quantified by the external quantum efficiency (EQE) which is defined as the ratio between the flux of collected electrons and the flux of incident photons. Since the final aim of this work is to improve the c-Si cell efficiency, the EQE of the complete device (c-Si cell and upconverter) noted UC-PV similarly to [8] was calculated @ 1.54 μm. The results obtained for the investigated materials as a function of the excitation power are represented on Figure 9. On contrary to classical PV devices, a slight dependence of EQE on the incident power is observed. As already discussed in [8], a particular behaviour of EQE with the incident power is obtained when using an upconverter. The photons responsible for the creation of electron-hole pairs in the semiconductor result from a non linear process as briefly described in the introduction. The number of electron-hole pairs is proportional to the number of NIR photons ($\phi_{e-h} \propto (\phi_p)^n$) as it is observed in Figure 5a in most of the power range. In particular for the IR (1.54 μm) → NIR (0.98 μm) conversion with Er^{3+}, two IR photons are required for one NIR photon. Then EQE in a UC-PV device is expressed as [8]:

$$EQE = \frac{(\phi_p)^n}{\phi_p} \propto (\phi_p)^{n-1} \text{ with } n = 2 \text{ in our case.} \quad (3)$$

conversion efficiency is not constant as shown on Figure 4c and the following calculations were done using the η^{UC} determined as a function of the excitation density.

First, the experimentally measured short circuit current measured as a function of the distance exhibits a variation which is perfectly parallel to the intensity of the NIR upconverted emission recorded in the same conditions but without the cell between the end fiber and the sample as shown on Figure 8(a) for the CYF sample. For a better comparison, the two set of data were normalized to the maximum value (i.e. for the shortest distance) and the excitation density corrected from the reduced transmission of the c-Si cell for curve (2) on Figure 8(a). This result confirms the role of the NIR upconverted emission to the short circuit current. The model has also been applied to calculate the I_{sc} value under these experimental conditions. First, the incident power was calculated from a reduction factor taking into account the ratio between the excitation surface at the minimum distance and the

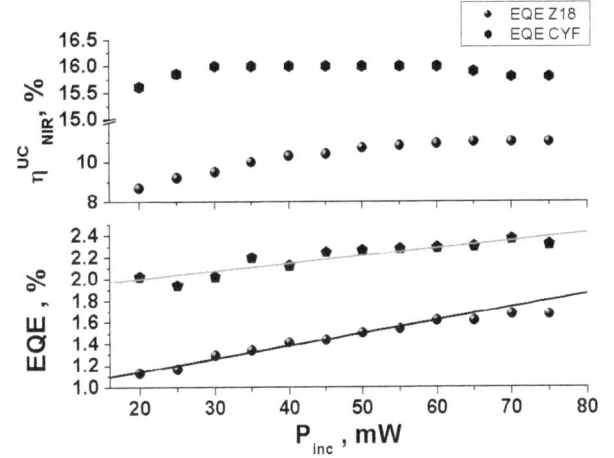

Thus, a linear dependence of EQE is expected with such a UC-PV device. This explains the EQE behaviour with the incident power represented on Figure 9. Moreover, the deviation from a linear dependence is observed at high incident power especially for Z18. Considering the absolute IR → NIR conversion efficiency versus the excitation power, represented on Figure 9 (upper curves), η^{UC} is quite constant for CYF over the incident power range and EQE increases linearly from 20 to 75 mW while η^{UC} increases from 20 to 30 mW, then stays stable up to 60 mW and then decreases for higher excitation power. This results are in agreement with the dependence of the conversion efficiency with the excitation power due to competing mechanisms occurring at high pump power in the upconversion processes as described in [22]. The EQE power dependence obtained for Z18 reflects quite well the η^{UC} dependence with the excitation power as observed in Figure 9.

An absolute increase of the quantum efficiency is found at 1.54 μm (2.4%, 1.7% for CYF, Z18 respectively, at maximum flux) by adding an upconverter at the rear face of the c-Si cell. These results are lower than that reported in [8] but still in the same range of order.

Although the UC systems reported are among the most efficient ones (a photon to photon quantum efficiency of 10.8% corresponding to a photon to photon power efficiency of 16.7%, see Table 2), the global system quantum efficiency is plagued by 3 issues that could be rather easily improved on:

- The transmission of the Si Cell in the IR (at 1.54 μm) is only about 60%. These losses come from the degraded AR coating (a good AR coating should yield less than 10% losses from reflection), a non optimal grid with shadowing from both grids and a doping level in the n/p parts that yields free carrier absorption.

- The rather low EQE of the cell at 980 nm, slightly less than 40%, due to the degraded AR coating. With texturing and improved AR coating, the EQE of the cell could well approach 80% in this region for such a thick cell.

- The non-optimized optical coupling between the cell and the UC crystals (with a very small air gap) as well as between UC and the mirror. From our evaluation, using a transfer matrix formalism to take into account all internal reflections, this induces losses that can be up to 50% of the UC photons absorbed in the Si cell.

With a UC quantum efficiency equal to 16.7% (Table 2), this yields the system EQE close to 2.4%. In terms of power efficiency, assuming a cell output voltage of 700–900 mV (this depends on solar concentration); this EQE translates into a 1.3 to 1.8% efficiency of power conversion of the 1.5 μm wavelength. With a somewhat optimized system, especially with optical coupling and NIR response of the Si cell, a factor of 3 to 5 of improvement could be obtained: 5 to 10% power conversion efficiency of the IR photons in the 1.5 μm band that can be absorbed by Er ions seems therefore an achievable target.

6 Conclusion

In this work, we could compare internal photon to photon conversion in UC compounds (something that is rarely measured) to the total photovoltaic system efficiency. We could show that quite efficient (up to 16.7% efficiency powerwise) compounds do exist for the 1.5 μm to 0.98 μm conversion that is well suited for use in combination with Si solar cells. In such a non optimal system, the power conversion efficiency was found around 1−2%, but a careful analysis of the losses showed that 5% is a realistic goal.

Of course, these systems have the drawback that they work only under very high concentration ($\times$ 2000 to $\times$ 20 000) at the present state of the art. This is something that can hardly be improved unless the absorption coefficient of the UC material can be improved, but still, with present day technology, this represents a few W/m^2 more that could be harvested.

This work has been supported by the ANR Solaire Photovoltaïque (project "THRI-PV"). G. Conibeer (ARC Photovoltaics Centre of Excellence, University of New South Wales, Sydney, Australia) is acknowledged for providing the c-Si solar cell. The authors would like to thank D. Lincot (IRDEP, UMR 7174 – CNRS/EDF/Chimie ParisTech) for helpful discussions. F. Pellé and J.-F. Guillemoles would like to dedicate this paper to Dr O. Guillot-Noël, disappeared with 227 other people in the AF447 Paris-Rio crash.

Appendix A

Model for the calculation of the photogenerated current by upconverter materials

1. Simple model neglecting internal reflections in the upconverter (Fig. A.1)

The used formalism is very close to that proposed in [23] but some modifications are needed due to the configuration of the experiments performed with an upconverter. The short circuit current generated by upconverted emission extending between λ_0 to λ_1 from the samples excited at 1.54 μm is given by the following expression:

$$I_{sc}\left(\lambda_{em}\right) = q \int_{\lambda_0}^{\lambda_1} \eta^{cell}\left(\lambda\right) F\left(\lambda_{em}\right) d\lambda, \qquad (A.1)$$

where q is the electron charge (Coulombs), $\eta^{cell}\left(\lambda\right)$ the quantum efficiency of the cell at the emission wavelength, $F\left(\lambda_{em}\right)$ the emission photon flux from the upconverter material.

First, the emission flux is calculated by taking into account for the NIR emission since UV-visible upconverted

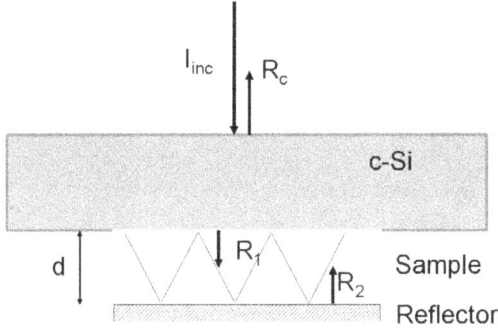

Fig. A.1. Schematic diagram of the incoming and outgoing incident excitation and converted emission in the complete system (solar cell + upconverter + reflector) (For details see text).

emissions observed between 400 nm and 800 nm contribute only for a minor part. Considering that the internal quantum efficiency of the cell $\eta^{cell}(\lambda)$ is only slightly dependent on the wavelength in the NIR emission range, Equation (A.1) can be simplified as:

$$I_{sc}(\lambda_{av}) = q\eta^{cell}(\lambda_{av}) \int_{\lambda_0}^{\lambda_1} f(\lambda)\,d\lambda$$
$$= q\eta^{cell}(\lambda_{av})\,F(\lambda_{av}). \qquad (A.2)$$

The integrated emission photon flux from the upconverter material $F(\lambda_{av})$ depends on the incident flux on the upconverter and on the IR → NIR conversion efficiency. First of all, the incident flux on the material is reduced from the incident flux on the cell due to reflections at the different interfaces (entrance and rear faces of the cell, cell-upconverter). In addition, if the semiconductor is transparent at 1.54 μm, the presence of contacts on both faces of the cell induces losses in the 1.54 μm transmission.

The incident flux $\phi_{cell}^{inc}(\lambda_{exc})$ on the cell is calculated from the excitation power at the excitation wavelength (i.e. 1.54 μm) following:

$$\phi_{cell}^{inc}(\lambda_{exc}) = \frac{P_{cell}^{inc}(\lambda_{exc})\,\lambda_{exc}}{hc}. \qquad (A.3)$$

In a first approximation, the incident flux on the upconverter material $\phi_{Mat}^{inc}(\lambda_{exc})$ is obtained from $\phi_{cell}^{inc}(\lambda_{exc})$ and reflections at the different interfaces following:

$$\phi_{Mat}^{inc}(\lambda_{exc}) = (1-R_c)(1-R_1)\,\phi_{cell}^{inc}(\lambda_{exc}) \qquad (A.4)$$

with R_c, R_1 the reflection coefficients at the air-cell interface, at the cell-upconverter interface respectively, given by Fresnel's equations:

$$R_C = \frac{(n_a - n_c)^2}{(n_a + n_c)^2} \qquad (A.5a)$$

$$R_1 = \frac{(n_{UC} - n_c)^2}{(n_{UC} + n_c)^2} \qquad (A.5b)$$

n_a, n_c, n_{UC} are the refractive index of the air (equal to 1), the cell and the upconverter respectively.

To achieve the calculation of the short-circuit current, n_c has to be evaluated. To this aim, we consider the cell as a whole system to calculate an effective index since we do not know exactly the thickness of the coating layers. The transmission of the system has been measured at 1.54 μm, out of the absorption range of the device. The loss of the excitation power is due to reflections from the entrance face, and internal reflections at interfaces coating-semiconductor. From the measured transmission behind the cell it is possible to calculate the effective refractive index of the system by taking into account reflections at the different interfaces.

$$I_{Trans} = (1-R_C)^2\,I_{inc} \qquad (A.6)$$

with R_C is given by equation (A.5a). The squared transmission coefficient is due to two interfaces air-cell in this configuration.

The transmitted power was found to be half of the incident one. Solving equations (A.5a) and (A.6) give rise to an effective index equal to 3.35. This value seems quite reasonable considering the refractive indices of the coating and the semiconductor equal to 2.1 and 3.48 respectively at 1.5 μm.

Absorption in the upconverter is due first from the direct absorption along the optical path d and in the second part due to the non absorbed photons which are reflected by the mirror and absorbed travelling in the opposite direction (see Fig. A.1), thus the rate of absorption of the excitation photons in the upconverter $\phi_{Mat}^{abs}(\lambda_{exc})$ is expressed as:

$$\phi_{Mat}^{abs}(\lambda_{exc}) = \phi_{Mat}^{inc}(\lambda_{exc})\,[(1-\exp(-\alpha d))$$
$$\times (1 + R_2\exp(-\alpha d))] \qquad (A.7)$$

where α is the absorption coefficient of the upconverter at the excitation wavelength, d is the upconverter thickness, and R_2 is the reflection coefficient of the mirror.

The complete expression of the absorbed flux by the upconverter is given by:

$$\phi_{Mat}^{abs}(\lambda_{exc}) = (1-R_c)(1-R_1)\,\phi_{cell}^{inc}(\lambda_{exc})$$
$$[(1-\exp(-\alpha d))(1+R_2\exp(-\alpha d))]. \qquad (A.8)$$

Neglecting internal reflections in the upconverter slab, that is neglecting R_I, the emission photon flux $F(\lambda_{av})$ is then expressed as:

$$F(\lambda_{av}) = \phi_{Mat}^{abs}(\lambda_{exc})\,\eta^{UC} \qquad (A.9)$$

with η^{UC} the absolute efficiency of the 1.54 μm → 0.98 μm conversion.

2. Model including internal reflections in the upconverter material

If the incident light is reflected on the interface upconverter/cell, the non absorbed part of the incident flux will be reflected and then absorbed travelling through the upconverter to the mirror and so on (Fig. A.1). This can be expressed by the following relationships between the reflection coefficients (R_1 and R_2) of the different interfaces, the absorption coefficient α and d the thickness of the upconverter

$$
\begin{aligned}
\phi_1^{abs} &= \phi_{Mat}^{inc} \ (1 - \exp(-\alpha d)) \\
\phi_2^{abs} &= \phi_{Mat}^{inc} \ (1 - \exp(-\alpha d)) \ R_2 \exp(-\alpha d) \quad \text{(A.10)} \\
\phi_3^{abs} &= \phi_{Mat}^{inc} \ (1 - \exp(-\alpha d)) \ R_1 R_2 \exp(-2\alpha d) \\
\phi_4^{abs} &= \phi_{Mat}^{inc} \ (1 - \exp(-\alpha d)) \ R_1 R_2^2 \exp(-3\alpha d) \ldots
\end{aligned}
$$

with ϕ_{Mat}^{inc} defined by equation (A.4).

For a large number of internal reflections, we get

$$
\begin{aligned}
\phi_n^{abs} = \phi_{Mat}^{inc} \ (1 - \exp(-\alpha d)) \left[\sum_{i=0}^{i=\infty} R_1^i R_2^i \exp(-2i\alpha d) \right. \\
\left. + \sum_{i=0}^{i=\infty} R_1^i R_2^{i+1} \exp(-(2i+1)\alpha d) \right] \quad \text{(A.11)}
\end{aligned}
$$

where the first and second terms of the above expression can be considered as geometrical series which sum is:

$$
A = \frac{1}{1 - R_1 R_2 \exp(-2\alpha d)} \text{ for the first term, and for}
$$

the second term $B = \dfrac{R_2 \exp(-\alpha d)}{1 - R_1 R_2 \exp(-2\alpha d)}$.

Finally, the absorbed flux within the upconverter is given by:

$$
\phi_{Mat}^{abs} = \phi_{Mat}^{inc} \ (1 - \exp(-\alpha d)) \frac{1 + R_2 \exp(-\alpha d)}{1 - R_1 R_2 \exp(-2\alpha d)} \tag{A.12}
$$

that gives for the emission flux:

$$
\begin{aligned}
F(\lambda_{av}) = \phi_{cell}^{inc}(\lambda_{exc})(1 - R_C)(1 - R_1) \\
\times (1 - \exp(-\alpha d)) \frac{1 + R_2 \exp(-\alpha d)}{1 - R_1 R_2 \exp(-2\alpha d)} \eta^{UC}. \tag{A.13}
\end{aligned}
$$

The final expression for the photogenerated current is obtained using equation (A.2) with the emission flux given by equation (A.13).

References

1. M.A. Green, K. Emery, Y. Hishikawa, W. Warta, Progress in Photovoltaics **17**, 85 (2009)
2. T. Trupke, M.A. Green, P. Würfel, J. Appl. Phys. **92**, 4117 (2002)
3. T. Trupke, M.A. Green, P. Würfel, J. Appl. Phys. **92**, 1668 (2002)
4. W. Shockley , H.J. Quiesser, J. Appl. Phys. **32**, 510 (1961)
5. T. Trupke, A. Shalav, B.S. Richards, P. Würfel, M.A. Green, Sol. Energy Mater. Sol. Cells **90**, 3327 (2006)
6. S. Baluschev, P.E. Keivanidis, G. Wegner, J. Jacob, A.C. Grimsdale, K. Mullen, T. Miteva, A. Yasuda, G. Nelles, Appl. Phys. Lett. **86**, 061904 (2005)
7. A. Shalav, B.S. Richards, T. Trupke, K.W. Krämer, H.U. Güdel, Appl. Phys. Lett. **86**, 13505 (2005)
8. B.S. Richards, A. Shalav, IEEE Trans. Elect. Dev. **54**, 2679 (2007)
9. A. Shalav, B.S. Richards, M.A. Green, Sol. Energy Mater. Sol. Cells **91**, 829 (2007)
10. S. Fischer, J.C. Goldschmidt, P. Löper, G.H. Bauer, R. Brüggemann, K. Krämer, D. Biner, M. Hermle, S.W. Glunz, J. Appl. Phys. **108**, 044912 (2010)
11. C. Strumpel, M. McCann, G. Beaucarne, V. Arkhipov, A. Slaoui, V. Svrcek, C. del Canizo, I. Tobias, Sol. Energy Mater. Sol. Cells **91**, 238 (2007)
12. F. Auzel, Comptes Rendus de l'Académie des Sciences (Paris) **262**, 1016 (1966)
13. V.V. Ovsyankin, P.P. Feofilov, JETP letters **4**, 471 (1966)
14. F. Auzel, Chem. Rev. **104**, 139 (2004)
15. Q. Nie, L. Lu, T. Xu, S. Dai, X. shen, X. Liang, X. Zhang, X. Zhang, J. Phys. Chem. Solids **67**, 2345 (2006)
16. Z. Jin, Q. Nie, T. Xu, S. Dai, X. Shen, X. Zhang, Mater. Chem. Phys. **104**, 62 (2007)
17. D. Chen, Y. Wang, Y. Yu, P. Huang, F. Weng, J. Solid State Chem. **181**, 2763 (2008)
18. S. Ivanova, F. Pellé, J. Opt. Soc. Am. B **26**, 1930 (2009)
19. F. Auzel, D. Morin, French patent FR2755309 (A1), Patent B.F. N 96 13327 (1996)
20. A.M. Tkachuk, S.É. Ivanova, F. Pellé, Optika i Spektroskopiya, **106**, 907 (2009)
21. A.M. Tkachuk, S.É. Ivanova, F. Pellé, Condensed-Matter Spectroscopy **106**, 821 (2009)
22. M. Pollnau, D.R. Gamelin, S.R. Lüthi, H.U. Güdel, M.P. Hehlen, Phys. Rev. B **61**, 3337 (2000)
23. B.-C. Hong, K. Kawano, Sol. Energy Mater. Sol. Cells **80**, 417 (2003)
24. M. Born, E. Wolf, Pergamon, 1999

12

Determining diode parameters of illuminated current density-voltage curves considering the voltage-dependent photo current density

F. Obereigner[a] and R. Scheer[b]

Photovoltaics Group, Martin-Luther-University Halle-Wittenberg, 06120 Halle (Saale), Germany

Abstract Suitable procedures for the determination of diode parameters of illuminated and dark current density-voltage (j-V) curves of solar cells are investigated. Within these procedures we avoid misinterpretations by the commonly used shifting approximation which assumes a voltage-independent photo carrier collection. To this end, the voltage-dependent collection efficiency is determined by j-V measurements under incremental illumination. After that, illuminated curves can be converted into dark curves taking the voltage-dependent photo current density into account. The converted dark curve next is consigned into a program that numerically solves the implicit equation of the current density and derives the diode parameters by means of a weighted least-squares minimization. Only if fluctuations of the illumination are noticeably high this procedure cannot be used, since the converted dark curve also reveals a high noise level. Then, the determination of diode parameters is performed stepwise with the help of commonly used methods.

1 Introduction

The characterization of illuminated solar cells is often restricted to the quest for first and second level parameters: energy conversion efficiency as well as short-circuit current density j_{sc}, open-circuit voltage V_{oc} and fill factor. However, illuminated ($j_L(V)$) curves may provide also third level metrics information, which is assignable to microscopic processes, for example the recombination mechanism (value of diode quality factor) or carrier lifetimes (value of saturation current density). Referring to the common diode model, a fit of the dark ($j(V)$) curve unveils the diode parameters: saturation current density j_{01}, diode quality factor A_1, shunt conductance G_{sh} and series resistance R_s. Since these parameters can change due to illumination, it is common practice to measure the dark and illuminated curves and compare the derived parameter sets. Light induced differences of j_{01} and A_1 for instance are observed in Cu(In,Ga)Se$_2$ solar cells, which can be explained by the so-called red-light metastability [1]. The interpretation of illuminated curves provides the general pitfall of a voltage-dependent current collection [2]. If the photo current density is voltage-dependent, the shift-

ing approximation, this is $j^{1\,sun} = j_L - j_{sc}$[1], leads to inaccurate interpretations. Hence, the voltage-dependent collection efficiency $\eta(V) = j_{ph}(V)/j_{sc}$ has to be introduced. The collection efficiency can be obtained experimentally with the help of illuminated curves measured at different illumination intensities j_E [3].

In this work suitable procedures for the determination of diode parameters of illuminated and dark curves are proposed. Since the one- and two-diode models are implicit in current density, a program has been established being able to solve numerically the implicit equation of the current density. Additionally, it performs a weighted least-squares minimization to achieve the diode parameters of the dark curve. This approach is explained in the following section (Sect. 2). Illuminated curves are converted into dark curves with the help of the collection efficiency. The determination of the collection efficiency is demonstrated in Section 3. Furthermore, we give advises to cope with some technical and physical problems. The calculation of the collection efficiency is impeded by fluctuations of the illumination anyway. In the case of a high noise level all diode parameters have to be computed stepwise with the help of commonly used methods, which are

---

[a] e-mail: `florian.obereigner@physik.uni-halle.de`
[b] e-mail: `roland.scheer@physik.uni-halle.de`

[1] $j^{1\,sun}$ represents the dark current density measured at illuminated conditions.

illustrated in Section 4. A practical application of our presented approaches is shown in the following. At first, we explain the experimental details in Section 5. Afterwards, we demonstrate the results on the basis of dark and illuminated curves of a Cu(In,Ga)Se$_2$ solar cell in Section 6.

2 Fit of dark curves

A fit of a dark curve shall determine the diode parameters of a measured data set $j^0(V_i)$, $i = 1, \ldots, N$ by means of the well-known two-diode model

$$
j(V) = j_{01} \left(\exp\left(\frac{\beta(V - j(V)R_s)}{A_1} \right) - 1 \right)
$$
$$
+ j_{02} \left(\exp\left(\frac{\beta(V - j(V)R_s)}{A_2} \right) - 1 \right)
$$
$$
+ G_{sh}(V - j(V)R_s), \tag{1}
$$

where $\beta = e/(k_B T)$ is the fraction of the elementary charge e and the thermal energy $k_B T$. It is obvious that there is no analytic solution for the current density with given diode parameters and a fixed voltage, in general. Nevertheless, a numerical computation can easily be achieved with the help of Newton's method. Since it is an iterative approach one needs appropriate initial diode parameters, which are calculated using the two-diode model without any series resistances ($R_s = 0$ in Eq. (1)). For this task the convergence of Newton's method is very good, as long as the series resistance has a minor influence on the curve.

The degree of deviation between measured and computed current density is defined by the vertical distance in a j-V diagram. In contrast to orthogonal distances, vertical ones can be mathematically described more easily and require less numerical effort. The fit shall minimize the degree of deviation by varying the diode parameters. This approach is known as method of least squares. It is indispensable to use the weighted least squares sum

$$
\chi^2 = \sum_{i=1}^{N} \left(\frac{j(V_i)}{j^0(V_i)} - 1 \right)^2, \tag{2}
$$

to be sure that low and high absolute magnitudes of current densities have the same significance. The minimization routine is the most important step for a successful establishment of a nonlinear fit. The centerpiece is a non-commercial nonlinear optimization routine "lmder" being part of the Minpack library [4]. It is based on the higher-dimensional Newton's method for nonlinear optimization problems enhanced by the Levenberg-Marquardt approach. Since initial diode parameters are crucial for nonlinear optimization problems we propose an initialization routine for typical Cu(In,Ga)Se$_2$ solar cells to expand the feasible interval of sufficient initial diode parameters.

Initialization routine:

1. *At first define standard parameters:*
   $j_{01} = 10^{-10}$ A cm^{-2}, $A_1 = 1.5$,
   $j_{02} = 10^{-6}$ A cm^{-2}, $A_2 = 2.5$.

2. *Approximate the shunt conductance and the series resistance with the slope respectively the inverse slope of linear fits in the appropriate intervals.*

3. *Now, perform a fit with an approximated two-diode model*

$$
j(V_i) = j_{01} \left(\exp\left(\frac{\beta(V_i - j^0(V_i)R_s)}{A_1} \right) - 1 \right)
$$
$$
+ j_{02} \left(\exp\left(\frac{\beta(V_i - j^0(V_i)R_s)}{A_2} \right) - 1 \right)
$$
$$
+ G_{sh}(V_i - j^0(V_i)R_s) \tag{3}
$$

using the measured current densities $j^0(V_i)$ on the right-hand side of equation (1).

4. *Use the results of step 3 as initial diode parameters for the fit with numerically solved equation (1).*

The resulting diode parameters in step 3 are in general not equal to the fit results of the exact model, but the magnitudes of diode parameters are sufficient to use them as initial diode parameters for the final fit [5]. This entire procedure is able to fit j-V curves from a broad parameter space with respect to the one or two-diode model. If the fit is successful, the standard deviation

$$
\sigma = \sqrt{\frac{\chi^2}{N}} \tag{4}
$$

can be calculated, which defines the average deviation between measured and computed current density for one point.

3 Determination of the collection efficiency

The collection efficiency has to be taken into account whenever a solar cell exhibits a voltage dependent photo current density. Note that the shifting approximation assumes a constant collection efficiency. Reasons for this voltage dependency are the width of the space charge region or interface recombination in a heterojunction thin-film solar cell. Both quantities depend on the applied voltage in an a priori unknown fashion. Hence, we are forced to determine the voltage-dependent photo current density of each solar cell with experimental input. At first sight this seems to be easy using two illuminated curves j_L^n, j_L^m measured at different illumination intensities j_E^n, j_E^m, as proposed by Hegedus [3]. The collection efficiency turns out to be

$$
\eta^{n,m}(V) = \frac{j_L^n - j_L^m}{j_{sc}^n - j_{sc}^m}. \tag{5}
$$

This approach is sufficient but can lead to a variety of physical and technical problems. In contrast to Hegedus [3] we use a very small range of j_E to evaluate equation (5) under similar conditions. The comparison of the dark j-V curve with the curve under one sun illumination leads to the assumption that both recombination current densities are the same. In general, this is not the case for $Cu(In,Ga)Se_2$ solar cells since the defect density can change by illumination [6]. This small range of j_E in turn leads to large noise levels on $\eta^{n,m}(V)$ due to fluctuating illumination and other noise sources. The natural solution is averaging a large number of measurements. This can be done by current averaging under constant voltage bias or by $\eta^{n,m}(V)$ averaging using several illumination levels j_E^x; $x = k, l, m, n, \ldots$ The first averaging scheme can lead to errors, when solar cells show time-dependent effects as being observed, e.g. for $Cu(In,Ga)Se_2$ [6,7] and CdTe [8] solar cells. Such a voltage bias metastability leads to a modification of the solar cell. That is why averaging data points in j-V measurements shall be avoided, whenever a drift of the current density occurs under persistent voltage bias at timescales of j-V measurements. Hence, we propose to apply the second averaging scheme and to record a convenient number M of illuminated curves with different illumination intensities, which should not be less than 4. Note that in this case we obtain $\frac{1}{2}(M-1)M = 6$ different collection efficiencies. A further technical problem arises from $j(V)$ interpolations in equation (5). Significant failures in the upper range of the curve can be excluded by applying the same voltages for all illuminated curves. Measuring j_{sc} avoids further interpolation problems. With these notes the collection efficiencies should be calculated to a satisfactory level. The mean collection efficiency $\bar{\eta}(V)$ emerges from averaging of all $\eta^{n,m}$. After that, the voltage-dependent photo current density results from the product of the mean collection efficiency and the short-circuit current density, which can easily be measured or be obtained from a linear fit.

The smoothness of the mean collection efficiency is crucial for the determination of illuminated diode parameters. Whenever $\bar{\eta}$ has a low noise level the fit of the converted dark curve should be successful. However, it is impossible to convert the illuminated into dark curves adequately in our setup because of fluctuations of the illumination intensity provided by our solar simulator. In general, illuminated curves show a non-vanishing magnitude of fluctuations around $V = 0$. In the high-voltage region the relative fluctuation amplitude is smaller, since the photo current density usually decreases at higher voltages. On this account, the mean collection efficiency as well as the converted dark curves indicate the same noise level because their computations rely on illuminated curves. In order to give a quantitative limit for the noise level we calculate the standard deviation of the current densities near $V = 0$ by a linear fit of one curve (see inset in Fig. 1). The standard deviation $\delta j = \sqrt{N^{-1} \sum_{i=1}^{N} j_{corr}(V_i) - j_{linear}(V_i)^2}$ is estimated by summing up the squared absolute deviations of current densities from the line $j_{linear}(V)$. δj has to be small in comparison to the difference of short-circuit current densities Δj_{sc} of two illuminated curves. The ratio $\delta j / \Delta j_{sc}$ gives the amount of noise level, which disturbs the calculation of the collection efficiency. We achieve a mean deviation $\delta j = 0.04$ mA cm^{-2} in our setup (see inset in Fig. 1), which leads to $\delta j / \Delta j_{sc} \approx 2.8\%$ by a given $\Delta j_{sc} = 1.5$ mA cm^{-2}. Another setup using a flash light solar simulator turns out to be more suitable for this procedure.

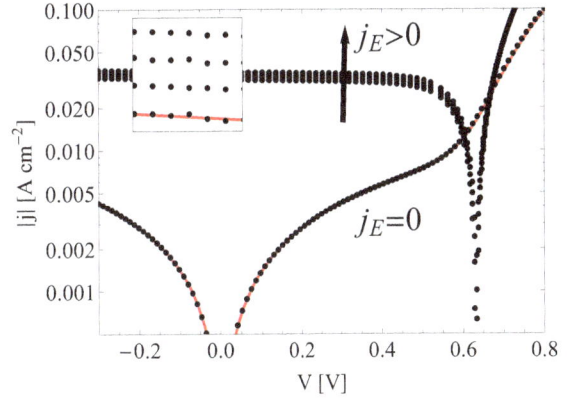

Fig. 1. Illuminated (blue) curves for different illumination intensities j_E and dark curve (black) in combination with the computed current density of the fit (red). The inset at the upper left side magnifies the illuminated curves being very close to each other and clarifies the procedure for determining the noise level of j_L. The red line is $j_{linear}(V)$.

We estimate a mean deviation of $\delta j = 0.004$ mA cm^{-2} in this case. In combination with the difference of the photo current densities $\Delta j_{sc} = 1.7$ mA cm^{-2} one obtains $\delta j / \Delta j_{sc} \approx 0.26\%$, which is one order of magnitude lower than the first case. If so, we can sufficiently convert the illuminated into a dark curve and perform a fit with the proposed procedure in Section 2. Since the flash light setup is not at our disposal we find another solution to retrieve the diode parameters. In this procedure the diode parameters are determined stepwise.

4 Stepwise determination of diode parameters from illuminated curves

At first, the difference between the slopes of the linear intervals near $V = 0$ of the illuminated curve and the mean collection efficiency are assigned to the shunt conductance [9]

$$G_{sh} \approx \left(\frac{dj_L(V)}{dV} - j_{sc} \frac{d\eta(V)}{dV} \right)\Big|_{V=0}. \tag{6}$$

The short-circuit current densities of each curve can be achieved in this step, too. Next, we determine the so-called suns-V_{oc} curve, i.e. the photo current densities and open-circuit voltages $(j_{ph}(V_{oc}), V_{oc})$ for all illuminated curves. Analysis of this curve yields the diode-quality factor A_1 and the saturation current density j_{01}. An exponential graph without any influence of the series resistance is expected. The use of $j_{ph}(V_{oc}) = j_{sc}\eta(V_{oc})$ is important, whenever the photo current density at open-circuit condition is much lower than the short-circuit current density. Unfortunately, the diode parameters j_{02} and A_2 cannot be determined by this method in general, because the typically chosen illumination intensities are close to $j_E = 1$ sun. Note that the use of an enlarged illumination range would allow to achieve those parameters, but this can lead to misinterpretations because of illumination-dependent changes. Therefore, the suns-V_{oc} graph is restricted to the upper part, which corresponds to A_1 and j_{01}. Last but not least the series resistance is calculated with the help of the converted dark curve by the method of the signal conductance, which has been introduced by Werner [10]. Here, we make use of the smooth behavior of the collection efficiency at higher voltages (see Sect. 3). The corrected current density $j_{corr} = j^{1\,sun} - G_{sh}V$ and the signal conductance $G = dj_{corr}/dV$ are drawn in a G/j_{corr}-G graph. A linear fit for high magnitudes of G is connected with the series resistance R_s as well as the diode quality factor A_1. It was shown by Eron and Rothwarf [2] that the diode quality factors of both methods using the shifting approximation can lead to differences. Hence, the results of the whole procedure are most reliable, if the outcomes of the differently determined A_1 are about identical.

5 Experimental details

Our measurements were made in a measurement chamber from Kurt J. Lesker. The used solar simulator SF 1000 W is supplied from the company Sciencetech. The light beam was distributed with a semi-transparent polka dot mirror to the device under test and to a reference cell to measure the light intensity in parallel. Dark and illuminated j-V curves were taken by a Digital Sourcemeter 2400 from Keithley.

The device under test was a ZnO/CdS/Cu(In,Ga)-Se$_2$/Mo solar cell with the following parameters. The Cu(In,Ga)Se$_2$ absorber layer (3 μm thickness) was prepared in a three-stage co-evaporation process. The CdS buffer layer (50 nm thickness) was deposited in a chemical bath and the ZnO window (0.5 μm thickness) was sputtered.

6 Practical application

The sample exhibits a time-dependent drift of the current density, hence constant bias during current averaging should be avoided. The illuminated (blue) and dark (black) curves are drawn in Figure 1. We use a number of measurement points being between 50 and 100. The fit of the dark (red) curve is also presented in Figure 1 and demonstrates a very low degree of deviation between measured and computed current densities. The narrow differences between the illuminated curves illustrate that the chosen range for illumination intensities is sufficiently small. However, we find that small fluctuations of j_E lead to high fluctuations in the collection efficiencies as the approach for computing $\eta^{n,m}$ is based on differences between the curves. These problems can be seen in Figure 2, which shows all possible collection efficiencies $\eta^{n,m}$ (black) according to the four illuminated curves as well as the mean collection efficiency $\bar{\eta}$ (red). There are obvious fluctuations in all graphs for voltages lower than 0.55 V. Nevertheless, the graphs seem to be more smooth for higher voltages. If one converts the illuminated curve at $j_E = 1$ sun by the

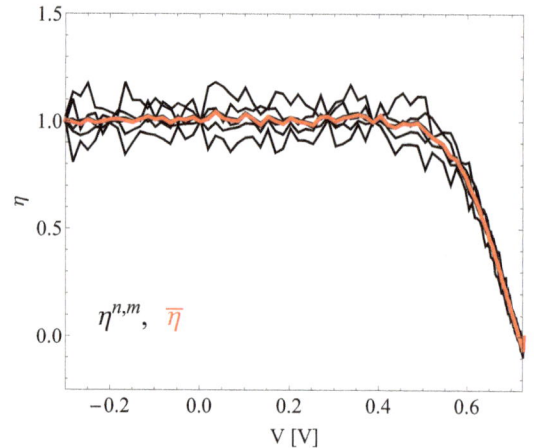

Fig. 2. All possible collection efficiencies (black) and the mean collection efficiency (red) of the sample being extracted from four different illuminated curves.

help of $\bar{\eta}$ and j_{sc}, the converted dark curve $j^{1\,sun}$ would exhibit the same fluctuations. This is the reason why diode parameters can only be investigated in the high-voltage range of $j^{1\,sun}$ being demonstrated in Figure 3. The figure shows the expected higher degree of fluctuations in comparison to the dark or illuminated curves (see Fig. 1). We have no choice but extracting the diode parameters of the illuminated curve stepwise.

To calculate the shunt conductance G_{sh} the slope of the mean collection efficiency yields

$$\frac{d\bar{\eta}(V)}{dV}\bigg|_{V=0} = -0.0085 \text{ V}^{-1}.$$

Combining this with the slopes of the illuminated curves and the short-circuit current densities the average shunt conductance $G_{sh} = 4.28 \times 10^{-3}$ S cm^{-2} of all illuminated curves is obtained. Next, we fit the suns-V_{oc} curve by a

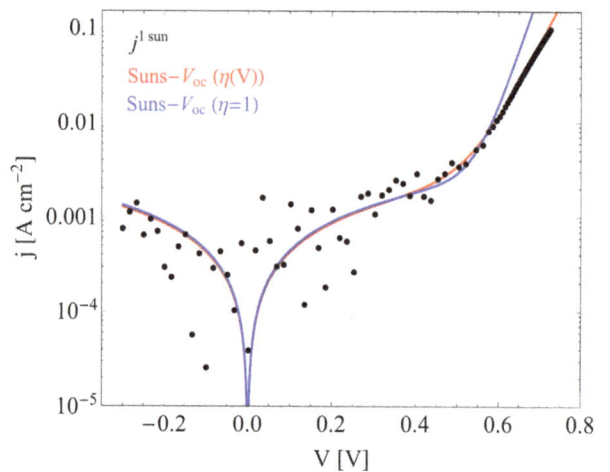

Fig. 3. The converted dark curve $j^{1\,sun}$ at an illumination intensity $j_E = 1$ sun and the converted dark curve without any series resistance proposed by the outcomes of the suns-V_{oc} analysis (voltage dependent (red) and independent (blue) photo current).

Table 1. Diode parameters by the proposed procedures for dark and illuminated curves of a typical Cu(In,Ga)Se$_2$ solar cell. In addition the influence of a voltage-dependent and -independent ($\eta = 1$) photo current density is compared during the suns-V_{oc} analysis. The standard deviation σ is always calculated by a numerical solution of equation (1).

Condition	j_{01} [A cm^{-2}]	A_1	G_{sh} [S cm^{-2}]	R_s [Ω cm^2]	σ
dark	1.35×10^{-8}	1.79	1.97×10^{-3}	0.19	0.0362
illuminated:					
with $\eta(V)$	6.90×10^{-8}	1.98	4.28×10^{-3}	0.16	1.84
with $\eta = 1$	2.28×10^{-10}	1.31	4.47×10^{-3}	0.30	1.95

simple non-implicit function. The diode parameters can be used to calculate a dark curve without any series resistance, which is illustrated in Figure 3. Obviously, there is a good agreement between the suns-V_{oc} curve (red line) and the converted dark curve, whereas the suns-V_{oc} curve (blue line) proposed by the shifting approximation shows an overestimated slope. Since the mean collection efficiency of the sample drops down to 0.5 at the open-circuit voltage, this observable difference is expected when the shifting approximation or the voltage-dependent collection efficiency is applied. The results of the non-implicit fits are shown in Table 1. The shifting approximation underestimates both diode parameters, which was explained by the in general non-positive slope of the collection efficiency [2]. Interpreting the results of the shifting approximation would lead to the conclusion that the recombination mechanism and recombination rate change drastically by white light illumination in comparison to dark conditions. In contrast, the outcomes of the voltage-dependent collection efficiency reveal a nearly unchanged recombination. The series resistance and the diode quality factor are calculated now with the help of the signal conductance G by applying difference quotients to the converted dark curve. Note that $j^{1\,sun}$ is smooth at high voltages, which correspond to high signal conductances. The linear fit for these high magnitudes of G yields $R_s = 0.16\ \Omega\,\mathrm{cm}^2$ and $A_1 = 1.90$ being very consistent with the outcome of the last step, which unveils nearly the same diode quality factor. All diode parameters are summarized in Table 1. The standard deviation σ being calculated by the numerical solution of equation (1) are also shown. σ of any illuminated parameter set is significantly higher than the dark one. The reasons are the high fluctuations in the lower range of the converted dark curve. However, in contrast to the shifting approximation the parameter set of voltage dependent photo current reveals a somewhat lower standard deviation. It becomes apparent that the illumination does neither distinctly influence the resistances nor the recombination. By these steps all diode parameters of the illuminated and dark curve are extracted in accordance with the one-diode model.

7 Conclusion

We have investigated procedures for the determination of diode parameters of illuminated and dark j-V curves. Dark curves are fitted using a numerical solution of the implicit one- or two-diode model. Illuminated curves are converted into dark curves by means of the voltage-dependent photo current density, which has to be determined by j-V measurements under incremental illumination. Considering the voltage-dependency of the collection efficiency can lead to substantially different interpretations of illuminated curves in contrast to the frequently-used shifting approximation. The presented procedures can easily be integrated in an electronic characterization software for solar cells. Such a software package was established by our group and was tested on many measured curves. It is embedded in a graphical user interface in Wolfram Mathematica for dark (PV-jV-$Dark.nb$) and illuminated (PV-jV-$Light.nb$) curves. It integrates a Fortran program (PV-jV-$DiodeFit.exe$), which performs the non-implicit and implicit fits for dark curves. This software package is offered at our homepage www.physik.uni-halle.de/fachgruppen/photovoltaik.

References

1. F. Engelhardt et al., Phys. Lett. A **245**, 489 (1998)
2. M. Eron, A. Rothwarf, Appl. Phys. Lett. **44**, 131 (1984)
3. S. Hegedus, Prog. Photovolt.: Res. Appl. **5**, 151 (1997)
4. Argonne National Laboratory, http://www.netlib.org/minpack/
5. F. Obereigner, Master thesis, Martin-Luther-University Halle-Wittenberg, 2012
6. S. Lany, A. Zunger, J. Appl. Phys. **100**, 113725 (2006)
7. H. Kempa et al., Thin Solid Films **535**, 340 (2013)
8. M.A. Lourenco, J. Appl. Phys. **82**, 1423 (1997)
9. R. Scheer, H.W. Schock, *Chalcogenide Photovoltaics* (Wiley-VCH, Weinheim, 2011)
10. J.H. Werner, Appl. Phys. A **47**, 291 (1988)

13

Epitaxial growth of silicon and germanium on (100)-oriented crystalline substrates by RF PECVD at 175 °C

M. Labrune[1,2,a], X. Bril[1], G. Patriarche[3], L. Largeau[3], O. Mauguin[3], and P. Roca i Cabarrocas[1]

[1] LPICM, CNRS-École Polytechnique, 91128 Palaiseau Cedex, France
[2] TOTAL S.A., Gas & Power, R&D Division, Courbevoie, France
[3] Laboratoire de Photonique et de Nanostructures, CNRS, Marcoussis, France

Abstract We report on the epitaxial growth of crystalline Si and Ge thin films by standard radio frequency plasma enhanced chemical vapor deposition at 175 °C on (100)-oriented silicon substrates. We also demonstrate the epitaxial growth of silicon films on epitaxially grown germanium layers so that multilayer samples sustaining epitaxy could be produced. We used spectroscopic ellipsometry, Raman spectroscopy, transmission electron microscopy and X-ray diffraction to characterize the structure of the films (amorphous, crystalline). These techniques were found to provide consistent results and provided information on the crystallinity and constraints in such lattice-mismatched structures. These results open the way to multiple quantum-well structures, which have been so far limited to few techniques such as Molecular Beam Epitaxy or MetalOrganic Chemical Vapor Deposition.

1 Introduction

In the field of solar energy, there is a continuous search for ways to increase the cost-effectiveness of solar cells. This is particularly the case of crystalline silicon solar cells which is the leading technology and covers more than 80% of the PV market. For this technology to keep its advantage, reducing the thickness of expensive c-Si wafers is mandatory. Various approaches have already been used to produce efficient and thin c-Si solar cells resulting in efficiencies above 22% for heterojunction c-Si solar cells on a wafer thinned down to 98 μm [1]. However this approach still requires to grow ingots and to slice them into wafers. Another way to cut costs is to grow the mono or multi crystalline silicon directly on a foreign substrate, using for instance Chemical Vapor Deposition [2], or on a polycrystalline seed layer obtained by the crystallization of an amorphous silicon layer, using a catalyst in the case of aluminium induced crystallization [3] or in a catalyst-free approach using solid phase crystallization [4]. However, these processes usually lead to relatively low solar cell efficiencies since there is a non-monotonic relationship between grain size and solar cell efficiency that implies that in the case of multicrystalline silicon one should have very large grains, as reported by Bergmann [5].

More recently the epitaxial growth of Si films on c-Si substrates has been achieved by various techniques such as Plasma Enhanced Chemical Vapor Deposition (PECVD) [6], Atmospheric Pressure CVD [7], Inductively Coupled PECVD [8] as well as by Hot Wire CVD [9]. In most of these cases, except in reference [6], the substrate was kept at a relatively high temperature to favor the epitaxial growth ($T \geqslant 700$ °C). These high temperature approaches have resulted in solar cell efficiencies of 17% for a 50 μm thick free-standing c-Si base material epitaxially grown on a porous Si substrate before being detached [10], or over 15% by growing a 20 μm thick epitaxial Si base on a seed substrate and also using diffusion processes [11], and about 7% when growing a 2 μm thick epitaxial layer by HWCVD at 700 °C [12].

Alike, epitaxial growth of Ge films was obtained by PECVD on (100) NaCl substrates kept at 450 °C during growth [13] or by Molecular Beam Epitaxy on (100) GaAs [14]. This is of great interest since simulations demonstrated the feasibility of efficient structures combining Si and Ge materials [15].

Previous results obtained in our laboratory have shown that it is possible to obtain epitaxial layers of Ge on (100) gallium arsenide (GaAs) substrates [16] and of Si on (100) silicon substrates [17–19], both by RF PECVD and at substrate temperature as low as 175 °C. By doing so, we have been able to obtain solar cells with an efficiency as high as 7% for an intrinsic absorber layer of 2.4 μm grown at 165 °C [19].

2 Experiments

All the samples of this study were deposited in a multiplasma monochamber reactor PECVD reactor operated at

[a] e-mail: martin.labrune@mines-paris.org

Table 1. Process conditions used to grow the epitaxial films of Si and Ge at 175 °C.

Film on (111) c-Si	Film on (100) c-Si	Pressure (Torr)	SiH$_4$ (sccm)	GeH$_4$ in H$_2$ (sccm)	H$_2$ (sccm)	RF power (mW/cm^2)	Deposition rate (Å s^{-1})
μc-Ge:H	Epitaxial	1.4	0	5	200	31	0.15
pm-Si:H	Epitaxial	1.2	12	0	200	25	0.5

a frequency of 13.56 MHz [20]. We used (100) and (111)-oriented Si substrates and (100)-oriented GaAs substrates for the TEM experiments. All the crystalline substrates were submitted to a 30 s dip in hydrofluoric acid, to remove the native oxide, prior to being loaded in the reactor, which was pumped down to a base pressure below 7×10^{-7} Torr. We emphasize that this procedure allows us to achieve heterojonction solar cells with efficiencies up to 17% [18] and even homojonction solar cells when the deposited layer is epitaxial [17]. This is a good indication of the excellent surface passivation and of the high quality of the epitaxial doped layers respectively, when we use this cleaning procedure and this PECVD reactor. The depositions were performed using silane (SiH$_4$) and hydrogen (H$_2$) gas mixtures for the silicon films and H$_2$ and germane (GeH$_4$, 2%-diluted in H$_2$) gas mixtures for the germanium films. All the samples were characterized via spectroscopic ellipsometry using a phase modulated ellipsometer (UVISEL from HORIBA Jobin-Yvon). The Raman spectrometer used in this work is a DILOR Jobin Yvon XY with a He-Ne laser excitation at 632.8 nm. The TEM microscope is a JEOL 2200 FS, being able to operate in the STEM (scanning TEM) or TEM mode. High angle X-ray diffraction and grazing incidence X-ray diffraction measurements have been performed using a Rigaku Smartlab high resolution diffractometer equipped with a 9 kW rotating anode and a 7-axes goniometer.

3 Results

Previous studies in our laboratory have shown that plasma conditions leading to hydrogenated microcrystalline germanium films on glass can eventually lead to an epitaxial growth when applied on a (100)-oriented GaAs substrate [16]. We obtained the same results for silicon films, for which conditions known to lead to hydrogenated polymorphous silicon (pm-Si:H) on glass lead to epitaxial growth when applied to (100)-oriented Si substrates [17,18,21]. Other research groups also found that a rather broad range of experimental parameters would eventually lead to unwanted epitaxial films on (100) Si substrates [22,23]. In the case of heterojunction solar cells, this epitaxy is unwanted since this crystalline layer does not provide any surface passivation [22]. However, one can take benefit of this capability of PECVD to grow thick epitaxial layers and use them as the active material in solar cells [19] or as the emitter of c-Si solar cells [17].

Table 1 summarizes the plasma conditions used to achieve epitaxy on (100)-oriented substrates for the Si and Ge films respectively. Figure 1 shows the SE spectra of a multilayer stack co-deposited on (100) GaAs, (100) Si and

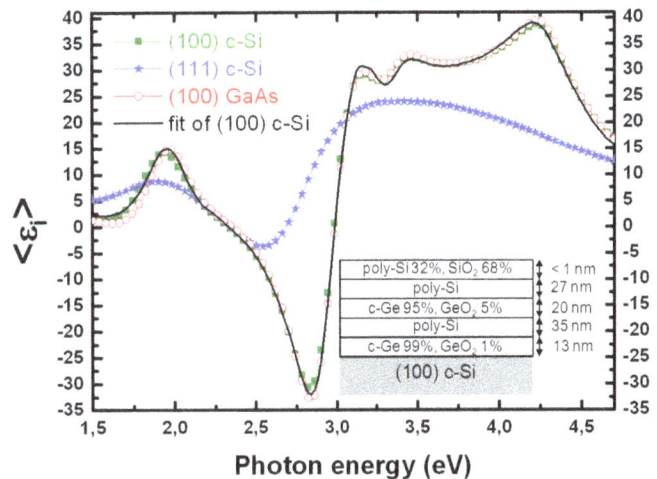

Fig. 1. Imaginary part of the pseudo-dielectric function of the multi layer stack (905071) co-deposited onto various substrates as deduced from SE measurements, the black line corresponds to the fit obtained by modeling the stack deposited on the (100) Si substrate using the optical model described in the inset.

(111) Si. The high photon energy part (3–5 eV) of the SE spectrum is more sensitive to the top and also bulk part of the films. The last deposited layer being silicon, in Figure 1, we can see the characteristic spectrum of c-Si, which has two peaks around 3.4 and 4.2 eV on both GaAs and Si (100)-oriented substrates, whereas on (111) c-Si substrate, the silicon films are amorphous (a similar spectrum was measured for the film co-deposited on glass, not shown for clarity). The lower photon energy part of the spectrum is sensitive to the bulk and thickness of the film, the interference fringes providing information on the thickness of the whole stack. The Bruggeman Effective Medium Approximation (BEMA) model [24] used for the film grown on (100) Si is shown in the inset of Figure 1 and the spectrum obtained from the model is also plotted with a dark line. Interestingly, the measurement performed on the film grown on the (100) GaAs substrate, can be fitted using the same model as the one used to fit the measurement performed on the (100) Si. This means that we obtained the same stack on both (100) substrates. Even though a fit based on crystalline silicon and germanium materials provided a reasonable match with the experimental data (with a figure of merit χ^2 of 2.3), we could improve the fit ($\chi^2 = 1.5$) by using the dielectric function of large grains polysilicon material reference file obtained by Jellison et al. [25] for the silicon layers and a mixture of crystalline germanium and a small fraction of germanium oxide (1–5%), as obtained by Aspnes and Studna [26] for the germanium layers. There are at least

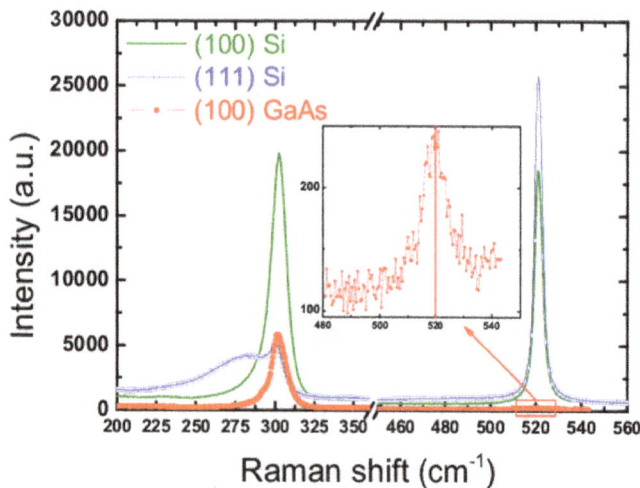

Fig. 2. Raman scattering intensity as a function of the Raman shift for the multi layer samples (905071) co-deposited on various substrates ((100) GaAs, (100) Si and (111) Si).

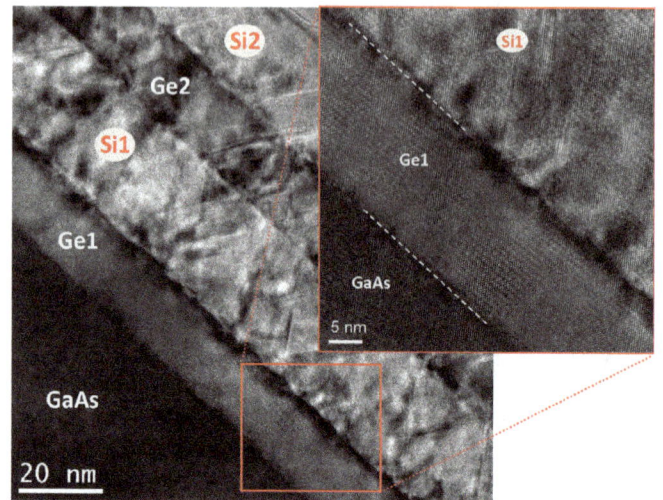

Fig. 3. HRTEM image of the multilayer stack on GaAs. The inset zooms on the two first Ge and Si layers.

two reasons that may explain the better fit when considering polycrystalline dielectric function and introducing GeO$_2$ in the Ge layers. The first one is that we cannot expect to have films with no roughness so that among our four interfaces, none is perfectly flat (as shown by the presence of a surface roughness of about 1 nm in the model of Fig. 1). Introducing a rough interface between each layer could even further reduce χ^2 but this is at the cost of drastically increasing the number of parameters of the model. The second one is that these films are produced at 175 °C in a standard RF PECVD reactor without load-lock nor special precaution concerning gas purity (no gas purifiers) so that we can expect our films to contain carbon and oxygen impurities as well as a significant amount of hydrogen. Those may slightly alter the dielectric functions of the materials as compared to their calibrated crystalline counterparts.

We also investigated the structure of the films by Raman spectroscopy. The Raman spectra of the stack deposited on the three substrates are shown in Figure 2, the substrates being (100) GaAs, (100) Si and (111) Si. The Raman spectra of the multilayer films deposited on GaAs and (100) Si show a sharp peak around 300 cm^{-1}, which is consistent with fully crystallized Ge layers [27, 28]. It has been shown that a peak at 300 cm^{-1} could also originate from Si substrates [29], but in our case, a comparison between a Si substrate and a Si substrate capped with a thin epitaxial Ge layer showed that no signal from the Si substrate alone at 300 cm^{-1} was distinguishable whereas a very sharp peak would appear in the presence of this thin Ge layer. On the other hand, the film deposited on (111) c-Si substrate displays a shoulder towards lower wavenumbers, indicating that the film is partially crystallized and contains an amorphous phase, since the hydrogenated amorphous Ge has a TO mode at 278 cm^{-1} [28]. The peak around 520 cm^{-1} can be assigned to crystalline silicon [27], but does not give much information on Si substrates where this peak is due to the substrate and

masks the small contribution of the amorphous film (at 480 cm^{-1}) to the Raman spectrum. However, on the GaAs substrate, we can detect the signal from the c-Si film despite of its small thickness (see inset in Fig. 2). Moreover, one can see that the film is fully crystallized as there is no shoulder at 480 cm^{-1}, as it would be the case if there were an amorphous silicon phase.

We used the sample deposited on (100) GaAs for TEM measurements. In Figure 3, we show an example of the High Resolution TEM (HRTEM) images of the stack we obtained. The red square in Figure 3 indicates the area on which we zoomed in the inset to focus on the two first layers and the two first interfaces. Based on such images, we can get a reasonable estimate of the thickness of each layer. The thicknesses are approximately 20 nm for the Ge layers and 30 nm for the Si layers. These results are in good agreement with the ones obtained by SE even though some discrepancy exists for the first Ge layer, the closest layer to the substrate. The fact is that the accuracy of the SE fit is not very sensitive to the value of the thickness of the first Ge layer so that the value we obtained had a significant error bar ($\approx$1.5 nm). On the inset of Figure 3, it appears that they are very few defects in the first Ge layer and that some start to appear at the interface between the first silicon layer and the first germanium layer. In all these layers we could observe some dislocations but it must be noted that that they do not propagate systematically from one layer to the other and that these dislocations do not prevent our layers to keep a monocrystalline structure, which supports the data obtained from spectroscopic ellipsometry and Raman spectroscopy.

Moreover, the microscope could also be operated in the Scanning TEM mode and by doing so we could obtain High Angle Annular Dark Field (HAADF) images of our sample as the one we show in Figure 4. For such images, the contrast in the gray scale comes from a difference in the Z value of the elements since we only collect the electrons that have scattered to high deviation angles and therefore the ones that are the most sensitive to the

Table 2. Spacings between adjacent lattices and calculated lattice parameters obtained by XRD for Si and Ge.

Diffraction planes	Distance between Si planes (Å)	Si lattice parameter (Å)	Distance between Ge planes (Å)	Ge lattice parameter (Å)
220	1.9183 ± 0.001	$a_\parallel = 5.4257 \pm 0.003$	1.97974 ± 0.001	$a_\parallel = 5.5996 \pm 0.003$
400	1.358 ± 0.001	$a_\parallel = 5.432 \pm 0.004$	–	$a_\perp < 5.65325$
004	1.3609 ± 0.001	$a_\perp = 5.444 \pm 0.004$	–	$a_\parallel > 5.65325$

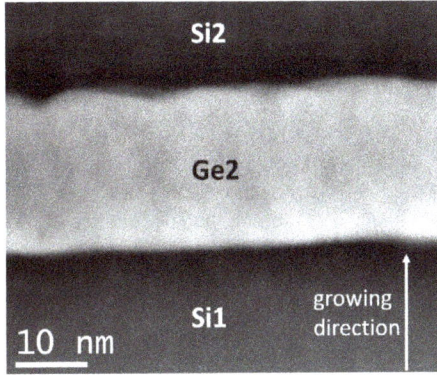

Fig. 4. HAADF image obtained by STEM of the multilayer sample on GaAs.

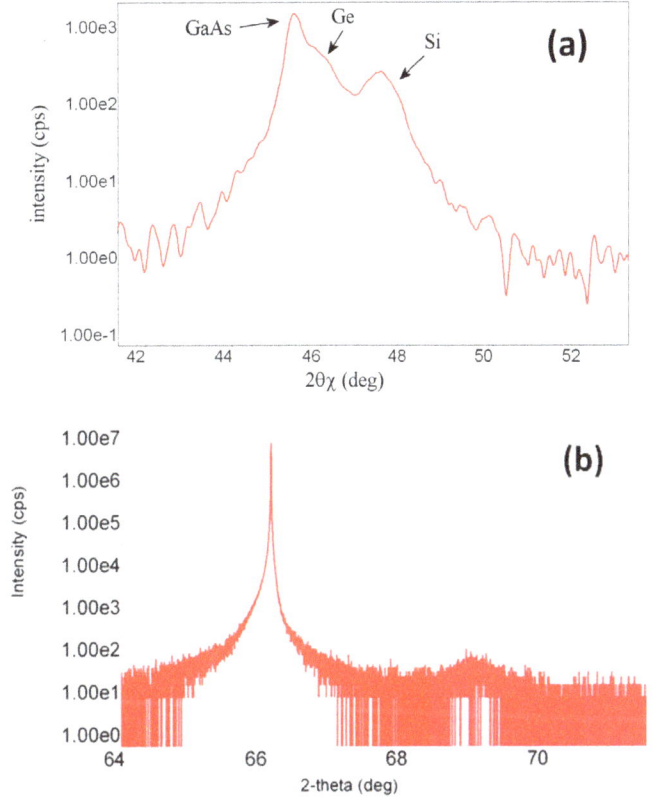

Fig. 5. (a) XRD measurement of the 220 planes perpendicular to the surface of the film. (b) High resolution XRD measurement of the {004} planes parallel to the surface of the film.

Z value. This means that we somehow get a "chemical picture" of the stack. Even if the atoms remain in a very crystallized arrangement the elements are not always the same. At the interface between Si and Ge in Figure 4 we can observe that there is a pronounced "chemical roughness". This means that we do not have perfectly flat interfaces and that we can still have Ge elements in the very first nanometers of the following Si layer. This complements the observation of Figure 3 and can be explained by the deposition technique itself rather than any diffusion process since we used low temperature deposition but we used the same chamber for Ge and Si films growth.

We coupled high angle $2\theta/\omega$ X-ray diffraction and grazing incidence X-ray diffraction (GIXRD) to study our sample in order to get information about the crystallinity of the films and also to get more quantitative information regarding the lattice parameters. The Rigaku Smartlab diffractometer, based in the LPN laboratory, allowed us to study the diffraction from the crystallographic planes parallel to the surface (lattice parameter: $a_\perp$) by scanning the 2θ and ω angles in the direction normal to the surface. 2θ is the angle between the diffracted beam and the surface and ω is the angle between the incident beam and the surface of the sample. We also studied the diffraction of the crystallographic planes perpendicular to the surface (lattice parameter: $a_\parallel$) by scanning $2\theta\chi$ and φ angles in grazing incidence (φ corresponds to the rotation of the sample on itself while $2\theta\chi$ is the sample in-plane angle between the crystallographic planes and the detection). This $2\theta\chi/\varphi$ in-plane configuration corresponds to a $2\theta/\omega$ configuration for crystallographic planes normal to the surface. Figure 5a shows the GIXRD of the {220} planes of the GaAs substrate and the Ge and Si layers. Knowing the interplanar distance of the 220 GaAs substrate planes, we can deduce the 220 interplanar distance

for Ge and Si by measuring the mismatch between the Ge, Si and GaAs peaks. These values are summarized in the first row of Table 2.

We also performed GIXRD of the {400} planes normal to the surface of GaAs, Ge and Si (not shown here). This allows us to calculate the distance between planes in the 400 direction for Si (second row of Tab. 2). Unfortunately we were not able to deconvolve the peaks due to GaAs and Ge but based on the fact the graph showed a shoulder towards higher angles around the 400 peak of GaAs, we can assume that we have a lattice parameter lower than the one of GaAs.

Finally we operated the XRD set-up in the high angle $2\theta/\omega$ configuration in order to obtain the high resolution diffraction from the planes parallel to the surface. We obtained diffraction from the {004} planes that we show in Figure 5b. Even though we could not observe a distinct peak for Ge we could still notice the dissymmetry of the GaAs peak towards smaller angles, indicating a Ge $a_\perp$ lattice parameter higher than the GaAs $a_\perp$ lattice parameter.

Nevertheless, we could again calculate the parameters for Si and we summarized them in the third row of Table 2.

Based on these results, averaging the value of $a_{\parallel}$ for Si, and using the formula in equation (1) we could calculate the bulk lattice parameter of Si. We used for Si the following parameters: $C_{11} = 16.6 \times 10^{11}$ dyn cm^{-2} and $C_{22} = 6.4 \times 10^{11}$ dyn cm^{-2}.

$$\frac{a_{\perp} - a_{bulk}}{a_{bulk}} = -2\frac{C_{12}}{C_{11}}\frac{a_{\parallel} - a_{bulk}}{a_{bulk}}. \qquad (1)$$

We obtained $a_{\text{bulk}} = 5.4374 \pm 0.004$ Å. This parameter is slightly higher that the theoretical value of 5.431 Å. Furthermore, we measure $a_{\parallel} < a_{\perp}$ indicating a slight compressive constraint for Si.

For Ge, $a_{\perp} > 5.65325$ Å (GaAs lattice parameter) and $a_{\parallel} = 5.5996 \pm 0.003$ Å, that corresponds to an important compressive strain. It is so far rather speculative to explain these results but we expect the process conditions to play a major role on the constraints of the material. Indeed, in the case of Ge, the layers were deposited from GeH$_4$ which was 2%-diluted in H$_2$ so that the plasma was mostly a hydrogen plasma resulting in much more aggressive deposition conditions compared to a-Si:H deposition case.

4 Conclusions

In this paper we have demonstrated the epitaxial nature of Si and Ge films grown on (100) crystalline Si as well as the growth of Si on epitaxially grown Ge layers in a PECVD reactor at 175 °C. Raman spectroscopy, Transmission Electron Microscopy, X-ray diffraction and Spectroscopic Ellipsometry consistently showed that the films were made up of monocrystalline layers. Further studies are needed to get a better understanding of the growth mechanisms and also to study the influence of the lattice parameter of the substrate on the constraints in the resulting films.

References

1. T. Mishima, M. Taguchi, H. Sakata, E. Maruyama, Sol. Energy Mater. Sol. Cells **95**, 18 (2011)
2. T. Yamazaki, Y. Uraoka, Takashi Fuyuki, Thin Solid Films **487**, 26 (2005)
3. W. Fuhs, S. Gall, B. Rau, M. Schmidt, J. Schneider, Sol. Energy **77**, 961 (2004)
4. P.A. Basore, *Proceedings of the 4th World Conference on Photovoltaic Energy Conversion, Waikoloa, Hawaii, 2006*
5. R.B. Bergmann, Appl. Phys. A **69**, 187 (1999)
6. R. Shimokawa, M. Yamanaka, I. Sakata, Jpn J. Appl. Phys. **46**, 7612 (2007)
7. E. Schmich, N. Schillinger, S. Reber, Surf. Coat. Technol. **201**, 9325 (2007)
8. M. Kambara, H. Yagi, M. Sawayanagi, T. Yoshida, J. Appl. Phys. **99**, 074901 (2006)
9. H.M. Branz, C.W. Teplin, D.L. Young, M.R. Page, E. Iwaniczko, L. Roybal, R. Bauer, A.H. Mahan, Y. Xu, P. Stradins, T. Wang, Q. Wang, Thin Solid Films **516**, 743 (2008)
10. J.H. Petermann, D. Zielke, J. Schmidt, F. Haase, E.G. Rojas, R. Brendel, Prog. Photovolt.: Res. Appl. **20**, 1 (2012)
11. I. Kuzma-Filipek, K.V. Nieuwenhuysen, J.V. Hoeymissen, M.R. Payo, E.V. Kerschaver, J. Poortmans, R. Mertens, G. Beaucarne, E. Schmich, S. Lindekugel, S. Reber, Prog. Photovolt.: Res. Appl. **18**, 137 (2010)
12. K. Alberi, I.T. Martin, M. Shub, C.W. Teplin, M.J. Romero, R.C. Reedy, E. Iwaniczko, A. Duda, P. Stradins, H.M. Branz, D.L. Young, Appl. Phys. Lett. **96**, 073502 (2010)
13. R.A. Outlaw, P. Hopson Jr., J. Appl. Phys. **55**, 6 (1984)
14. B. Jenichen, V.M. Kaganer, R. Shayduk, W. Braun, A. Trampert, Phys. Stat. Sol. A **206**, 1740 (2009)
15. C.-H. Lin, Thin Solid Films **518**, S255 (2010)
16. E.V. Johnson, G. Patriarche, P. Roca i Cabarrocas, Appl. Phys. Lett. **92**, 103108 (2006)
17. M. Labrune, M. Moreno, P. Roca i Cabarrocas, Thin Solid Films **518**, 2528 (2010)
18. J. Damon-Lacoste, P. Roca i Cabarrocas, J. Appl. Phys. **105**, 063712 (2009)
19. R. Cariou, M. Labrune, P. Roca i Cabarrocas, Sol. Energy Mater. Sol. Cells **95**, 2260 (2011)
20. P. Roca i Cabarrocas, J.B. Chévrier, J. Huc, A. Lloret, J.Y. Parey, J.P.M. Schmitt, J. Vac. Sci. Technol. A **9**, 2331 (1991)
21. P. Roca i Cabarrocas, R. Cariou, M. Labrune, J. Non-Cryst. Solids **358**, 2000 (2012)
22. J.J.H. Gielis, P.J. van den Oever, B. Hoex, M.C.M. van de Sanden, W.M.M. Kessels, Phys. Rev. B **77**, 205329 (2008)
23. C.W. Teplin, D.H. Levi, E. Iwaniczko, K.M. Jones, J.D. Perkins, H.M. Branz, J. Appl. Phys. Lett. **97**, 103536 (2005)
24. D.A.G. Bruggeman, Ann. Phys. **416**, 636 (1935)
25. G.E. Jellison Jr., F. Chisholm, S.M. Gorbatkin, Appl. Phys. Lett. **62**, 25 (1993)
26. D.E. Aspnes, A.A. Studna, Phys. Rev. B **27**, 985 (1983)
27. J.H. Parker Jr., D.W. Feldman, M. Ashkin, Phys. Rev. B **155**, 3 (1967)
28. D. Bermejo, M. Cardona, J. Non-Cryst. Solids **32**, 405 (1979)
29. A.V. Kolobov, J. Appl. Phys. **87**, 2926 (1999)

14

X-Ray diffraction and Raman spectroscopy for a better understanding of ZnO:Al growth process

C. Charpentier[1,2,a], P. Prod'homme[1], I. Maurin[3], M. Chaigneau[2], and P. Roca i Cabarrocas[2]

[1] TOTAL SA – Gas & Power R&D Division Tour Lafayette, 2 place des Vosges La Défense 6, 92400 Courbevoie, France
[2] LPICM-CNRS – Laboratoire de Physique des Interfaces et Couches Minces, École Polytechnique, 91128 Palaiseau, France
[3] LPMC-CNRS – Laboratoire de Physique de la Matière Condensée, École Polytechnique, 91128 Palaiseau, France

Abstract ZnO:Al thin films were prepared on glass substrates by radio frequency RF magnetron sputtering from a ceramic ZnO target mixed with Al_2O_3 (1 wt%) in pure argon atmosphere with a power of 250 W. Two series of samples were deposited, the first one as a function of the substrate temperature (between 20 and 325 °C) at 0.12 Pa, the second one as a function of the working pressure (between 0.01 and 2.2 Pa) at room temperature. The influence of these deposition parameters was studied by a detailed microstructural analysis using X-Ray diffraction and Raman spectroscopy. Their electrical properties were characterized by Hall effect measurements. As the substrate temperature is increased, crystallite size increases while the strain and the electrical resistivity decrease. An improvement of the crystallinity is deduced from Raman spectra and X-Ray diffraction patterns. From the pressure series, an electrical optimum is observed at 0.12 Pa. Above 0.12 Pa, there is a degradation of the crystallinity, attributed to the low mobility of the adatoms and below 0.12 Pa, high-energy sputtered particles damage the growing films. The trends given by Thornton structure zone model are also consistent with X-Ray diffraction results.

1 Introduction

In μc-Si:H solar cells, doped and intrinsic layers are packed between a transparent conductive oxide (TCO) front contact and a highly reflective back contact. In the PIN configuration light enters through the glass and TCO layer and is trapped into the intrinsic layer by the light scattering properties of the TCO layer and by reflection on back contacts. TCOs are a key point for improvement of μc-Si:H solar cells [1–4]. The TCO front contact must meet a number of requirements. This layer must be highly transparent in the active wavelength range of the absorber layer, between 400 and 1100 nm for μc-Si:H, and also highly conductive to avoid ohmic losses. In the PIN structure, the front contact additionally has to possess a rough surface to provide efficient light scattering and must feature favourable physical and chemical properties for an efficient growth of μc-Si:H. For example, TCO has to be inert to hydrogen-rich plasmas and a high work function is required in order to provide an ohmic TCO/p-contact for cell performance.

A promising TCO material is aluminium doped zinc oxide because of its low cost, non-toxicity, high stability against hydrogen plasma and high surface texturability [5,6]. Highly conductive and transparent in the visible range Al-doped ZnO thin films have been deposited by RF magnetron sputtering. This deposition technique enables a high deposition rate, a low process temperature and a dense layer formation [7–9]. An optimized light trapping is achieved by a wet-chemical etching step after vacuum deposition.

Many studies have addressed the correlation between deposition parameters and quality of the films [8–16]. A deeper understanding of the relationship between different sputter process conditions, growth process and microstructure is still required to further optimize electrical and optical properties of the films. In this work, we study the influence of the deposition temperature and working pressure on the microstructural and electrical properties of ZnO:Al thin films prepared by RF magnetron sputtering on glass substrates. A wide variety of characterization techniques is used, including atomic force microscopy (AFM) and Hall effect measurements. Moreover, a detailed analysis of X-Ray diffraction data using Williamson-Hall plots [17–19] complemented by Raman spectroscopy [19–23] was performed to obtain additional information on microstructure, crystallinity and defects in the films.

2 Experiments

2.1 ZnO:Al film deposition

ZnO:Al films were deposited on glass substrates by RF magnetron sputtering using a ZnO/1 wt% Al_2O_3 as

---

[a] e-mail: `coralie.charpentier@polytechnique.edu`

a ceramic target of 152 mm in diameter, and a RF power of 250 W. The base pressure was 4×10^{-5} Pa and the working pressure was controlled by the flow rate of argon regulated by a mass flow controller.

For the first set of samples grown at different substrate temperatures, depositions were carried out between room temperature around 20 °C (± 5 °C for temperature without set point) and 325 °C (± 1 °C for temperature regulated by set point value) at a working pressure of 0.12 Pa in pure argon atmosphere. For the second set of samples prepared at room temperature, depositions were carried out in the working pressure range of 0.01 Pa to 2.2 Pa. A non-intentional increase of the substrate temperature during the sputtering can be observed up to a maximum of 10 °C independently of the working pressure. The X-Ray diffraction (XRD) patterns were recorded with a PANalytical X'Pert powder diffractometer equipped with a Cu Kα radiation ($\lambda = 1.5418$ Å). The Raman experimental setup consists of a high-resolution (0.1 cm^{-1}) Raman spectrometer (Labram HR800 from HORIBA Jobin Yvon) in a confocal microscope backscattering configuration. The accommodate objective used in this study was a 100× (NA = 0.9) objective from Olympus. A tunable Ar laser (514 nm wavelength) was employed in this confocal configuration. For all the Raman Stokes analysis, the collection time was 10 s with 6 consecutive spectrum accumulations. Surface morphology was characterized by AFM in tapping mode (Agilent Technologies 5500). A HMS5000 from Microworld was used to obtain the Hall mobility, carrier concentration and resistivity of the films by Hall effect.

All studies were carried out on films of approximately the same thickness (1000 $\pm$ 100 nm) estimated by ellipsometric analysis. This thickness corresponds to a good compromise for the use of this layer as front contact in μc-Si:H solar cells, between the highest transmission in the visible range and the lowest resistivity.

2.2 Microstructural analyses

2.2.1 X-Ray diffraction

All the films exhibit a strong c-axis preferred orientation revealed by the high intensity of the (002) reflection of ZnO:Al Würtzite structure (Fig. 1) [12–17, 19]. These strongly oriented grains will be referred to *crystallites* in the following, with a given vertical coherence length called from now *crystallite size*. Crystallite size and strain can be obtained from the distribution of the scattered intensity in reciprocal space. The crystallite size is correlated with a reciprocal lattice point broadening in the q_Z direction perpendicular to the substrate plane. In the radial-scan direction (θ–2θ scans) of the symmetric reflections of (002), (004), and (006), a reduced correlation length normal to the substrate plane and a heterogeneous strain along the c-axis are responsible for the broadening of the Bragg reflections. Broadening in reciprocal space due to finite size effects is independent of the scattering vector q_Z, but the broadening associated with strain is proportional to the scattering order. A graphical separation of these

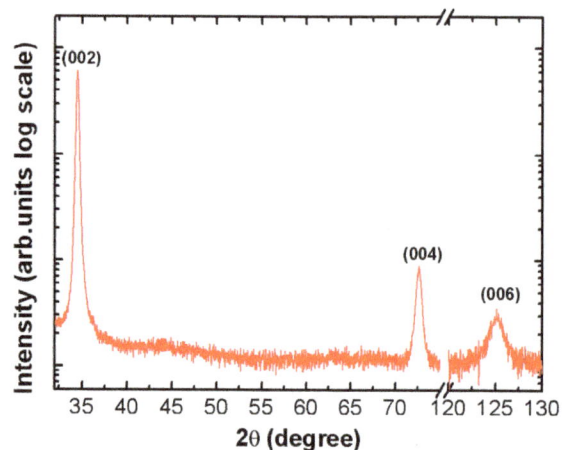

Fig. 1. (Color online) XRD profile of ZnO:Al film deposited on glass substrate at room temperature, RF power of 250 W and Ar pressure of 0.12 Pa.

two effects owing to their different dependence on q_Z is possible by recording high-order reflections and using Williamson-Hall plots [17,18]. The (002), (004) and (006) peaks were fitted to a linear combination of Gauss and Cauchy functions (Pseudo-Voigt function) to determine the Full Width at Half Maximum FWHM called β_{exp} in the following.

FWHM values have to be corrected from instrumental broadening following equation (1) for a Gauss profile, or equation (2) for a Cauchy profile:

$$\beta^2 = \beta_{exp}^2 - \beta_{instr}^2 \tag{1}$$

$$\beta = \beta_{exp} - \beta_{instr}. \tag{2}$$

For Williamson-Hall plots, $(\beta \cos\theta/\lambda)^2$ is plotted against $(2\sin\theta/\lambda)^2$ for each reflection for Gauss profiles or alternatively, $\beta \cos\theta/\lambda$ is plotted against $2\sin\theta/\lambda$ for Cauchy profiles. λ and θ are the wavelength and the incident angle of the X-rays, respectively. The crystallite size D and strain ε can be respectively estimated from the intercept and slope of the linear Williamson-Hall plots following equation (3) for a Gauss profile or equation (4) for a Cauchy one.

$$(\beta \cos\theta/\lambda)^2 = (1/D)^2 + (\varepsilon 2 \sin\theta/\lambda)^2 \tag{3}$$

$$\beta \cos\theta/\lambda = 1/D + \varepsilon 2 \sin\theta/\lambda. \tag{4}$$

In the present case, the Cauchy and Gauss contributions in the Pseudo-Voigt function are obtained by the fit of the FWHM for the (002), (004) and (006) peaks and are equal, so none of these equations are strictly valid for Williamson-Hall. Nonetheless, we observed that the strain extracted either from equation (3) or from equation (4) are similar ($\pm 3\%$). The values presented in the following correspond to the average. In contrast, a Cauchy approximation systematically underestimates the width of the reflections after correction of the instrumental resolution and that of the transverse coherence length. In the following, values obtained from Gauss analyses were used to qualitatively estimate the crystallite size.

2.2.2 Raman spectroscopy

Raman spectroscopy gives information on vibrational properties of ZnO. In hexagonal structures with C_{6v}^4 symmetry like ZnO, six sets of phonon normal modes at the Γ point (centre of the Brillouin zone) are optically active modes [19, 20, 23]. The phonons for Würtzite ZnO belong to the following irreducible representation:

$$\Gamma = A_1 + E_1 + 2B_1 + 2E_2. \quad (5)$$

Where both A_1 and E_1 modes are polar, and split into transverse optical modes (A_1-TO and E_1-TO), resulting from beating in the basal plane, and longitudinal optical modes (A_1-LO and E_1-LO), resulting from beating along the c-axis, with different frequencies due to the macroscopic electric fields associated with LO phonons. For lattice vibrations with A_1 and E_1 symmetry, the atoms move parallel and perpendicular to the c-axis, respectively. The A_1 and E_1 branches are both Raman and infrared active, the two nonpolar E_2 branches are Raman active only, and the B_1 branches are inactive [23].

In addition to the selection rules mentioned above for Raman-active phonon modes, the modes to be detected by Raman scattering spectroscopy also depend on the scattering geometry. The selection rules and the backscattering geometry of our Raman system imply that the two E_2 modes and the A_1-LO modes are expected while all other modes are forbidden. The E_2 modes at low and high frequency and the A_1-LO mode are respectively at 101 cm^{-1}, 437 cm^{-1} and 574 cm^{-1}. The E_2 (low-frequency) mode is associated with the vibration of the heavy Zn sublattice, while the E_2 (high-frequency) mode involves only oxygen atoms [23].

3 Results and discussion

3.1 Influence of the substrate temperature

3.1.1 Microstructural properties

The Raman spectra of this series of ZnO:Al thin films are dominated by the A_1-LO vibrations around 578 cm^{-1} (Fig. 2). This band, which is highly asymmetric, arises from two contributions, a low wavenumber contribution and a high wavenumber one.

The Raman spectra provide qualitative information about the crystallinity of the films. Indeed, the relative area of the disordered (low wavenumber contribution) and crystalline (high wavenumber contribution) peaks enable to compare the degree of crystallinity corresponding to different spectra. For the results listed in Table 1, peaks were fitted to pure Gauss profiles, with a linear baseline from 525 to 625 cm^{-1} which reproduces quite well the glass substrate contribution. The two peaks used for the deconvolution are attributed to different phases present in the sample: the narrow peak centred around 578 cm^{-1} is attributed to ZnO:Al crystallites while the broad peak around 560 cm^{-1} is characteristic of a disordered phase

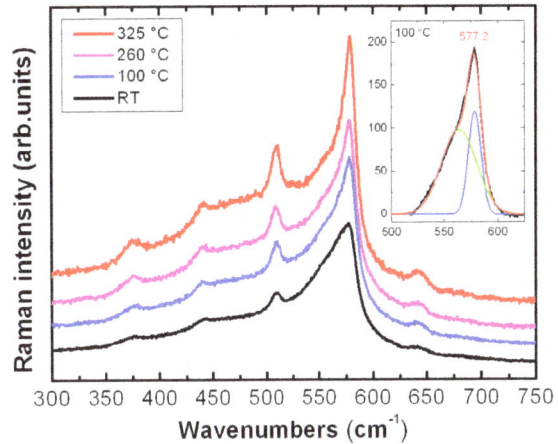

Fig. 2. (Color online) Raman spectra of ZnO:Al films deposited at various temperatures. The asymmetric A_1-LO vibration is fitted to two Gaussian contributions (example given in the inset).

Table 1. Relative amount of the crystalline and disordered contributions derived from the deconvolution of the A_1-LO Raman mode as a function of the substrate temperature.

T_s (°C)	RT	100	260	325
Peak position (cm^{-1})	576.1	577.2	577.7	578.2
% Area Crystalline/ Disordered	24	30	30	36

(Fig. 2). For the present time, values of the Table 1 have to be considered as semi-quantitative as the fitting procedure of the A_1-LO peak has not been optimized.

With increasing temperature, the high wavenumber contribution increases compared to the low wavenumber contribution (Tab. 1). This trend suggests an improvement of the crystallinity and a reduction in defects of the ZnO:Al films.

Some additional weak peaks are observed at 380 cm^{-1} and 501 cm^{-1}. The first one can be assigned to the A_1-TO mode. This mode is not expected to appear in Raman backscattering geometry for single crystalline ZnO and could arise from the polycrystalline character of the films. The second peak at 501 cm^{-1} is possibly related to the doping of the zinc oxide layer and to the charge carrier concentration [20, 21].

The improvement of crystallinity observed in Raman spectra with increasing temperature is also observed by X-Ray diffraction, from the increase of the intensity of the (002) peak. It is correlated with an increase of the crystallite size deduced from X-Ray diffraction analyses (Fig. 3), which is attributed to an increase in surface diffusion of the adsorbed species. The crystallite size varies from 70 nm for a sample deposited at room temperature to 100 nm at a substrate temperature of 325 °C. In parallel, the strain decreases with increasing substrate temperature, because of thermal annealing. The improvement of the crystallinity

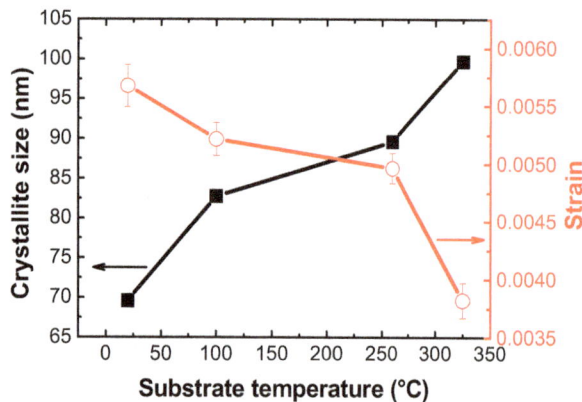

Fig. 3. (Color online) Crystallite size (■) and strain (○) estimated from Williamson-Hall plots for samples deposited at various substrate temperatures.

leads to a reduction in defect density inside the films which involves a network relaxation [15, 17, 19]. The strain is also an indication of the crystalline quality of the layer. A decrease in strain indicates an improvement of the crystallinity of the film.

3.1.2 Morphological studies

The growth of thin films can be described by a process involving various steps. The first one is nucleation, when the dispersed individual crystalline nuclei are randomly oriented on the amorphous substrate. The second step consists of the island growth. Coalescence of islands leads to the last step of the process, the column growth [24,25]. During the deposition of ZnO:Al on glass substrate, the initial nuclei may contain all possible orientations, and crystals growing from these nuclei are randomly oriented. The complete coalescence of the islands induces the formation of polycrystalline areas and channels which leads to the development of polycrystalline films by columnar growth of grains perpendicularly to the substrate plane. The columnar texture of the film is ascribed to the competitive growth of crystals, due to the different growth rates in various crystallographic directions. In ZnO case, the growth competition favours the (002) orientation over the others (Fig. 1).

Film microstructure is consistent with the trend given by Thornton's structure zone model where T and Tm are respectively the substrate temperature and ZnO melting point (Fig. 4) [26,27]. With the increase of the substrate temperature, a transition from tapered crystallites separated by voids zone (Zone I) to a densely packed fibrous grains zone (Zone T) is observed. The increase of the crystallite size along the growth direction with substrate temperature (Fig. 3) is accompanied with an increase of the lateral dimension of the columns observed on the top of the film in AFM images (Fig. 5). The two structure zones have not the same X-Ray diffraction profiles. All the films have a c-axis preferred orientation but additional (101), (102), (103) and (112) peaks appear with increasing substrate temperature (Fig. 6) [14].

Table 2. Relative amount of the crystalline and disordered contributions derived from the deconvolution of the A_1-LO Raman mode as a function of the substrate temperature.

Working pressure (Pa)	0.01	0.12	0.32	1.3	2.2
Peak position (cm^{-1})	576.7	576.1	568.7	553.8	552.2
% Area Crystalline/ Disordered	28	24	12	0	0

At low substrate temperature, in zone I, the film consists of tapered crystallites separated by voids (Fig. 4), the surface of the glass substrate is covered by a high density of small nuclei. Crystalline grains grow out of these primary nuclei until the top of the film and the (002) oriented grains rapidly dominate over the other orientations. The nucleation density determines the lateral size of the fibres which are growing uninterruptedly side by side. No minor orientations are observed in X-Ray diffraction diagrams (Fig. 6).

At high substrate temperature, in zone T, the film consists of densely packed fibrous grains (Fig. 4), atoms on the surface of the glass substrate are highly energetic. The surface is covered of a low density of large islands, leading to a competitive growth of differently oriented neighbouring islands and thus V-shaped columns. The lateral growth of the grains is faster than the vertical growth. Minor orientations like (101), (102), (103) and (112) are observed, coming from the region between glass and ZnO:Al, at the interface (Fig. 6). This hypothesis is supported by the X-Ray diffraction profiles of ZnO:Al with different thicknesses. For the same deposition conditions, the area of minor orientations increases comparing to the area of (002) orientation when thickness decreases.

3.1.3 Electrical properties

With the increase of the substrate temperature, an increase of the carrier concentration and of the Hall mobility, resulting in a decrease of the resistivity, are observed (Fig. 7). This improvement of the electrical properties is related to an increase of the layer density and an increase in crystallite size along the growth direction for films deposited at high substrate temperature which results in a decrease in grain boundaries that act as scattering centres for charge carriers.

3.2 Influence of the working pressure

3.2.1 Microstructural properties

The working pressure has a strong influence on film microstructure and on the Raman spectra (Fig. 8). The position of the A_1-LO band gradually varies from 552 cm^{-1} at 2.2 Pa to 577 cm^{-1} at 0.01 Pa (Tab. 2). The apparent shift of LO modes towards lower wavenumbers is due to the increase of the low wavenumber contribution compared to

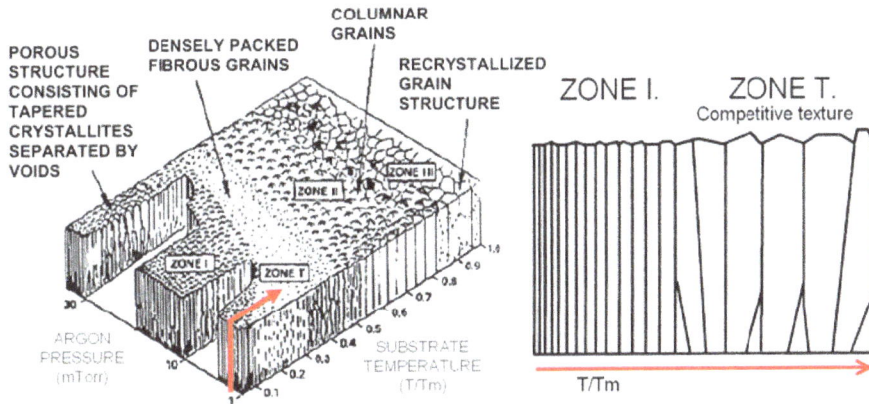

Fig. 4. (Color online) Thornton structure zone model correlating the argon pressure and the substrate temperature to the morphological properties of the films. When increasing the substrate temperature from room temperature to 325 °C, at 0.12 Pa, the structure of the film transits from zone I to zone T [24, 26, 27].

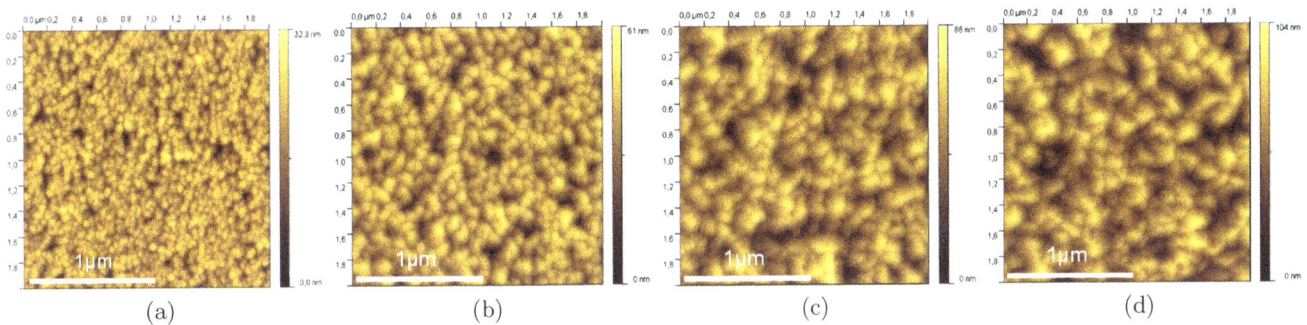

Fig. 5. (Color online) AFM images of the ZnO:Al films as function of substrate temperature: (a) room temperature, (b) 100 °C, (c) 260 °C and (d) 325 °C.

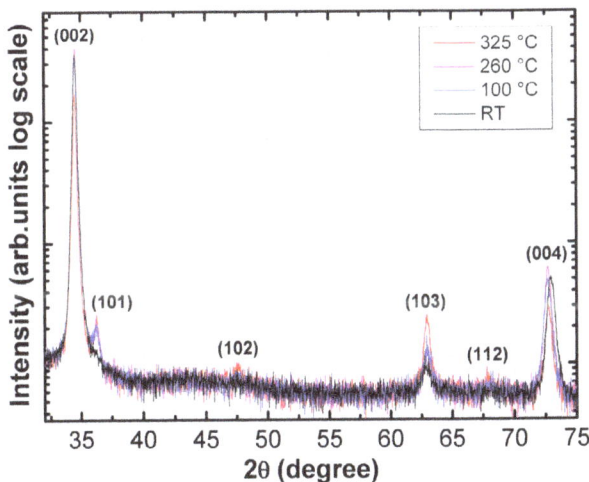

Fig. 6. (Color online) XRD profiles of ZnO:Al films for various substrate temperatures.

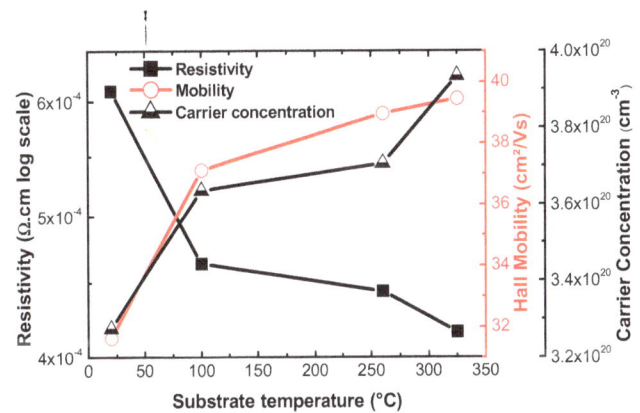

Fig. 7. (Color online) Resistivity (■), Hall mobility (○) and carrier concentration (▲) as functions of the substrate temperature.

Above 0.12 Pa, the degradation of the crystallinity results from the increase of strain and low surface mobility of atoms at the surface of the films (Fig. 10).

3.2.2 Morphological properties

As for the influence of the substrate temperature, experiments as a function of the working pressure are in

the high wavenumber contribution with increasing working pressure. This effect is attributed to an increase in defects. This increase correlates well with the significant increase in strain and the decrease in crystallinity estimated from the decrease of the intensity of (002) reflection at high deposition gas pressure (Fig. 9).

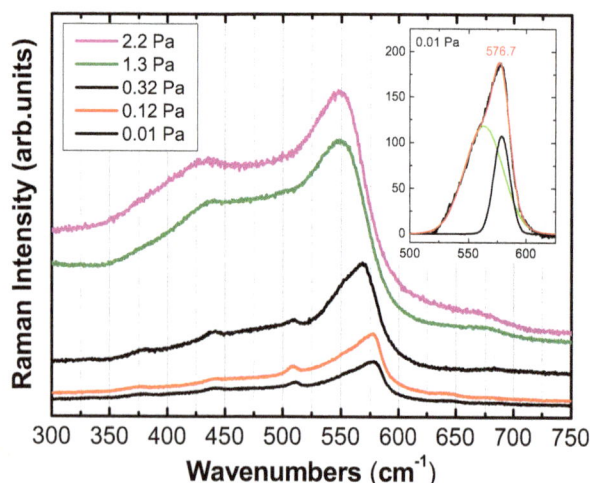

Fig. 8. (Color online) Raman spectra of ZnO:Al films deposited at various working pressures. The asymmetric A_1-LO vibration is fitted to two Gaussian contributions (example given in the inset).

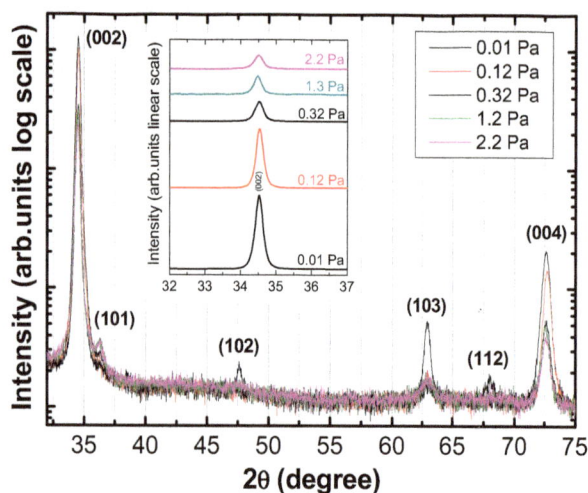

Fig. 9. (Color online) XRD profiles of ZnO:Al films for various working pressures.

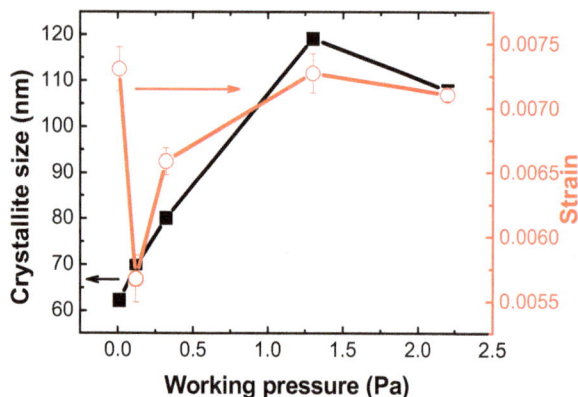

Fig. 10. (Color online) Crystallite size (■) and strain (○) estimated from Williamson-Hall plots for samples deposited at various working pressures.

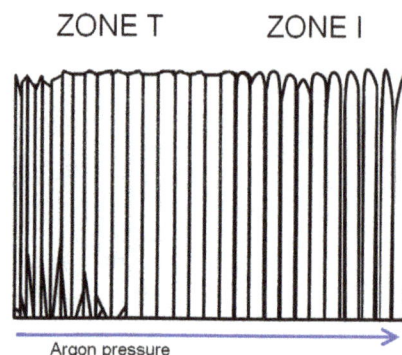

Fig. 11. (Color online) Structure zone model depending of the argon pressure. When increasing the working pressure from 0.01 Pa to 2.2 Pa, at room temperature, the structure of the film transits from zone T to zone I.

good agreement with the Thornton's structure zone model (Fig. 11). At low working pressure, the surface structure corresponds to the zone consisting of densely packed fibrous grains and relatively smooth domed surfaces. At low pressure as well as at high substrate temperature, the densely packed fibrous grains are revealed by the observation of (101), (102), (103) and (112) peaks (Fig. 9), when the growth competition is tougher because of the high mobility of the atoms at the surface of the growing film.

With the increase of the working pressure, the mean free path of the sputtered species decreases. The sputtered species undergo many collisions and as a result of their low energy these impinging particles have low surface mobility, causing the network compression (Fig. 10). The structure zone of films undergoes a transition from densely packed fibrous grains to a porous zone consisting of tapered crystallites separated by voids. (101), (102), (103) and (112) peaks are no longer observed (Fig. 9).

The microstructure predicted by the Thornton's structure zone model is confirmed by AFM images (Fig. 12). At low pressure, below 0.12 Pa, the surface morphology is not distinguishable, while at high pressure, above 0.12 Pa, crystallites are well defined, surrounded by voids. Low energy sputtered particles limit the mobility of surface atoms and self-shadowing effect becomes pronounced at high argon pressure, leading to an increase of the surface roughness. But below 0.12 Pa, in densely packed fibrous grains zone, there is also an increase of the roughness, this time due to the bombardment by high energy ions. This bombardment introduces defects inside the films, leads to an increase in strain and a degradation of electrical properties. Surprisingly, the crystallinity of the films, as measured by Raman spectroscopy (Tab. 2) and X-Ray diffraction analysis (Fig. 9), seems not affected by the bombardment of sputtered particles.

3.2.3 Electrical properties

Above 0.12 Pa, with the increase of the working pressure, a degradation of the electrical properties is observed, revealed by an increase in resistivity and a decrease in Hall

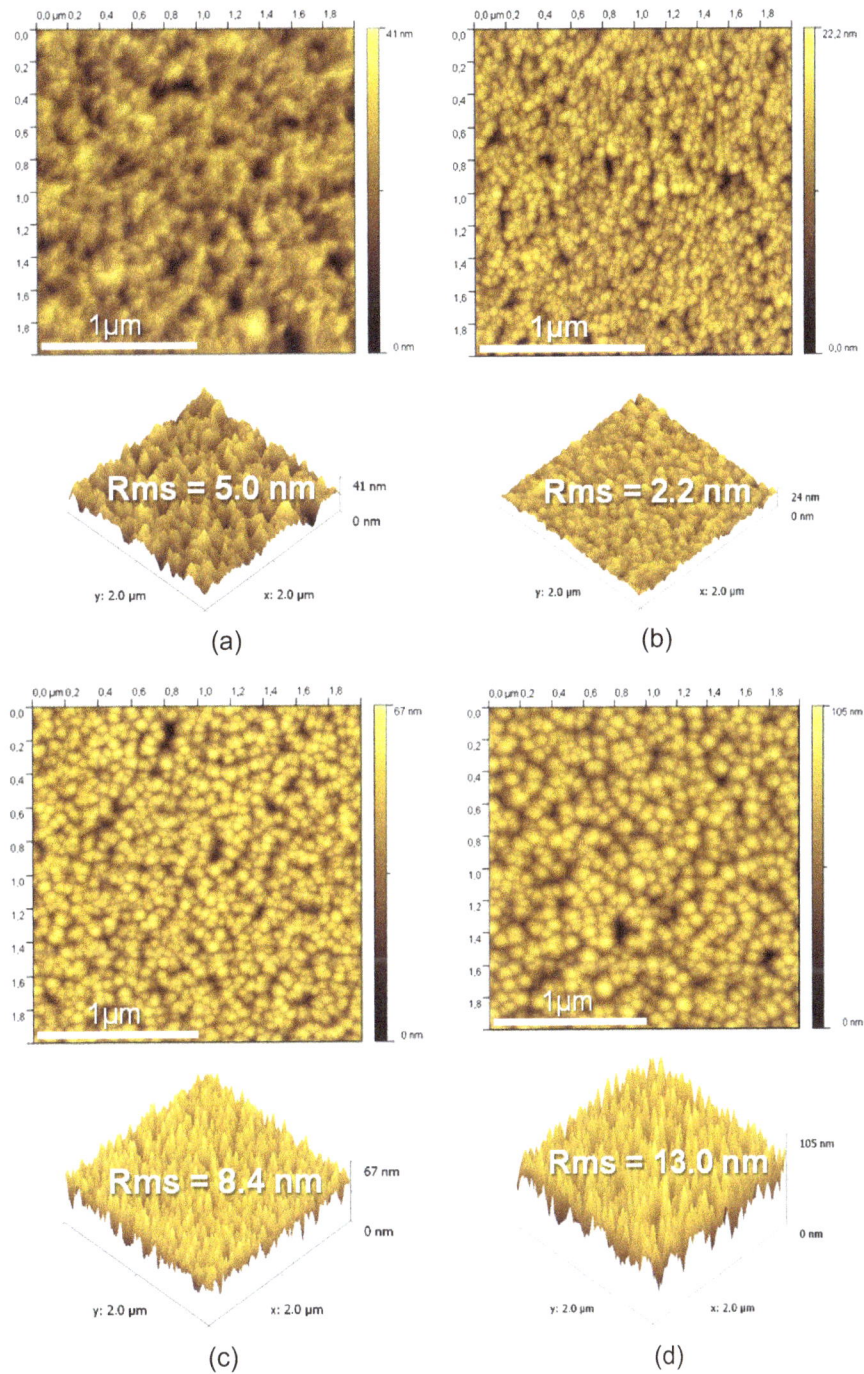

Fig. 12. (Color online) AFM images of the ZnO:Al films as function of working pressure: (a) 0.01 Pa, (b) 0.12 Pa, (c) 0.32 Pa, (d) 1.3 Pa.

mobility (Fig. 13). At high pressure, the increase in surface roughness increases the effective surface area of the film and then enhances the number of adsorption sites for oxygen. The highly porous film structure intensifies the physisorption of oxygen after deposition at the grain boundaries which are scattering centres for carriers. Oxygen adsorbed on the surface of crystallites traps electrons, decreases the carrier concentration and also decreases the Hall mobility by increasing the potential barrier at the surface of crystallites.

4 Conclusion

By using X-Ray diffraction and Raman spectroscopy, a correlation between microstructural, morphological, electrical and optical properties is possible. Depending on substrate temperature and working pressure, two structure zones are observed: a densely packed fibrous grains zone at high temperature and low pressure, and a porous structure consisting of tapered crystallites separated by voids at high pressure. But the knowledge of the microstructure

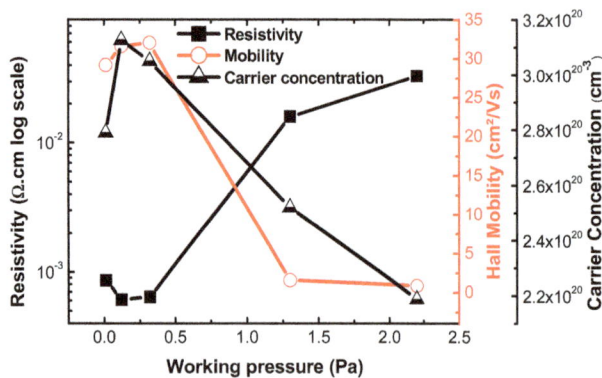

Fig. 13. (Color online) Resistivity (■), Hall mobility (○) and carrier concentration (▲) as functions of the working pressure.

is not sufficient to find the electrical optimum. Zones of densely packed fibrous grains exhibit an improvement of the crystallinity revealed by Raman spectroscopy and X-Ray diffraction analysis, and a low strain. However, at low pressure, the high energy of sputtered particles enables a high mobility of atoms at the surface of the films but also leads to a high bombardment which creates defects, and degrades electrical properties of the films. Therefore the electrical optimum is in densely packed fibrous grains zone at high temperature, but at intermediate pressure.

References

1. B. Rech, T. Repmann, M.N. van den Donker, M. Berginski, T. Kilper, J. Hüpkes, S. Calnan, H. Stiebig, S. Wieder, Thin Solid Films **511**, 548 (2006)
2. W. Beyer, J. Hüpkes, H. Stiebig, Thin Solid Films **516**, 147 (2007)
3. J. Springer, B. Rech, W. Reetz, J. Muller, M. Vanecek, Solar Energy Mater. Solar Cells **85**, 1 (2005)
4. J. Müller, B. Rech, J. Springer, M. Vanecek, Sol. Energy **77**, 917 (2004)
5. J. Yoo, J. Lee, S. Kim, K. Yoon, I. Jun Park, S.K. Dhungel, B. Karunagaran, D. Mangalaraj, J. Yi, Thin Solid Films **480**, 213 (2005)
6. O. Kluth, B. Rech, L. Houben, S. Wieder, G. Schöpe, Thin Solid Films **351**, 247 (1999)
7. K. Ellmer, J. Phys. D **33**, 17 (2000)
8. O. Kluth, G. Schöpe, J. Hüpkes, C. Agashe, J. Müller, B. Rech, Thin Solid Films **442**, 80 (2003)
9. C. Agashe, O. Kluth, G. Schöpe, H. Siekmann, J. Hüpkes, B. Rech, Thin Solid Films **442**, 167 (2003)
10. C. Agashe, O. Kluth, J. Hüpkes, U. Zastrow, B. Rech, J. Appl. Phys. **95**, 1911 (2004)
11. Y. Igasaki, H. Kanma, Appl. Surf. Sci. **169**, 508 (2001)
12. S. Rahmane, M.A. Djouadi, M.S. Aida, N. Barreau, B. Abdallah, N. Hadj Zoubir, Thin Solid Films **519**, 5 (2010)
13. D. Song, A.G. Aberle, J. Xia, Appl. Surf. Sci. **195**, 291 (2002)
14. S.-S. Lin, J.-L. Huang, Surf. Coat. Technol. **185**, 222 (2004)
15. R. Wen, L. Wang, X. Wang, G.-H. Yue, Y. Chen, D.-L. Peng, J. Alloys Compd. **508**, 370 (2010)
16. J. Woo, J. Lee, S. Kim, K. Yoon, I. Jun Park, S.K. Dhungel, B. Karunagaran, D. Mangalaraj, J. Yi, Phys. Stat. Sol. **2**, 1228 (2005)
17. S. Singh, T. Ganguli, R. Kumar, R.S. Srinivasa, S.S. Major, Thin Solid Films **517**, 661 (2008)
18. R. Chierchia, T. Böttler, H. Heinke, S. Einfeldt, S. Figge, D. Hommel, J. Appl. Phys. **93**, 8918 (2003)
19. S. Singh, R.S. Srinivasa, S.S. Major, Thin Solid Films **515**, 8718 (2007)
20. D. Song, Appl. Surf. Sci. **254**, 4171 (2008)
21. M. Tzolov, N. Tzenov, D. Dimova-Malinovska, M. Kalitzova, C. Pizzuto, G. Vitali, G. Zollo, I. Ivanov, Thin Solid Films **379**, 28 (2000)
22. Y.-C. Lee, S.-Y. Hu, W. Water, Y.-S. Huang, M.-D. Yang, J.-L. Shen, K.-K. Tiong, C.-C. Huang, Solid State Commun. **143**, 250 (2007)
23. H. Morkoç, Ü. Özgür, in *Zinc Oxide, Fundamentals, Materials and Device Technology* (Wiley VCH, 2009)
24. J.B. Barna, M. Adamik, Thin Solid Films **317**, 27 (1998)
25. F.C.M. Van De Pol, F.R. Blom, T.J.A. Popma, Thin Solid Films **204**, 349 (1991)
26. J.A. Thornton, J. Vac. Sci. Technol. **11**, 666 (1974)
27. J.A. Thornton, Annu. Rev. Mater. Sci. **7**, 269 (1977)

15

Recent advances in small molecular, non-polymeric organic hole transporting materials for solid-state DSSC

Thanh-Tuan Bui[a] and Fabrice Goubard[b]

Laboratoire de Physicochimie des Polymères et des Interfaces, Université de Cergy-Pontoise, 5 mail Gay Lussac, Neuville-sur-Oise, 95031 Cergy-Pontoise Cedex, France

Abstract Issue from thin-film technologies, dye-sensitized solar cells have become one of the most promising technologies in the field of renewable energies. Their success is not only due to their low weight, the possibility of making large flexible surfaces, but also to their photovoltaic efficiency which are found to be more and more significant (>12% with a liquid electrolyte, >7% with a solid organic hole conductor). This short review highlights recent advances in the characteristics and use of low-molecular-weight glass-forming organic materials as hole transporters in all solid-state dye-sensitized solar cells. These materials must feature specific physical and chemical properties that will ensure both the operation of a photovoltaic cell and the easy implementation. This review is an english extended version based on our recent article published in *Matériaux & Techniques* **101**, 102 (2013).

1 Construction and performance of dye-sensitized solar cell

The energy-scenario analysis in 2050 plans an increase until 300% of the world energy consumption. Such a need cannot be exclusively satisfied by the fossil fuels, humanity has to turn to renewable energies and notably towards the potentiality of solar energy. Photovoltaics is a promising renewable energy technology that converts sunlight to electricity, with broad potential to contribute significantly to solving the future energy problem that humanity faces. Historically, in 1954, Bell Labs created the first silicon-based solar cells with 6% efficiency [1]. To date, inorganic semiconductor solar cells dominate commercial markets, with crystalline Si having an 80% share; the remaining 20% is mostly thin film solar technology, such as CdTe and CuInGaSe [2]. However, the use of this conventional silicon involves a non-negligible production cost, in particular because of the silicon purification process to reach the solar-grade silicon. This cost strikingly reduces the competitiveness of these silicon cells compared with the traditional sources of energy for the ground applications.

Since the last two decades, low-cost solutions have emerged mainly using thin-film solar photovoltaic technologies. Among them, a new type of cell, known as dye-sensitized solar cells (DSSC) [3], has been studied to develop low-cost photovoltaic devices. In this type of cell,

a light-harvesting material, generally a molecular dye absorbs photons and is photoexcited. The photogenerated electrons and holes from the dye are quickly separated and transferred into two different transporting media (metallic oxide and electrolyte) reducing strongly the electron-hole recombination in the absorber material. Moreover, the device is based on the superposition of active layers whose thicknesses are ten to twenty-folds inferior to that of crystalline silicon wafers. In addition, the requested purity of materials is 10 to 100 times less than for a silicon device.

The DSSC is composed of a photo-anode and a photoinert counter electrode (cathode) sandwiching a redox mediator (Fig. 1). It consists of five materials: (1) a fluorine-doped SnO_2 (FTO) coated glass substrate, (2) a nanocrystalline TiO_2 semiconductor thin film, (3) a photosensitizer, organic dye or metal coordination complex, (4) an electrolyte containing redox mediator, and (5) a platinum-coated glass substrate (Fig. 1).

A schematic presentation of the operating principles of DSSC is shown in Figure 1 [4]. Upon absorption of photons, dye molecules anchored on the TiO_2 surface are excited and an electron passes from the highest occupied molecular orbital (HOMO) to the lowest unoccupied molecular orbital (LUMO). The excited dyes inject electrons from their LUMOs into the conduction band of the oxide. The oxidized dyes are then regenerated by electron donation from the redox mediator, such as the iodide/triiodide couple. The iodide is regenerated, in turn, by the reduction of triiodide at the counter electrode, with

[a] e-mail: thanh-tuan.bui@u-cergy.fr
[b] e-mail: fabrice.goubard@u-cergy.fr

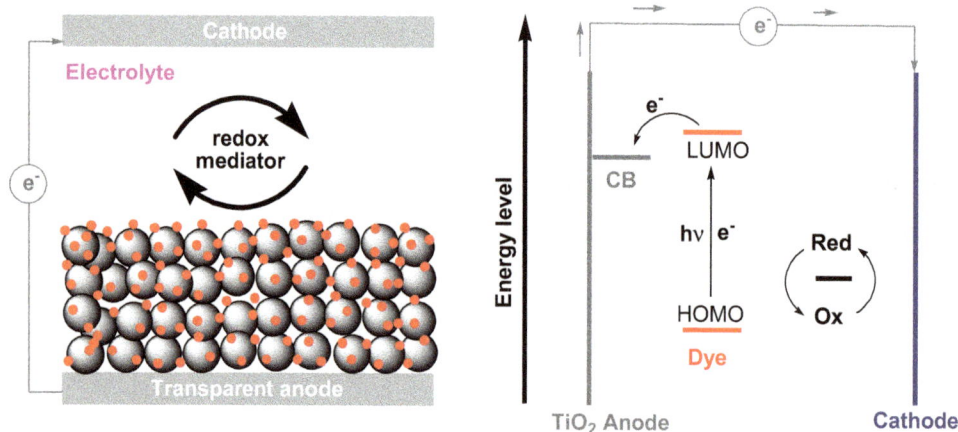

Fig. 1. Schematic presentation of liquid electrolyte-based DSSC and its operation principle.

the circuit being completed via electron migration through the external load.

In 1991, O'Regan and Grätzel [5] published a seminal paper on DSSC based on the mechanism of a fast regenerative photoelectrochemical process. The overall efficiency of this new type of solar cell was 7.1–7.9% (under simulated solar light) with I^-/I_3^- as redox mediator and Ru complex as dye. The originality and the device performances are mainly governed by (i) the high surface area and the controlled nanoporosity of the oxide materials, and (ii) the high efficiency regeneration of the photo-oxidized dye molecules induced with an iodide/triiodide redox electrolyte.

Since the last two decades, many research groups have been focusing their efforts into developing and optimizing all the processes and materials constituting a DSSC [3,6]. Nowadays, the photoconversion efficiencies reach over 12% for small single cells [7,8] and about 10% for modules [9], connected with high stability upon sunlight exposure. One of the manufacturing challenges for DSSCs is the need for a robust sealing process that would prevent the liquid electrolyte in the cells from leakage and evaporation. Moreover, the electrolyte, the iodide/triiodide redox couple, is corrosive to the surroundings. These problems create difficulties for the large scale production and greatly curb the commercialization of DSSC [10–12]. For this reason, "all solid-state" DSSCs have emerged in an attempt to replace the liquid electrolytes with small organic or inorganic hole conductors, π-conjugated polymers, or gel electrolytes. These molecules, except polymeric materials, by their small size can infiltrate into the pores of metallic oxide to replace the liquid electrolyte. The structure and the operation principle of the device are thus modified (Fig. 2). Here, hole transfer occurs directly from the dye to the hole-transporting materials (HTM), with the holes being transported via hopping between electronic states on the organic molecules to a metallic counter electrode.

To well ensure the operation cycle of a solid-state DSSC, several characteristics are essential for any organic p-type semiconductor.

(1) It must be able to be deposited within the porous nanocrystalline layer: an amorphous state is required

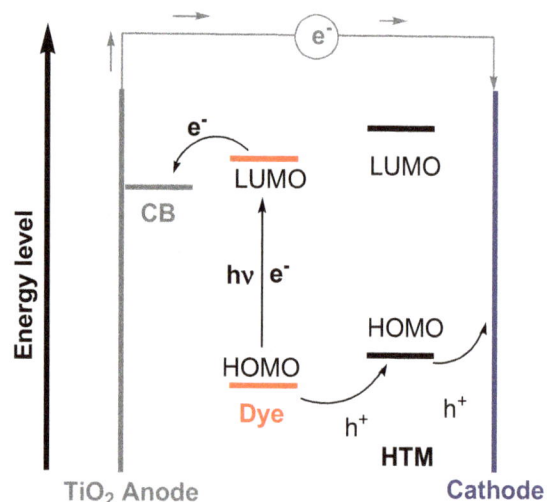

Fig. 2. Operation principle of solid state DSSC device.

with a glass transition superior to room temperature. The amorphous phase, with small molecular size and high solubility, facilitate pore filling, allowing good contact with the dye and hence the best efficiency of the cell.

(2) It must be able to collect holes from the sensitizing dye after the dye has injected electrons into the TiO$_2$; that is why the HOMO energy level of p-type semiconductors must be located above the ground state level of the dye.

(3) A method must be available for depositing the p-type semiconductors without dissolving or degrading the monolayer of dye on TiO$_2$ nanocrystallites.

(4) It must be transparent in the visible spectrum with no absorption screen effect compared to dye.

(5) The mobility of charge carriers in the organic p-type semiconductors must be higher and substantially equal than that in the nanoporous TiO$_2$ layer avoiding a charge excess at the interface between active layer and electrode.

A solid-state DSSC with cyanidin as natural dye and CuI as inorganic hole transporter was first demonstrated by Tennakone et al. [13] in 1995. In direct sunlight (ca. 800 Wm^{-2}) the cell generated a J_{SC} of 2.5 mA cm^{-2}, a V_{OC} of 375 mV with the maximum energy conversion efficiency equal to 0.8%. Since, the conversion efficiencies of cells employed inorganic p-type semiconductor electrolyte based solar cells have significantly increased. Recently, Chung et al. [14] have reported new efficient solution processable solid state DSSCs with p-type inorganic semiconductor CsSnI$_3$ as hole conductor with a high hole mobility ($\mu_h = 585$ cm^2 V^{-1} s^{-1} at room temperature). With a bandgap of 1.3 eV, CsSnI$_3$ enhances light absorption in visible domain to outperform the typical dye-sensitized solar cells. The cell sensitized with N719 using pristine CsSnI$_3$ gave interesting photocurrent density-voltage characteristics: $V_{OC} = 0.638$ V, $FF = 66.1\%$, $J_{SC} = 8.82$ mA cm^{-2}, $\eta = 3.72\%$. Moreover, the authors showed that doping of CsSnI$_3$ with fluor and SnF$_2$ dramatically improves these characteristics. At an optimum molar concentration of 5% F and 5% SnF$_2$, the cell exhibits the highest efficiency: $\eta = 10.2\%$ ($\eta = 8.51\%$ with a mask) under the standard AM 1.5 irradiation (100 mW cm^{-2}).

The potential of ionic liquids as solvents for electrolytes for DSSCs has been also investigated during the last decade. The non-volatility, good solvent properties and high electrochemical stability of ionic liquids make them attractive solvents in contrast to volatile organic counterparts. However, the problem of leakage of the liquid electrolyte during the long-term operation is always persistent [11,15]. Then, polymeric ionic liquids, combining both features of ionic liquids and organic polymeric materials, have been used as a promising approach leading to high efficient devices. Recently, it was reported I$_2$-free solid state DSSCs based on a solid state polymerized ionic liquids, i.e., poly(1-alkyl-3-(acryloyloxy)hexylimidazolium iodide) and poly((1-(4-ethenylphenyl)methyl)-3-butylimidazodium iodide) that can sufficiently penetrate the TiO$_2$ nanopores. These devices exhibited cell efficiencies of 5.29 and 5.93%, respectively, with an excellent long-term stability of the devices [16,17]. Intrinsically, conducting polymers are well known as good hole transporting material, carrying current densities of several mA cm^{-2}. Thus, these materials are potential candidates to use as HTM in solid state DSSCs. Nevertheless, polymers cast from solutions must penetrate into the pores of the nanoparticles and should form a good contact with the adsorbed sensitizer. Semiconducting polymer as poly(3,4-ethylenedioxythiophene) (PEDOT) [18,19] and poly(3-hexylthiophene) (P3HT) [20] are mainly used in solid state DSSC as solid hole conductor [21]. However, the conversion efficiencies reported with conducting polymers as HTM are lower in comparison to the cells using a liquid electrolyte or small organic hole conductor molecules, mainly due to the incomplete filling of the TiO$_2$ nanopores by the HTM [22], which leads to poor electronic contact between the dye molecules and the hole conductor. To overcome this drawback, in situ polymerized HTMs have been developed. The polymerization of monomers has taken place directly inside the TiO$_2$ networks leading to high efficient solid

state DSSCs with η ranges from ~4–7% [18,19,23–26]. A breakthrough was recently reported in solid-state hybrid solar cells by using organolead iodide perovskite (CH$_3$NH$_3$PbI$_3$) nanoparticles [27] as a light absorber and spiro-MeOTAD as HTM, which shows an impressive PCE of 9.7% with superior stability over 500 h. Etgar et al. [28] have described the photovoltaic device comprised of a mesoscopic CH$_3$NH$_3$PbI$_3$/TiO$_2$ heterojunction with CH$_3$NH$_3$PbI$_3$ nanoparticles as both a light harvester and a HTM, which shows a high efficiency of 5.5% under 1 Sun illumination. Moreover, when the n-type semiconductor TiO$_2$ was recently replaced by an insulating Al$_2$O$_3$ mesoporous scaffold, the device displayed a high PCE of 10.9%, where a CH$_3$NH$_3$PbI$_2$Cl thin absorber layer functions as both a light absorber and a charge transport materials [29].

In this short review, we limited our study on solid small molecular, non-polymeric organic hole conducting materials for solid state DSSCs.

2 Molecular glasses based solid state DSSCs

Historically, the first study concerning the use of molecular glasses as hole conductors in solid state DSSC was realized by Hagen et al. [30] in 1997. The solid state cells were prepared by evaporation of the (N,N-9-diphenyl-N,N-9-bis(3-methylphenyl)4,49-diamine (TPD) transport layer (Fig. 3) on a ruthenium complex sensitized nanoporous TiO$_2$ (N3, Fig. 4). External quantum efficiencies of up to 0.2% was achieved with a low open circuit voltage equal to 300 mV (Tab. 1). Poor titania pore filling has been reported as a major performance limiting factor in this study: the vacuum evaporation process involved a low interface between the dye and TPD yielding a low charge transfer between sensitized TiO$_2$ and TPD.

Senadeera et al. [31] elaborated solid-state dye-sensitized photocell based on pentacene (Fig. 3) as a hole collector, with N3 as sensitizer (Fig. 4). Pentacene was thermally evaporated in vacuum on top of the dye-coated TiO$_2$ film to form films ca. 50 nm thick. Conductivity measurements on these iodine doped pentacene films revealed that conductivity is in the order of 1.25×10^{-2} S cm^{-1}, which was lower than that of pentacene films doped either with iodine or bromine reported in the literature (electrical conductivity of up to 100 S cm^{-1}) [32,33]. Different film thickness, doping concentration and doping method might be reasons for this low value. Using an outer surface graphite-painted gold-plated FTO glass as counter electrode, the cell gave photovoltaic characterizations under illumination: $J_{SC} = 3.6$ mA cm^{-2}, $V_{OC} = 415$ mV, $FF = 49\%$ and $\eta = 0.8\%$. The low value of the efficiency can be explained by a partial absorption in visible light of the pentacene involving an overlap in absorption with the dye and cut off a significant amount of light reflected by the back contacts. Additionally, a low pore filling and a poor injection of the holes to the pentacene might decrease the photovoltaic efficiency.

In 1998, Salbeck et al. [34] and Bach et al. [35] reported the first dye-sensitized heterojunction of TiO$_2$

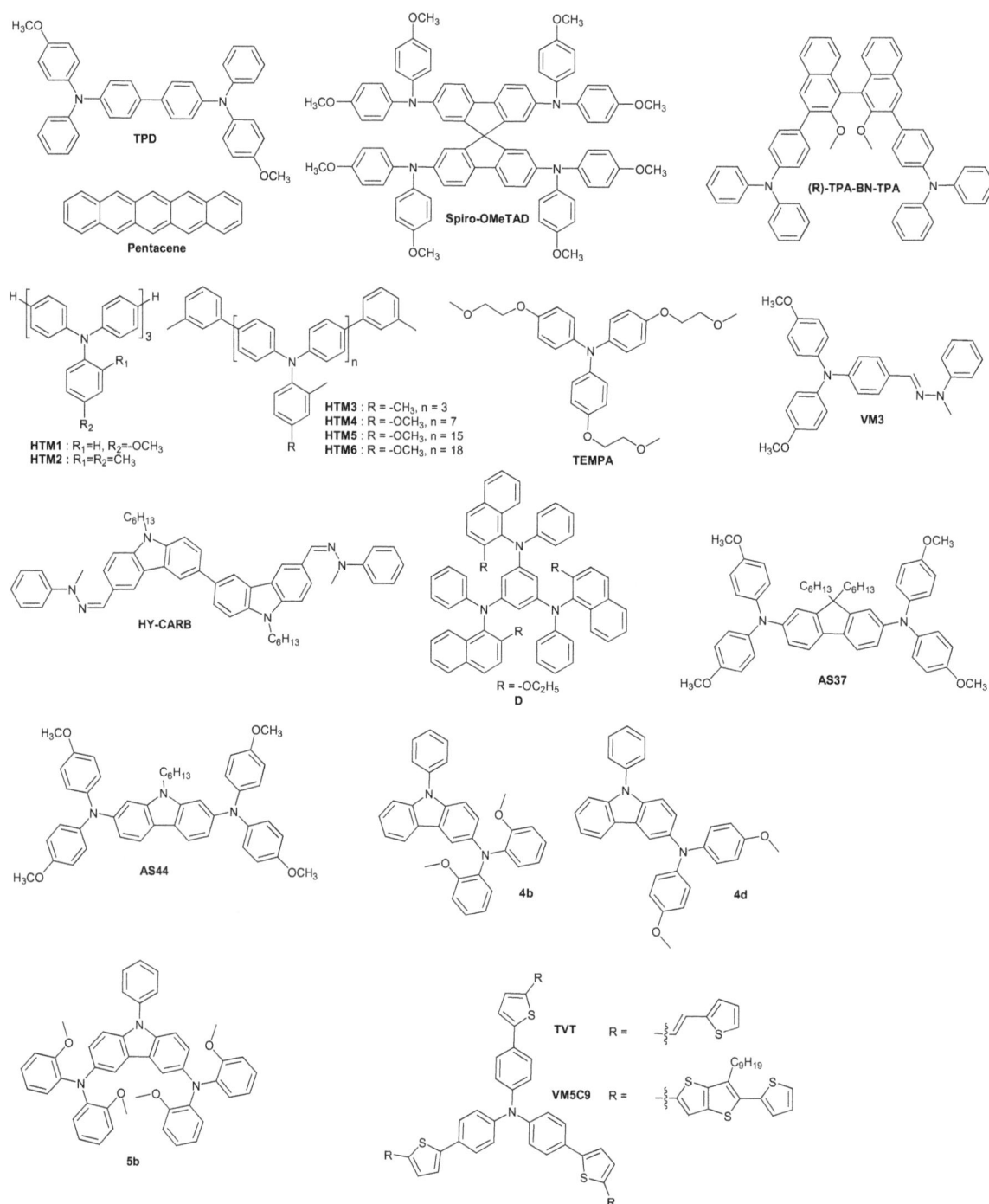

Fig. 3. Chemical structures of organic hole transporters discussed in this article.

with the amorphous organic hole-transport material 2,2',7,7'-tetrakis(N,N-di-p-methoxyphenyl-amine) 9,9'-spirobifluorene (Spiro-OMeTAD) (Fig. 3), with an overall conversion efficiency of 0.74%. Spiro-OMeTAD contains a spiro-centre (a tetrahedral carbon linking two aromatic moieties) which is introduced in order to improve the glass-forming properties and prevent crystallization of the organic material [36]. Its glass transition temperature of $T_g = 120$ °C is higher that of the widely used hole conductor TPD. Moreover, the methoxy groups are

introduced in order to (i) match the oxidation potential of the hole-transport medium to that of the dye, (ii) move the absorption towards the UV domain thanks to their effect donor and (iii) make pore filling easier with a more hydrophilic molecule.

Usually, the photovoltaic efficiencies in solid state DSSC are dependent on the choice of the dye-HTM couple: a low interface or a screen effect in absorption between both molecules decreases considerably the overall conversion efficiency. In the case of the spiro-OMeTAD, the

Table 1. Photovoltaic properties of hole transporters discussed in this article.

Compound	μ_h (cm^2 V^{-1} s^{-1})	E_{HOMO} (eV)	Sensitization dye	J_{SC} (mA/cm^2)	V_{OC} (mV)	FF (%)	η (%)	Ref.
TPD	10^{-3}	−5.06	N3 (Ru)	n/a	300	n/a	0.2	[30]
Spiro-OMeTAD	2×10^{-4}	−4.77	Y123 (OD)	9.5	986	76	7.2	[37]
Pentacene	n/a	n/a	N3 (Ru)	3.6	415	49	0.8	[31]
HTM1	4.68×10^{-4}	−4.97	N719 (Ru)	3.8	789	67.1	2.01	[48]
HTM2	8.07×10^{-4}	−5.0	N719 (Ru)	1.69	750	34.9	0.44	[48]
HTM3	1.76×10^{-3}	−4.99	N719 (Ru)	1.96	778	59.2	0.9	[48]
HTM4	2.06×10^{-4}	−4.96	N719 (Ru)	0.87	779	51.8	0.35	[48]
HTM5	8.23×10^{-4}	−4.98	N719 (Ru)	2.29	815	56.6	1.05	[48]
HTM6	1.04×10^{-3}	−4.96	N719 (Ru)	2.51	797	49.5	0.99	[48]
(R)-TPA-BN-TPA	n/a	n/a	N719 (Ru)	1.07	887	58	0.55	[49]
HY-CARB	6.4×10^{-5}	−5.06	N719 (Ru)	0.42	500	35	0.07	[50]
TMEPA	n/a	−5.34	K51 (Ru)	4.60	770	70	2.4	[51]
VM3	n/a	−5.25	N719 (Ru)	0.332	521	43	0.075	[52]
D	n/a	−5.70	Red sandal (OD)	0.9	250	n/a	0.39	[53]
AS37	5×10^{-5}	−4.98	Z907 (Ru)	5.5	730	62	2.48	[54]
AS44	1×10^{-5}	−4.99	Z907 (Ru)	6.0	730	67	2.94	[54]
4d	n/a	−4.82	D102 (OD)	2.63	630	32	0.54	[55]
4b	7.4×10^{-5}	−4.92	D102 (OD)	0.75	573	28	0.12	[55]
5b	1×10^{-3}	−4.86	D102 (OD)	1.72	531	35	0.32	[55]
TVT	n/a	n/a	SQ (OD)	0.64	480	64	0.19	[56]
VM5C9	n/a	−5.21	D102 (OD)	1.83	680	38	0.47	[57]

n/a: not available; (OD): organic dye; (Ru): ruthenium complex.

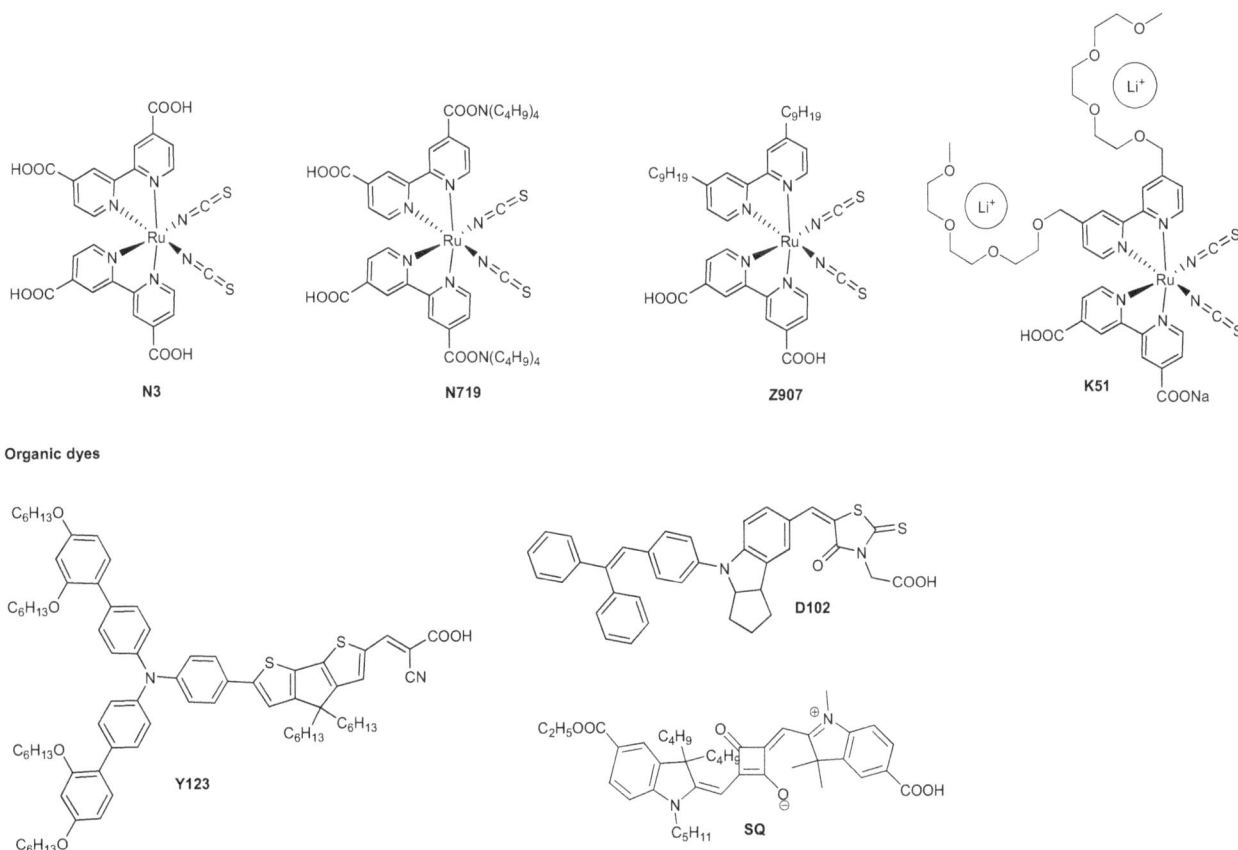

Fig. 4. Chemical structures of photosensitizers associated with hole transporters discussed in this article.

evolution of both the nanostructuration of TiO_2 and the synthesis of new dyes (mainly organic dyes) allowed a considerable increase of photovoltaic efficiencies with a record in 2011 in 7.2% thanks to the organic dye Y123 (Fig. 4) by Burschka et al. [37]. The inorganic dye based solar devices have attracted considerable attention in recent years as they offer high and interesting overall conversion efficiency: recently, Kim et al. [27] reported on solid-state mesoscopic heterojunction solar cells employing nanoparticles of methyl ammonium lead iodide $(CH_3NH_3)PbI_3$ as light harvesters yielding a power conversion efficiency of 9.7%. Moreover, Lee et al. [29] replaced TiO_2 anode by mesoporous alumina as an inert scaffold in sensitized solar cells: with intense visible to near-infrared absorptivity by the highly crystalline perovskite as absorber, the power conversion efficiency of 10.9% in a single-junction device under simulated full sunlight is observed. Electron transfer is then carried out on the oxide surface and through the perovskite.

Moreover, the hole conduction of spiro-OMeTAD increased by adding lithium bistrifluoromethylsulfonyl imide, $(Li[(CF_3SO_2)_2N])$ as an electrolytic salt and tert-butylpyridine (tBP) as solvent. This salt addition improves the performances in terms of photocurrent, open circuit voltage [38] and hole mobility [39,40]. Upon the addition of lithium salts to the spiro-OMeTAD hole-transporter matrix, Snaith et al. [39] observe a 100-fold increase in conductivity through spiro-MeOTAD within a dye-sensitized TiO_2 mesoporous network. By making "hole-only" diodes of pure spiro-MeOTAD and those doped with lithium salts, the authors calculate that the hole mobility increases from 1.6×10^{-4} to 1.6×10^{-3} cm^2/V s. Using a molecular hole conductor $(4,4",4"',4""-ter-N,N-diphenyl[4-(methoxymethyl)phenyl]-amine)$, Snaith and Grätzel [41] have also shown that under illumination, the hole density increases, resulting in striking enhancements in film conductivity, up to 10^6 times, and charge carrier mobility, up to 10^3 times (hole mobility increases from 2.0×10^6 to 1.2×10^{-3} cm^2/V s when increasing the incident illumination intensity from dark to 100 mW cm^2 at room temperature. Moreover, 4-trimethylsilylpyridine as the replacement for tBP involved a 10% increase of overall conversion efficiency [42].

The rate of pore filling by the molecular glass is also an important parameter for optimal overall conversion efficiency. Infiltration of molecular glasses into mesoporous TiO_2 film is usually accomplished by depositing molecular glass solution on the film and soaking the film for 1 minute, followed by spin coating to remove excess solution. By this way and with spiro-OMeTAD the filling fraction is as high as 60–70% for films with thickness <3 μm; however, as the film thickness increases, the filling fraction goes down, which may be the cause of poor efficiency of these devices [43,44]. Moreover, the coating method could influence the pore filling: Doctor blade coating is more adapted for the higher thickness [45]. Melhem et al. [46] demonstrated also different filling fraction depending on crystalline morphology of TiO_2.

The other structures of studied molecular glasses are mainly arylamine derivatives [47]: the hydrogen atoms of the phenyls near the nitrogen atom create between them a steric hindrance inducing the three-dimensional structure and consequently an amorphous state over the ambient temperature like spiro-OMeTAD. Kroeze et al. [48] studied oligomers HTM with core triphenylamine of variable molecular masses (HTM1 to HTM6, Fig. 3), and showed in particular, higher efficiencies when the number of entity decrease. Furthermore, the valuable difference of mobility values, contrary to the filling, is not correlable in the performances of cells. Also let us note that these values of mobility are the same order as those of spiro-OMeTAD but the performances are much lower. It can be the consequence of a difficult penetration of these oligomers in the porous anode.

As described above, the active layer of solid-state DSSC is comprised of a nanostructured TiO_2 film coated with a monolayer of photosensitizer and infiltrated with an organic electron-donating/hole-transporting semiconductor. The most successful approach is to spin-coat a highly concentrated solution of hole transporter on top of the nanoporous film, allowing the favorable interaction between the hole transporter and the dye-coated surface to act as a driving force to aid material infiltration during solvent evaporation. However, this later step is far from perfect and it is still difficult to obtain material infiltration through more than a few micrometers thickness. Snaith et al. [51] have, therefore, suggested that a liquid organic semiconductor may offer the ideal solution, enabling infiltration without the need for subsequent solvent evaporation. They have synthesized tris-[4-(2-methoxy-ethoxy)-phenyl]-amine (TMEPA, Fig. 3) which absorbs only in the UV region. Due to the methoxy-ethoxy side-chain's characteristics, TMEPA is in a liquid state at room temperature ($T_g = -14$ °C, $T_m = -7$ °C). DSSC fabricated from $NOBF_4$ chemically doped TMEPA and K51-sensitized porous anode delivered a power conversion efficiency of 2.4% with a quantum efficiency of over 50%.

In 2007, Zhao et al. [49] reported a newly chiral small organic hole conductor, (R)-2,20-dimethoxyl-3, 3'-di(phenyl-4-yl-diphenyl-amine)-[1,10]-binaphthyl ((R)-TPABN- TPA, Fig. 3). This molecule features two triphenylamine and two methoxynaphthalene moieties. Introduced into dye-sensitized photoelectrical cell as solid-state electrolyte, the device converted light to electricity with an efficiency of 0.07% without any additives in electrolyte. The solid-state devices showed an attractive conversion efficiency of up to 0.55% with the assistance of some functional small-molecules in the electrolyte. This amelioration is attributed to the positive effects of functional additives (tBP, $Li[(CF_3SO_2)_2N]$, $N(PhBr)_3SbCl_6$) such as inhibiting interface charge recombination, improving hole-transporting properties and penetration of solid-state electrolyte.

In 2008, our group reported a new class of hole transporting molecular glasses based on hydrazone derivatives (HY-CARB, Fig. 3) [50, 58]. Thermal and optoelectrochemical investigations revealed interesting properties suitable for solid-state dye-sensitized solar cell application. With an optical gap of 3.1 eV, which is large enough to avoid screening effect in absorption with dye, and a HOMO energy level close to -5.06 eV, HY-CARB is well adapted for the regeneration of the photooxidized ruthenium dye ($E_{HOMO} = -5.45$ eV). Indeed, differential scanning calorimetric characterizations confirm the metastable amorphous properties of HY-CARB with glass transition temperature at 76 °C. Without any additives, the solid-state DSSC based on N719-sensitized TiO$_2$ anode give the following photovoltaic characteristics: $V_{OC} = 230$ mV, $FF = 29\%$ and $J_{SC} = 1$ μA/cm^2. Upon adding NOBF$_4$ and Li[(CF$_3$SO$_2$)$_2$N] into electrolyte, photovoltaic output was significantly improved ($J_{SC} = 0.42$ mA cm^{-2}, $V_{OC} = 500$ mV et $FF = 35\%$). However, it is still far from satisfactory. The low hole mobility (6.4×10^{-5} cm^2 V^{-1} s^{-1}), as well as the poor infiltration into the nanoporous oxide layer because of its hydrophobic structure seem to be the cause of this unsatisfactory performance.

Juozapavicius et al. [52] have reported a molecular glass featuring hydrazone and triphenylamine moieties (VM3, Fig. 3). Designed for the development of solvent-free, melt-infiltration process, VM3 has relatively low melting point (134 °C). Based on transient absorption spectroscopy measurement, the authors have proved a good efficiency of dye regeneration suggesting a good hole transporter infiltration. However, low photocurrent lead to low light to electricity conversion yield ($\eta = 0.075\%$ with electrolyte doped by iodine vapor).

Mathew and Haridas [53] have recently reported a starburst naphthylamine-based molecular glass hole conductor ($T_g = 80$ °C), named N,N,N'-tris-(2-ethoxy-naphthalene-1-yl)N,N,N'-triphenylbenzene1,3,5-triamine ($\mathbf{D}$, Fig. 3). Using this compound as solid electrolyte, natural-organic-dye-sensitized solar cell was fabricated and the performance was analyzed. Despite unsatisfactory photovoltaic output performance (0.39%), the authors highlight the economical aspect of these cells: the total cost is down to nearly 30–40 % of a similar type of fabricated cell.

Recently, Leijtens et al. [54] have reported two new hole transporters, both containing double di(4-methoxyphenyl)amino moieties (AS37 et AS44, Fig. 3). Conceptually, the new HTMs were designed to have similar functional groups and energy levels close to that of spiro-OMeTAD but differ in molecular size, solubility, glass transition temperature, and melting point. AS37 can be considered as a half of Spiro-OMeTAD, where the central spiro-link was replaced by two hexyl chains. In the other hand, replacement of the spiro-bridge by carbazole core, which is more electron donating than fluorene core, lead to AS44 In addition to possessing high solubility, their optoelectrochemical and electrical properties well suited for solid-state DSSC. These molecules have relatively low glass transition temperature and melting point than that of Spiro-OMeTAD (AS37: $T_g = 43$ °C, $T_f = 106$ °C, AS44: $T_g = 59$ °C, $T_f = 132$ °C). Using

standard device fabrication methods with Z907-sensitized TiO$_2$ nanoporous anode, interesting power conversion efficiencies of 2.94% (AS44) and 2.48% (AS37) in 2-μm-thick cells were achieved. In 6-μm-thick cells, the device performance is shown to be higher than that obtained using spiro-OMeTAD in the same condition despite their lower hole mobility, making these new HTMs promising for preparing high-efficiency solid-state DSSCs. The authors suggested that the high solubility of the new materials result in significantly improved pore filling in thick devices, thus photovoltaic performance.

Tomkeviciene et al. [55] have recently synthesized a series of molecular glasses based on 9-phenylcarbazole (4b, 4d, and 5b, Fig. 3). All of these compounds contain bis(methoxyphenyl)amino functional groups with methoxy groups in the different position of diphenylamino. These bis(methoxyphenyl)amino moieties modulate the optoelectronic and thermal properties of targeted molecules. In addition to possessing high solubility, high thermal stability ($T_d = 344$–475 °C), stable amorphous state ($T_g = 65$–89 °C), their optical, photoelectrical and electrochemical properties are well suited for photovoltaic application. The derivatives were tested as hole transport materials in solid-state D102-sensitized solar cells and showed overall conversion efficiency of up to 0.54%.

Recently, Unger et al. [56] has reported a star-shaped hole transporting molecular glass absorbing in the visible spectral domain, named tris(thienyl-vinyl-thienyl)-triphenylamine (TVT, Fig. 3). Associated with a near infrared harvesting squaraine dye (SQ, Fig. 4) in solid-state DSSC, it has been demonstrated that TVT contributes to the photocurrent. Energy transfer from TVT to SQ was investigated by photoluminescence emission and excitation measurements and the excitation energy transfer efficiency from TVT to SQ was determined to be 26%. The low power conversion efficiency (0.19%) may be due to the slow regeneration of SQ by TVT and narrow absorption band of SQ.

Our group has recently reported a starbust triphenylamine-based molecular glass (VM5C9, Fig. 3) absorbing strongly in the visible spectral domain ($\lambda_{max} = 428$ nm/$\varepsilon = 131\,000$ L mol^{-1} cm^{-1}) [57, 59], which was successfully employed in solid-state DSSCs. The device made from a spin-coated thin film of VM5C9 onto a TiO$_2$ nanoporous anode (no dye-coated) delivered an overall energy conversion yield of 0.3% after cathode (Au) evaporation. It is noteworthy to note that, in this device, VM5C9 played a double role as photosensitizer and also as electron-donating/hole transporter. In the other hand, VM5C9 was not chemically absorbed onto TiO$_2$ surface by covalent bonds. By using a D102-sensitized anode, the performance of the obtained device increased to 0.47%. Our first investigations suggest that poor pore filling and strong screen effect in absorption between VM5C9 and D102 caused this unsatisfactory performance. We are actively working on the device optimization and concerning results will be published elsewhere shortly.

From the analysis of photovoltaic properties of different hole transporters cited in the article, it appears that rational design of new hole transporter need some requirements. However some criteria are more relevant than other ones.

- High hole mobility is desirable but not vital. The hole drift speed should be at least in the same magnitude to that of the electrons in inorganic semiconductor porous network. Best results in PCE with TiO_2/dye/ spiro-OMeTAD were obtained with charge mobilities of 10^{-4}–10^{-3} cm^2/V s.
- The energy level of the HOMO of hole conductor must be slightly higher than that of the photosensitizer. A difference of 0.3–0.4 eV is requested in order to ensure good hole injection from excited photosensitizer toward hole transporter.
- The screen effect in absorption between the electrolytic molecular glass and the photosensitizer should be avoided. Competing absorption between these two components conducts to a bad impact on the photovoltaic efficiency of devices. Ideally, the hole transporter should only absorb in the UV spectral domain to ensure a good compatibility with larger range of photosensitizer.
- Good penetration of hole conductor into the mesoporous titanium dioxide networks is vital. Contrary to liquid-based electrolyte DSSC where liquid electrolyte can penetrate easily and deeply into TiO_2 anode, the pore filling of solid organic hole transporters stands only around 60–70%. The insufficient pore-filling of HTM causes lower photocurrent and poorer performance under full sunlight and faster recombination as well. The pore filling should be taken into account in designing new molecules as dye or HTM for sDSSC application. For example: dyes having similar functional group to methoxyphenyl amine group on spiro-OMeTAD could have a favorable contact between dye molecules and HTM, and this good contact can lead to better pore-filling. Additionally, the long alkyl chains help to promote pore-filling of the HTM due to their strong hydrophobicity. Moreover, the presence of a long alkyl chain retarded charge recombination and increased electron lifetime [60].
- Finally, in terms of morphological stability, glass transition should occur at high temperature, ideally higher than 80 °C. The cell temperature can increase under continuous full sunlight illumination, which leads to morphological changes in the layer of hole conductor if its T_g is too low.

3 Conclusion and outlook

Since one decade, the performance of solid-state dye-sensitized solar cells has been made remarkable progress. In this article, we review current progress in the use of organic molecular glasses as hole transporting materials in these devices. Contrary to liquid-based electrolyte DSSCs, interfacial issues of mesoporous TiO_2/colorant/ hole transporter are more critical and greatly influent

the photovoltaic performance of devices. New hole transporter's conception and chemical molecular engineering focus primarily on the delicate balance of HOMO-LUMO energy levels/no screen effect in absorption with photosensitizer/infiltration into the mesoporous metal oxide network. Many molecular glasses hole transporters have been reported in the literature. Structurally, these compounds are all tertiary arylamine derivatives. The introduction of methoxy hydrophilic units in the molecular structure seems to be primordial. These groups help to modulate the optoelectronic and thermal properties of the targeted molecule. On the other hand, the penetration of molecular glasses into porous networks is favored leading to more efficient devices. Currently, hole transporting materials based on arylamines are dominant and the most efficient is spiro-OMeTAD.

But still several critical issues should be resolved in solid DSSC: incomplete pore-filling of the spiro-OMeTAD, stability of the cell, and low light harvesting efficiency. As well as, new hole transporting materials should be developed in order to enhance a photovoltaic performance, the device efficiency can be improved by applying new light absorbing materials with high molar extinction coefficient or modifications of the mesoporous TiO_2 layer.

References

1. D.M. Chapin, C.S. Fuller, G.L. Pearson, J. Appl. Phys. **25**, 676 (1954)
2. J. Bisquert, Chem. Phys. Chem. **12**, 1633 (2011)
3. A. Hagfeldt, G. Boschloo, L. Sun, L. Kloo, H. Pettersson, Chem. Rev. **110**, 6595 (2010)
4. M. Grätzel, Inorg. Chem. **44**, 6841 (2005)
5. B. O'Regan, M. Grätzel, Nature **353**, 737 (1991)
6. M. Grätzel, Acc. Chem. Res. **42**, 1788 (2009)
7. Y. Chiba, A. Islam, Y. Watanabe, R. Komiya, N. Koide, L. Han, Jpn J. Appl. Phys. **45**, L638 (2006)
8. A. Yella, H.-W. Lee, H.N. Tsao, C. Yi, A.K. Chandiran, M.K. Nazeeruddin, E.W.-G. Diau, C.-Y. Yeh, S.M. Zakeeruddin, M. Grätzel, Science **334**, 629 (2011)
9. M.A. Green, K. Emery, Y. Hishikawa, W. Warta, E.D. Dunlop, Prog. Photovolt.: Res. Appl. **20**, 606 (2012)
10. J.B. Baxter, J. Vac. Sci. Technol. A **30**, 020801 (2012)
11. S.M. Zakeeruddin, M. Grätzel, Adv. Funct. Mater. **19**, 2187 (2009)
12. L.M. Peter, J. Phys. Chem. Lett. **2**, 1861 (2011)
13. K. Tennakone, G.R.R.A. Kumara, A.R. Kumarasinghe, K.G.U. Wijayantha, P.M. Sirimanne, Semicond. Sci. Technol. **10**, 1689 (1995)
14. I. Chung, B. Lee, J. He, R.P.H. Chang, M.G. Kanatzidis, Nature **485**, 486 (2012)
15. M. Gorlov, L. Kloo, Dalton Trans. **20**, 2655 (2008)
16. W.S. Chi, J.K. Koh, S.H. Ahn, J.-S. Shin, H. Ahn, D.Y. Ryu, J.H. Kim, Electrochem. Commun. **13**, 1349 (2011)
17. G. Wang, L. Wang, S. Zhuo, S. Fang, Y. Lin, Chem. Commun. **47**, 2700 (2011)
18. J.K. Koh, J. Kim, B. Kim, J.H. Kim, E. Kim, Adv. Mater. **23**, 1641 (2011)
19. J. Kim, J.K. Koh, B. Kim, S.H. Ahn, H. Ahn, D.Y. Ryu, J.H. Kim, E. Kim, Adv. Funct. Mater. **21**, 4633 (2011)

20. W. Zhang, R. Zhu, F. Li, Q. Wang, B. Liu, J. Phys. Chem. C **115**, 7038 (2011)

21. W. Zhang, Y. Cheng, X. Yin, B. Liu, Macromol. Chem. Phys. **212**, 15 (2011)

22. L. Yang, U.B. Cappel, E.L. Unger, M. Karlsson, K.M. Karlsson, E. Gabrielsson, L. Sun, G. Boschloo, A. Hagfeldt, E.M.J. Johansson, Phys. Chem. Chem. Phys. **14**, 779 (2012)

23. X. Liu, Y. Cheng, L. Wang, L. Cai, B. Liu, Phys. Chem. Chem. Phys. **14**, 7098 (2012)

24. I.Y. Song, S.-H. Park, J. Lim, Y.S. Kwon, T. Park, Chem. Commun. **47**, 10395 (2011)

25. B. Kim, J.K. Koh, J. Kim, W.S. Chi, J.H. Kim, E. Kim, ChemSusChem. **5**, 2173 (2012)

26. X. Liu, W. Zhang, S. Uchida, L. Cai, B. Liu, S. Ramakrishna, Adv. Mater. **22**, E150 (2010)

27. H.-S. Kim, C.-R. Lee, J.-H. Im, K.-B. Lee, T. Moehl, A. Marchioro, S.-J. Moon, R. Humphry-Baker, J.-H. Yum, J.E. Moser, M. Grätzel, N.-G. Park, Sci. Rep. **2**, 591 (2012)

28. L. Etgar, P. Gao, Z. Xue, Q. Peng, A.K. Chandiran, B. Liu, M.K. Nazeeruddin, M. Grätzel, J. Am. Chem. Soc. **134**, 17396 (2012)

29. M.M. Lee, J. Teuscher, T. Miyasaka, T.N. Murakami, H.J. Snaith, Science **338**, 643 (2012)

30. J. Hagen, W. Schaffrath, P. Otschik, R. Fink, A. Bacher, H.-W. Schmidt, D. Haarer, Synth. Met. **89**, 215 (1997)

31. G.K.R. Senadeera, P.V.V. Jayaweera, V.P.S. Perera, K. Tennakone, Sol. Energy Mater. Sol. Cells **73**, 103 (2002)

32. T. Minakata, I. Nagoya, M. Ozaki, J. Appl. Phys. **69**, 7354 (1991)

33. T. Minakata, H. Imai, M. Ozaki, K. Saco, J. Appl. Phys. **72**, 5220 (1992)

34. J. Salbeck, N. Yu, J. Bauer, F. Weissortel, H. Bestgen, Synth. Met. **91**, 209 (1997)

35. U. Bach, D. Lupo, P. Comte, J.E. Moser, F. Weissortel, J. Salbeck, H. Spreitzer, M. Grätzel, Nature **395**, 583 (1998)

36. R. Pudzich, T. Fuhrmann-Lieker, J. Salbeck, Adv. Polym. Sci. **199**, 83 (2006)

37. J. Burschka, A. Dualeh, F. Kessler, E. Baranoff, N.-L. Cevey-Ha, C. Yi, M.K. Nazeeruddin, M. Grätzel, J. Am. Chem. Soc. **133**, 18042 (2011)

38. J. Krüger, R. Plass, L. Cevey, M. Piccirelli, M. Grätzel, U. Bach, Appl. Phys. Lett. **79**, 2085 (2001)

39. H.J. Snaith, M. Grätzel, Appl. Phys. Lett. **89**, 262114 (2006)

40. G. Boschloo, L. Haggman, A. Hagfeldt, J. Phys. Chem. B **110**, 13144 (2006)

41. H.J. Snaith, M. Grätzel, Phys. Rev. Lett. **98**, 177402 (2007)

42. K. Kakiage, T. Tsukahara, T. Kyomen, M. Unno, M. Hanaya, Chem. Lett. **41**, 895 (2012)

43. H.J. Snaith, R. Humphry-Baker, P. Chen, I. Cesar, S.M. Zakeeruddin, M. Grätzel, Nanotechnology **19**, 424003 (2008)

44. I.-K. Ding, N. Tetreault, J. Brillet, B.E. Hardin, E.H. Smith, S.J. Rosenthal, F. Sauvage, M. Grätzel, M.D. McGehee, Adv. Funct. Mater. **19**, 2431 (2009)

45. I.K. Ding, J. Melas-Kyriazi, N.-L. Cevey-Ha, K.G. Chittibabu, S.M. Zakeeruddin, M. Grätzel, M.D. McGehee, Org. Electron. **11**, 1217 (2010)

46. H. Melhem, P. Simon, L. Beouch, F. Goubard, M. Boucharef, C. Di Bin, Y. Leconte, B. Ratier, N. Herlin-Boime, J. Bouclé, Adv. Energy Mater. **1**, 908 (2011)

47. C.-Y. Hsu, Y.-C. Chen, R.Y.-Y. Lin, K.-C. Ho, J.T. Lin, Phys. Chem. Chem. Phys. **14**, 14099 (2012)

48. J.E. Kroeze, N. Hirata, L. Schmidt-Mende, C. Orizu, S.D. Ogier, K. Carr, M. Grätzel, J.R. Durrant, Adv. Funct. Mater. **16**, 1832 (2006)

49. Y. Zhao, W. Chen, J. Zhai, X. Sheng, Q. He, T. Wei, F. Bai, L. Jiang, D. Zhu, Chem. Phys. Lett. **445**, 259 (2007)

50. R. Aich, F. Tran-Van, F. Goubard, L. Beouch, A. Michaleviciute, J.V. Grazulevicius, B. Ratier, C. Chevrot, Thin Solid Films **516**, 7260 (2008)

51. H.J. Snaith, S.M. Zakeeruddin, Q. Wang, P. Pechy, M. Grätzel, Nano Lett. **6**, 2000 (2006)

52. M. Juozapavicius, B.C. O'Regan, A.Y. Anderson, J.V. Grazulevicius, V. Mimaite, Org. Electron. **13**, 23 (2012)

53. S. Mathew, K.R. Haridas, Bull. Mater. Sci. **35**, 123 (2012)

54. T. Leijtens, I.K. Ding, T. Giovenzana, J.T. Bloking, M.D. McGehee, A. Sellinger, ACS Nano **6**, 1455 (2012)

55. A. Tomkeviciene, G. Puckyte, J.V. Grazulevicius, M. Degbia, F. Tran-Van, B. Schmaltz, V. Jankauskas, J. Bouclé, Synth. Met. **162**, 1997 (2012)

56. E.L. Unger, A. Morandeira, M. Persson, B. Zietz, E. Ripaud, P. Leriche, J. Roncali, A. Hagfeldt, G. Boschloo, Phys. Chem. Chem. Phys. **13**, 20172 (2011)

57. N. Metri, X. Sallenave, C. Plesse, L. Beouch, P.-H. Aubert, F. Goubard, C. Chevrot, G. Sini, J. Phys. Chem. C **116**, 3765 (2012)

58. F. Goubard, R. Aîch, F. Tran-Van, A. Michaleviciute, F. Wünsch, M. Kunst, J. Grazulevicius, B. Ratier, C. Chevrot, Proc. Estonian Acad. Sci. Eng. **12**, 96 (2006)

59. N. Metri, X. Sallenave, L. Beouch, C. Plesse, F. Goubard, C. Chevrot, Tetrahedron Lett. **51**, 6673 (2010)

60. Z.-S. Wang, N. Koumura, Y. Cui, M. Takahashi, H. Sekiguchi, A. Mori, T. Kubo, A. Furube, K. Hara, Chem. Mater. **20**, 3993 (2008)

16

Simulations of geometry effects and loss mechanisms affecting the photon collection in photovoltaic fluorescent collectors

L. Prönneke[1,a], G.C. Gläser[1], and U. Rau[2]

[1] Institut für Photovoltaik, Universität Stuttgart, Pfaffenwaldring 47, 70569 Stuttgart, Germany
[2] IEK5-Photovoltaik, Forschungszentrum Jülich, 52425 Jülich, Germany

Abstract Monte-Carlo simulations analyze the photon collection in photovoltaic systems with fluorescent collectors. We compare two collector geometries: the classical setup with solar cells mounted at each collector side and solar cells covering the collector back surface. For small ratios of collector length and thickness, the collection probability of photons is equally high in systems with solar cells mounted on the sides or at the bottom of the collector. We apply a photonic band stop filter acting as an energy selective filter which prevents photons emitted by the dye from leaving the collector. We find that the application of such a filter allows covering only 1% of the collector side or bottom area with solar cells. Furthermore, we compare ideal systems in their radiative limits to systems with included loss mechanisms in the dye, at the mirror, or the photonic filter. Examining loss mechanisms in photovoltaic systems with fluorescent collectors enables us to estimate quality limitations of the used materials and components.

1 Introduction

Fluorescent collectors (FCs) use organic dye molecules or inorganic fluorescent quantum dots surrounded by a dielectric material to trap and concentrate solar photons. The dye absorbs incoming photons with energy E_1 and emits photons due to Stokes shift with $E_2 < E_1$. The emission occurs with a randomized direction. Total internal reflection traps part of the radiation in the system and guides the photons to the collector sides. In a photovoltaic system, solar cells applied to the collector sides or the back side collect these photons and convert them into electrical energy. Already in the late 1970s and early 1980s Goetzberger, Wittwer and Greubel described the technological potential of FCs in photovoltaic systems [1,2]. Recently, the basic idea has regained some interest in the context of building photovoltaic structures which exceed the classical efficiency limitations by using up- and down-converting dyes [3–5]. Theoretical tools to describe FCs thermodynamically have been developed [6–8]. Numerical approaches analyzing the FC behavior gain more interest [9,10]. In order to estimate theoretical limitations, the photovoltaic systems with FCs have been highly idealized. However, realistic setups show loss mechanisms which need to be considered. The classic idea of assembling an FC in a photovoltaic system is based on its behavior of guiding emitted photons to the sides. Therefore, the classical setup mounts solar cells to the collector sides [2,11–17].

Technically, it seems less expensive to apply and connect solar cells at the bottom side of the collector. Experimentally, fluorescent collectors and photonic structures on top of solar cells prove to raise the output current by 95% compared to a non-fluorescent glass on top of the cell [18].

The present paper uses Monte-Carlo ray-tracing simulations for a comparison of the classical side-mounted system to a system where solar cells cover the FC back side. We see that the side-mounted system performs better in most cases, especially at larger collector sizes. However, for both systems the maximum collection probability for photons $p_c = 97\%$ is only achieved in the presence of a back side mirror and a photonic band stop (PBS) filter at the collector top surface acting as an energy selective filter. This maximal photon collection occurs in the statistical limit. Here, numerous small solar cells cover the FC in close proximity. Such a small-scale system is more favorable than a system with few large-scaled solar cells taking the same coverage fraction. The maximal number of collected photons is equivalent in a photovoltaic system with neither FC nor PBS, but our assembly saves us 99% of solar cell area.

In order to describe FCs in photovoltaic systems, we use numerical and analytical approaches based on the principle of detailed balance. Starting from ideal systems in their radiative recombination limitation, we also examine the influences of non-radiative recombination in the fluorescent dye, of non-perfect reflection at the mirrors, and of non-perfect reflection conditions at the PBS filter. The results point out that reflection losses at the back

---
[a] e-mail: `liv.proenneke@ipv.uni-stuttgart.de`

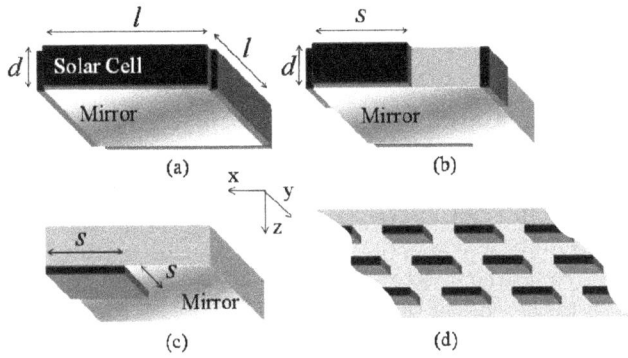

Fig. 1. Sketch of the fluorescent collector geometries compared in the present paper (seen from the bottom). (a) Classical design with the solar cells mounted at each side of the collector with length l and thickness d. (b) Modified classical system where only a fraction of the respective collector sides is covered with a solar cell area $A_{cell} = ds$. (c) Collector with solar cells with an area $A_{cell} = s^2$ mounted at the bottom. In all cases the (remaining) back side is covered with a mirror. (d) Systems (b) and (c) are assumed to be periodic in space. Exemplary, a detail of the bottom-mounted system is shown.

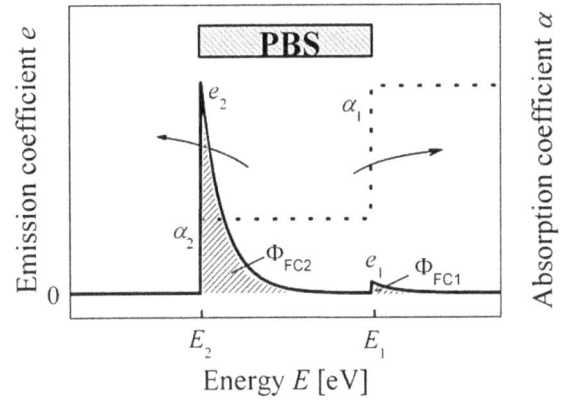

Fig. 2. Sketch of the absorption and emission behavior as assumed in this paper. The dye absorption is given by a step function. Incoming photons have the energy E_1 and a high absorption coefficient α_1. The lower absorption coefficient α_2 holds for the lower energy E_2 and leads with Kirchhoffs law (Eq. (1)) to a high emission coefficient e_2. The model also features the possibility of an energy selective photonic band stop (PBS) that keeps the emitted photons in the FC system.

surface cause a higher decrease than losses due to non-radiative recombination in the dye. Compared to a system without applied PBS non-radiative losses induce higher decreases of photon collection in the PBS covered system.

2 Collector geometries and dye properties

This section describes the three photovoltaic systems discussed in this paper. A characterization of the FC dye properties as well as an explanation of the functionality of a PBS filter follows.

Figure 1a shows an FC in the classical configuration with an acrylic plate of length l and thickness d doped with fluorescent dye. The collecting solar cells are mounted at the sides of the plate. Let us define the coverage fraction $f = A_{cell}/A_{coll}$ as the ratio between the area $A_{cell} = 4dl$ of the solar cells in the system and the illuminated collector area $A_{coll} = l^2$. For the configuration in Figure 1a, we have $f = 4dl/l^2 = 4d/l$, hence the coverage fraction depends only on the ratio between the collector thickness d and the side length l. A perfect mirror covers the FC back side. Figure 1b features a variant of the side-mounted FC where only a part of each side is covered with a solar cell. The system is repeated periodically in x- and y-direction. Therefore, photons hitting a collector side experience periodic boundary conditions and enter the opposite side. In an alternative but equivalent perception perfect mirrors cover the remainders. The coverage fraction for the system in Figure 1b is $f = 4ds/l^2$ with the side length $s \leq l$ of the solar cells. Thus, coverage fraction f and collector length l are decoupled and this geometry offers an additional degree of freedom for the collector design. Again, the FC back side is covered by a mirror. The collector design of Figure 1c uses a square solar cell with a side length s at the back side of the FC. Thus, the solar cells in this

bottom-mounted system cover a fraction $f = s^2/l^2$ of the surface. Figure 1d shows a detail of the bottom-mounted system which is also assumed to be periodically repeated in x- and y-direction. As shown, square solar cells occupy the back surface of the collector with a period length l. The remaining parts of the back side are covered with a mirror.

We model an FC consisting of an acrylic plate with the refractive index $n_r = 1.5$ and embedded fluorescent dye molecules. Figure 2 depicts the absorption/emission behavior of the fluorescent dye used in the following. We assume a stepwise increase of the absorption constant α from zero at energies $E < E_2$ to a value α_2 for $E > E_2$ and a further increase to α_1 for energies $E > E_1$. The emission coefficient e is linked to the absorption coefficient α via Kirchhoffs law

$$e(E) = \alpha(E)n_r\phi_{bb}(E) \tag{1}$$

with the black body spectrum

$$\phi_{bb}(E) = \frac{2}{h^3c^2}\frac{E^2}{e^{E/kT}-1} \approx \frac{2E^2}{h^3c^2}e^{-E/kT} \tag{2}$$

where n_r is the refractive index of the collector material, h is Planck's constant, c the speed of light, and kT the thermal energy corresponding to the temperature T of the collector and its surroundings ($T = 300$ K, throughout this paper).

The absorption/emission dynamics used in the following is given by a two-level scheme as used earlier to describe the detailed balance limit of FCs [11, 12, 19]. The choice of this simple approach ensures a certain generality of our results such that the trends caused by the collector geometries or by the introduction of loss mechanisms should be equally found in real systems with more complex spectral absorption/emission properties. For the present two-level system we consider the emission probabilities

$$p_1 = \frac{\alpha_1}{p} \int\limits_{E_1}^{\infty} E^2 \exp\left(-\frac{E}{kT}\right) dE = \frac{\alpha_1 p_\infty \left(E_1\right)}{p} \quad (3)$$

and

$$p_2 = \frac{\alpha_2}{p} \int\limits_{E_2}^{E_1} E^2 \exp\left(-\frac{E}{kT}\right) dE = \frac{\alpha_2 \left[p_\infty \left(E_2\right) - p_\infty \left(E_1\right)\right]}{p} \quad (4)$$

for photon emission by the fluorescent dye in the range of photon energies $E > E_1$ and $E_1 > E > E_2$, respectively. In equations (3) and (4), we use the definition

$$p_\infty \left(E_x\right) = \int\limits_{Ex}^{\infty_1} E^2 \exp\left(-\frac{E}{kT}\right) dE$$

$$= kT \left[2 \left(kT\right)^2 + 2 E_x kT + E_x^2\right] \exp\left(-\frac{E}{kT}\right) \quad (5)$$

and the normalization factor p such that $p_1 + p_2 = 1$.

The choice of the energies E_1, E_2, and the absorption coefficients α_1, α_2 leads to the emission probabilities $p_2 \gg p_1$, in contrast to the absorption coefficients $\alpha_1 \gg \alpha_2$. Due to the dominance of the exponential factor in equation (5) we approximate

$$\frac{p_1}{p_2} \approx \frac{\alpha_1}{\alpha_2} \exp\left(\frac{E_2 - E_1}{kT}\right). \quad (6)$$

Thus, a choice of an energy difference $\Delta E = E_1 - E_2 = 200$ meV and of absorption coefficients $\alpha_1 = 100\alpha_2$ still ensures $p_2 \approx 20 p_1$. In the following, we assume $E_1 = 2.0$ eV, $E_2 = 1.8$ eV and absorption coefficients $\alpha_1 = 3/d$, $\alpha_2 = 0.03/d$. Therefore, the system provides a high emission coefficient e_2 for photons with energy E_2 and a significantly lower emission coefficient e_1 for photons with high energies.

Figures 3a–3e sketch the functionality of the PBS filter. Incoming photons have the energy E_1. The dye absorbs these photons and emits spatially randomized photons with angles θ, ϕ defined in Figure 3a at a lower energy E_2. Figure 3b shows that emitted photons impinging at the top surface with an incident angle θ higher than the critical angle θ_c for total internal reflection are guided to the collector sides. Whereas photons with $\theta < \theta_c$ leave the collector as shown in Figure 3c. The application of a PBS filter avoids this loss mechanism. Figure 3d shows the ideal PBS filter which is energy selective only $(\theta_{pbs} = \theta_c)$. The filter has a reflection $R = 1$ and a transmission $T = 0$ for photon energies $E < E_{th}$. For the other part of the spectrum $R = 0$ and $T = 1$ is assumed. Therefore, E_{th} denotes the upper cut-off energy of the filter. We choose $E_{th} = E_1$ throughout this paper. Two- and three-dimensional photonic crystals [20–22] are promising materials which might be used as omnidirectional PBS

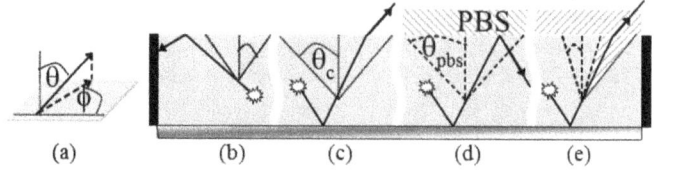

Fig. 3. Light guiding behavior of a fluorescent collector covered with solar cells at the sides and a mirror at its back side. (a) Definition of photon ray angle θ. (b) Absorbed photons are reemitted spatially randomized. The system leads rays with $\theta > \theta_c$ to the sides of the collector. (c) Rays with angle $\theta < \theta_c$ for total internal reflection leave the top surface. (d) Applying a photonic band structure (PBS) keeps rays with $\theta < \theta_{pbs}$ in the system as well. This PBS is energy selective with $\theta_{pbs} = \theta_c$. Therefore, rays with energies $E \leq E_1$ are kept in the system only. (e) For an energy and angular selective PBS a reflection cone is assumed such that only rays with $E \leq E_1$ and $\theta < \theta_{pbs}$ are kept in the system.

in FC systems. However, technological developments have led to dielectric mirrors used as band pass filters with almost rectangular cut-off characteristics for normal incident photons [23,24]. These rugate-filters show a high angular dependency by blocking only photons with almost perpendicular incidence. In order to examine the influence of this angular selectivity, we vary the reflection cone of the filter. Figure 3e depicts that a PBS with $\theta_{pbs} < \theta_c$ reflects photons with $E \leq E_1$ and $\theta < \theta_{pbs}$. Thus, rays with $\theta_c > \theta > \theta_{pbs}$ hitting the collector surface within the striped angle cone are neither reflected by the PBS nor subject to total internal reflection and leave the system.

3 Simulation method

The Monte-Carlo simulation calculates the collection probability for photons p_c for the different collector geometries shown in Figures 1a–1c with varied collector dimensions and component quality. In order to allow also the comparison between systems with and without PBS, we provide only incoming photons with energy $E = E_1$. All photons enter into the collector perpendicular with random coordinates (x, y) with $0 < x < l$ and $0 < y < l$. Their statistical absorption occurs following Beer's absorption law after a path length

$$w = -\frac{1}{\alpha_1} \ln \left(p_w\right) \quad (7)$$

where p_w is a random number $0 \leq p_w \leq 1$. After its absorption a photon is re-emitted with a probability $p_e = 1 - p_{nr}$ with the non-radiative recombination probability p_{nr} of the fluorescent dye. According to equations (3) and (4) the energy of the re-emitted photons lies with the probability p_1 in the energy range $E \geq E_1$ and with p_2 in the range $E_1 > E \geq E_2$. After re-emission the photon also obtains a pair of spherical angles (θ, ϕ) with $0 < \theta < \pi$ and $0 < \phi < 2\pi$ using the probabilities $p_\theta = \sin(\theta)/2$ and $p_\phi = 1/2\pi$ (for the definition of θ and ϕ, see Fig. 3a).

Subsequently, either the dye molecules reabsorb the re-emitted radiation or the photons hit one of the six collector surfaces at a coordinate (x_s, y_s, z_s). At the *top surface* $(x_s, y_s, 0)$, the photon is reflected if $\theta > \theta_C$ with $\sin(\theta_C) = 1/n_r$. In the presence of an omnidirectional PBS, the photon is reflected for all photon energies $E < E_1$. An assumed angular selectivity sets as a reflection condition $E < E_1$ and $\theta < \theta_{pbs}$ as defined in Figure 3e. Non-reflected photons are lost and the number N_{lost} of lost photons is increased accordingly. At the *bottom surface* (x_s, y_s, d) a mirror perfectly reflects the photons with a probability $p_r = 1$. We use $p_r < 1$ for the analysis of loss mechanisms. For the system shown in Figure 1c, the bottom-mounted solar cells collect photons with $x_s \leq s$ and $y_s \leq s$. In this case, the photons add to the number N_{coll} of collected photons. A special case discussed below is the *statistical limit* where we simply assume that a photon hitting the bottom of the collector enters a solar cell with the probability f, the solar cell coverage fraction. Throughout this paper, we assume a collection probability of 100% for photons hitting the solar cell area with energy E higher than the solar cell band gap E_{gap}. In order to analyze the principle limitations of applying FCs to photovoltaic systems, we choose $E_{gap} = 1.8$ eV which corresponds to the emission peak of the FC.

If photons hit the *collector sides*, for instance at (l, y_s, z_s) for the right collector side, side-mounted solar cells collect the photons for the geometry shown in Figure 1a. The geometry depicted in Figure 1b collects photons hitting the collector right surface with $y_s \leq s$. Otherwise, the photon is subject to periodic boundary conditions. For the bottom-mounted system, we apply periodic boundary conditions on each collector side, i.e. the photon is re-injected at the respective facing side with unchanged spherical angles (θ, ϕ).

Our ray tracing program, that typically handles a number $N_{in} = 5 \times 10^4$ photons in parallel, runs until all photons are collected by the solar cells, lost by re-emission from the collector surface, by non-radiative recombination in the dye, by reflection losses at the mirrors or at the PBS. With $N_{in} = N_{lost} + N_{coll}$, we obtain the collection probability for photons $p_c = N_{coll}/N_{in}$ as the final result.

4 Simulation results

In this section, we describe the simulation results. First, the classical FC system with side-mounted solar cells as shown in Figure 1a is modeled. Here, the variation of the collector dimensions modifies the coverage fraction of solar cells. Second, photon collection probabilities for the bottom-mounted system as depicted in Figure 1c are calculated. In these simulations, we already include the loss mechanism of non-radiative recombination in the dye. The influences of collector dimension and coverage fraction are de-coupled in this system. In the third part, the approach of de-coupling these two aspects also in the side-mounted system as sketched in Figure 1b allows the comparison between side-mounted and bottom-mounted

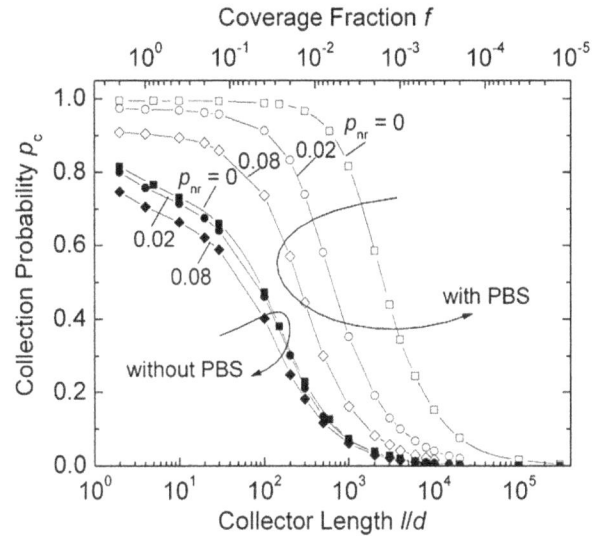

Fig. 4. The collection probability p_c of the classical fluorescent collector geometry (Fig. 1a) fully covered with solar cells at its sides depends on the collector length l normalized to the thickness d. A perfect mirror covers the back side. The systems with a photonic band stop filter (PBS, open symbols) have higher collection probabilities than those without PBS (filled symbols). For the system with PBS a value of p_c close to unity remains up to a normalized collector length $l/d \approx 500$ and non-radiative recombination probability $p_{nr} = 0$ in the dye. In contrast, the system without PBS has a maximum p_c only slightly above 80% even in the radiative case with $p_{nr} = 0$. With increasing collector length l/d the collection probability p_c decreases immediately. The consideration of non-radiative recombination in the dye, i.e. $p_{nr} > 0$, leads to a deterioration of p_c in all cases. The cell coverage fraction f (top axis) is directly linked to the collector size for the chosen geometry.

system. We compare both systems, first at constant collector lengths for varied coverage fractions. Secondly, at constant coverage fractions under the inclusion of three loss mechanisms: a non-radiative recombination in the dye, a non-perfect mirror at the collector back side, and an angular selectivity at the photonic structure lying on top of the collector.

4.1 Classical collector geometry

This subsection studies the influence of the collector and cell geometry on the photon collection properties of a classical collector system as shown in Figure 1a. Figure 4 shows the dependence of the collection probability p_c on the collector length l/d normalized to the collector thickness d calculated for systems with and without PBS at the top surface. We have also considered non-radiative recombination in the dye by assuming $p_{nr} = 0.02$ and 0.08.

The most important feature in Figure 4 is that the systems with PBS have a considerably higher collection probability p_c than those without PBS. This is because the PBS decreases the emission of photons through the surface of the collector as shown in Figure 3c. For systems without PBS this non-radiative loss occurs whenever a photon falls into the critical angle θ_c of total reflectance. For the system with PBS the photon additionally must be emitted at an energy $E \geq E_1$. This emission probability is low, but non-zero for reasons of detailed balance. For the same reason systems with PBS obtain the high values of p_c also for larger collector lengths l/d, whereas for the systems without PBS p_c drops considerably already upon slight increases of l/d. Furthermore, with increased l/d also the number of photons absorbed by the dye a second or third time increases. Each absorption event leads to θ-randomization of the re-emitted photon and, in consequence, to a certain probability that the photon is lost by emission from the collector surface.

Also shown in Figure 4 are curves that reflect non-radiative recombination in the dye ($p_{nr} = 0.02$ and 0.08). We see that the system with PBS is especially sensitive to non-radiative recombination in the range of large l/d. At low values of l/d a photon is most likely absorbed only once before collected by the side-mounted solar cells. The maximum of p_c decreases therefore only proportionally to $(1 - p_{nr})$. For higher values of l/d, repeated re-absorption of photons not only increases the risk for radiative but now also for non-radiative losses. As radiative losses are low in the systems with PBS the relative importance of non-radiative losses is higher. Whereas in the radiative case a value of $p_c > 90\%$ remains up to a normalized collector length $l/d \approx 500$, for $p_{nr} = 0.02$ and 0.08 we have $p_c > 90\%$ only for $l/d \leq 100$ and $l/d \leq 6$, respectively. The changes that occur by non-radiative recombination in the case of systems without PBS over the whole range of l/d are less significant due to the high emission losses that are present in the system anyway.

4.2 Scaling effect

The classical side-mounted collector geometry has a strict relation between the collector length l and the cell coverage fraction $f = A_{cell}/A_{coll} = 4d/l$. In this subsection we examine the bottom-mounted system shown in Figure 1c where both quantities can be treated independently. For Figure 5 the coverage fraction $f = 0.01$ is fixed and we vary the collector length l. Because of $f = s^2/l^2$, we adjust the cell side length s to $s = l\sqrt{f}$.

Figure 5 demonstrates that the collection probability p_c at constant coverage fraction f drastically depends on the collector length l/d. The data in Figure 5 display asymptotic behavior in both limits, for small and large ratios l/d. We observe a wide transition regime between the two limiting cases where the collection probability p_c changes from high values at small l/d to significantly smaller values at large l/d. Such a behavior is typical for spatially extended inhomogeneous systems. If the characteristic feature length (here the collector length l) is large with respect to the length scale that is characteristic for interactions within the system (here the mean free path of photons), the system can be looked at as a parallel connection of spatially separated subsystems without interaction. The collection probability $p_{c,ls}$ in this *large scale limit* [25] is then the weighted average of a portion f that

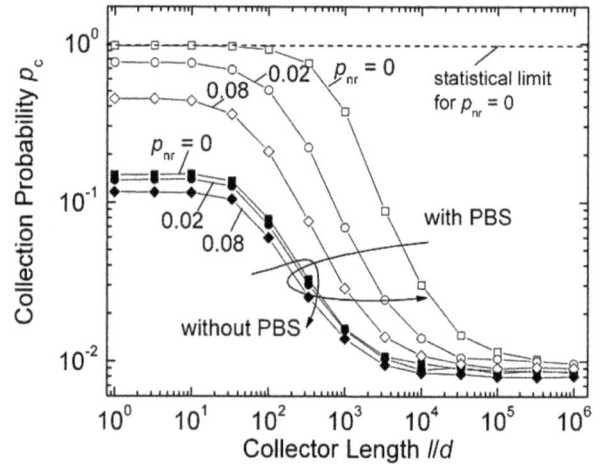

Fig. 5. Collection probability p_c of the fluorescent collector geometry with solar cells mounted at the bottom of the collector (Fig. 1c). The coverage fraction f is kept constant at $f = 0.01$ and the normalized collector length l/d is varied. All data feature a transition between two asymptotic situations at low and high ratios l/d. Only the systems with a photonic band pass (PBS, full symbols) achieve $p_c > 0.5$ at low ratios of l/d. For these systems, the maximum p_c as well as the transition from the small-scale to the large-scale behaviour is strongly dependent on the non-radiative recombination probability p_{nr}. The systems without PBS (open symbols) have already a relatively low $p_c < 20\%$ even in the more favorable case of $l/d < 10$. Accordingly, the sensitivity to the introduction of non-radiative recombination in the dye ($p_{nr} > 0$) is less pronounced.

has a local p_c of a collector with full back coverage, i.e. $p_c \approx 1$, for the system with $p_{nr} = 0$ and a portion $(1 - f)$ with $p_c = 0$. This is why we observe in this limit that p_c of the system with PBS approaches the coverage fraction f.

In contrast, approaching the *small scale* limit ($l < 1/\alpha_2$, marked in Figure 5 with dashed line [26]), any ray, reflected forth and back within the collector, has often the possibility to hit a cell at the collector back side. In this situation, the system might be looked at as spatially homogeneous with statistical cell coverage at its back. In fact, the value of $p_{c,ss}$ in the small scale limit is consistent with quasi-one-dimensional computations that simulate the bottom-mounted solar cells by a probability $p_c = f$ for a photon to be collected by cells at the back side. Therefore, we denote this case also as the *statistical limit*.

The introduction of non-radiative recombination ($p_{nr} = 0.02, 0.08$) in Figure 5 leads to a deterioration of the collection probability in all cases. As we have already seen in Figure 4, the system with PBS is much more sensitive to losses in the dye because of its overall high collection probability in the radiative limit. Especially important is the influence of a finite p_{nr} on the transition from the small-scale to the large-scale limit. Whereas with $p_{nr} = 0$ all systems with $l/d \leq 100$ yield the same high collection probability, the limit for $p_{nr} = 0.08$ reduces to $l/d \leq 10$.

All data in Figure 5 represent the same coverage $f = 0.01$, i.e. the same amount of solar cell area per unit collector area. Nevertheless, the collection probability strongly depends on the chosen size of collector and solar cell. A proper scaling of these quantities is therefore necessary to tune the collection probability and, finally, the collector performance to its optimum.

4.3 Statistical limit

The following section derives an analytical description for the statistical limit marked in Figure 5 with the dashed line [27]. This analytical description only holds for the statistical limit in the radiative case and for systems with applied photonic structure. As discussed by Markvart [28] and Rau et al. [29], using photon fluxes only describes systems with equal chemical potential μ for the incoming photons. Applying a PBS equalizes μ because the absorption coefficient for all incoming photons is now α_1. In the Monte-Carlo simulation, we excite the system with a monochromatic beam with $E = 2.22$ eV. The results fit the calculation of the analytical expression because both cases fulfill the condition of equalized μ. In contrast, the system without PBS does not provide a spectrally equal absorption for all incoming photons leading to an inhomogeneous μ. Additionally, photons experience a spatial inhomogeneity for systems beyond the statistical limit as depicted in the former section. This also leads to an inhomogeneous μ. Considering these limitations, we compare in the following a side- and a bottom-mounted system in the statistical limit ($l/d = 1$) with applied PBS.

The Monte-Carlo simulation gives us the expressions for the photon flux incident on the collector surface

$$\varphi_{sun}^{MC} = \frac{N_{in}}{A_{coll}}, \tag{8}$$

the photon flux which is absorbed in the collector by the solar cells

$$\varphi_{FC}^{MC} = \frac{N_{coll}}{A_{cell}}, \tag{9}$$

and the photon flux leaving the collector without hitting a solar cell

$$\varphi_{out}^{MC} = \frac{N_{in} - N_{coll}}{A_{coll}}. \tag{10}$$

We propose that the ratio between the incident photon flux Φ_{sun} and the photon flux Φ_{FC} kept in the collector by total internal reflection gives the expression

$$c_{TIR}^{max} = \Phi_{FC}/\Phi_{sun} \tag{11}$$

which is at the same time the maximum concentration of a concentrator based on total internal reflection only. A collector with a refractive index of $n = 1.5$ yields a concentration $c_{TIR} = 2.25$. Inside a FC with a dielectric material doped with a spectral shifting dye, the overall flux

$$\Phi_{FC} = \Phi_{FC1} + \Phi_{FC2} \tag{12}$$

is composed of two fluxes Φ_{FC1} and Φ_{FC2} as depicted in Figure 2. Here,

$$c_{p,max} = (\Phi_{FC1} + \Phi_{FC2})/\Phi_{sun} \tag{13}$$

denotes the maximum concentration inside the FC. By integrating over the corresponding sections on the energy axes, the analytical expressions for the two fluxes inside the collector

$$\Phi_{FC1} = \frac{2n^2}{h^3 c^2} \int_{E_1}^{\infty} E^2 e^{-E/kT} dE \tag{14}$$

and

$$\Phi_{FC2} = \frac{2n^2}{h^3 c^2} \int_{E_2}^{E_1} E^2 e^{-E/kT} dE \tag{15}$$

are derived. The PBS reflecting all photons with energy $E_2 < E < E_1$ limits the incident photon flux

$$\Phi_{sun} = \frac{2}{h^3 c^2} \int_{E_1}^{\infty} E^2 e^{-E/kT} dE. \tag{16}$$

With equations (14) to (16) it is possible to calculate the maximum concentration analytically

$$c_{p,max} = n^2 \frac{\left(E_2^2 + 2kTE_2 + 2(kT)^2\right) e^{-E_2/kT}}{\left(E_1^2 + 2kTE_1 + 2(kT)^2\right) e^{-E_1/kT}} = 4251 \tag{17}$$

with $kT = 0.0258$. In order to reach $c_{p,max}$, the system has to be in open circuit condition, thus, no solar cells are present at the bottom of the collector.

In order to derive values for the fraction dependent concentration $c_p^{an}(f)$ analytically, we follow the interpretation of the FC system of Glaeser and Rau [12], Meyer and Markvart [30]. Here, the collection probability p_c follows the expression

$$p_c^* = \frac{A_{cell} \Phi_{FC}^*}{A_{cell} \Phi_{FC}^* + A_{coll} \Phi_{out}^*} \tag{18}$$

with the photon flux absorbed by the solar cells

$$\Phi_{FC}^* = \frac{2n^2}{h^3 c^2} e^{\mu/kT} \int_{E_1}^{\infty} E^2 e^{-E/kT} dE \tag{19}$$

and the emitted photon flux leaving the collector without hitting a solar cell

$$\Phi_{out}^* = \frac{2}{h^3 c^2} e^{\mu/kT} \int_{E_2}^{\infty} E^2 e^{-E/kT} dE. \tag{20}$$

Note, that Markvart proposes an analytical approximation by introducing uniform chemical potentials μ for all photons in the system. This assumption is exact as long as the system is in open circuit condition. Then the system is in thermal equilibrium and the chemical potential for all photons is equal. This is the same condition for the system Rau et al. [29] describe. As derived in equation (11), the concentration is the ratio between the flux

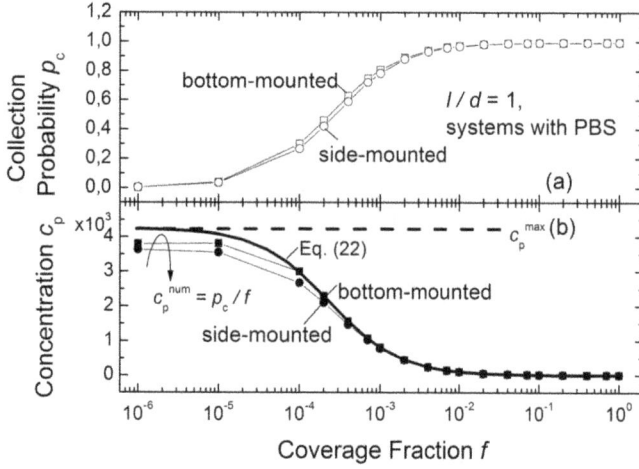

Fig. 6. (a) Collection probabilities p_c for photons derived with a Monte-Carlo simulation for a side- and a bottom-mounted system in the statistical limit and with applied PBS. (b) Symbol-lines are the values for the concentration c_p derived from p_c in (a). Solid lines show the analytically calculated $c_p(f)$ from equation (22). Side-mounted system shows less agreement between the analytically and the numerically derived concentration because photons entering the system close to a solar cell are absorbed with a higher probability. This leads to an inhomogeneous chemical potential μ for the incoming photons which violates the assumptions for the calculation. But, the approach μ = const. is a good approximation in the statistical limit for the bottom-mounted solar cells [27].

inside the collector Φ_{FC} and the incident flux Φ_{sun}. Following the approximately uniform μ proposed by Markvart et al. and comparing equation (13) to the result of dividing equation (19) by equation (20), we also understand the maximum concentration as

$$c_{p,max}^* = \frac{\Phi_{FC}^*}{\Phi_{out}^*}. \tag{21}$$

Linking the descriptions of the system via equalizing the photon fluxes in both descriptions with the values derived from the Monte-Carlo simulation ($\varphi_i^{MC} = \Phi_i = \Phi_i^*$) yields the analytical expression for the concentration

$$
\begin{aligned}
c_p^{an}(f) = c_{p,max} &= \frac{\varphi_{FC}^{MC}}{\varphi_{sun}^{MC}} = \frac{N_{coll}}{A_{cell}}\frac{A_{coll}}{N_{in}} = \frac{p_c}{f} \\
&= \frac{1}{f}\frac{A_{cell}N_{coll}/A_{cell}}{A_{cell}N_{coll}/A_{cell} + (N_{in} - N_{coll})} \\
&= \frac{1}{f}\frac{A_{cell}\varphi_{FC}^{MC}}{A_{cell}\varphi_{FC}^{MC} + A_{coll}\varphi_{out}^{MC}} = \frac{p_c^*}{f} \\
&= \frac{A_{coll}}{A_{cell}}\frac{A_{cell}\varphi_{FC}^{MC}/\varphi_{out}^{MC}}{A_{cell}\varphi_{FC}^{MC}/\varphi_{out}^{MC} + A_{coll}\varphi_{out}^{MC}/\varphi_{out}^{MC}} \\
&= \frac{c_{p,max}^*}{c_{p,max}^*f + 1}
\end{aligned}
\tag{22}
$$

with the coverage fraction $f = A_{cell}/A_{coll}$. In Figure 6b we derive the numerical concentration $c_p^{num}(f) = p_c/f$ from the numerical simulated collection probability p_c of Figure 6a and compare this with the analytically calculated

concentration $c_p^{an}(f)$ from equation (22). The analytical solutions excellently fit the statistically derived values for the bottom-mounted system. As described above, the system is described in thermal equilibrium and with same chemical potential μ for all incoming photons. In particular it holds $qV_{oc} = \mu$ for the cell at the side of the collector. Yet, under short circuit conditions, the voltage V of the cell equals zero and at the solar cell the chemical potential of the photons is $\mu = 0$. Therefore, μ cannot be constant throughout the system. (This finding is also derived by the consideration that in short circuit condition photons enter the system and there they are transported to the solar cells.) A net flux of photons requires local differences in the chemical potential. In a more coarse resolution of the solar cells, at which the period length l exceeds the mean free path of the photons as it is the case for the side-mounted system, the results are not valid any more. Photons entering the system close to a solar cell are absorbed with a higher probability, whereas photons in areas with no solar cell are most likely reabsorbed by the dye and, with a higher probability, reemitted from the collector. Systems in Figures 1a and 1b become identical. Thus, in Figure 6b the side-mounted system shows less agreement between the analytically and the numerically derived concentration. However, the approach μ = constant is a good approximation in the statistical limit for the bottom-mounted solar cells [27]. We achieve the open-circuit condition by reducing the coverage fraction f of solar cell area to photovoltaic unreasonable low values. The collection probability p_c decreases with decreasing f, and reduces to nearly zero at $f < 10^{-5}$. Concurrently, the concentration approaches the theoretical maximum $c_{p,max} = 4251$. Note, that c_p is reaching the maximum $c_{p,max}$ for coverage fractions f at which p_c is almost zero. The thermodynamic limit of the concentration lies therefore beyond photovoltaic useful collector dimensions.

4.4 Comparison of side-mounted and bottom-mounted FC

This and the following section compare FCs with bottom-mounted to FCs with side-mounted solar cells. In order to compare the bottom-mounted FC, where the coverage fraction and the collector length are decoupled, with a side-mounted system, we use the modified side-mounted system displayed in Figure 1b. Keeping a constant coverage fraction f upon variation of the collector length l requires the adjustment of the solar cell side length s to

$$s = f\frac{l^2}{4d}, \tag{23}$$

likewise $s/d = f(l/d)^2/4$ for the normalization of all quantities to the collector thickness d. The maximum coverage fraction f_{max} for the side-mounted system is given by

$$f_{max} = \frac{4d}{l} \tag{24}$$

because in this case the side length s of the solar cell in equation (23) equals the collector length, and the systems in Figures 1a and 1b become identical.

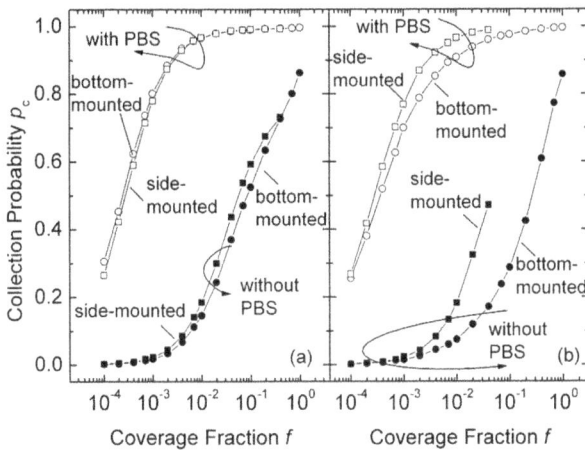

Fig. 7. Comparison of FC systems with solar cells at the sides or at the back as sketched in Figures 1b and 1c. (a) Systems at collector length $l = 10d$. Without PBS side-mounted solar cells provide slightly higher collection probabilities for $10^{-3} < f < 4 \times 10^{-1}$. Applying PBS eliminates the difference in collection probability p_c for high coverage fractions. Both systems achieve approximately $p_c = 1$ for coverage fractions $f \approx 1$. For coverage fractions $f < 10^{-2}$ mounting solar cells at the FC back side is of slight advantage. (b) Systems at collector length $l = 100d$. Mounting solar cells at collector sides leads to higher collection probabilities for all cases. Compared to Figure 7a, the side-mounted system obtains the same values. Therefore, this system works still in the small-scale limit, whereas for this collector length the bottom-mounted system is already in the transition regime to the large-scale limit which is also shown in Figure 5.

Figures 7a, 7b compare the collection probabilities of side- and bottom-mounted systems for fixed collector length $l = 10d$ (Fig. 7a) and $100d$ (Fig. 7b). Therefore, for the side-mounted system coverage fractions up to $f_{max} = 0.4$ and 0.04 respectively are modeled.

Figure 7a presents a comparison between bottom-mounted and side-mounted system, both with a collector length $l = 10d$. The results outline that without applied PBS the side-mounted system performs better for low coverage fractions. The application of PBS on top of the collector yields higher photon collection for the bottom-mounted system in this region. In the region of $f > 10^{-2}$ both systems reach collection probabilities close to 100%. As shown in Figure 5 a collector with bottom-mounted solar cells works in the statistical limit. Without the application of a PBS side-mounted solar cells provide slightly higher collection probabilities for $10^{-3} < f < 4 \times 10^{-1}$.

Figure 7b shows the comparison of the two systems with collector length $l = 100d$. Compared to Figure 7a, the side-mounted system obtains the same values for p_c. This implies that this system still works in the statistical limit whereas the bottom-mounted system passes into the large-scale region. This is accordingly seen in Figure 5. Therefore, in Figure 7b the FC system with solar cells covering the sides performs better at all coverage fractions with or without PBS.

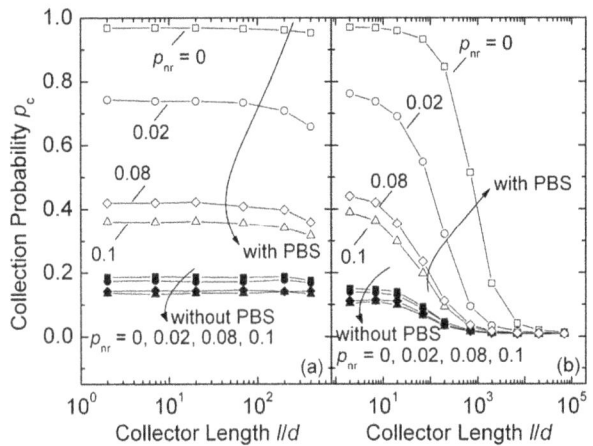

Fig. 8. Comparison of FC systems depicted in Figures 1b and 1c with a constant coverage fraction $f = 0.01$ analyzing influence of non-radiative losses in the dye. (a) Side-mounted system is only modeled up to a collector length $l \leq l_{max} = 4d/f$ corresponding to a full coverage of all collector sides. Inclusion of non-radiative losses deteriorates the photon collection p_c in all cases. The effect is more significant in the system with applied PBS. (b) Bottom-mounted system. The simulation covers a wider range of collector lengths. With PBS the bottom-mounted system in its radiative limit shows a significant drop in p_c for $l/d \geq 100$. The inclusion of non-radiative losses reduces this limit to $l/d \geq 1$. In comparison: for $p_{nr} = 0$ (only radiative losses) the side-mounted system leads to a better photon collection p_c, especially at large values of l. Inclusion of non-radiative losses leads to the same behavior in the largescale region, but for collector lengths in the statistical limit, the bottom-mounted system performs better. Without PBS the side-mounted system yields higher p_c at all collector lengths even with inclusion of non-radiative losses.

4.5 Loss mechanisms

This section analyzes the systems in the radiative limit as well as the influence of non-radiative losses in the dye, of a non-perfect mirror at the back side, and of a non-ideal reflection cone at the PBS filter. Unlike in Figures 7a, 7b, we now vary the collector length l/d while assuming a constant coverage fraction $f = 0.01$ for all simulations and an additional $f = 0.9$ for Figures 9a, 9b. Thus, the side-mounted system is modeled up to a maximum collector length $l = l_{max} = 4d/f = 400d$ and $4.44d$, respectively, where the side length s of the solar cells equals the collector length l.

4.5.1 Non-radiative losses in the dye

Figures 8a, 8b compare the side- and the bottom-mounted systems in their radiative limit with a probability $p_{nr} = 0$ for non-radiative losses and under the influence of non-radiative recombination in the dye ($p_{nr} = 0.02, 0.08$ and 0.1). Figure 8a shows the results for the side-mounted system. In the radiative limit the application of a PBS increases the photon collection from $p_c \approx 19\%$ to $p_c \approx 95\%$. These values remain constant up to the maximal simulated collector length $l_{max} = 400d$. As discussed above,

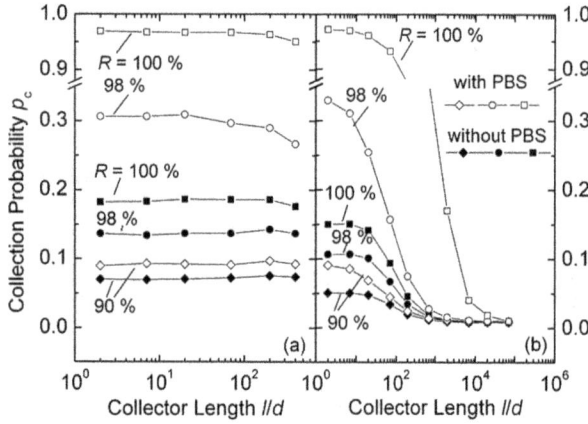

Fig. 9. Influence of non-perfect mirror at the back side of the FC in systems with a coverage fraction $f = 0.01$. (a) Side-mounted solar cells. Inclusion of non-radiative losses in the dye decreases the photon collection significantly. The relative deterioration in a system with applied PBS is higher than in the system without PBS. (b) Bottom-mounted solar cells. At small l/d the system shows a similar behavior as the side-mounted system.

this means that the re-absorption length in this system lies considerably under the collector length.

Figure 8b presents the results for the bottom-mounted system. Note, that the simulation covers a larger range of collector lengths l/d. As explained above, enlarging the systems degrades p_c even more than the inclusion of non-radiative recombination in the dye. However, the lowering in the photon collection occurs with $p_{nr} = 0$ at $l/d = 100$, but this limit reduces to $l/d \leq 10$ for $p_{nr} = 0.1$.

Under the application of PBS in the radiative limit both systems display a similar $p_c \approx 96.5\%$ in the limit of small collector lengths $l/d < 10$. Whereas the data for the bottom-mounted system considerably decay at $l/d > 100$, a high collection probability of $p_c > 95\%$ is maintained by the side-mounted system up to the maximum collector length $l_{max}/d = 400$. The inclusion of non-radiative losses in the calculation deteriorates the collection probability in both cases significantly. Also, the transition to large-scale limit occurs at lower collection lengths. Interestingly, in the small-scale limit $l/d < 10$ the bottom-mounted system achieves a better photon collection than the side-mounted system. Here, the collection probability decreases by $\Delta p_c \approx 52\%$ from $p_c \approx 98\%$ to $p_c \approx 46\%$ whereas in the side-mounted system it decreases by $\Delta p_c \approx 55\%$.

Comparing these two systems without the application of PBS shows that the side-mounted option displays a collection probability that is consistently higher than that of the bottom mounted systems. In addition, the transition towards the large-scale limit that occurs at $l/d \approx 20$ for the cells at the bottom is absent for the side mounted system. A collection probability $p_c > 18\%$ is maintained up to l_{max}. With included non-radiative loss $p_{nr} = 0.08$ the over-all collection probability drops by about 3% for both options. The difference between side- and bottom-mounted systems remains the same.

4.5.2 Reflection losses at the back side mirror

Mirrors cover the collector back sides either fully in the side-mounted system or partly in the bottom-mounted system. Optical losses due to non-ideal reflection are a major loss mechanism especially for large collector systems. Note that other losses, especially parasitic absorption in the polymer matrix (not considered in the present paper), are expected to have comparable effects. In the following, we study the influence of reflection coefficients $R < 1$ in comparison to the perfect case ($R = 1$). We do not consider explicitly the case of an air-gap between mirror and collector which would restrict the losses to angles larger than the critical angle of total internal reflection. The present approach uses R-values which are independent of the direction of the photons and could be looked at as directional averages.

Figures 9a, 9b depict the influences of reflection losses on the photon collection p_c of both systems. Figure 9a presents the results for the side-mounted system. Since less radiative losses occur in the system with PBS the contribution of non-radiative reflection losses at the back side mirror is more significant. Therefore, the decrease in the photon collection is higher.

Figure 9b shows the photon collections for the system with bottom-mounted solar cells. For small l/d the system behaves similar to the side-mounted system. The larger the system the more often a reflection takes place, thus, increasing the number of lost photons. Therefore, included reflection losses decrease the photon collection already at $l/d = 10$, whereas for a mirror with $R = 1$ this drop occurs at $l/d = 100$.

In systems with coverage fraction $f = 0.01$ at any of the calculated collector length the re-absorption is a rare event compared to the number of reflections at the FC back side as the following example clarifies. Most photons emitted by the dye experience the absorption coefficient $\alpha_2 = 0.03/d$ as mentioned above. With $d = 3$ cm, a typical thickness for industrialized fluorescent collectors, the photons are re-absorbed after 100 cm. A photon emitted with an angle $\theta = 45°$ hits the collector sides every 4.2 cm. Since the coverage fraction $f = 0.01$ is very low, most likely the photon is re-absorbed before it hits a solar cell. Such, a photon with $\theta = 45°$ is reflected 23 times before it is re-absorbed. Therefore, a non-perfect mirror has a stronger influence than an equally high non-radiative recombination loss in the dye. A reflection $R = 98\%$ at the back side mirror causes a drop $\Delta p_c \approx 70\%$ for small l/d which is higher than the drop $\Delta p_c \approx 25\%$ caused by the non-radiative recombination loss as shown in Figures 8a, 8b. Therefore, the mirror applied in the system has to feature a superior quality compared to the fluorescent dye.

4.5.3 Non-perfect photonic structure

In the simulations the application of a photonic band stop as an energy selective filter increases the photon collection in all cases. However, in realistic filters blocking the photons depends not only on the energy but also on the angle of incidence of the photon as schematically shown in

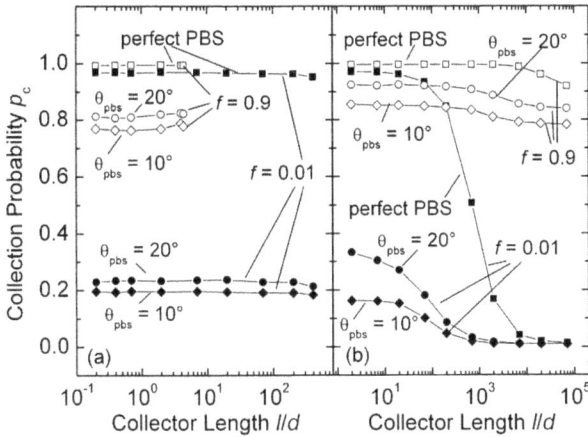

Fig. 10. Influence of a photonic band stop filter with angular selectivity. (a) Side-mounted system. Coverage fraction $f = 0.9$ achieves higher photon collections p_c than $f = 0.01$ for all θ_{pbs}. (b) Bottom-mounted system. For small scales the system shows a better performance than the side-mounted system because this system benefits from the randomization of photon direction during re-emission.

Figure 3e. Figures 10a, 10b outline the influence of smaller reflection cones ($\theta_{pbs} = 10°$ and $20°$) on side- and bottom-mounted systems with coverage fractions $f = 0.01$ and 0.9.

Figure 10a depicts the results for the side-mounted system. The application of a non-perfect PBS filter decreases the photon collection at all collector lengths. However, a relatively higher drop occurs for the system with $f = 0.01$ than for the system with $f = 0.9$. Here, a higher re-absorption rate occurs due to longer distances between the cells. Therefore, photons are more often re-emitted with their direction spatially randomized. The frequent randomization carried out also in unfavorable angles contributes to the number of lost photons.

Figure 10b presents the calculation results for the bottom-mounted system. At small l/d this system collects more photons under the application of a non-perfect PBS filter than the side-mounted system. This is due to the effect that the disadvantageous angles for the PBS are very favorable angles for solar cells at the FC back side. Therefore, the collection of a photon is more likely than in the side-mounted system. As in the side-mounted system, the collection in the system with $f = 0.01$ decreases more at large l/d than in the system with $f = 0.9$.

4.5.4 Realistic values

In order to get a more realistic picture, we perform a system simulation which includes several loss mechanisms simultaneously. While the absorption/emission in the fluorescent collector is as ideal as presented in Figure 2, the non-radiative recombination rate lies at 5%, which is a value typically achieved in industrial FCs. For the back side mirror, we assume a reflectivity $R = 92\%$, which corresponds to the reflectivity of Aluminium. Furthermore, we assume dielectric layers as photonic structures which show transmittances $T > 95\%$ if manufactured carefully.

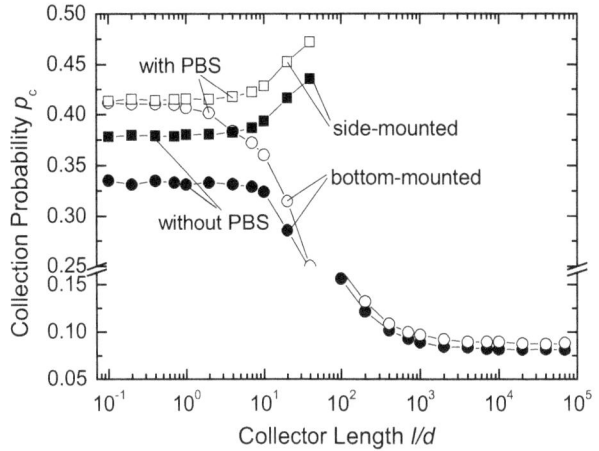

Fig. 11. Side-mounted and bottom-mounted solar cells on fluorescent collectors including realistic values for the back side mirror, the photonic structure and the fluorescent dye. The bottom-mounted solar cells are in the small-scale limit for $l/d < 1$ and in the large-scale limit from $l/d > 10^3$. For a clear acrylic glass on top of the solar cells, the coverage fraction $f = 0.1$ leads to a collection probability $p_c = 0.1$. Thus, a fluorescent collector system with the assumed losses increases the collection probability by 23%, with a PBS on top by 27%. The side-mounted solar cells show an interesting behavior for $l/d > 11$. Instead of descending into the large-scale limit, p_c increases up to 0.47 at $l/d > 40$ (with PBS on top).

However, they always show a blue shift in their transmittance spectrum for oblique incident angles. Thus, adjusting the transmittance spectrum properly should lead to a full band stop for all ray directions. Nevertheless, we assume $\theta_{pbs} = 30°$. The coverage fraction is $f = 0.1$. Note, that for a clear acrylic glass on top of the solar cells, the coverage fraction $f = 0.1$ leads to collection probability $p_c = 0.1$.

Figure 11 shows that bottom-mounted solar cells perform as expected from the previous results: the system is in the small-scale limit for $l/d < 1$ and in the large-scale limit from $l/d > 10^3$. Thus, a FC system with the assumed losses increases the collection probability by 23%, with a PBS on top even by 27% compared to clear acrylic glass. The side-mounted solar cells show an interesting behavior for $l/d > 11$. Instead of descending into the large-scale limit, p_c increases up to 0.47 at $l/d > 40$ (with PBS on top). We explain this effect, which also is already slightly indicated in the data of Figure 10a, as follows: the larger l/d, the more collector edge area is covered with solar cells until they are fully covered for $l/d = 40$. Thus, the photon path length is larger for small l/d than for large l/d, because the probability to hit solar cell area is smaller. The longer the path length, the higher is the probability for the photon to be subject to loss mechanisms.

5 Conclusion

Monte-Carlo simulations compare photovoltaic fluorescent collectors in side-mounted and bottom-mounted systems as well as the systems with or without the application of a photonic band stop filter on top. The filter greatly enhances the overall collection probability by suppressing emission of converted photons from the surface of the collector. Also, the sensitivity of the collector to repeated reabsorption and re-randomization of the fluorescent photons is reduced. We find that the collection probability of systems with identical coverage fraction, i.e., the same amount of solar cell area per unit collector area is heavily influenced by the scaling of the solar cell size. Many small solar cells generally perform better than few large solar cells with the same overall area (scaling-effect). The systems in their radiative limits are compared to systems which include loss mechanisms. The comparison of collectors with solar cells mounted at the collector side to a system with the cells at the bottom shows that in most cases the side-mounted system displays a higher collection efficiency. We find that a non-perfect reflection at the back side mirror causes a higher deterioration in the photon collection than a comparable non-radiative recombination in the dye. This is because in the analyzed systems the reabsorption is a rare event compared to a reflection at the back side. Therefore, the quality of the back side mirror merits especial care. Assuming a restricted reflection cone of the photonic band stop filter causes higher losses in the side-mounted system than in the bottom-mounted system which is caused by the randomization of the photon angles during the emission by the dye. All loss effects are especially significant for systems without a filter. We see that a good quality of the filter is especially necessary in systems with low coverage fraction of solar cells. In comparison to the classical side-mounted system, the system with solar cells at the back side performs equally well for small scales. Therefore, applying fluorescent collectors technically less expensive *on top* of solar cells is a promising approach, if the solar cells are properly scaled in size and distance.

This work was supported by a grant of the Deutsche Forschungsgemeinschaft (DFG, contract PAK88, 'nanosun'). The authors wish to thank G. Bilger, C. Ulbrich, and T. Kirchartz for numerous discussions as well as J.H. Werner for continuous support.

References

1. A. Goetzberger, V. Wittwer, Sol. Cells **4**, 3 (1981)
2. A. Goetzberger, W. Greubel, Appl. Phys. A **14**, 123 (1977)
3. T. Trupke, M.A. Green, P. Würfel, J. Appl. Phys. **92**, 4117 (2002)
4. T. Trupke, M.A. Green, P. Würfel, J. Appl. Phys. **92**, 1668 (2002)
5. J.A.M. van Roosmalen, Semiconductors **38**, 970 (2004)
6. A.J. Chatten, K.W.J. Barnham, B.F. Buxton, N.J. Ekins-Daukes, M.A. Malik, Solar Energy Mater. Solar Cells **75**, 363 (2003)
7. E. Yablonovitch, J. Opt. Soc. Am. **70**, 1362 (1980)
8. G. Smestad, H. Ries, R. Winston, E. Yablonovitch, Solar Energy Mater. Solar Cells **21**, 99 (1990)
9. L.H. Slooff, E.E. Bende, A.R. Burgers, T. Budel, M. Pravettoni, R.P. Kenny, E.D. Dunlop, A. Büchtemann, Phys. Stat. Sol. PRL **2**, 257 (2008)
10. M. Bendig, J. Hanika, H. Dammertz, J.C. Goldschmidt, M. Peters, M. Weber, in *IEEE/EG Symposium on Interactive Ray Tracing, California, USA, 2008*, p. 93
11. U. Rau, F. Einsele, G.C. Glaeser, Appl. Phys. Lett. **87**, 171101 (2005)
12. G.C. Glaeser, U. Rau, in *Proc. SPIE*, edited by A. Gombert (2006), Vol. 6197, p. 143
13. J.C. Goldschmidt, M. Peters, L. Prönneke, L. Steidl, R. Zentel, B. Bläsi, A. Gombert, S. Glunz, G. Willeke, U. Rau, Phys. Stat. Sol. A **205**, 2811 (2008)
14. M. Peters, J.C. Goldschmidt, P. Löper, B. Bläsi, A. Gombert, J. Appl. Phys. **105**, 014909 (2009)
15. S. Knabe, N. Soleimani, T. Markvart, G.H. Bauer, Phys. Stat. Sol. PRL **4**, 118 (2010)
16. J.C. Goldschmidt, M. Peters, A. Bösch, H. Helmers, F. Dimroth, S.W. Glunz, G. Willeke, Solar Energy Mater. Solar Cells **93**, 176 (2009)
17. R. Reisfeld, Opt. Mater. **32**, 850 (2010)
18. L. Prönneke, G.C. Glaeser, Y. Uslu, U. Rau, in *24th European Photovoltaic Solar Energy Conference Hamburg, Germany, 2009*, p. 385–387
19. T. Markvart, J. Appl. Phys. **99**, 026101 (2006)
20. E. Yablonovitch, Phys. Rev. Lett. **58**, 2059 (1987)
21. E. Yablonovitch, J. Opt. Soc. Am. B **10**, 283 (1993)
22. E. Yablonovitch, Opt. Lett. **23**, 1648 (1998)
23. L. Martinu, D. Poitras, J. Vac. Sci. Technol. A **18**, 2619 (2000)
24. D.N. Chigrin, C.M. Sotomayor Torres, Opt. Spectrosc. **91**, 484 (2001)
25. M. Schöfthaler, U. Rau, J.H. Werner, J. Appl. Phys. **76**, 4168 (1994)
26. P. Kittidachan, L. Danos, T.J.J. Meyer, N. Aldermann, T. Markvart, Chimia **61**, 780 (2007)
27. G.C. Gläser, Ph.D. thesis, University of Stuttgart, 2007
28. T. Markvart, Appl. Phys. Lett. **88**, 176101 (2006)
29. U. Rau, F. Einsele, G.C. Glaeser, Appl. Phys. Lett. **88**, 176102 (2006)
30. T.J.J. Meyer, T. Markvart, Appl. Phys. **105**, 063110 (2009)

17

Effect of thermal annealing in vacuum on the photovoltaic properties of electrodeposited Cu_2O-absorber solar cell

T. Dimopoulos[1,a], A. Peić[1,b], S. Abermann[1], M. Postl[2], E.J.W. List-Kratochvil[2,3], and R. Resel[3]

[1] AIT-Austrian Institute of Technology, Energy Department, Giefinggasse 2, 1221 Vienna, Austria
[2] NanoTecCenter Weiz Forschungsgesellschaft mbH, Franz-Pichler-Str. 32, 8160 Weiz, Austria
[3] Graz University of Technology, Institute of Solid State Physics, Petergasse 16, 8010 Graz, Austria

Abstract Heterojunction solar cells were fabricated by electrochemical deposition of p-type, cuprous oxide (Cu_2O) absorber on sputtered, n-type ZnO layer. X-ray diffraction measurements revealed that the as-deposited absorber consists mainly of Cu_2O, but appreciable amounts of metallic Cu and cupric oxide (CuO) are also present. These undesired oxidation states are incorporated during the deposition process and have a detrimental effect on the photovoltaic properties of the cells. The open circuit voltage (V_{OC}), short circuit current density (j_{SC}), fill factor (FF) and power conversion efficiency (η) of the as-deposited cells are 0.37 V, 3.71 mA/cm^2, 35.7% and 0.49%, respectively, under AM1.5G illumination. We show that by thermal annealing in vacuum, at temperatures up to 300 °C, compositional purity of the Cu_2O absorber could be obtained. A general improvement of the heterojunction and bulk materials quality is observed, reflected upon the smallest influence of the shunt and series resistance on the transport properties of the cells in dark and under illumination. Independent of the annealing temperature, transport is dominated by the space-charge layer generation-recombination current. After annealing at 300 °C the solar cell parameters could be significantly improved to the values of: $V_{OC} = 0.505$ V, $j_{SC} = 4.67$ mA/cm^2, $FF = 47.1\%$ and $\eta = 1.12\%$.

1 Introduction

Cuprous oxide (Cu_2O) is one of the first known semiconducting materials. Although its photovoltaic (PV) property was very early recognized [1], the technological advancements towards a Cu_2O-based solar cell were only sporadic. Presently, Cu_2O attracts increased attention, driven by the need for abundant and environmental-friendly materials for PV and by newly introduced technological approaches for solar cell fabrication. In a recent study, Cu_2O appears as one of the most intriguing among 23 inorganic absorbers for abundant and affordable electricity supply, based on the dual constraints of material supply and lowest cost per watt [2].

Indeed, Cu_2O is abundant, non-toxic and can be prepared by non-vacuum techniques, such as thermal oxidation of Cu sheets and electrochemical deposition (ECD). It is a native p-type semiconductor, due to a high concentration of negatively-charged copper vacancies [3]. Its direct bandgap of $\sim$2 eV is quite large for an ideal match of the solar spectrum, but still allows for a theoretical maximum power conversion efficiency of $\sim$20% for a single junction under AM1.5 illumination [4]. This is combined with a high absorption coefficient for energies above the bandgap (10^3 up to more than 10^5 cm^{-1}) [5], high majority carrier mobility (in the range of 100 cm^2/Vs) [6,7] and large minority carrier diffusion length (up to several micrometers) [8,9]. Early investigations showed that Cu_2O Schottky barrier-type solar cells lead to poor PV performance [8] and pointed out the need for heterojunction or homojunction architectures. Since the n-type doping of Cu_2O is extremely difficult due to the mechanism of dopant self-compensation [10], research efforts focused inevitably on heterojunctions. Many n-type window layers have been tested for this purpose, such as In_2O_3 [11,12], SnO_2 [11], CdO [11,13], Cu_xS [14], TiO_2 [15], Ga_2O_3 [16] and ZnO, with the last two being the most promising.

Indeed, in 2006, an efficiency of more than 2% was achieved by sputter-depositing a thin ZnO layer on a bulk Cu_2O substrate, produced by thermal oxidation of a Cu

[a] e-mail: `theodoros.dimopoulos@ait.ac.at`
[b] *Present address*: EV Group E. Thallner GmbH, DI Erich Thallner Str. 1, 4782 St. Florian am Inn, Austria.

Table 1. List of the deposition parameters for each material, along with the corresponding sputter rates.

Target	Power density (W/cm^2)	Gas	Pressure (Pa)	Dep. rate (nm/sec)
Al:ZnO (AZO)	1.97 (DC)	Ar	0.1	0.78
ZnO	0.49 (DC)	Ar/O$_2$: 80/20	1	0.08
NiO	2.47 (RF)	Ar/O$_2$: 80/20	0.5	0.04
Au	0.43 (DC)	Ar	0.2	0.32

sheet at temperatures exceeding 1100 °C and process duration of several hours [7]. Through optimization of the thermal oxidation process and appropriate ZnO doping with Mg, the efficiency increased to 4.3% for bulk Cu$_2$O cells [17, 18]. Recently, record efficiency of 5.38% was reported through the use of Ga$_2$O$_3$ n-layer [16]. However, the demand is to turn from bulk to thin film absorbers, which require significantly less material and lower fabrication cost. To this end, ECD of Cu$_2$O has a large potential as it takes place close to room temperature (about 50−60 °C) and from aqueous solution. In addition, ECD is easily up-scalable, high throughput and roll-to-roll compatible. However, the maximum achieved efficiency of solar cells with thin film, ECD-grown Cu$_2$O in combination with ZnO remained for long limited at 1.28% [19]. Only very recently did the efficiency jump to 2.65% by employing a tin-doped ZnO layer, formed by atomic layer deposition and a substrate-type of solar cell [20].

The need remains to identify the factors that limit the efficiency of ECD Cu$_2$O solar cells and introduce technological processes to address them. The heterojunction interface and the absorber bulk are of paramount importance for the charge carrier separation and collection. Enhanced recombination at the ZnO/Cu$_2$O interface arises due to structural defects and the large conduction band misalignment [21], leading to the degradation of the open circuit voltage and fill factor of the solar cell [22]. Further degradation is caused by current shunts related to a defect-rich heterojunction interface, as well as by the carrier recombination and large series resistance contributions associated with the absorber's bulk.

Thermal annealing can generally improve the quality of layers' interfaces and bulk. For potentiostatically-grown Cu$_2$O on Cu and SnO$_2$, annealing in air was shown to promote crystallization, to lower the film's resistivity and to improve the conversion efficiency of photoelectrochemical cells [23]. On the other hand, the annealing of Cu$_2$O films in air at elevated temperatures was reported to lead to the formation of CuO (cupric oxide) with poor PV properties [23].

In this paper we show that by thermal annealing ZnO/ECD-Cu$_2$O solar cells in vacuum, the crystallinity and composition of the Cu$_2$O absorber film are improved, eliminating the amounts of metallic Cu and CuO in the layer, which are incorporated during the electrochemical deposition. Furthermore, vacuum conditions prevent the oxidation of Cu$_2$O to CuO that arises during annealing in air. Annealed solar cells present increased shunt resistance

and lower series resistance, suggesting improvement of the heterojunction interface and absorber's bulk. The combination of these effects leads to the enhancement of the short circuit current, open circuit voltage and fill factor of the heterojunction solar cells.

2 Experimental

2.1 Fabrication

The solar cells were deposited on 1 mm-thick glass slides. Apart from the Cu$_2$O absorber, all other films were deposited by sputtering. During the sputter process, the targets and substrate holder were water-cooled at 25 °C. The target-substrate distance was 10 cm and the sputter chamber base pressure in the range of $7 \times 10^{-6} - 1.2 \times 10^{-5}$ Pa. All oxide films were sputtered from stoichiometric oxide targets. The deposition parameters of the films are shown in Table 1. The film thicknesses and the corresponding sputter rates were extracted from step-height measurements using a surface profilometer.

Firstly, a 400 nm-thick AZO layer was deposited on the glass slide, followed by the deposition of the intrinsic ZnO with a thickness of 40 nm. ZnO covered the whole AZO surface except from a small window at the sample's edge, which is needed to establish a good contact to the sample during the ECD and later for the electrical measurements. AZO has a resistivity of 1.5×10^{-3} Ωcm, while the i-ZnO is highly resistive (>10^5 Ωcm).

For the ECD of the Cu$_2$O film, an aqueous solution with reagent-grade purified and deionized $Milli$-Q water was prepared, containing 0.2 M CuSO$_4$x5H$_2$O, 3 M lactic acid and 125 mg of ZnO powder, which was stirred for several hours until all grains were fully dissolved. A pH of 12.5 was reached by gradually adding NaOH to the solution. Lactic acid is used to stabilize Cu (II) ions at bath pH higher than 7, while the introduced ZnO powder served as sacrificial material in order to prevent the dissolution of the underlying ZnO layer during ECD and the formation of unwanted impurities, as proposed by Musselman et al. [24]. The reaction solution volume was 200 ml. The aqueous solution was heated under permanent mild stirring at the reaction temperature of 50 °C prior to the immersion of the AZO/ZnO-covered glass substrate (cathode), which was placed vertically in the reservoir. A wound platinum wire with a diameter of 0.5 mm was used as the counter electrode. A negative DC potential of −0.6 V relative to an Ag/AgCl reference electrode was applied

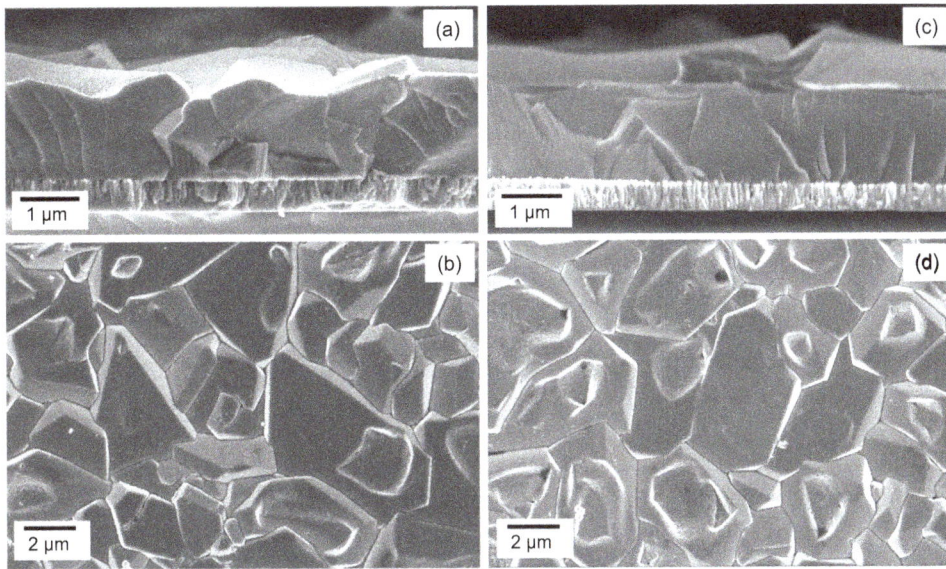

Fig. 1. Cross section and top view SEM images of the as-deposited heterojunction (a, b) and after annealing at 300 °C (c, d).

for 70 min, using an AUTOLAB potentiostat/galvanostat. After the deposition the sample was removed from the solution and thoroughly rinsed in flowing deionized water to eliminate residual salts and unreacted products from the surface. Then the sample was transferred to the sputtering chamber for the deposition of 20 nm of NiO, followed by Au deposition through a contact mask, to define solar cell areas from 1 to 9 mm^2.

2.2 As-deposited cell

Figures 1a and 1b show scanning electron microscopy (SEM) images of the cleaved edge and of the surface of the as-deposited heterojunction, respectively (without the top Au contact). The sputtered AZO and ZnO layers have a characteristic columnar structure, while the Cu_2O layer, of approximately 1.7 μm thickness, is compact, rough and features large, flat-top grains, with sizes up to several micrometers. Large grain-size is important in order to keep the charge carrier scattering low and thus improve the absorber's conductivity and the minority carrier transport length by diffusion or drift. Two groups estimated the latter to be approximately 160 nm for electrodeposited Cu_2O solar cells, much lower than for thermally oxidized Cu_2O [25, 26]. It is admitted, though, that this value is sensitive to the structural properties of the absorber (crystallinity, grain size, defect density) and of the heterojunction quality.

The specular X-ray diffraction (XRD) pattern of the as-deposited multilayer in Figure 2 shows the typical hexagonal ZnO (002) plane reflection at $2\theta = 34.25°$, with a secondary peak stemming from the (103) planes at 62.6°. In the case of Cu_2O, its structure is cubic with the dominant reflection coming from the (111) family of planes at 36.5°. The same preferred Cu_2O texture was reported by potentiostatic [27–29] or galvanostatic [19, 30] deposition, for solutions with high pH. Indeed, the increase of hydroxyl ion concentration for high bath pH,

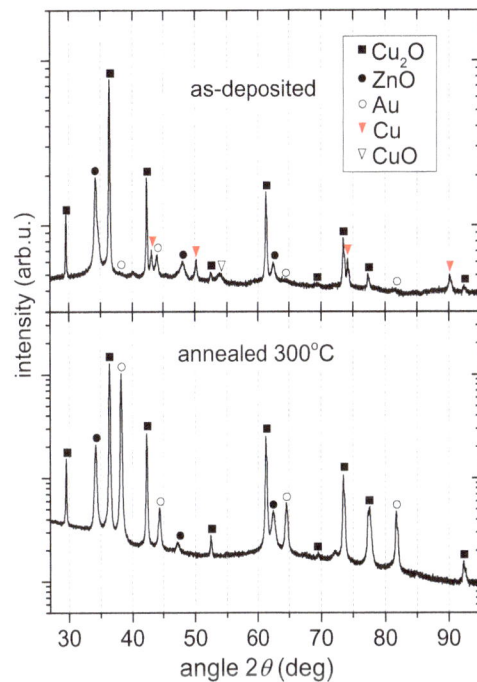

Fig. 2. Specular XRD measurements of the as-deposited and annealed cell at 300 °C.

leads to increased oxygen supply in the film. This favors the growth of planes with high oxygen ion density, such as the (111) family of planes [31]. The XRD pattern also contains Cu_2O peaks of smaller intensities at $2\theta = 42.3$, 61.4, 29.6, 73.5, 77.6, 52.4 and 92.4° (in descending intensity order), corresponding to the (200), (220), (110), (311), (222), (211) and (400) planes, respectively.

Most importantly, we also observe reflections from metallic Cu at $2\theta = 43.2$, 50.2, 74.2 and 90.1°, which correspond to the fcc (111), (200), (220) and (311) planes, respectively. Finally the peak at $2\theta = 53.9°$ is ascribed to the monoclinic CuO (020) plane reflection. It is therefore

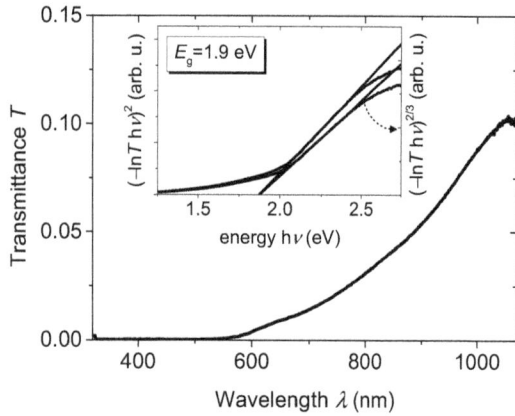

Fig. 3. Transmittance spectrum (excluding the glass substrate) and Tauc plot (inset) from where the direct bandgap of Cu_2O is extracted (as-deposited sample).

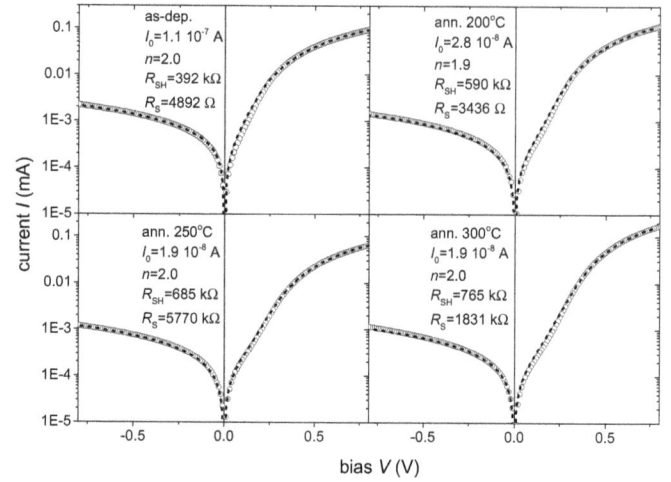

Fig. 4. Measured dark I-V curves (open circles) for a solar cell with 6.5 mm^2 area at the as-deposited state and after annealing at the marked temperatures. Dashed lines represent fits to equation (1), from where the saturation current, diode ideality factor, shunt and series resistance values are extracted.

shown that the as-deposited absorber may predominantly consist of Cu_2O, but appreciable amounts of metallic Cu and CuO are present. Indeed, the oxidation state of the deposit is very sensitive on the applied ECD potential. For a fixed bath pH, a more negative potential favors the deposition of Cu, a moderate potential the deposition of Cu_2O and the least negative the formation of CuO [32]. The limits between the different potential windows are not sharply defined, leading to the possibility that unwanted oxidation states are present in the film. In line with our findings, Septina et al. [29] reported the incorporation of Cu^{2+} and Cu^0 impurities during potentiostatic ECD for a wide range of potential values. Also, it is pointed out that traces of these impurities may well be present, even if the corresponding phases cannot be resolved in the X-ray diffractogram [19]. The presence of metallic Cu in the absorber was shown to degrade the PV characteristics [19], while CuO also leads to reduced PV performance compared to Cu_2O [28,33].

Figure 3 shows the optical transmittance, T, spectrum of the AZO/n-ZnO/p-Cu_2O heterojunction, excluding the glass substrate, measured with a spectrophotometer. From this, the Tauc plot is constructed, i.e. the quantity $(-\ln T\,h\nu)^m$ versus $h\nu$, where $h\nu$ is the photon energy and m an exponent, the value of which depends on the type of optical transition: $m = 2$ for direct allowed transition and $m = 2/3$ for direct forbidden. The inset of Figure 4 shows the curves for both exponent values, leading to equally good linear dependence at the absorption edge window. The linear extrapolation to the $h\nu$-axis yields in both cases a bandgap of 1.9 eV, in accordance with the anticipated value for Cu_2O.

Following the electrical characterization of the as-deposited solar cells, the sample was annealed in vacuum (base pressure in the 10^{-4} Pa regime) at temperatures of 200, 250, 300 and 350 °C for 1 h. The ramp-up time to the targeted temperature was 5–10 min, while the ramp-down time to room temperature was 4–6 h. The cells were electrically characterized after each annealing step in dark and under AM1.5G illumination, before proceeding to the next temperature. The presented measurements

concern a cell area of 6.5 mm^2 (as defined by the size of the Au electrode). The current density was, nevertheless, practically independent of the cell area, showing that the current flow is homogeneous throughout the surface of the cell and that the influence of the cell's edges can be neglected. Also, the photovoltaic properties did not depend on whether a shadow mask was used for illumination or not, showing that the collection of current from outside the cell area is negligible. This is a consequence of the very resistive absorber, restricting lateral carrier transport.

2.3 Annealed cell

Figures 1c and 1d show cross section and surface SEM images of the sample after annealing at 300 °C. No significant changes can be observed as opposed to the as-deposited sample (Figs. 1a and 1b). On the other hand, the XRD pattern of the annealed sample in Figure 2 is free of metallic Cu and CuO reflections, i.e. the absorber has acquired the pure Cu_2O phase. In parallel, the intensities of the secondary Cu_2O peaks relative to the principal (111) peak are notably higher. This is especially the case for the (211), (311), (222) and (400) plane reflections. We assume that this gain in intensity arises from the transformation of the Cu and CuO phases to Cu_2O.

The effects of vacuum annealing on the cells' electrical properties were investigated through dark and illuminated current-voltage (I-V) measurements. After each annealing step the sample was kept in ambient conditions for 2 days before being measured. This period was necessary in order to obtain stable electrical properties from the solar cells. The dark I-Vs obtained for a 6.5 mm^2 area cell as a function of the annealing temperature, are presented in Figure 4 in $\log I$ vs. V plots. They are fitted to the one-diode model expression:

$$I = I_0 \left[\exp\left(q(V - IR_S)/nkT\right) - 1\right] + (V - IR_S)/R_{SH}$$
(1)

Table 2. Extracted device parameters from the I-V measurements of the solar cells in dark and under AM1.5G illumination.

	As-dep.	200 °C	250 °C	300 °C
n	2.0	1.9	2.0	2.0
R_{SH} (kΩ) – dark	392	590	685	765
R_S (Ω) – dark	4892	3436	5770	1831
I_0 (10^{-8} A)	11.0	2.8	1.9	1.9
R_{SH} (Ω) – illum.	4425	5867	6920	14 689
R_S (Ω) – illum.	889	1000	861	599
R_{CH} (Ω)	1386	1282	1432	1494
R_{SH}/R_{CH}	3.2	4.6	4.8	9.8
j_{SC} (mA/cm^2)	3.71	3.91	4.61	4.67
V_{OC} (V)	0.37	0.44	0.47	0.51
FF (%)	35.7	35.9	37.4	47.1
η (%)	0.49	0.62	0.81	1.12

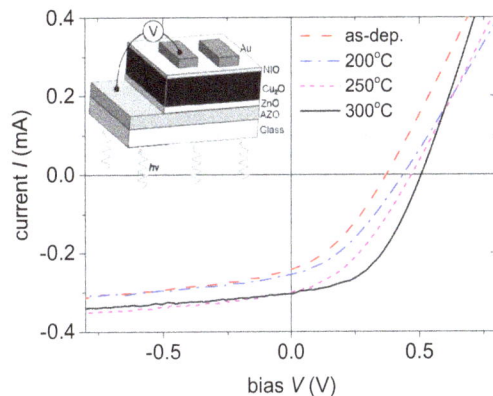

Fig. 5. I-V curves measured under simulated solar spectrum (AM1.5G) for a cell of 6.5 mm^2 area after being annealed at different temperatures. Inset shows schematic of the solar cell structure.

where n the diode's ideality factor, q the elementary charge, k the Boltzmann's constant, T the temperature in Kelvin, I_0 the diode's saturation current, R_{SH} the shunt resistance and R_S the series resistance. The fit yields in all cases ideality factors close to $n = 2.0$ (Tab. 2), which suggests that transport is dominated by the space-charge layer generation-recombination current as described by the Sah-Noyce-Shockley model [34], with the generation-recombination level located at mid-gap. The parasitic shunt resistance has an important contribution on the transport characteristics at reverse and low forward bias (<0.2 V). It stems from defect-rich junction regions with nearly ohmic behavior, resulting in a practically linear increase of the current with the bias. The value of R_{SH} rises constantly with increasing annealing temperature (Tab. 2). Since low shunt resistance is related to poor heterojunction quality, we conclude that thermal annealing has, indeed, improved the ZnO/Cu$_2$O interface. The improvement of the electronic quality of the materials forming the junction is also reflected upon the continuous decrease of the saturation current I_0 with the annealing temperature (Tab. 2), which can be related to the reduction of charge carrier recombination. Finally, the R_S values, related to the resistance of the semiconductor films and contacts and manifesting itself at large forward bias, also appear in Table 2. Although R_S varies for intermediate annealing temperatures, there is a clear decrease of its value after annealing at 300 °C, as compared to the as-deposited case. This R_S reduction can be partly attributed to the reduction of the Cu$_2$O resistivity. Indeed, the resistivity of Cu$_2$O after annealing at 300 °C is found to moderately decrease to 7.2×10^4 Ωcm from the 11.6×10^4 Ωcm level of the as-deposited state.

The parasitic resistance values were also extracted from the illuminated I-V curves in Figure 5. The R_{SH} is obtained from the inverse of the curve's slope at $V = 0$ (short circuit condition). We observe again a continuous increase of R_{SH} with the annealing temperature, which becomes especially large after annealing at 300 °C (Tab. 2). The fact that R_{SH} is much lower under illumination than in dark suggests that the shunting paths are photoactive. The cell's series resistance, as extracted from the I-V slope at $I = 0$ (open circuit condition), significantly reduces at 300 °C. The difference between dark and illuminated values is not as strong here as with the case of R_{SH}.

Also important is the fact that the characteristic cell resistance, R_{CH}, at the point of maximum power, is not significantly modified by the annealing, resulting to a steady improvement of the R_{SH}/R_{CH} ratio (Tab. 2). As a consequence of the considerable increase of R_{SH} and the decrease of R_S, an improvement of the fill factor from 37.4 to 47.1% is achieved after annealing at 300 °C (Tab. 2). In parallel, the open circuit voltage, V_{OC}, of the cell reaches 505 mV after annealing at 300 °C, steadily increasing from the 370 mV value of the as-deposited state. Likewise, the short circuit current density, j_{SC}, increases with the temperature from 3.71 to 4.67 mA/cm^2. As a result, the power conversion efficiency, η, improves from 0.49% for the as-deposited case to 1.12% after annealing at 300 °C (Tab. 2). We note here that further annealing at 350 °C for 1 h resulted to linear I-V characteristics for the majority of the cells, which strongly suggests the formation of short-cuts at the heterojunction.

In conclusion, we showed that thermal annealing in vacuum can considerably improve the photovoltaic performance of solar cells with electrochemically-deposited Cu$_2$O. The composition of the annealed absorber is purely Cu$_2$O, with no traces of CuO and Cu, as it is the case for the as-deposited films. The electrical characterization of the solar cells in dark and under AM1.5G illumination showed that the influence of the shunt and series resistance is decreased with annealing and all PV parameters are improved, obtaining a maximum power conversion efficiency of 1.12%.

The authors would like to acknowledge financial support from the Austrian Klima & Energiefonds projects "SAN-CELL" and "CopperHEAD".

References

1. B. Lange, *Photoelements* (Reinhold Publ. Corp., New York, 1938)
2. C. Wadia, A.P. Alivisatos, D.M. Kammen, Environ. Sci. Technol. **43**, 2072 (2009)
3. W. Brattain, Rev. Mod. Phys. **23**, 203 (1951)
4. W. Shockley, H.J. Queisser, J. Appl. Phys. **32**, 510 (1961)
5. C. Malerba, F. Biccari, C. Leonor Azanza Ricardo, M. D'Incau, P. Scardi, A. Mittiga, Sol. Energy Mater. Sol. Cells **95**, 2848 (2011)
6. A. Musa, T. Akomolafe, M. Carter, Sol. Energy Mater. Sol. Cells **51**, 305 (1998)
7. A. Mittiga, E. Salza, F. Sarto, M. Tucci, R. Vasanthi, Appl. Phys. Lett. **88**, 163502 (2006)
8. L.C. Olsen, F.W. Addis, W. Miller, Sol. Cells **7**, 247 (1982)
9. F. Biccari, C. Malerba, A. Mittiga, Sol. Energy Mater. Sol. Cells **94**, 1947 (2010)
10. Y. Tsur, I. Riess, Phys. Rev. B **60**, 8138 (1999)
11. L. Papadimitriou, N.A. Economou, D. Trivich, Sol. Cells **3**, 73 (1981)
12. C.J. Dong, W.X. Yu, M. Xu, J.J. Cao, C. Chen, W.W. Yu, Y.D. Wang, J. Appl. Phys. **110**, 073712 (2011)
13. Y. Hameş, S. Eren San, Sol. Energy **77**, 291 (2004)
14. R. Wijesundara, L.D.R. Perera, K. Jayasuriya, W. Siripala, K.T. De Silva, A. Samantilleke, I. Dharmadasa, Sol. Energy Mater. Sol. Cells **61**, 277 (2000)
15. S. Hussain, C. Cao, Z. Usman, Z. Chen, G. Nabi, W.S. Khan, Z. Ali, F.K. Butt, T. Mahmood, Thin Solid Films **522**, 430 (2012)
16. T. Minami, Y. Nishi, T. Miyata, Appl. Phys. Express **6**, 044101 (2013)
17. T. Minami, Y. Nishi, T. Miyata, J. Nomoto, Appl. Phys. Express **4**, 062301 (2011)
18. T. Minami, N. Yuki, T. Miyata, A. Shinya, in *Pacific Rim Meeting on Electrochemical and Solid-State Science (PRiME), 222nd Meeting of ECS, Honolulu, 2012*
19. M. Izaki, T. Shinagawa, K.-T. Mizuno, Y. Ida, M. Inaba, A. Tasaka, J. Phys. D **40**, 3326 (2007)
20. Y.S. Lee, J. Heo, S.C. Siah, J.P. Mailoa, R.E. Brandt, S.B. Kim, R.G. Gordon, T. Buonassisi, Energy Environ. Sci. **6**, 2112 (2013)
21. B. Kramm, A. Laufer, D. Reppin, A. Kronenberger, P. Hering, A. Polity, B.K. Meyer, Appl. Phys. Lett. **100**, 094102 (2012)
22. T. Minemoto, T. Matsui, H. Takakura, Y. Hamakawa, T. Negami, Y. Hashimoto, T. Uenoyama, M. Kitagawa, Sol. Energy Mater. Sol. Cells **67**, 83 (2001)
23. T. Mahalingam, J.S.P. Chitra, J.P. Chu, S. Velumani, P.J. Sebastian, Sol. Energy Mater. Sol. Cells **88**, 209 (2005)
24. K.P. Musselman, A. Marin, A. Wisnet, C. Scheu, J.L. MacManus-Driscoll, L. Schmidt-Mende, Adv. Funct. Mater. **21**, 573 (2011)
25. Y. Liu, H.K. Turley, J.R. Tumbleston, E.T. Samulski, R. Lopez, Appl. Phys. Lett. **98**, 162105 (2011)
26. K.P. Musselman, Y. Ievskaya, J.L. MacManus-Driscoll, Appl. Phys. Lett. **101**, 253503 (2012)
27. T. Mahalingam, J.S.P. Chitra, J.P. Chu, S. Velumani, P.J. Sebastian, Sol. Energy Mater. Sol. Cells **88**, 209 (2005)
28. S. Hussain, C. Cao, G. Nabi, W.S. Khan, Z. Usman, T. Mahmood, Electrochimica Acta **56**, 8342 (2011)
29. W. Septina, S. Ikeda, M.A. Khan, T. Hirai, T. Harada, M. Matsumura, L.M. Peter, Electrochimica Acta **56**, 4882 (2011)
30. S.S. Jeong, A. Mittiga, E. Salza, A. Masci, S. Passerini, Electrochimica Acta **53**, 2226 (2008)
31. L. Wang, Ph.D. thesis, The University of Texas at Arlington, pp. 55–58, 2006
32. L. Chen, S. Shet, H. Tang, H. Wang, T. Deutsch, Y. Yan, J. Turner, M. Al-Jassim, J. Mater. Chem. **20**, 6962 (2010)
33. T. Dimopoulos, A. Peić, P. Müllner, M. Neuschitzer, R. Resel, S. Abermann, M. Postl, E.J.W. List, S. Yakunin, W. Heiss, H. Brückl, J. Renew. Sustain. Energy **5**, 011205 (2013)
34. C. Sah, R. Noyce, W. Shockley, Proc. IRE **45**, 1228 (1957)

18

Solution-processed In_2S_3 buffer layer for chalcopyrite thin film solar cells

Lan Wang[1,a], Xianzhong Lin[1], Ahmed Ennaoui[2], Christian Wolf[1], Martha Ch. Lux-Steiner[1,3], and Reiner Klenk[1]

[1] Helmholtz-Zentrum Berlin für Materialien und Energie, Hahn-Meitner-Platz 1, 14109 Berlin, Germany
[2] Qatar Environment and Energy Research Institute and Hamad Bin Khalifa University, Education City, Doha, Qatar
[3] Freie Universität Berlin, Fachbereich Physik, Arnimallee 14, 14195 Berlin, Germany

Abstract We report a route to deposit In_2S_3 thin films from air-stable, low-cost molecular precursor inks for Cd-free buffer layers in chalcopyrite-based thin film solar cells. Different precursor compositions and processing conditions were studied to define a reproducible and robust process. By adjusting the ink properties, this method can be applied in different printing and coating techniques. Here we report on two techniques, namely spin-coating and inkjet printing. Active area efficiencies of 12.8% and 12.2% have been achieved for In_2S_3-buffered solar cells respectively, matching the performance of CdS-buffered cells prepared with the same batch of absorbers.

1 Introduction

While the industrial production of chalcopyrite solar modules relies on vacuum-based deposition for most of the layers, there are in addition efforts to implement alternative vacuum-free methods. Among them, printing technologies are deemed attractive because of superior utilization of raw materials and the potential for high throughput roll-to-roll fabrication. Developments in this area have been concentrating mainly on the chalcopyrite absorber [1,2]. The buffer layer, either CdS or, preferably, a Cd-free material such as Zn(O,S) [3,4] or In_2S_3 [5–10], can be prepared by dry as well as solution-based processes. The latter are typically implemented by chemical bath deposition and could be combined with a printed absorber to implement vacuum-free manufacturing of the core components of the cell. Nevertheless, printing the buffer layer could offer additional advantages in terms of material usage, in-line integration and with respect to the amount of (liquid) waste generated. In addition, certain printing technologies (such as inkjet printing [11]) provide highly localized deposition, direct patterning (without lithography) of materials in atmospheric environment and excellent raw material utilization. This is very attractive for the implementation of advanced cell concepts such as micro-concentrator cells [12].

In this work, precursor inks (molecular inks) were developed and combined with drying and annealing steps for the fabrication of In_2S_3 buffers. By tuning the rheological properties of the inks, both, spin-coating as well as inkjet printing could be implemented. By optimizing the composition of the precursor ink as well as the processing conditions, cells with more than 12% active area efficiency were achieved for both process variants.

2 Experimental

The precursor-based processing of indium sulphide thin films consists of four steps (see Fig. 1): (i) formulation of metal salt precursor inks, (ii) spin coating/inkjet printing, (iii) pre-heating of deposited inks and (iv) annealing in H_2S/Ar atmosphere. Firstly, a precursor ink was formulated by adding 1 mmol $In(NO_3)_3$ (Sigma-Aldrich, 99.99%) and 1.5 mmol thiourea ($SC(NH_2)_2$, Merck, 99%) to an ethanol-based solvent (containing 9 mL ethanol and 1 mL ethylene glycol). Assuming that each thiourea molecule results in one free sulphur atom, the In/S ratio was then in nominal accordance with the stoichiometry of the In_2S_3 compound. However, different ratios were used in the experiments to study the influence on the cell performance. The ink was subjected to continuous stirring until it became transparent. Secondly, the precursor ink was deposited by either spin coating or inkjet printing onto substrates. For inkjet printing, a PiXDRO LP50 inkjet printer (Roth & Rau B.V.) and an industrial-grade Trident 256Jet^{TM} printhead (Trident, ITW) were used. The Trident print head has 256 piezoelectric nozzles, each with a diameter of 50 μm and a nominal drop volume

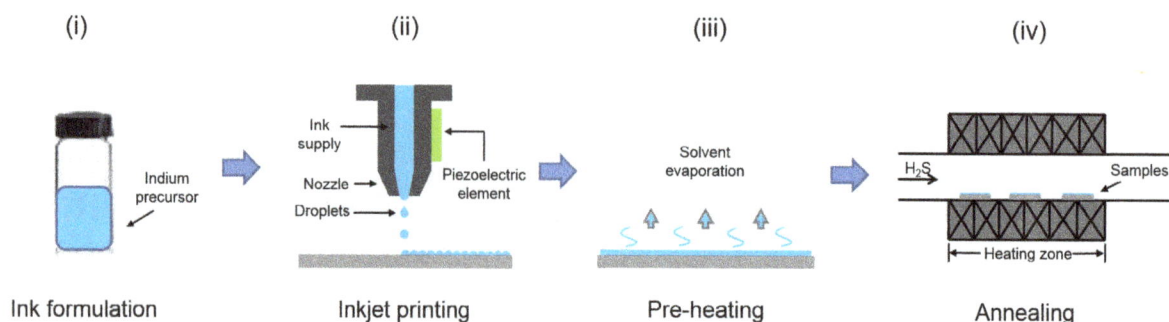

Fig. 1. Schematic diagram of In_2S_3 thin film formation process.

of $15-20$ pL. The applied voltage and pulse duration of each nozzle can be tuned to obtain a well-jetted string of droplets. By increasing the resolution (dots per inch, dpi) of the printed pattern, deposited droplets can merge into a continuous film. The resolution used in our experiment was 500 dpi, in both X and Y directions. Then the sample was pre-heated in air at 150 °C for 1 min to evaporate the solvents and dry the film. Finally the pre-heated dry film was annealed in a quartz-tube furnace at 225 °C for 10 min under 5% H_2S in Ar atmosphere to drive the formation of an In_2S_3 thin film. The quartz tube was evacuated and filled with argon three times before starting annealing.

The structure of In_2S_3 thin films deposited onto glass substrates was characterized by X-ray diffraction (XRD) operated in a 2θ range of $10-70°$ on a Bruker D8-Advance X-ray diffractometer with Cu $K\alpha1$ radiation at an incident angle of 0.5° (grazing incidence mode), using a step size of 0.02° and step time of 5 s. The thickness and morphology of buffer layers were analyzed in a LEO 1530 GEMINI scanning electron microscope (SEM) of Zeiss. The SEM images were recorded at an acceleration voltage of 10 kV.

Test solar cells were fabricated by depositing the In_2S_3 buffers onto soda lime glass/Mo/absorber thin film stacks (cut down to a sample size of 2.5×2.5 cm^2) from our in-house large-area baseline. The sequential processing of $Cu(In,Ga)Se_2$ absorbers relies on the chalcogenization of a multi-layered sputtered metal CuGa/In precursor, which is performed in nitrogen at atmospheric pressure using elemental Se [13]. Reference cells with chemical bath deposited CdS buffers were made with absorbers from the same batch. All devices were completed with an i-ZnO/ZnO:Al window layer and Ni/Al grids. Mechanical scribing was applied on each sample to define 8 solar cells with an area of 0.5 cm^2. Solar cells were measured under simulated AM 1.5 illumination and standard conditions. The cells were not annealed or light-soaked before the measurements. The quantum efficiency was recorded using Xe arc and halogen lamps and a monochromator and referenced to calibrated Si and Ge solar cells. The short-circuit current densities measured with the sun simulator were slightly too high due to spectral mismatch. Unless stated otherwise, the values given below were calculated from the measured quantum efficiency (EQE) measurement using a tabulated AM 1.5 reference spectrum and refer to the active area of the devices [14].

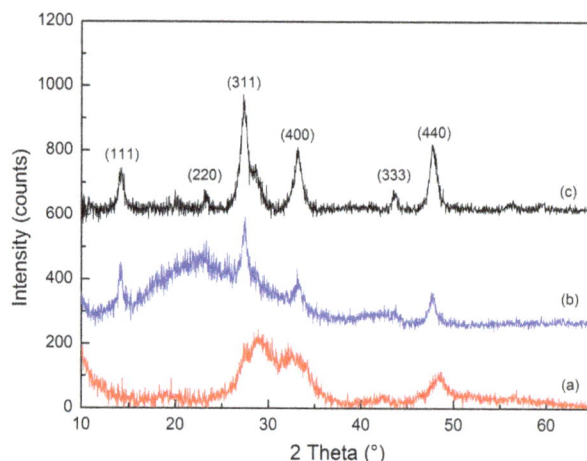

Fig. 2. XRD patterns of indium sulfide films prepared from precursor inks drop-cast onto glass substrates, (a) thiourea-containing precursor annealed in air, (b) thiourea-free precursor annealed in H_2S and (c) thiourea-containing precursor annealed in H_2S. The annealing time was 30 min for all samples.

3 Results

To confirm that the formation of In_2S_3 is possible in principle, precursor ink was drop-cast onto rinsed glass substrates followed by a pre-heating step in air at 150 °C for 1 min to remove residual solvents. XRD patterns of this pre-heated film showed no peak and further annealing in air for 30 min led to broad peaks with low intensity (see Fig. 2a). Instead, when sulfurized in Ar/H_2S at 225 °C for 30 min, the film showed pronounced XRD peaks (Fig. 2c) corresponding to β-In_2S_3(JCPDS 00-025-0390) phases, which indicates the formation of crystalline film at this moderate annealing temperature. To identify the influence of thiourea, as a sulphur source, on the formation of In_2S_3, precursor ink without thiourea was also used for structural characterization. However, the XRD pattern of the obtained film shown in Figure 2b indicates inferior crystallinity. Compared to Figure 2c, the intensities of (311) and (400) reflections were decreased and embedded in a background presumably stemming from an amorphous phase.

By adjusting the spin coating parameters, a buffer layer with a thickness of ca. 30 nm (after annealing) could

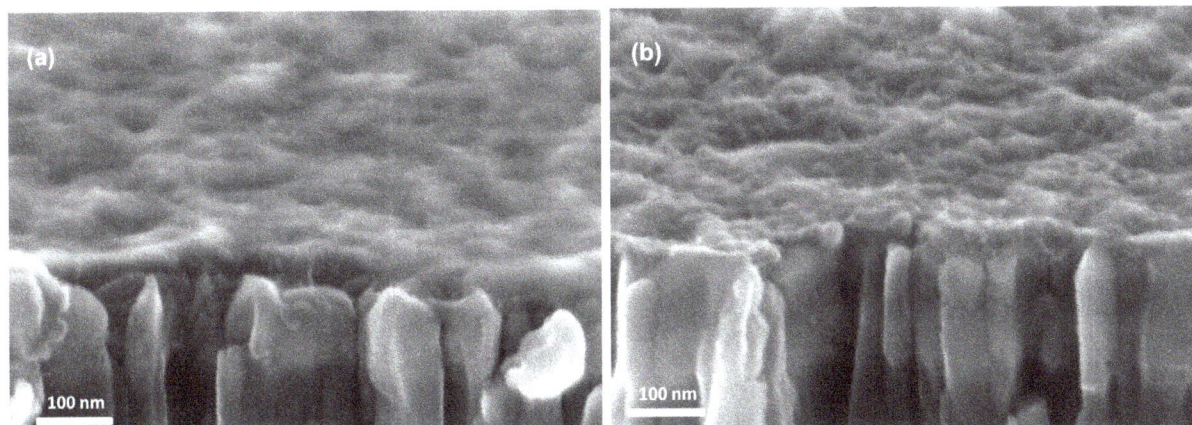

Fig. 3. SEM images of spin-coated films on glass/Mo substrates: (a) precursor film after pre-heating and (b) after annealing.

Table 1. Device parameters (best cells) of CIGSe solar cells with standard CdS buffer, spin-coated (C-03, C-04, C-05 and C-06) and inkjet-printed (C-72, C-77) In_2S_3 buffers. Sample C-03 was prepared from the thiourea-free precursor, while C-04, C-05 and C-06 were spin-coated from precursors with varied In-S ratio 1:1, 1:2 and 1:3 respectively. C-72 was prepared with the buffer annealed only in argon atmosphere (without H_2S), while C-77 was annealed in H_2S-containing atmosphere.

Sample	In/S ratio	V_{oc} (mV)	J_{sc} (mA/cm^2)	FF (%)	η (%)
CdS	–	511	39.1	63	12.5
C-03	–	389	39.6	49	7.6
C-04	1:1	478	39.9	55	10.4
C-05	1:2	491	39.2	66	12.8
C-06	1:3	485	38.6	61	11.5
C-72	1:2	458	37.5	63	10.8
C-77	1:2	481	38.6	65	12.2

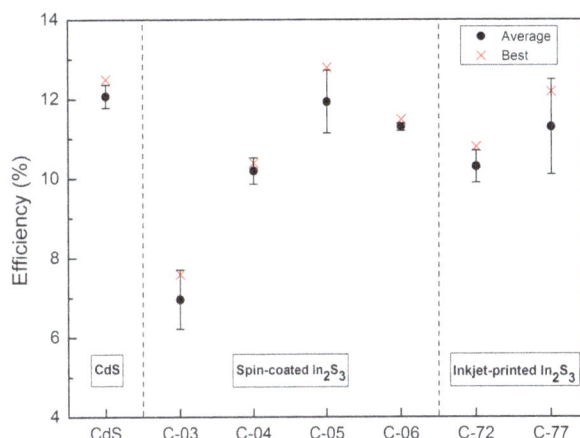

Fig. 4. Average (16 cells) and best efficiencies of solar cells buffered with CdS, spin-coated In_2S_3 and inkjet-printed In_2S_3. The error bars represent the standard deviation.

be achieved. The cross-sections of spin-coated films on a glass/Mo substrate before and after annealing are shown in Figure 3. The thickness of the film after annealing was slightly reduced (from 34 nm to 28 nm in the given SEM images). Inkjet-printed films were similar in morphology on a microscopic level but slightly thinner (25 nm before and 20 nm after annealing, respectively). In terms of material utilization, this drop-on-demand technique is superior to spin coating. For instance, only 4 µL ink is needed to inkjet print a buffer layer on a standardized sample, in comparison with 100 µL ink required for spin coating.

Photovoltaic parameters of solar cells from a typical test run (best cells) are summarized in Table 1. To achieve the optimal solar cell efficiency, variations of the In/S ratio (1:1, 1:2 and 1:3) in the precursor were applied. The highest efficiency (12.8%) was obtained with the In/S ratio of 1:2 (C-05). The sample fabricated from the thiourea-free precursor (C-03) showed lower open-circuit voltage and fill factor compared to samples produced with thiourea-containing precursors. The transfer of the processing route from spin-coating to inkjet printing was successfully implemented, with the best cell efficiency of 12.2% (C-77).

Solar cells with a buffer annealed in pure argon (C-72) showed only slightly lower efficiency. For each experimental condition, two 2.5 × 2.5 cm^2 samples have been prepared identically, each containing 8 cells. The average and best efficiencies of the cells are depicted in Figure 4. The trend of the average of cell efficiencies upon different processing conditions corresponds well with the best values and all the data are within a reasonable deviation range.

The current-voltage (J-V) characteristics and external quantum efficiencies (EQE) of selected best cells from Table 1 are presented in Figure 5. The J-V curves of both cells with solution-processed In_2S_3 demonstrate diode properties (shunt and series resistance, diode quality factor) leading to fill factors that are even slightly higher than that of the CdS reference. The EQE measurements show a good photocurrent collection with a maximum of about 90% at a wavelength between 650 nm and 680 nm.

4 Discussion

The best results were achieved by annealing in a sulphur containing atmosphere but reasonable performance

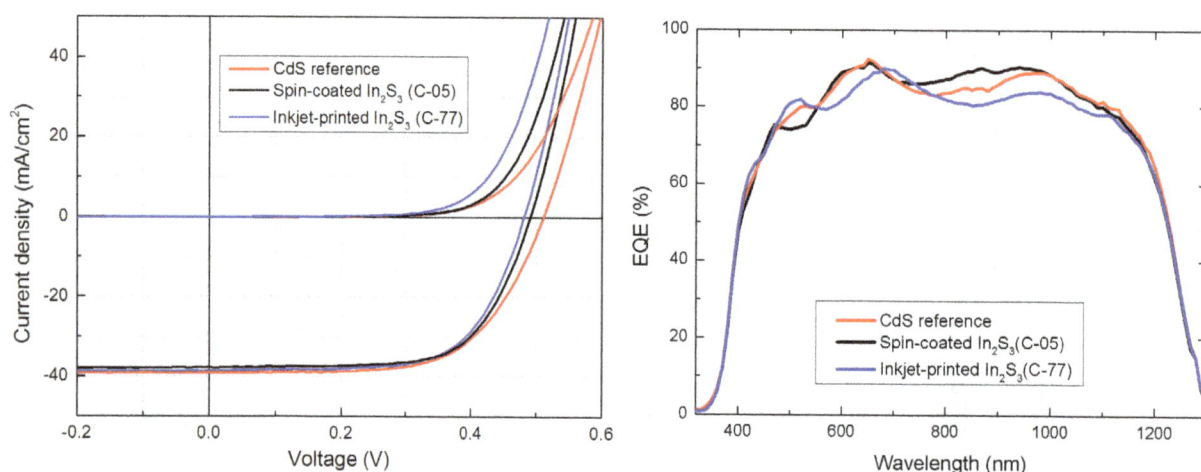

Fig. 5. Comparison of dark and illuminated current-voltage (J-V) characteristics (left, current densities as measured, without correction for spectral mismatch) and external quantum efficiencies (right) of three best solar cells buffered with standard CdS, spin-coated In_2S_3 and inkjet-printed In_2S_3, respectively.

was also achieved using an inert, sulphur-free atmosphere (sample C-72 in Tab. 1). In contrast, adding a source of sulphur (preferably with S/In = 2) to the ink appears to be crucial, as judged from the XRD patterns as well as cell data. The influence of thiourea on crystallization, phases, and final structure of the In_2S_3 buffer needs to be investigated in more detail, also taking into account the possible out diffusion of sodium and copper from the absorber. A universally applicable ink should be non-corrosive (pH) and stable, i.e. should not produce precipitations even with prolonged storage. Besides the potential formation of $In_x(OH)$, the required sulphur content may lead to $In_x(OH,S)_y$ precipitation. Certain solutions can indeed be used to grow layers (chemical bath deposition of $In_x(OH,S)_y$ [15]. It is therefore not surprising that initial efforts to formulate stable and non-corrosive water-based inks were not successful. Different metal salts and organic solvents have then been tested and lead to the simple approach used here, i.e. $In(NO_3)_3$ and thiourea in ethanol. Addition of ethylene glycol allows tuning the viscosity of the ink according to the requirements of the deposition method. This solvent system was observed to be stable under ambient temperature over months. A single coating step is sufficient to achieve a thickness suitable for buffer layers. This is different from the deposition of absorbers which requires multi-coatings with intermediate preheating steps in order to reach a micrometer-scale thickness [11]. Therefore the as-deposited buffer layer can be annealed directly without the pre-heating/drying step (confirmed in experiments not shown here). Besides annealing in inert (pure argon) atmosphere, this is another possible simplification of the process that could lower production costs. In this context, note that the annealing step used here to synthesize the compound appears to eliminate the need for post-deposition annealing of the completed device typically required for evaporated In_2S_3 buffers [16]. Higher annealing temperatures than those used in our work are assumed to further increase the crystallinity of the films. However, it is known from In_2S_3

buffers prepared by other technologies that the junction properties are degrading when the annealing temperature is too high [17] or the annealing time too long, due to Cu, Ga and In interdiffusion between the absorber and the buffer. A possible disadvantage of the process is contamination (C, O, N) stemming from the organic solvents, drying in air, decomposition of thiourea, or the In-salt. Also the use of toxic H_2S, if not avoidable, may counterbalance the reduced amount of liquid waste.

Our In_2S_3 buffer, even in preliminary experiments, showed almost identical results to the CdS buffer in terms of homogeneity and cell parameters, suggesting that this route is stable with a wide process window. The potential with respect to different types of absorbers (including those capable of higher efficiency) needs to be verified in further work. Judging from the red response the current transport properties of the absorber are not adversely affected by the buffer layer preparation. Based on the experience with other deposition methods, the blue response with a thin In_2S_3 buffer is often slightly higher than that of CdS-buffered devices [10,18]. This is not clearly observed here, partly because the blue response of the reference cell is already quite high (thin CdS). The blue response is influenced not only by the buffer layer transparency, but also by interference fringes and the collection probability of charge carriers generated close to the interface. The latter can be reduced when the there is an extended inversion zone (low absorber doping) [19]. A comparison of the optical absorption of films deposited on glass and the quantum efficiency of the cells suggests that this may play a role here but more experiments are needed to judge the blue response achievable with our printed buffers.

5 Summary

A precursor ink based process for preparing In_2S_3 buffer layers was proposed. By tuning the solution compositions, stable inks with suitable rheological properties were formulated. In_2S_3 thin films could be obtained by annealing the precursor films in a H_2S-containing atmosphere at 225 °C as verified by XRD. The addition of thiourea to the precursor ink was found to improve the crystallinity of the films as well as cell performance. Successful implementation of ink deposition by spin coating or inkjet printing on CIGSe absorbers led to working solar cells with 12.8% and 12.2% efficiency respectively, which is comparable to CdS-buffered solar cells prepared from the same batch of absorbers.

The research leading to these results has received funding from the European Union Seventh Framework Programme (FP7/2007-2013) under grant agreement no. 609788 (CHEETAH project). The authors thank C. Kelch, M. Kirsch, M. Hartig for completion of the devices.

References

1. S.M. McLeod, C.J. Hages, N.J. Carter, R. Agrawal, Prog. Photovolt.: Res. Appl. **23**, 1550 (2015)
2. T.K. Todorov, O. Gunawan, T. Gokmen, D.B. Mitzi, Prog. Photovolt.: Res. Appl. **21**, 82 (2013)
3. R. Klenk, A. Steigert, T. Rissom, D. Greiner, C.A. Kaufmann, T. Unold, M.C. Lux-Steiner, Prog. Photovolt: Res. Appl. **22**, 161 (2014)
4. T.M. Friedlmeier, P. Jackson, A. Bauer, D. Hariskos, O. Kiowski, R. Wuerz, M. Powalla, IEEE J. Photovolt. **5**, 1487 (2015)
5. C. Hönes, J. Hackenberg, S. Zweigart, A. Wachau, F. Hergert, S. Siebentritt, J. Appl. Phys. **117**, 094503 (2015)
6. R. Sáez-Araoz, J. Krammer, S. Harndt, T. Koehler, M. Krueger, P. Pistor, A. Jasenek, F. Hergert, M.C. Lux-Steiner, C.-H. Fischer, Prog. Photovolt: Res. Appl. **20**, 855 (2012)
7. T. Todorov, J. Carda, P. Escribano, A. Grimm, J. Klaer, R. Klenk, Sol. Energy Mater. Sol. Cells **92**, 1274 (2008)
8. M. Bär, N. Allsop, I. Lauermann, C.H. Fischer, Appl. Phys. Lett. **90**, 132118 (2007)
9. B. Yahmadi, N. Kamoun, R. Bennaceur, M. Mnari, M. Dachraoui, K. Abdelkrim, Thin Solid Films **473**, 201 (2005)
10. N.A. Allsop, A. Schönmann, H.J. Muffler, M. Bär, M.C. Lux-Steiner, C.H. Fischer, Prog. Photovolt: Res. Appl. **13**, 607 (2005)
11. X. Lin, J. Kavalakkatt, M.C. Lux-Steiner, A. Ennaoui, Adv. Sci. **2**, 1500028 (2015)
12. M. Paire, L. Lombez, J.-F.O. Guillemoles, D. Lincot, J. Appl. Phys. **108**, 034907 (2010)
13. B. Rau, F. Friedrich, N. Papathanasiou, C. Schultz, B. Stannowski, B. Szyszka, R. Schlatmann, Photovolt. Int. **17**, 99 (2012)
14. ASTM G173-03 (2012), *Standard Tables for Reference Solar Spectral Irradiances: Direct Normal and Hemispherical on 37° Tilted Surface*, (W.C. ASTM International, PA, 2012), www.astm.org
15. D. Hariskos, M. Ruckh, U. Rühle, T. Walter, H.W. Schock, J. Hedström, L. Stolt, Sol. Energy Mater. Sol. Cells **41-42**, 345 (1996)
16. P. Pistor, N. Allsop, W. Braun, R. Caballero, C. Camus, C.H. Fischer, M. Gorgoi, A. Grimm, B. Johnson, T. Kropp, I. Lauermann, S. Lehmann, H. Mönig, S. Schorr, A. Weber, R. Klenk, Phys. Stat. Sol. A **206**, 1059 (2009)
17. D. Abou-Ras, G. Kostorz, D. Hariskos, R. Menner, M. Powalla, S. Schorr, A.N. Tiwari, Thin Solid Films **517**, 2792 (2009)
18. P. Pistor, A. Grimm, D. Kieven, F. Hergert, A. Jasenek, R. Klenk, in: *Proceedings 37th IEEE Photovoltaic Specialists Conference (PVSC), Seattle, 2011*, p. 002808
19. R. Klenk, H.W. Schock, in: *Proceedings 12th European Photovoltaic Solar Energy Conference (EU PVSEC), Amsterdam, 1994*, p. 1588

19

Comparison of silicon oxide and silicon carbide absorber materials in silicon thin-film solar cells

Cordula Walder[1,a], Martin Kellermann[1], Elke Wendler[2], Jura Rensberg[2], Karsten von Maydell[1], and Carsten Agert[1]

[1] NEXT ENERGY · EWE Research Centre for Energy Technology at the University of Oldenburg, Carl-von-Ossietzky-Straße 15, 26129 Oldenburg, Germany

[2] Institut für Festkörperphysik, Friedrich-Schiller-Universität Jena, Helmholtzweg 3, 07743 Jena, Germany

Abstract Since solar energy conversion by photovoltaics is most efficient for photon energies at the bandgap of the absorbing material the idea of combining absorber layers with different bandgaps in a multijunction cell has become popular. In silicon thin-film photovoltaics a multijunction stack with more than two subcells requires a high bandgap amorphous silicon alloy top cell absorber to achieve an optimal bandgap combination. We address the question whether amorphous silicon carbide (a-SiC:H) or amorphous silicon oxide (a-SiO:H) is more suited for this type of top cell absorber. Our single cell results show a better performance of amorphous silicon carbide with respect to fill factor and especially open circuit voltage at equivalent Tauc bandgaps. The microstructure factor of single layers indicates less void structure in amorphous silicon carbide than in amorphous silicon oxide. Yet photoconductivity of silicon oxide films seems to be higher which could be explained by the material being not truly intrinsic. On the other hand better cell performance of amorphous silicon carbide absorber layers might be connected to better hole transport in the cell.

1 Introduction

In silicon thin-film photovoltaics multijunction solar cells are a promising concept for efficiency enhancement by reducing thermalization losses in the absorber material and increasing the overall light absorption. For this purpose subcells with optimal absorber layer bandgaps should be combined in a layer stack. Yunaz et al. [1] have shown that in a triple cell structure with a bottom cell bandgap of 1.1 eV the optimal bandgap of the top cell absorbing material exceeds the standard value of amorphous silicon. Consequently high bandgap amorphous silicon alloys containing carbon, oxygen or both are considered as absorber materials for the top cell. Their single layer properties have been studied for some time [2–8] and recently their applicability in single and multijunction cells has been confirmed [9–13]. Yet it is unclear which amorphous silicon alloy is preferable as top cell absorber. Haga et al. [5] and Fujikake et al. [6] observed the photoconductivity of a-SiO:H to be higher by orders of magnitude compared to a-SiC:H at similar optical bandgaps. However photoconductivity does not always relate to cell performance if the absorber material is not truly intrinsic [14]. Therefore the objective of our work is to compare single layer and single cell properties of a-SiO:H and a-SiC:H to see which material shows better results as a high bandgap absorber layer.

2 Experimental methods

All amorphous silicon materials were deposited by PECVD (plasma enhanced chemical vapor deposition) at 13.57 MHz at the cluster tool from Von Ardenne where separate chambers are available for intrinsic and doped layers. We created amorphous silicon single cells according to the layer stack in Figure 1 with 1 cm^2 cell area defined by the size of the silver back contact. High bandgap absorber layers were achieved by introducing either carbon dioxide or methane as additional source gas to silane and hydrogen in the intrinsic layer. All process conditions (see Tab. 1) were kept constant except the methane or carbon dioxide flow which was varied in order to compare the influence of carbon and oxygen in the absorber material. For each single cell we also produced single layers at the process conditions used for the cell absorber. Cells were deposited on commercial rough NSG (Nippon Sheet Glass) which comprises SnO_2:F as TCO electrode. Single

[a] e-mail: cordula.walder@next-energy.de

Table 1. Summary of the deposition parameters used in this work.

	p-a-SiO:H	i-a-Si:H	i-a-SiO:H	i-a-SiC:H	n-a-Si:H
ϕ (H$_2$) [sccm]	200	200	900	900	200
ϕ (SiH$_4$) [sccm]	25	40	30	30	40
ϕ (CO$_2$) [sccm]	50		3–20		
ϕ (CH$_4$) [sccm]				7–30	
ϕ (B$_2$H$_6$) [sccm]	32				
ϕ (PH$_3$) [sccm]					25
P [W]	10	10	10	10	15
p [mbar]	0.3	1	1	1	1
d_{el} [mm]	25	15	15	15	20
T_{sub} [°C]	220	220	180	180	220

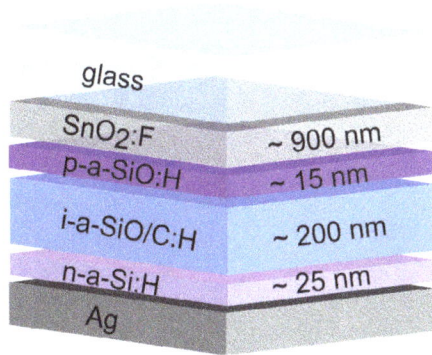

Fig. 1. Layer stack of a-SiO:H and a-SiC:H single cells.

layers were produced on commercial Schott Eco glass and on monocrystalline silicon wafer pieces polished on both sides to enable infrared transmission measurements.

Single cells were characterized by illuminated current voltage measurements (IV) using a WACOM dual lamp solar simulator under standard test conditions (AM1.5G spectrum, 1000 W/m^2, 25 °C cell temperature). The optical Tauc bandgap [15] and the corresponding Tauc slope parameter B of single layers were determined from reflection and transmission measurements with the UV-VIS-NIR spectrometer Cary 5000 from Varian. Fourier transformed infrared spectra of single layers on silicon wafers were taken with the FTIR spectrometer Spectrum 400 from Perkin Elmer. The strength of the resulting absorption peaks was analysed with the software Scout/Code by W. Theiss Hard- and Software. The hydrogen concentration was calculated from the 2000 cm^{-1} band of Si-H stretching modes according to the relation given by Lucovsky et al. [16]:

$$conc.(\mathrm{H}) = 0.77 \frac{\mathrm{at.\%}}{\mathrm{eV\ cm}^{-1}} \int \alpha \, dE. \qquad (1)$$

The microstructure factor was obtained from the absorption strength of polyhydrogen modes divided by that of all hydrogen modes around the wavenumber $\nu = 2000$ cm^{-1}. For conductivity measurements single layer samples were evaporated with coplanar aluminum pads having 1 cm length and 1 mm gap in between. These samples were placed in a vacuum atmosphere of ca. 10^{-6} mbar and their current voltage characteristics were taken with the electrometer 6517B from Keithley. Photoconductivity was measured the same way using a blue LED lamp. Constant photocurrent measurements (CPM) were conducted with the same sample geometry to calculate the Urbach energy from the exponential decay of the absorption coefficient at low photon energies. Oxygen and carbon concentrations were obtained from elastic (non-Rutherford) backscattering spectrometry (EBS). In contrast to conventional Rutherford backscattering spectrometry this method utilizes the enhancement of the scattering cross-section of He$^{+/++}$ ions on ^{12}C (^{16}O) by a factor of more than 120 (20) at a He$^{+/++}$ ion energy of 5700 keV (3035 keV). The carbon (oxygen) content can be determined by comparing the backscattering yield of each sample with the backscattering yield of a reference sample of known carbon (oxygen) concentration after subtracting the silicon background. Here, glassy carbon and a 1 μm thick thermal SiO$_2$ layer on a Si substrate were used as references.

3 Results

3.1 Single cell results

Amorphous silicon alloy single cells were produced with varying flows of either CO$_2$ or CH$_4$ in the intrinsic layer. Figure 2 shows open circuit voltage, fill factor, short circuit current and efficiency of these solar cells as a function of the Tauc bandgap which was measured on equivalent single layers. The addition of either CO$_2$ or CH$_4$ leads to a rise in open circuit voltage compared to the reference cells without carbon or oxygen in the intrinsic layer (Fig. 2a). Cells deposited with CH$_4$ reach higher values of open circuit voltage (V_{OC} up to 1000 mV) than cells deposited with CO$_2$ (V_{OC} up to 920 mV). While there is a rising tendency of open circuit voltage with Tauc bandgap for CH$_4$ in the case of CO$_2$ the open circuit voltage remains approximately constant after a first increase. Since CO$_2$ and CH$_4$ introduce defects into the intrinsic material [7,8] the fill factor decreases with rising Tauc bandgap in both cases but takes higher values for cells with CH$_4$ (Fig. 2b). The short circuit current drops significantly at the first addition of CO$_2$ or CH$_4$ and starts to decrease with the bandgap as expected (Fig. 2c). CO$_2$ and CH$_4$ have a similar influence on the short circuit current with

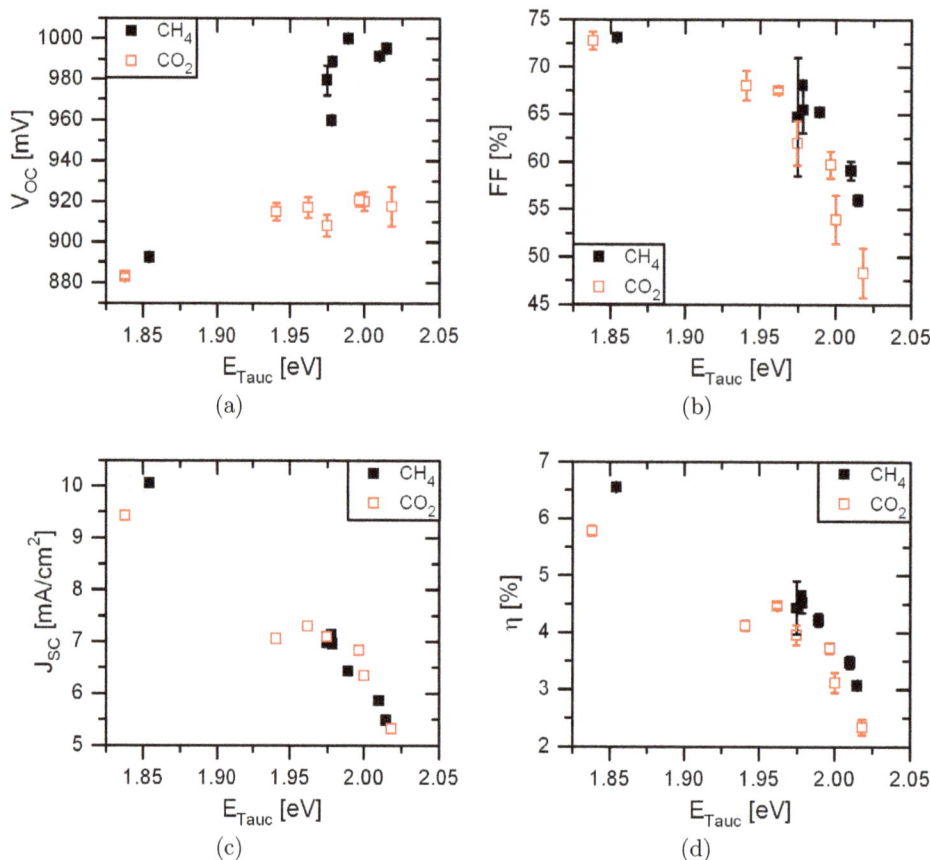

Fig. 2. IV parameters of amorphous silicon alloy single cells produced with CO_2 or CH_4 in the intrinsic layer against Tauc bandgap.

respect to the Tauc bandgap. Although the difference between the highest efficiencies reached with CO_2 and CH_4 is relatively small (0.2%) the results of the current voltage characteristics clearly favor CH_4 because open circuit voltage and fill factor are predominant factors in the development of high voltage top cells while the current can be adjusted by the cell thickness and light management.

3.2   Single layer results

Single layers corresponding to the intrinsic cell absorbers were deposited on Schott Eco glass and double sided polished monocrystalline silicon wafers. In Figure 3 the oxygen and carbon concentrations determined by EBS are shown as a function of the CO_2 or CH_4 gas flow. With increasing gas flow the concentrations of carbon and oxygen rise approximately linearly. The increase of the oxygen concentration with CO_2 flow is almost twice as high as that of the carbon concentration with an equivalent CH_4 flow. When CO_2 is used the incorporated carbon concentration even for the highest flow remains below 1 at.%.

Figure 4 shows the optical Tauc bandgap E_{Tauc} of the films prepared with CO_2 or CH_4 as a function of the oxygen or carbon concentration. The Tauc bandgap first rises steeply and then more gently with the concentration. At

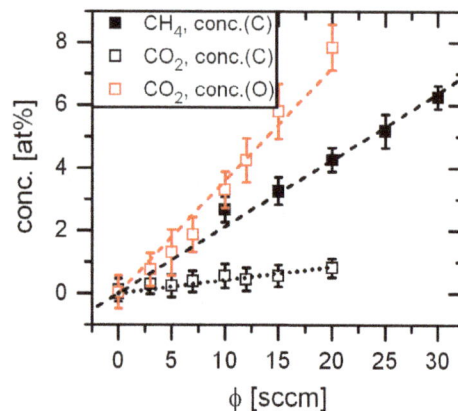

Fig. 3. C and O concentration from EBS of amorphous silicon alloy single layers against CH_4 or CO_2 flow.

equivalent oxygen and carbon concentrations there is no significant difference in the values of the Tauc bandgap for the films prepared with CO_2 or CH_4. This is an interesting result, since oxygen and carbon have different binding energies with silicon.

Infrared transmission measurements show the typical absorption band of Si-H stretching vibrations around the wavenumber of $\nu = 2000$ cm^{-1}. In Figure 5 the hydrogen

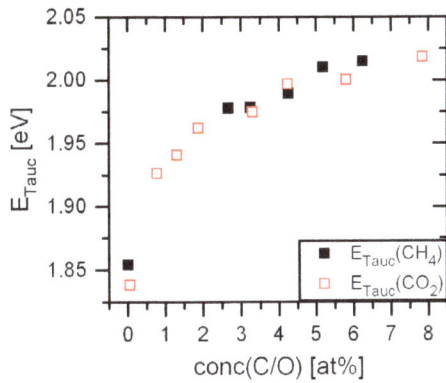

Fig. 4. Tauc bandgap of amorphous silicon alloy single layers against C or O concentration from EBS.

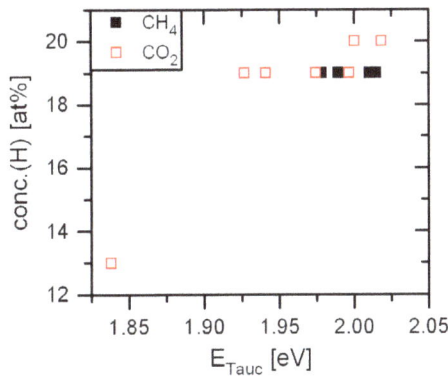

Fig. 5. Hydrogen concentration deduced from Si-H stretching modes around the wavenumber $\nu = 2000$ cm^{-1} for amorphous silicon alloy single layers produced with CO_2 or CH_4 on c-Si wafers.

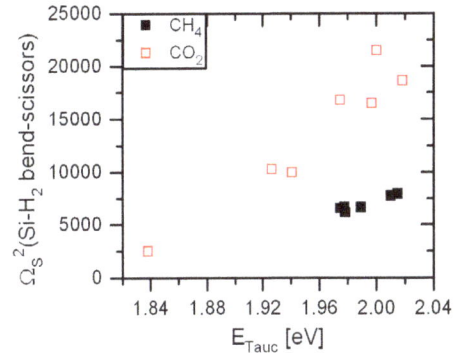

Fig. 6. Squared oscillator strength of Si-H$_2$ bend-scissors IR mode at $\nu = 880$ cm^{-1} for amorphous silicon alloy single layers produced with CO_2 or CH_4 on c-Si wafers.

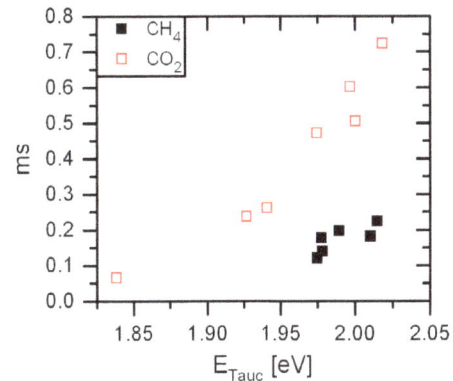

Fig. 7. Microstructure factor ms of amorphous silicon alloy single layers produced with CO_2 or CH_4 on c-Si wafers.

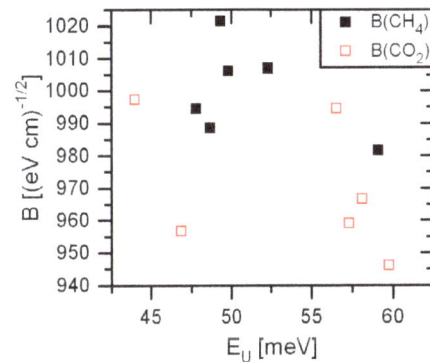

Fig. 8. Tauc slope parameter B of amorphous silicon alloy single layers produced with CO_2 or CH_4 against Tauc bandgap.

concentration according to formula (1) is depicted for single layers prepared with CO_2 or CH_4 in dependence on the Tauc bandgap. The alloys with oxygen or carbon are prepared at a 6 times higher hydrogen dilution than the reference sample and consequently show a higher hydrogen content of about 20 at.% compared to 13 at.% for the reference. Almost no difference in hydrogen content can be detected for all silicon carbide and silicon oxide samples regardless of the source gas or the Tauc bandgap. In Figure 6 the squared oscillator strength of the Si-H$_2$ bend-scissors IR mode at $\nu = 880$ cm^{-1} is presented. It is proportional to the integrated absorption coefficient of this mode and therefore is a measure for the amount of Si-H$_2$ bonds which are known to promote degradation effects. For CO_2 and CH_4 the amount of Si-H$_2$ bonds rises with increasing bandgap indicating a deterioration of material quality. Yet for CH_4 it is considerably lower than for CO_2 at equivalent bandgaps. Figure 7 shows the microstructure factor for films prepared with CO_2 or CH_4 as a function of the Tauc bandgap. The microstructure factor mirrors the trend of the amount of Si-H$_2$ bonds as expected. It rises steeply from 0.24 to 0.72 for films prepared with CO_2 and more moderately from 0.12 to 0.22 for films prepared with CH_4.

The Tauc slope parameter B represents the steepness of the absorption coefficient over energy and therefore is an indicator of material quality. Figure 8 illustrates the Tauc slope parameter against the Urbach energy for samples prepared with CO_2 or CH_4. Samples prepared with CH_4 tend to achieve higher B parameters and lower Urbach energies compared to samples prepared with CO_2. This suggests less disorder and less bond angle distortion in the case of CH_4 compared to CO_2. Otherwise it is difficult to make out a clear relationship between the Tauc slope parameter and the Urbach energy. Ambrosone et al. [17]

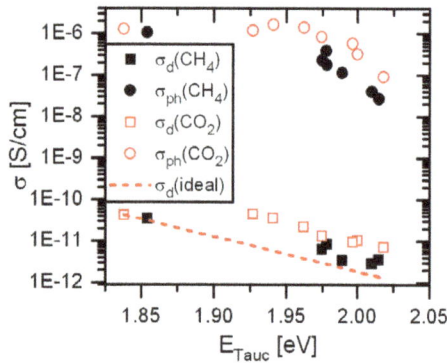

Fig. 9. Dark and photoconductivity of amorphous silicon alloy single layers produced with CO_2 or CH_4 against Tauc bandgap at room temperature.

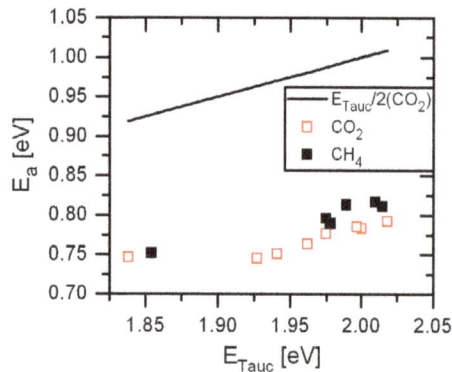

Fig. 10. Activation energy of amorphous silicon alloy single layers produced with CO_2 or CH_4 against Tauc bandgap at room temperature.

observe how the Tauc slope parameter decreases from 788 to 663 $(eV\ cm)^{-1/2}$ with rising Urbach energy in a range from 72 to 170 meV. This trend is not clearly demonstrated by our results since the variation in Urbach energy is quite small.

In Figure 9 dark and photoconductivity measurements for films prepared with CO_2 or CH_4 are depicted. As expected dark and photoconductivity both decrease at higher bandgaps. Yet the films prepared with CO_2 show higher photoconductivity by half an order of magnitude compared to CH_4 which seems to contradict our remaining results. Figure 10 shows the activation energy calculated from the dark conductivity at room temperature with a constant conductivity prefactor of $\sigma_0 = 150$ S/cm as proposed in reference [18]. The activation energy increases with the Tauc bandgap and takes slightly lower values for samples prepared with CO_2 compared to those prepared with CH_4. Half the Tauc bandgap is also depicted for comparison since it should be close to the activation energy of truly intrinsic layers.

4 Discussion

The addition of either CH_4 or CO_2 as source gases for high bandgap amorphous silicon alloys leads to the incor-

poration of carbon or oxygen into the amorphous silicon network. Although CO_2 contains carbon almost no carbon is incorporated into the layer in agreement with previous results [6]. As can be seen in Figure 3 oxygen is incorporated more easily into the network since higher oxygen than carbon concentrations are reached at equivalent flows of CO_2 and CH_4. This can be explained by the higher energy required for the dissociation of CH_4 compared to CO_2 [19,20]. Furthermore Bullot and Schmidt [21] suggest that in the low power regime CH_4 is only dissociated by secondary reactions with silicon species but not directly by electron impact.

The Tauc bandgap rises first steeply and then more gently with oxygen or carbon concentration (see Fig. 4). The initial steep increase is promoted by much higher hydrogen dilution and therefore higher hydrogen incorporation in the alloyed samples compared to the a-Si:H reference (see Tab. 1 and Fig. 5). Hydrogen is well-known to increase the bandgap of amorphous silicon and its alloys [22]. Further increase in the bandgap with oxygen or carbon concentration is explained by higher binding energies and backbonding effects of oxygen or carbon with silicon [7]. Yet oxygen has a higher bond strength with silicon than carbon and therefore should produce a higher bandgap at equivalent concentrations [23] which is not observed in our results. Possibly this is due to the silicon carbide layers having a higher hydrogen content than the silicon oxide films. While CH_4 is known to promote hydrogen incorporation [24] CO_2 has been observed to suppress it [25]. If hydrogen is bonded in CH_3 groups in the silicon carbide films the difference in hydrogen concentration to the silicon oxide films is not observable in the FTIR mode at $\nu = 2000\ cm^{-1}$ used in Figure 5. Moreover the determination of the hydrogen concentration with formula (1) introduces large uncertainties especially with respect to amorphous silicon alloys instead of just a-Si:H.

Single cell results reveal much higher open circuit voltage and slightly higher fill factor in dependence on the Tauc bandgap for CH_4 than for CO_2. Furthermore the open circuit voltage rises with the Tauc bandgap in case of CH_4 while for CO_2 it stays almost constant. This can be explained by better material quality of the silicon carbide layers since the microstructure factor and the amount of Si-H_2 bonds suggest a more compact material compared to the layers prepared with CO_2. In contrast Beyer found a bigger void structure for amorphous silicon alloys with 18 at.% carbon concentration than for those with 23 at.% oxygen concentration from effusion measurements [26]. This indicates that either process conditions change the observed trends considerably or that low carbon and oxygen concentrations lead to a different behavior, since the investigated samples show concentrations below 8 at.% which is about a factor of 2-3 lower than in the case of Beyer.

Unlike our remaining results conductivity measurements seem to indicate better performance of samples prepared with CO_2 than with CH_4 in agreement with Haga et al. [5] and Fujikake et al. [6]. One explanation could be an unintentional doping effect especially in the silicon oxide films. According to Beck et al. [14] photo-conductivity only relates well to cell performance if the absorber material is truly intrinsic. Figure 9 shows that dark conductivity of silicon oxide only starts to decrease at higher Tauc bandgaps and not immediately as expected. The dashed line represents the ideal dark conductivity at room temperature if the Fermi level energy is half the Tauc bandgap energy:

$$\sigma_d(\text{ideal}) = \sigma_0 \exp\left\{-0.5\frac{E_{\text{Tauc}}}{kT}\right\}.$$

The conductivity prefactor σ_0 was kept constant and calibrated so that σ_d (ideal) matches the dark conductivity values of the reference samples at low Tauc bandgaps. Obviously the dark conductivity of silicon carbide is closer to the ideal curve than that of silicon oxide. Both materials are expected to be rather n-type and oxygen is known to act as a donor in the form of O_3^+ impurities [27]. Figure 10 suggests that CO_2 produces slightly lower activation energy than CH_4. Maybe unintentional n-doping leads to increased conductivity of the silicon oxide films but to worse cell performance compared to silicon carbide. Another explanation for the discrepancy between photoconductivity and cell results could be that cell performance also depends on hole carrier transport which is not monitored by our conductivity measurements. Wang et al. report that hole carrier collection is strongly deteriorated with increased oxygen concentration of amorphous silicon oxide absorber layers in single cells [18]. So it is possible that silicon carbide allows better hole carrier transport than silicon oxide at equivalent bandgaps when used as absorber layers in amorphous silicon single cells.

5 Conclusion

The objective of this work was to compare CO_2 and CH_4 as source gases for high bandgap amorphous silicon alloy absorber layers. Single cell results reveal higher fill factors and especially higher open circuit voltages for CH_4 in contrast to CO_2 at equivalent Tauc bandgaps. The microstructure factor from infrared transmission measurements indicates that the reason for this might be less voids in the structure of silicon carbide and consequently better material quality of layers produced with CH_4. Curiously photoconductivity shows higher values in the case of CO_2. One reason for this discrepancy could be higher unintentional n-doping in silicon oxide samples by O_3^+ impurities. Another explanation could be better hole carrier transport when silicon carbide absorber layers are used.

We would like to thank Ulrich Barth for his support with EBS measurements as well as Tim Möller and Ulrike Kochan for their help with UV-VIS optical measurements.

References

1. I.A. Yunaz, A. Yamada, M. Konagai, Jpn J. Appl. Phys. **46**, L1152 (2007)
2. B.G. Yacobi, R.W. Collins, G. Moddel, P. Viktorovitch, W. Paul, Phys. Rev. B **24**, 5907 (1981)
3. Y. Tawada, K. Tsuge, M. Kondo, H. Okamoto, Y. Hamakawa, J. Appl. Phys. **53**, 5274 (1982)
4. A. Morimoto, T. Miura, M. Kumeda, Jpn J. Appl. Phys. **21**, L119 (1982)
5. K. Haga, K. Yamamoto, M. Kumano, Jpn J. Appl. Phys. **25**, L39 (1986)
6. S. Fujikake, H. Ohta, A. Asano, Y. Ichikawa, H. Sakai, Mater. Res. Soc. Symp. Proc. **258**, 875 (1992)
7. D. Das, S.M. Iftiquar, A.K. Barua, J. Non-Cryst. Solids **210**, 148 (1997)
8. A. Desalvo, F. Giorgis, C.F. Pirri, E. Tresso, P. Rava, J. Appl. Phys. **81**, 7973 (1997)
9. I.A. Yunaz, K. Hashizume, S. Miyajima, A. Yamada, M. Konagai, Sol. Energy Mater. Sol. Cells **93**, 1056 (2009)
10. I.A. Yunaz, H. Nagashima, D. Hamashita, S. Miyajima, M. Konagai, Sol. Energy Mater. Sol. Cells **95**, 107 (2011)
11. S. Inthisang, B. Janthong, P. Sichanugrist, M. Konagai, in *26th European Photovoltaic Solar Energy Conference and Exhibition, Hamburg, 2011*, p. 2392
12. K. Sriprapha, A. Hongsingthong, T. Krajangsang, S. Inthisang, S. Jaroensathainchok, A. Limmanee, W. Titiroongruang, J. Sritharathikhun, Thin Solid Films **546**, 398 (2013)
13. J. Sritharathikhun, S. Inthisang, T. Krajangsang, A. Limmanee, K. Sriprapha, Thin Solid Films **546**, 383 (2013)
14. N. Beck, N. Wyrsch, Ch. Hof, A. Shah, J. Appl. Phys. **79**, 9361 (1996)
15. J. Tauc, R. Grigorovici, A. Vancu, Phys. Stat. Sol. **15**, 627 (1966)
16. G. Lucovsky, J. Yang, S.S. Chao, J.E. Tyler, W. Czubatyj, Phys. Rev. B **28**, 3225 (1983)
17. G. Ambrosone, D.K. Basa, U. Coscia, P. Rava, Thin Solid Films **518**, 5871 (2010)
18. S. Wang, V. Smirnov, T. Chen, B. Holländer, X. Zhang, S. Xiong, Y. Zhao, F. Finger, Jpn J. Appl. Phys. **54**, 011401 (2015)
19. C.W. Bauschlicher Jr., S.R. Langhoff, Chem. Phys. Lett. **177**, 133 (1991)
20. L.F. Spencer, A.D. Callimore, Plasma Chem. Plasma Process. **31**, 79 (2010)
21. J. Bullot, M.P. Schmidt, Phys. Stat. Sol. B **143**, 345 (1987)
22. A. Singh, E.A. Davis, J. Non-Cryst. Solids **122**, 233 (1990)
23. T. Jana, S. Ghosh, S. Ray, J. Mater. Sci. **32**, 4895 (1997)
24. D. Kuhman, S. Grammatica, F. Jansen, Thin Solid Films **177**, 253 (1989)
25. A. Samanta, D. Das, Sol. Energy Mater. Sol. Cells **93**, 588 (2009)
26. W. Beyer, J. Non-Cryst. Solids **266-269**, 845 (2000)
27. T. Shimizu, T, Ishii, M. Kumeda, A. Masuda, J. Non-Cryst. Solids **227-230**, 403 (1998)

Grain boundary assisted photocurrent collection in thin film solar cells

Susanna Harndt[1], Christian A. Kaufmann[1], Martha C. Lux-Steiner[1], Reiner Klenk[1,a], and Reiner Nürnberg[2]

[1] Helmholtz-Zentrum Berlin für Materialien und Energie, Hahn-Meitner-Platz 1, 14109 Berlin, Germany
[2] Weierstraß-Institut für Angewandte Analysis und Stochastik, Mohrenstr. 39, 10117 Berlin, Germany

Abstract The influence of absorber grain boundaries on the photocurrent transport in chalcopyrite based thin film solar cells has been calculated using a two dimensional numerical model. Considering extreme cases, the variation in red response is more expressed than in one dimensional models. These findings may offer an explanation for the strong influence of buffer layer preparation on the spectral response of cells with small grained absorbers.

1 Introduction

The red response of a planar heterojunction solar cell is determined mainly by absorber properties, i.e., its absorption coefficient, doping, minority carrier diffusion length, and band gap grading. At longer wavelengths photons are absorbed at a greater distance from the hetero junction and the quantum efficiency may be reduced due to insufficient transport of minority carriers over the enlarged distance. The quantum efficiency can be calculated in analytical approximation [1,2] or by numerical device modeling. Figure 1 shows a set of quantum efficiencies which have been calculated with SCAPS [3] for a typical chalcopyrite-based solar cell consisting of a Cu(In,Ga)Se$_2$ (CIGSe) absorber, CdS buffer and ZnO window layer. Absorber doping and carrier lifetime have been varied covering the full parameter range that we expect for these cells. Absorption data (which strongly influence the result) have been taken from [4]. In contrast to this, Figure 2 shows experimental results for a set of cells where only the buffer layers were prepared differently (ILGAR with InCl$_3$ or In(acac) precursor [5], chemical bath deposition of CdS, PVD of In$_2$S$_3$ [6]) but all the other layers were each produced in the same process. These curves are not easily understood in terms of the 1D models described above. The buffer layer by itself should have only a minor influence on carrier collection. The variation in red response in the experimental data would imply an even larger variation of absorber properties as used for the calculation in Figure 1. This suggests that buffer layer preparation somehow drastically changes the absorber properties. However, buffer prepara-

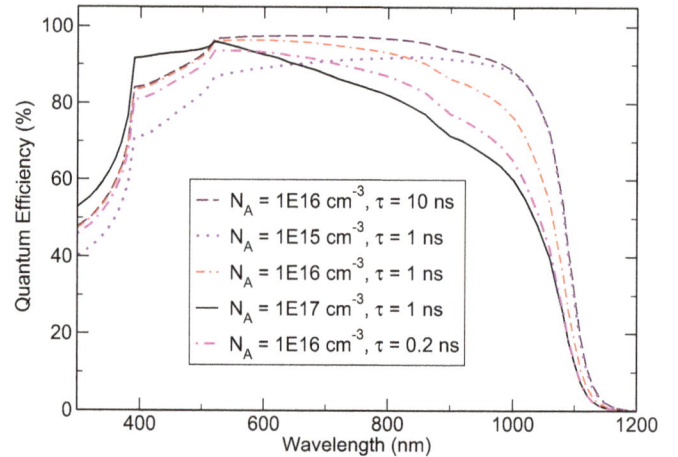

Fig. 1. Calculated quantum efficiency of CIGSe/CdS/ZnO solar cells with different shallow acceptor concentration N_A and lifetime τ of electrons and holes in the CIGSe absorber.

tion is carried out at much lower temperature (between room temperature and 225 °C) than the absorber preparation which should limit such post deposition changes in absorber properties.

A possible explanation of this discrepancy involves the role of grain boundaries within the absorber for carrier collection. Diffusion of impurities along grain boundaries is much faster than within the grains and therefore could be significant even at the lower temperature used for buffer layer preparation [7]. The influence of grain boundaries must be stronger in films with smaller grains. We note that the absorbers that yielded the experiential data shown in

^{a} e-mail: `klenk@helmholtz-berlin.de`

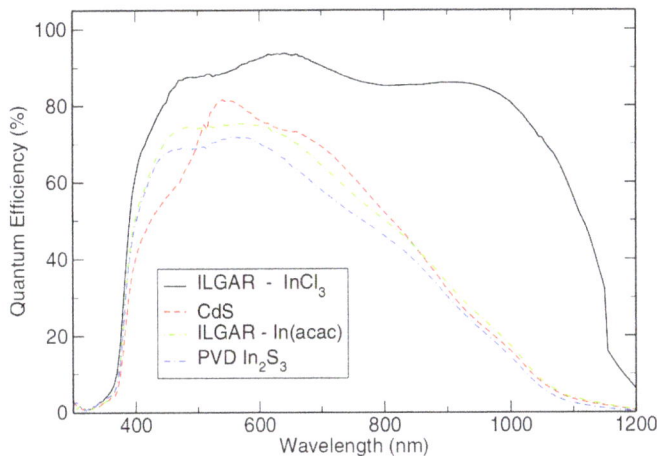

Fig. 2. Quantum efficiency of CIGSe/buffer/ZnO solar cells. All layers except the buffer were each grown in the same process. The buffer layers were deposited by ILGAR with two different precursors (ILGAR – InCl$_3$, ILGAR – In($acac$)), by chemical bath deposition (CdS) or evaporation (PVD In$_2$S$_3$).

Figure 2 were grown at lower than usual substrate temperature (<440 °C) as part of the optimization of processes for cells on polyimide foil [8] and they therefore have smaller grains. In this contribution we present 2D/3D numerical modeling of grain boundary assisted photocurrent collection and discuss the results in view of grain boundaries in chalcopyrite absorbers.

2 Calculation

For the electronic model the grain boundary is regarded as interface (ideal plane with zero thickness) between grains. Conduction and valence band edges are continuous across the interface. Defects at the interface within the band gap cause a certain recombination velocity. In addition, defects may carry a localized charge giving rise to a symmetrical band bending towards the grain boundary. This essentially follows Seto's model [9] for grain boundaries in polycrystalline silicon. Depending on the charge distribution within the grain boundary the band edges bend "upward" (accumulation) or "downward" (depletion) from their equilibrium position within the grain towards the grain boundary. In the first case the photogenerated electrons (minority carriers in the p-type absorber) are driven away from the grain boundary in analogy to back surface fields used for contact passivation. In the latter case electrons are driven towards the grain boundary, potentially increasing photo current losses. However, if the depletion is further increased the charged grain boundary will eventually create a region with conductivity type inversion. The electrons are then majority carriers within the inversion zone and their recombination probability becomes very small. Furthermore, if the inversion zone connects to the n-type part of the heterojunction (the buffer), photo electrons can be collected by transport to and within the grain boundary. The WIAS-TeSCA code [10–12], which solves the semiconductor drift-

diffusion equations in 2D or 3D, was used to calculate the effect of grain boundary inversion on current collection. A structural unit (grain) with cylindrical symmetry was chosen so that the calculation is performed in 2D parameter space but provides results for an idealized 3D structure. The grain has a grain boundary around its circumference extending from the heterojunction to close to the back contact. The small gap next to the ohmic back contact was introduced to prevent shunting of the device along a path from the junction via the grain boundary to the back contact. The defects at the grain boundary can have two main consequences: deep defects act as recombination centers and charged defects cause a band bending towards the grain boundary. Our model of the defect distribution reflects these considerations by a mid-gap defect which is an effective recombination center and a donor type defect with a concentration that is sufficiently high to pin the Fermi level. Only the energy position of the latter was varied to create different band bending towards the grain boundary. All other grain boundary defect parameters were kept constant. The material properties within the grain are the same as in the SCAPS calculations shown above. The absorber thickness was set to 3 μm to reduce the influence of back contact recombination. Full parameter sets are available on request from the authors.

3 Results

Figure 3 shows an example of a calculated potential distribution where the space charge region not only extends from the heterojunction into the absorber grain but also from its grain boundary. The direction of the electron current under illumination is given by the arrows and clearly shows the transport of electrons to the inverted grain boundary. The inversion is caused by the shallow donor type defect at the grain boundary. In Figure 4 the energy position of the donor type defect at the grain boundary has been varied. The band bending generally follows the energy of the donor. The calculated quantum efficiency at a wavelength of 1000 nm demonstrates the effect of grain boundaries on the red response of the solar cell. The red response is significantly enhanced for the fully inverted grain boundary (defect close to the conduction band, filled symbols), compared to the 1D case without a grain boundary (dashed lines). The quantum efficiency drops rapidly as the band bending is reduced, inversion is lost, and the hole density at the grain boundary increases. The transition occurs roughly where the band bending is equal to half of the band gap, i.e., when the defect is at mid-gap position (dotted vertical line in Fig. 4). However, close to the junction the band bending is also influenced by the buffer and window layer (Fig. 3), and a transition case exists where the grain boundary is inverted in the upper part of the grain but not in the lower part of the grain. A minimum (open symbols) is eventually reached when there still is significant band bending (depletion case of previous paragraph). As the defect level is moved deeper within the band gap, the resulting band bending does no longer follow the energy position of the defect in a 1:1

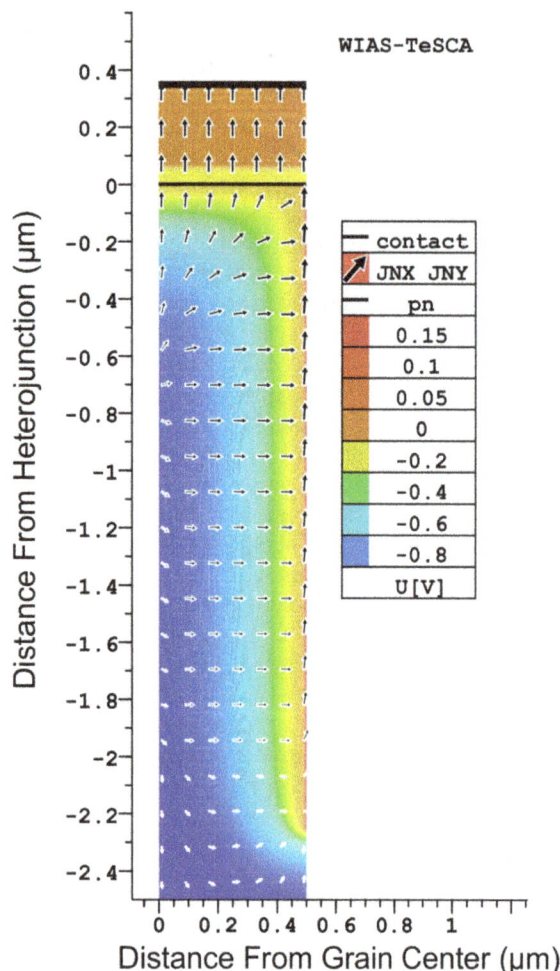

Fig. 3. Calculated potential (with respect to the equilibrium Fermi-level, color-coded) and electron current flow (arrows) in a CIGSe/CdS/ZnO structure with an inverted grain boundary in the absorber (only one half of the grain is shown). Parameters for this calculation are shallow acceptor density in the absorber $N_A = 10^{16}$ cm^{-3}, and carrier lifetimes τ of 1 ns. The cell is illuminated with monochromatic light (wavelength 1000 nm, power 100 mW/cm^2).

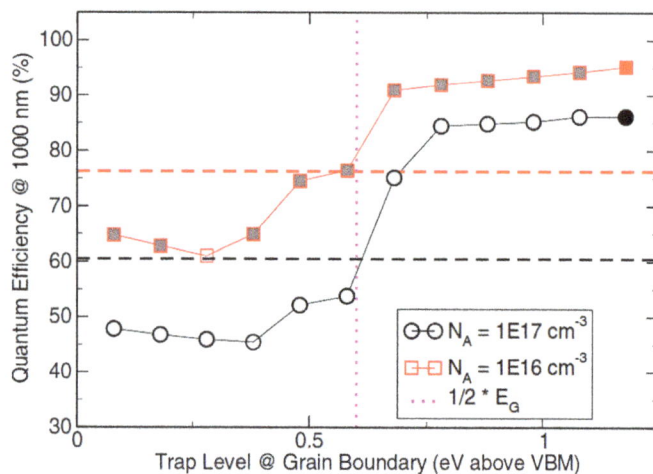

Fig. 4. Calculated quantum efficiency of polycrystalline CIGSe/CdS/ZnO solar cells at a wavelength of the incident light of 1000 nm. The energy position of the donor type defect at the grain boundary (trap level) has been varied to modify the band bending towards the grain boundary. The dashed horizontal lines indicate the quantum efficiency without grain boundary (1D case). The dotted vertical line marks the midgap position. The absorber grain radius is 1 μm and the carrier lifetimes are 1 ns.

for the extreme cases regarded here. Figure 6 shows quantum efficiencies over the full wavelength range (in analogy to Figs. 1 and 2) for the three grain boundary configurations (fully inverted grain boundaries, without grain boundary, maximum recombination). Results calculated with the WIAS-TeSCA code for the 1D case without grain boundaries are identical to those calculated with SCAPS. The discrepancy at lower wavelengths is only due to neglecting the optical absorption in the window and buffer layer in the WIAS-TeSCA model.

4 Discussion

The calculations confirm that the defect distribution (modeled here by just one mid-gap recombination center and one donor with varied energy) at grain boundaries is an additional and quite important parameter in the design of efficient thin film solar cells even when compared to other parameters such as grain size, absorber doping, and minority carrier lifetime. Measurements of device quality chalcopyrite films [13] have found depletion zones at grain boundaries and it was suggested that these may lead to improved carrier collection [14]. Current transport along grain boundaries has also been measured by conductive scanning probe microscopy [15, 16]. With respect to the current collection, our calculations show that there can be a negative as well as a positive influence depending on the band bending towards the grain boundary. The reported band bending would be too small to achieve the beneficial effect on current collection predicted by our model, however, it has been argued that the measurements underestimate the band bending at grain boundaries due

relation. This is due to incomplete pinning of the Fermi level by a single defect type. Consequently, the accumulation case is not reached in these calculations. Qualitatively, the results are very similar for two different levels of absorber doping, however, the effects are more significant for the higher doping level. The influence of grain boundaries on red response for varied absorber properties can be seen in Figure 5. Looking at, e.g., the emphasized black curves, the red response can be significantly higher (filled symbols) as well as lower (open symbols) as compared to the 1D case without grain boundaries (dashed lines). Furthermore, both gain and loss depend on the absorber bulk properties (doping and carrier lifetime). As expected, both maximum gain (inverted grain boundaries) as well as loss in quantum efficiency are reduced with increasing grain size. However, even at a grain radius of 2 μm there still is a significant influence of the grain boundaries, at least

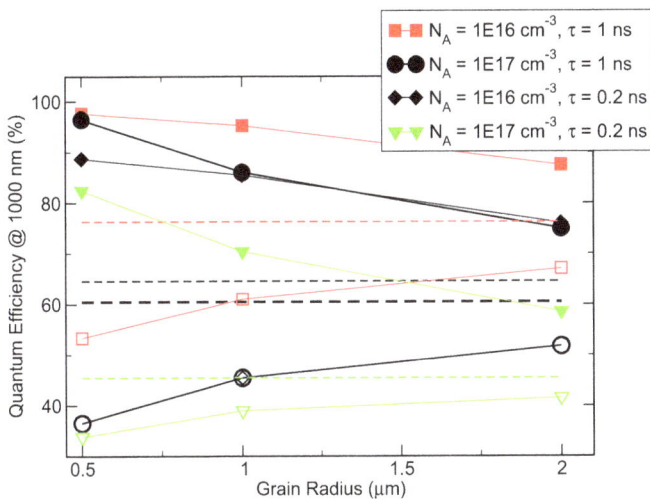

Fig. 5. Calculated quantum efficiency of polycrystalline CIGSe/CdS/ZnO solar cells at a wavelength of the incident light of 1000 nm as a function of absorber grain radius. Filled symbols indicate inverted grain boundaries whereas empty symbols refer to the minimum quantum efficiency that is reached for a specific trap level (see previous figure). The dashed lines indicate the quantum efficiency without grain boundary (1D case).

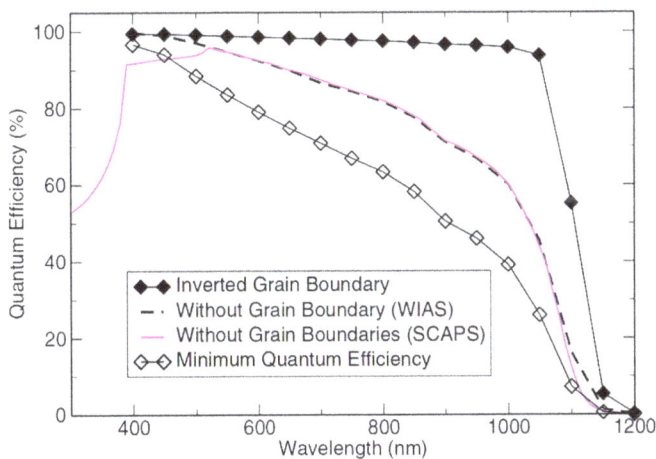

Fig. 6. Calculated spectral quantum efficiency of CIGSe/CdS/ZnO solar cells. Acceptor density and lifetime in the absorber are $N_A = 10^{17}$ and $\tau = 1$ ns, respectively. The absorber grain radius is 0.5 μm.

to surface band bending [13] and the averaging effect of the scanning probe [17].

Taretto et al. [18] limited their numerical calculations to the case of depleted grain boundaries and have thus shown only the negative influence of grain boundary recombination. Our finding of both, negative and positive influence, is in principle agreement with calculations described in reference [19]. However, the authors only reported the effect on the total photo current density under white light illumination and did not show the quantum efficiency at different wavelengths. An increase in white light induced photo current for inverted grain boundaries was

also found in reference [20]. A different electronic model of the grain boundary has been proposed where Cu depletion leads to a discontinuity of the valence band (hole barrier) [21,22]. The previous calculations in references [18,19] also show some results for this alternative model. It is clear that such a hole barrier would decrease the recombination of electrons at the grain boundaries even under the conditions which lead to poor carrier collection in our calculations. In summary, the influence of material properties on the spectral response would be closer to the 1D model (without grain boundaries) [20].

In terms of overall device performance, the optimum band bending towards the grain boundaries depends on other parameters such as carrier lifetime and doping. If the minority carrier diffusion length within the grains is sufficient for carrier collection, an accumulation layer towards the grain boundary is preferable to the inverted grain boundary. This is due to the decreased open circuit voltage of devices with inverted grain boundaries. The significance of this loss depends on the dominating recombination mechanism giving rise to the bucking current. Also, depending on how strongly the Fermi-level is pinned at the grain boundary, the inversion may vanish at higher forward bias, resulting in strongly voltage-dependent carrier collection (poor fill factor). In material with grains smaller than considered here the grain interior may become depleted and the band bending may be insufficient to establish inverted grain boundaries. For low quality material, the trade-off between improved current collection and reduced open circuit voltage may be optimized in the numerical model but difficult to achieve in the experiment. It was concluded in reference [20] that in most cases the performance of a CIGSe based cell cannot be improved by grain boundaries. The geometry of a device with grain boundary assisted carrier collection is in some respects similar to that of the ETA (Extremely Thin Absorber) solar cell. Analytical modeling [23] of this cell demonstrates both the potential as well as the complexity of adapting the geometry to achieve reasonable efficiencies with limited material quality.

If we consider again the large variation in red response in cells with different buffer layers (Fig. 1), it is indeed conceivable from the calculations that absorber grain boundaries could play a significant role. Together with the fast diffusion of impurities along grain boundaries, this suggests that process steps carried out after absorber preparation and at significantly lower temperature, such as the deposition of buffer and window layers, can still influence absorber and hence device properties. Here, a particularly good red response has been observed when the buffer layer was deposited with ILGAR using a chlorine containing precursor solution. It has been suggested that Cl may cause donor type defects (Cl_{Se}) in the bulk of $CuInSe_2$ [24,25]. We speculate that during buffer deposition Cl is diffusing along the grain boundaries, leads to inverted grain boundaries and superior carrier collection. The expected corresponding loss in open circuit voltage is also systematically observed in the experiment when using buffers prepared from the $InCl_3$ solution. Nevertheless, it should be understood that the experimental observation has mainly been the motivation to revisit the problem of grain boundary assisted current collection. Additional work would be required to establish the claim that ILGAR deposition of a buffer layer with the Cl containing precursor leads to inverted grain boundaries in the absorber. The additional work could include measurements of electron beam induced current to map the electron collection, assessment of the Cl concentration at the grain boundary, and direct measurement of the band bending by Kelvin probe microscopy.

5 Conclusions

Numerical calculations show that grain boundaries in the absorber can have a large influence on photocurrent collection. Hence, the defect distribution (recombination centers, charged defects) at grain boundaries is an additional and quite important parameter in the design of efficient thin film solar cells, even when compared to other parameters such as grain size, absorber doping, and minority carrier lifetime. Shallow donor type defects at the grain boundary can cause an inversion zone in which electrons are transported leading to very high quantum efficiency even in films with high doping and short carrier lifetimes. When the defect is located deeper in the band gap, the inversion and carrier transport are lost causing excessive recombination losses. These findings are a tentative explanation for the expressed differences in quantum efficiency which were found for cells with differently prepared buffer layers.

Financial support by the German Federal Ministry of Economy and Technology (BMWi), represented by the German Aerospace Center (DLR) under Contract 50RN1101 is gratefully acknowledged. The authors thank Adrien Bercegol for discussions.

References

1. R. Klenk, H.-W. Schock, in *Proceedings of the 12th European Photovoltaic Solar Energy Conference, Amsterdam, 1994*, pp. 1588–1591
2. M. Troviano, K. Taretto, Sol. Energy Mater. Sol. Cells **95**, 821 (2011)
3. M. Burgelman, P. Nollet, S. Degrave, Thin Solid Films **361**, 527 (2000)
4. M. Gloeckler, A. Fahrenbruch, J. Sites, in *Proceedings of the 3rd World Conference on Photovoltaic Energy Conversion, Osaka, 2003*, pp. 491–494
5. R. Sáez-Araoz, J. Krammer, S. Harndt, T. Koehler, M. Krueger, P. Pistor, A. Jasenek, F. Hergert, M.Ch. Lux-Steiner, Ch.-H. Fischer, Prog. Photovolt.: Res. Appl. **20**, 855 (2012)
6. P. Pistor, R. Caballero, D. Hariskos, V. Izquierdo-Roca, R. Wächter, S. Schorr, R. Klenk, Sol. Energy Mater. Sol. Cells **93**, 148 (2009)
7. M. Rusu, M. Bär, S. Lehmann, S. Sadewasser, L. Weinhardt, C.A. Kaufmann, E. Strub, J. Röhrich, W. Bohne, I. Lauermann, Ch. Jung, C. Heske, M.Ch. Lux-Steiner, Appl. Phys. Lett. **95**, 173502 (2009)
8. R. Caballero, C.A. Kaufmann, T. Eisenbarth, A. Eicke, Th. Unold, R. Klenk, H.-W. Schock, Mater. Res. Soc. Symp. Proc. **1165**, M02-10 (2009)
9. J.Y.W. Seto, J. Appl. Phys. **48**, 5247 (1975)
10. WIAS TeSCA – Modeling and Simulation of Semiconductor Devices, http://www.wias-berlin.de/software/tesca, last accessed on January 6th, 2015
11. H. Gajewski, Mitt. Ges. Angew. Math. Mech. **16**, 35 (1993)
12. H. Gajewski, H.-Chr. Kaiser, H. Langmach, R. Nürnberg, R.H. Richter, Mathematical Modeling and Numerical Simulation of Semiconductor Detectors, in *Mathematics – Key Technology for the Future*, edited by W. Jäger, H.-J. Krebs (Springer, New York, 2003), pp. 355–364
13. C.-S. Jiang, R. Noufi, J.A. AbuShama, K. Ramanathan, H.R. Moutinho, J. Pankow, M.M. Al-Jassim, Appl. Phys. Lett. **84**, 3477 (2004)
14. C.-S. Jiang, R. Noufi, K. Ramanathan, J.A. AbuShama, H.R. Mouthino, M.M. Al-Jassim, Appl. Phys. Lett. **85**, 2625 (2004)
15. D. Azulay, O. Millo, I. Baalberg, H.-W. Schock, I. Visoly-Fisher, D. Cahen, Sol. Energy Mater. Sol. Cells **91**, 85 (2007)
16. S. Sadewasser, D. Abou-Ras, D. Azulay, R. Baier, I. Balberg, D. Cahen, S. Cohen, K. Gartsman, K. Ganesan, J. Kavalakkatt, W. Li, O. Millo, Th. Rissom, Y. Rosenwaks, H.-W. Schock, A. Schwarzman, T. Unold, Thin Solid Films **519**, 7341 (2011)
17. R. Baier, C. Leendertz, D. Abou-Ras, M.C. Lux-Steiner, Sol. Energy Mater. Sol. Cells **130**, 124 (2014)
18. K. Taretto, U. Rau, J.H. Werner, Thin Solid Films **480-481**, 8 (2005)
19. M. Gloeckler, J.R. Sites, W.K. Metzger, J. Appl. Phys. **98**, 113704 (2005)
20. K, Taretto, U. Rau, Mater. Res. Soc. Symp. Proc. **1012**, 309 (2007)
21. M.J. Hetzer, Y.M. Strzhemechny, M. Gao, M.A. Contreras, A. Zunger, L.J. Brillson, Appl. Phys. Lett. **86**, 162105 (2005)
22. C. Persson, A. Zunger, Phys. Rev. Lett. **91**, 266401 (2003)
23. K. Taretto, U. Rau, Prog. Photovolt.: Res. Appl. **12**, 573 (2004)
24. T. Tanaka, T. Yamaguchi, T. Ohshima, H. Itoh, A. Wakahara, A. Yoshida, Sol. Energy Mater. Sol. Cells **75**, 109 (2003)
25. Y.-J. Zhao, C. Persson, S. Lany, A. Zunger, Appl. Phys. Lett. **85**, 5860 (2004)

21

Solar cell fabricated on welded thin flexible silicon

Maik Thomas Hessmann[1,a], Thomas Kunz[1], Taimoor Ahmad[1], Da Li[1], Stephan Wittmann[1], Arne Riecke[1], Jan Ebser[2], Barbara Terheiden[2], Kristian Cvecek[3], Michael Schmidt[1], Richard Auer[1], and Chistoph J. Brabec[1,4]

[1] Bavarian Center for Applied Energy Research (ZAE Bayern), Haberstr. 2a, 91058 Erlangen, Germany
[2] Department of Physics, University of Konstanz, Universitätsstr 10, Box 676, 78464 Konstanz, Germany
[3] BLZ-Bavarian Laser Center, Konrad-Zuse-Str. 2-6, 91052 Erlangen, Germany
[4] i-MEET: institute Materials for Electronics and Energy Technology, University of Erlangen-Nuremberg, Martensstr. 7, 91058 Erlangen, Germany

Abstract We present a thin-film crystalline silicon solar cell with an AM1.5 efficiency of 11.5% fabricated on welded 50 μm thin silicon foils. The aperture area of the cell is 1.00 cm^2. The cell has an open-circuit voltage of 570 mV, a short-circuit current density of 29.9 mA cm^{-2} and a fill factor of 67.6%. These are the first results ever presented for solar cells on welded silicon foils. The foils were welded together in order to create the first thin flexible monocrystalline band substrate. A flexible band substrate offers the possibility to overcome the area restriction of ingot-based monocrystalline silicon wafers and the feasibility of a roll-to-roll manufacturing. In combination with an epitaxial and layer transfer process a decrease in production costs can be achieved.

1 Introduction

Silicon solar cells are following the recipe of improving efficiency and reducing material usage to consolidate their position on the market. Thus silicon thin-film solar cells are in the focus of interest due to the low silicon usage as well as the potential lower production costs. In 2012, the porous silicon (PSI) layer transfer process has proved that high efficiency values can be obtained. Efficiency values of up to 19.1% (aperture area: 3.98 cm^2) have been reached at a thickness of 43 μm by Brendel and co-workers [1, 2]. This result was even excelled by the company Solexel, who reported also a 43 μm thick solar cell but with an area of 156 mm × 156 mm based on a PSI layer transfer process with a world record efficiency of 20.1% (aperture area: 242.60 cm^2) [3–5]. Reuse cycles of above 50 are already achieved without any degradation effects [4].

ZAE Bayern developed a method of producing a large-area substrate for solar cell production called "extended-monocrystalline-silicon-base-foil" (EMOSiB). This substrate offers the possibility of high efficiency solar cells while reducing the material usage at the same time. An illustration of the EMOSiB is depicted in Figure 1. This band substrate would be the first monocrystalline silicon band substrate, further details are published elsewhere [6–8]. In short, an endless thin and flexible band

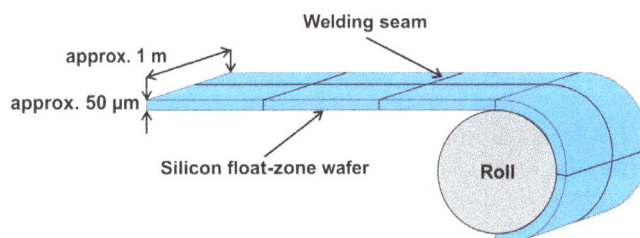

Fig. 1. Illustration of the EMOSiB, which consists of several individual silicon foils. The foils are welded together to the first monocrystalline band substrate. For further details for the solar cell manufacturing process see [8].

substrate of silicon is suggested as an epitaxial seed layer for the roll-to-roll production of crystalline solar cells. The lateral bonding of several thin single wafers to a band substrate are established by a laser welding process, for further details see [9,10]. Other approaches for lateral bonding are rare, only one other concept exists, and is published by Werner et al. [11]. Within this concept the gap between two silicon wafers are closed by lateral epitaxy. No results have been published since it was introduced in 2001.

The EMOSiB itself has to be manufactured by thin silicon wafers grown as ingots. A low thickness of the joined foils is essential for sufficient flexibility of the resulting substrate. By using an epitaxial process combined with

a e-mail: maik.hessmann@zae-bayern.de

Float-zone, (100), p-type, 0.5 Ω cm, 280 μm thick, 5 inch wafer
Laser cutting to 25 mm × 25 mm pieces
Chemical etching by KOH to a thickness of 50 μm
Keyhole welding process, 3 Si pieces (25 mm × 25 mm) to 1 Si foil (50 mm × 25 mm) at 1015 °C
RCA cleaning
5 μm Back side metallization (Al)
Spin-on doping on front side of phosphorus solution
RTP process, phosphorus diffusion, BSF creation
Removal of phosphosilicate glass by HF
Laser edge isolation process
Removal of residuals by HF
Front grid metallization (shadow mask)
Antireflective coating (SiN$_x$)

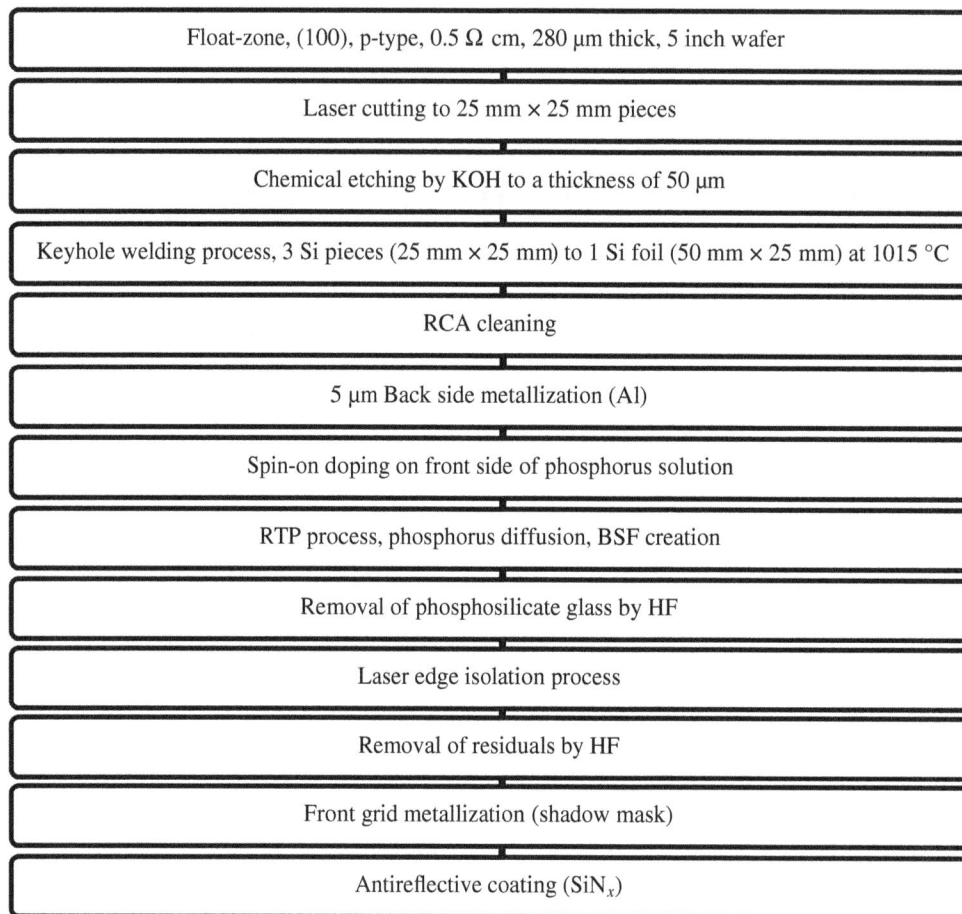

Fig. 2. Process flow diagram for solar cells on top of keyhole welded silicon foils as prepared in this work.

a layer transfer process such as PSI all necessary silicon layers for solar cell production can be produced from the gas phase afterwards [1, 12] and the initial silicon foil can be reused several times like in the PSI process [4]. Additionally, the EMOSiB offers the opportunity of an industrial roll-to-roll process for crystalline silicon solar cell fabrication.

This work shows the first solar cells manufactured on keyhole welded silicon thin foils. The flexible foils are obtained by chemical thinning of standard wafers. As the focus of this paper is on cell processing on thin laser-welded foils, no layer transfer or epitaxy is used so far. We discuss and analyze results from the AM1.5 measurement as well as quantum efficiency measurements. Cells at different areas of the welded sample are compared in order to study the influence of the area directly hit by the laser during welding.

2 Material and methods

2.1 Device fabrication

The process flow of the solar cell production is depicted in Figure 2 and started with 280 μm thick, 5 inch in diameter float-zone grown silicon wafers with an orientation

of (100), p-type, boron doped with a resistivity of 0.45–0.55 Ω cm. The wafers were laser cut into 25 mm × 25 mm pieces using a Nd:YVO$_4$ laser (Rofin-Sinar Laser GmbH, model: Power Line E20). The thickness of the pieces was decreased by etching in potassium hydroxide (KOH) solution to approx. 50 μm followed by a RCA cleaning using a system of Kufner Nassprozesstechnik GmbH. The KOH concentration was 22% and the solution was kept constantly at 85 °C.

Before the welding process began the silicon foils were placed on a specimen holder of fused quartz glass in a crucible furnace and the temperature was increased to 1015 °C in a nitrogen atmosphere. The laser beam was introduced through a fused quartz window in the furnace, for further details see [1]. The three silicon foils were keyhole welded to one silicon foil, two were placed in butt joint geometry and one in the middle at the back side of the other two as depicted in Figures 3 and 4. Further information about Keyhole welding are published elsewhere [9,10]. For welding a ytterbium single mode fiber laser model YLR-1000-SM made by the company IPG Photonics was used. This is a continuous wave laser with a wavelength of 1075 nm, a very high beam quality ($M^2 < 1.1$) and a maximum power of approx. 1000 W. Additionally a galvanometer scanner system with an objective focal length of 370 mm was used to focus and deflect the laser beam

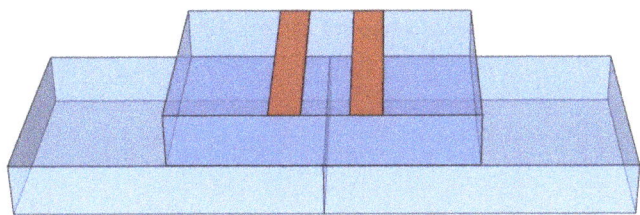

Fig. 3. Illustration of the keyhole welding geometry. Two silicon foils were placed in butt joint geometry and one in the middle at the back side of the other two. The area influenced by welding is colored red. Within the red colored area several keyhole welding lines were applied as depicted in Figure 4a.

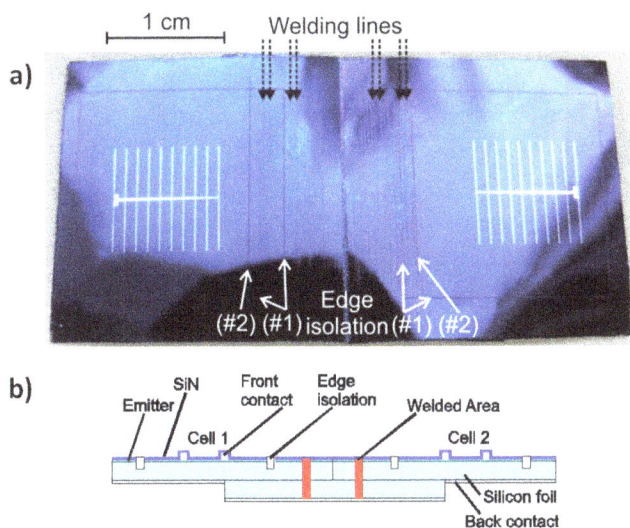

Fig. 4. (a) Photograph of two solar cells fabricated in the silicon foil. Two Si pieces (25 mm × 25 mm) can be seen at the front and one (25 mm × 25 mm) piece is underneath. Investigated cell areas were either with (edge isolation line #1) or without (additional isolation line #2) locations hit by the welding laser. (b) Schematic cross section of this solar cell design.

onto the sample surface. The resulting laser beam on the silicon foil surface had a spot diameter of 80 μm. Two times twenty lines with several keyhole welding spots were applied to increase the probability of welding through both silicon foils. Afterwards a RCA cleaning step was applied.

On the back side 5 μm aluminum was deposited by an electron beam evaporation system of Pfeiffer Vacuum (model: Classic 570). The phosphorus emitter was created by spin-on doping using the model Spin 150 from APT GmbH. Afterwards the welded foils were annealed in a nitrogen/oxygen atmosphere at 85 °C for 10 s in a rapid thermal processing (RTP) furnace (UniTemp GmbH model: UTP 1100). Also a back surface field (BSF) was established, for further information see [13,14]. Followed by the removal of the phosphosilicate glass by a 2% hydrofluoric acid (HF) etch step. For the electronic isolation, a laser edge isolation process (Rofin-Sinar Laser GmbH, model: Power Line E20) was applied. Created residuals were removed by 2% HF. Afterwards, the front contact grid, with

Ti 30 nm/Pd 30 nm/Ag 5 μm layers were formed by using shadow masks in an electron beam evaporator using a Pfeiffer Vacuum system model: Classic 570. A silicon nitride layer was deposited by plasma-enhanced chemical vapor deposition (PECVD) on the front side as antireflective coating (Roth & Rau GmbH, model: AK1000). A photograph of finished solar cells is depicted in Figure 4a.

2.2 Characterization

The sample was measured by an internally developed sun simulator at ZAE Bayern, the J-V curve was determined under AM1.5 illumination (1000 W/m^2) produced by halogen lamps and solar cell temperature of 25 °C. The height of the lamps were aligned according to the short-circuit current of a calibration sample. The value of parallel resistance (R_p) was determined from the gradient between -0.9 V to -0.7 V of the dark J-V characteristic and for the series resistance (R_s) from the comparison of the dark J-V characteristic and AM1.5 characteristic of the solar cell at V_{oc}.

A reflectance (R) and external quantum efficiency (EQE) measurement was performed at the physics department of the University of Konstanz using a solar cell analysis system LOANA, fabricated by pv-tools GmbH. The data was corrected for grid shading by using the software Lassie 7.5 of the company pv-tools and the internal quantum efficiency (IQE) was calculated from the reflectance and the external quantum efficiency:

$$IQE = EQE/(1 - R). \tag{1}$$

3 Results and discussion

The solar cell was measured three times under the sun simulator in order to study the influences of the welded area of the solar cell. The first measurement was performed after the cell processing. Afterwards an additional laser edge isolation line #2 was applied to exclude the welded area from the active area of the solar cell electrically as depicted in Figure 4a. The efficiency increased by a factor of 55.5% in between the first measurement (aperture area: 3.24 cm^2) and the second measurement (aperture area: 2.70 cm^2). For the last measurement the solar cell was masked with an opening of 1.0 cm^2, which corresponded to the optimum size with regard to the front grid. The efficiency increased by a factor of 24.7% in between second and third measurement.

The champion solar cell produced on keyhole welded silicon foils achieved an efficiency of 11.5% at an aperture area of 1.00 cm^2, more details are stated in Table 1 and depicted in Figure 5. Determined values of the champion cell were $V_{oc} = 570$ mV, $J_{sc} = 29.9$ mA/cm^2, $R_s = 0.61$ Ohm cm^2 and $R_p = 8.81$ kOhm cm^2. The values of the shunt resistance and the series resistance were both reasonable for a one-sun solar cell and did not severely affect the measured cell efficiency. Due to the thin thickness of the silicon foils and the lack of front site texturing

Table 1. Result of the three measurements of the champion solar cell on keyhole welded silicon foils (cell area A, fill factor FF, open-circuit voltage V_{oc}, short-circuit current density J_{sc} and efficiency η) determined by an internal developed sun simulator under AM1.5 illumination and 25 °C solar cell temperature.

Measurement	A [cm^2]	FF [%]	V_{oc} [mV]	J_{sc} [mA/cm^2]	η [%]
1	3.24	45.4	571	22.9	5.9
2	2.70	56.0	582	28.3	9.2
3	1.00	67.6	570	29.9	11.5

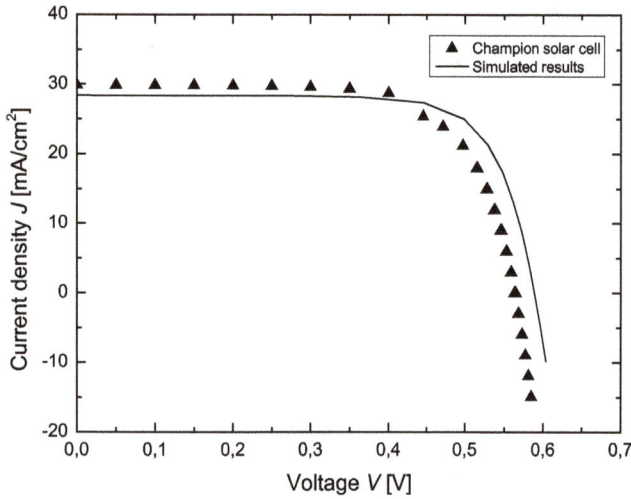

Fig. 5. In-house measured AM1.5 characteristic of the champion solar cell on keyhole welded silicon foils as well as the results of the PC1D simulation.

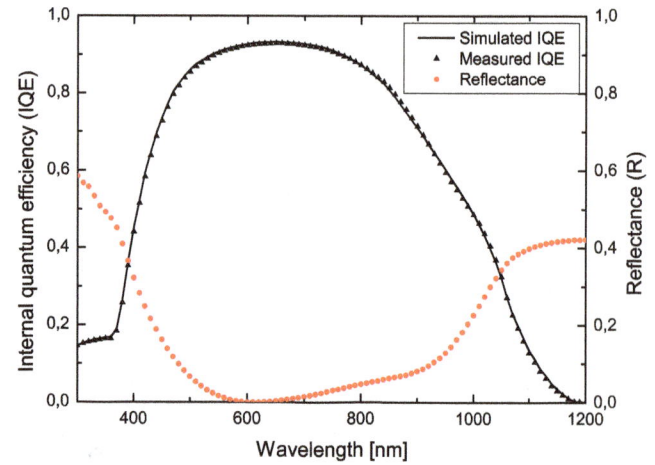

Fig. 6. Characteristics of the internal quantum efficiency (IQE) and reflectance of the champion solar cell on keyhole welded silicon foils as well as the results of the PC1D simulation.

the short-circuit current density was low. The fact that the efficiency was rising by excluding the welding area was attributed to changes within the silicon because of the welding process. Therefore, solar cells with a high efficiency values can only be build on the unirradiated areas.

Reference solar cells were fabricated in the same way except for the first three manufacturing steps in the process flow (laser cutting, KOH etching and keyhole welding) on top of 280 μm thick float-zone grown silicon wafers. Nine cells with an aperture area of 4.00 cm^2 each were fabricated on a single wafer. The mean efficiency value was 10.9% and the top efficiency value 12.4%.

Reflectance and quantum efficiency measurements revealed high reflectance values below 500 nm and above 900 nm as depicted in Figure 6. These values were due to the absence of surface texturing on the front side. Values above 900 nm were due to the reflectance on the back side of the silicon foils. The IQE measurement revealed high surface recombination velocity at the front and back side. Results of the back side were better than of the front side due to the BSF.

Using PC1D (Version 5.9) to simulate the champion solar cell and adjusting the model to the measured IQE characteristics values of the front surface recombination velocity of 1.60×10^6 cm/s and the rear surface recombination velocity of 1.50×10^4 cm/s were determined. More details of the PC1D simulation model are illustrated in Table 2.

The performance data determined by the PC1D simulation of the champion solar cell ($V_{oc} = 589$ mV, $J_{sc} = 28.4$ mA/cm^2, $\eta = 12.6$%) differed in comparison to the measured results. The open-circuit voltage as well as the efficiency was higher than the measured results and the short-circuit current density was lower. This discrepancy was attributed to the fact of unpassivated surfaces in the middle of the solar cell by stacking two silicon foils on top of each other in order to create a band substrate as illustrated in Figure 4b. In our case the active area of the solar cell was affected by stacking at measurement 1 up to 33.3%, measurement 2 up to 20.0% and measurement 3 up to 5.0%. Despite the surface the two foils were connected with each other at the welding points. This fact is not possible to simulate with an one dimensional software such as PC1D. An additionally issue was that maybe aluminum diffused too far into the silicon bulk material, which was observed sometimes during solar cell manufacturing. However, the measured IQE value were low at high wavelength, this showed that the BSF did not work as designed. This fact was inserted in the simulation model by setting the 1st rear diffusion to a very low value. This value is not realistic for a BSF simulation, but in our case necessary to fit our simulated data on the measured data.

The results of the solar cells are low in comparison to the established standard silicon solar cells, but this was a proof of concept with a simple cell process. Thus, higher efficiencies are possible by applying state of the art techniques like surface passivation and front side texturing.

Table 2. Selected input and output data of the PC1D simulation plus the comparison between simulated and measured data of the champion solar cell.

	Parameters	Simulated Data	Unit	Measured Data
Input	Device Area	1.00	cm^2	
	Emitter contact	0.61	Ohm	
	Thickness	50.00	μm	
	P-type background doping	3.25×10^{16}	cm^{-3}	
	Sheet Resistance	129.50	Ohm/square	
	1st rear diffusion	4.00×10^{16}	cm^{-3}	
	Bulk recombination (minority carrier lifetime)	50.00	μs	
	Front surface recombination	1.60×10^6	cm/s	
	Rear surface recombination	1.50×10^4	cm/s	
Output	V_{oc}	589	mV	570
	J_{sc}	28.4	mA/cm^2	29.9
	η	12.6	%	11.5

4 Conclusions

Flexible silicon foils were successfully welded together by keyhole welding. Solar cells at different locations of the welded sample were compared. For high efficiencies it is essential to exclude electrically the area directly hit by the welding laser e.g. by scribing the emitter. The best solar cell on 50 μm thin flexible keyhole welded silicon foils reached an efficiency of 11.5% at an aperture area of 1.00 cm^2, with promising values of $FF = 67.6\%$, $V_{oc} = 570$ mV and $J_{sc} = 29.9$ mA/cm^2 for a proof of concept. The process flow of the fabrication was simple without clean room environment, no front side texturing and surface passivation. Therefore, higher efficiencies are feasible by applying state of the art techniques. Further investigations with these techniques applied have to show the real potential of this concept.

We thank the Deutsche Forschungsgemeinschaft (No.: KU 2601/1-1 and KU 2601/1-2) for the financial support. Furthermore, the authors are very grateful to Bernhard Fischer from pv-tools GmbH for the software support.

References

1. R. Brendel, A novel process for ultrathin monocrystalline silicon solar cells on glass, in *14th European Photovoltaic Solar Energy Conference, 1997*, p. 1354
2. J.H. Petermann, D. Zielke, J. Schmidt, F. Haase, E.G. Rojas, R. Brendel, Prog. Photovolt. Res. Appl. **20**, 1 (2012)
3. M.A. Green, K. Emery, Y. Hishikawa, W. Warta, E.D. Dunlop, Prog. Photovolt. Res. Appl. **22**, 1 (2014)
4. M.M. Moslehi, World-Record 20.6% Efficiency 156 × 156 mm Full-Square Solar Cells Using Low-Cost Kerfless Ultrathin Epitaxial Silicon and Porous Silsicon Lift-off Technology for Industry-Leading High-Performance Smart PV Modules, in *PV Asia Pacific Conference, 2012*, p. 11
5. P. Kapur, M.M. Moslehi, A. Deshpande, V. Rana, J. Kramer, S. Seutter, H. Deshazer, S. Coutant, A. Calcaterra, S. Kommera, Y. Su, D. Grupp, S. Tamilmani, D. Dutton, T. Stalcup, T. Du, M. Wingert, A Manufacturable, Non-Plated, Non-Ag Metallization based 20.44% Efficient, 243 cm^2 Area, Back Contacted Solar Cell on 40 μm Thick Mono-Crystalline Silicon, in *28th European Photovoltaic Solar Energy Conference, 2013*, p. 2228
6. I. Burkert, T. Kunz, Ausgedehnte Siliziumsubstrate, German patent application DE102006037652A1, August 2006
7. T. Kunz, I. Burkert, R. Auer, M. Zimmermann, Towards extended free-standing crystalline silicon thin-films by laser joining, in *22nd European Photovoltaic Solar Energy Conference, 2007*, p. 1946
8. M.T. Hessmann, T. Kunz, I. Burkert, N. Gawehns, L. Schaefer, T. Frick, M. Schmidt, B. Meidel, R. Auer, C.J. Brabec, Thin Solid Films **520**, 595 (2011)
9. M.T. Hessmann, T. Kunz, M. Voigt, K. Cvecek, M. Schmidt, A. Bochmann, S. Christiansen, R. Auer, C.J. Brabec, Int. J. Photoenergy **2013**, 724502 (2013)
10. K. Cvecek, M. Zimmermann, U. Urmoneit, T. Frick, M. Heßmann, T. Kunz, Thermisches Prozessieren dünner Siliziumsubstrate für die solare Energieerzeugung, in *15th Laser Elektronikprod. Feinwerktech, 2012*, p. 91
11. J.H. Werner, R. Dassow, T.J. Rinke, J.R. Köhler, R.B. Bergmann, Thin Solid Films **383**, 95 (2001)
12. R. Brendel, Crystalline thin-film silicon solar cells from layer-transfer processes: a review, in *10th Workshop on Crystalline Silicon Solar Cell Materials and Processes, 2000*, p. 117
13. M. Mühlbauer, Dünne kristalline Silizium Wafer-Solarzellen mit Glasträger stabilisiert, Dissertation, FernUniversität Hagen, 2009
14. M. Mühlbauer, V. Gazuz, N. Gawehns, R. Weissmann, M. Scheffler, R. Auer, Novel heterojunction solar cell concept with thin monocrystalline silicon on low cost glass, in *21st European Photovoltaic Solar Energy Conference, 2006*, p. 1021

22

Resistance and lifetime measurements of polymer solar cells using glycerol doped poly[3,4-ethylenedioxythiophene]: poly[styrenesulfonate] hole injection layers

Emma Lewis[1], Bhaskar Mantha[2], and Richard P. Barber Jr.[1,a]

[1] Department of Physics and Center for Nanostructures, Santa Clara University, Santa Clara CA 95053, USA
[2] Department of Electrical Engineering, Santa Clara University, Santa Clara CA 95053, USA

Abstract We have performed resistivity measurements of poly[3,4-ethylenedioxythiophene]: poly[styrenesulfonate] (PEDOT:PSS) films with varying concentrations of glycerol. Resistivity is seen to decrease exponentially from roughly 3 Ω-cm for pure PEDOT:PSS to 3×10^{-2} Ω-cm for 35 mg/cm^3 glycerol in PEDOT:PSS. Beyond this concentration adding glycerol does not significantly change resistivity. Bulk heterojunction polymer solar cells using these variously doped PEDOT:PSS layers as electrodes were studied to characterize the effects on efficiency and lifetime. Although our data display significant scatter, lowering the resistance of the PEDOT:PSS layers results in lower device resistance and higher efficiency as expected. We also note that the lifetime of the devices tends to be reduced as the glycerol content of PEDOT:PSS is increased. Many devices show an initial increase in efficiency followed by a roughly exponential decay. This effect is explained based on concomitant changes in the zero bias conductance of the samples under dark conditions.

1 Introduction

Organic photovoltaics (OPVs) continue to attract interest as a low-cost and mechanically robust alternative to Si-based technologies. The primary challenges facing OPVs have been lower efficiency and lifetime, however significant progress has occurred over the last decade [1]. Furthermore, the primary advantage of OPVs, manufacturability, has been clearly demonstrated [2, 3]. A common approach to producing theses solar cells relies on the transparent conductive layer indium-tin-oxide (ITO). However the use of ITO imposes limits on the flexibility of the cells and hence the manufacturability [4, 5]. A potential replacement for the ITO layer is poly[3,4-ethylenedioxythiophene]: poly[styrenesulfonate] (PEDOT:PSS). PEDOT:PSS is already commonly used as a hole conducting planarizing layer over the ITO in many solar cell architectures [6]. To overcome the relatively low conductivity of PEDOT:PSS various studies have utilized additives [4, 5, 7–13] including glycerol [8, 10, 11] and carbon nanotubes [5]. The aging of the PEDOT:PSS electrodes has been investigated [4]. However, we are unaware of stability studies of actual solar cell devices based on these modified electrodes.

Given the importance of cell stability [1, 14, 15], the central motivation for this work is to find any effects that enhancing the PEDOT:PSS conductivity might have on device lifetime. Specifically, we report measurements of series resistance, efficiency and lifetime of solar cells using various glycerol/PEDOT:PSS (G-PEDOT:PSS) films as the transparent contact. These measurements are correlated with the resistivity of films cast from the same glycerol/PEDOT:PSS blends. In addition, previous work showed a dramatic decrease in the resistivity (2 orders of magnitude) of PEDOT:PSS when 30 mg per cm^3 was added [11]. We were also interested in whether this onset could be more finely tuned.

2 Experimental

A series of G-PEDOT:PSS (Aldrich 483095) solutions were prepared with glycerol concentrations of 0, 9.2, 22.3, 36.1, 70.7 and 95.0 mg per cm^3 of PEDOT:PSS. Films were spin cast from these solutions and annealed at 200 °C for two hours to produce electrical transport samples (Fig. 1a) or transparent contacts for solar cell measurements (Fig. 1b). All "wet" preparation, processing and annealing steps were conducted in an inert atmosphere glove box. Film resistance was measured using a variable current

a)

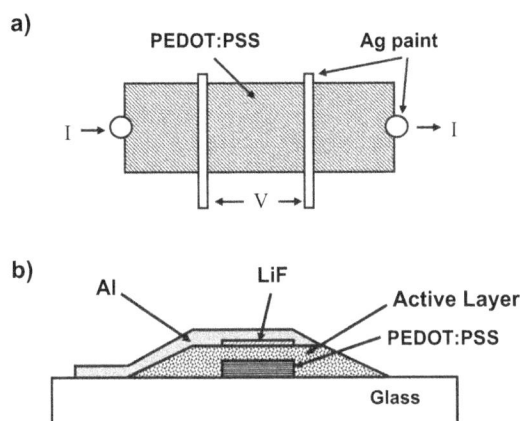

b)

Fig. 1. (a) Schematic sample layout for resistivity measurements and (b) solar cell architecture.

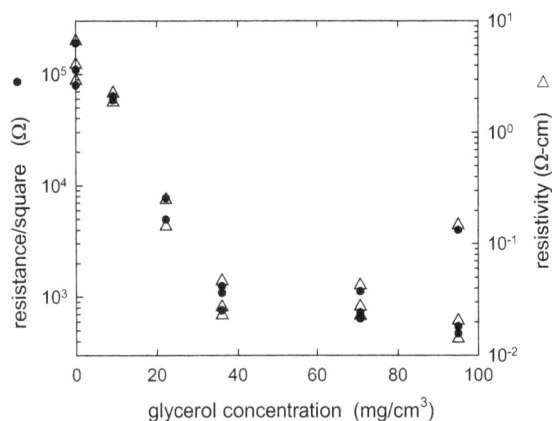

Fig. 2. Sheet resistance (left axis) and resistivity (right axis) for G-PEDOT:PSS films as a function of glycerol concentration.

source with separate voltage leads to derive current voltage (IV) curves (contacts schematically shown in Fig. 1a). Film resistance (R) was derived from $R = \lim_{I \to 0} \frac{\partial V}{\partial I}$, however no significant nonlinearity was observed in any of these samples. Given that resistance is given by $R = \frac{\rho L}{Wd}$, where L is the length of the film, W is the width, d is the thickness and ρ is the material's resistivity; we can also define the sheet resistance or resistance per square $R_{sq} = \frac{\rho}{d}$ since L/W is the number of squares. Finally, film thickness measurements obtained from a Gaertner L116B ellipsometer allow us to calculate the resistivity of our samples ($\rho = R_{sq}d$).

Solar cells were fabricated using a 0.16 mole fraction (equal weight) blend of [6,6]-Phenyl C61 butyric acid methyl ester (PCBM) in Poly(3-hexylthiophene-2,5-diyl) (P3HT) spin cast on to our G-PEDOT:PSS electrodes from a 1.5 weight percent solution in chlorobenzene. These samples were annealed for one hour at 190 °C and then transferred to a bell-jar evaporator system equipped with a quartz crystal thickness monitor where they were finished by evaporating ~1 nm of LiF followed by 100 nm of Al to form the top electrodes. The evaporator is not integrated into the glove box, so samples were transferred between the two using a vacuum tight vessel carrying dry nitrogen atmosphere. The elapsed time that samples were exposed to ambient air was typically under 5 min. Current-voltage (IV) characteristics of the devices were measured in ambient conditions alternately in darkness and illuminated by a PV Measurements, Inc. Small-Area Class-B Solar Simulator. Automated transport data collection utilized a MATLAB controlled routine via an IEEE 488 Bus interfaced Keithley 2400 Source Meter. IV curves were typically acquired at 15 min intervals. However as samples aged, this time was sometimes increased to 30 min, as changes became smaller.

3 Results and discussion

We have plotted the sheet resistance (left axis) for samples of G-PEDOT:PSS as glycerol concentration is

varied in Figure 2; these values are in reasonable agreement with previous work [11]. We observe that for glycerol doping below about 30 mg/cm³ the sheet resistance drops roughly exponentially with concentration, a more finely tuned result that was previously reported [11]. Beyond this value, there is little change. Overlaid in Figure 2 are our resistivity values from the same data (right axis). The nearly point-to-point correspondence between the sheet resistance and resistivity indicates consistent film thicknesses for all of the samples. The averaged thickness for the sixteen samples was 340 ± 39 nm, with only three outlier samples (outside of the error bars).

Figure 3 displays the results for three parameters of merit for the solar cell devices: power conversion efficiency η, series resistance R_S and characteristic time τ (lifetime) as plotted against the sheet resistance R_{sq} of the various G-PEDOT:PSS blends. The efficiencies are roughly two orders of magnitude below optimized PCBM/P3HT devices [16], however our focus is on the lifetime behavior and not the efficiency while using this ITO-free sample structure. It is important to note that we saw consistent device performance behavior throughout these measurements, and we believe that this reproducibility supports the validity of the results. We have so far been unable to find the cause of the poor device performance. We should note that in measurements of devices on the same substrate both with and without the ITO layer (but having the same PEDOT:PSS and active layer), we found that ITO-free samples performed as well or better than those with ITO. Series resistance is calculated by measuring the asymptotic behavior of the IV curve at the highest applied voltage. In order to measure τ, IV curves were taken in ambient atmospheric conditions at 15 min intervals for up to 24 h. From these data we derived the power conversion efficiency as a function of time $\eta(t)$ and fit those results to an exponential decay model to extract τ [17]. There is significant scatter in the results, however the trends are apparent (regression lines shown) and probably meaningful. As the transparent G-PEDOT:PSS resistance is lowered we see both an increase in efficiency and a decrease in R_S. These results

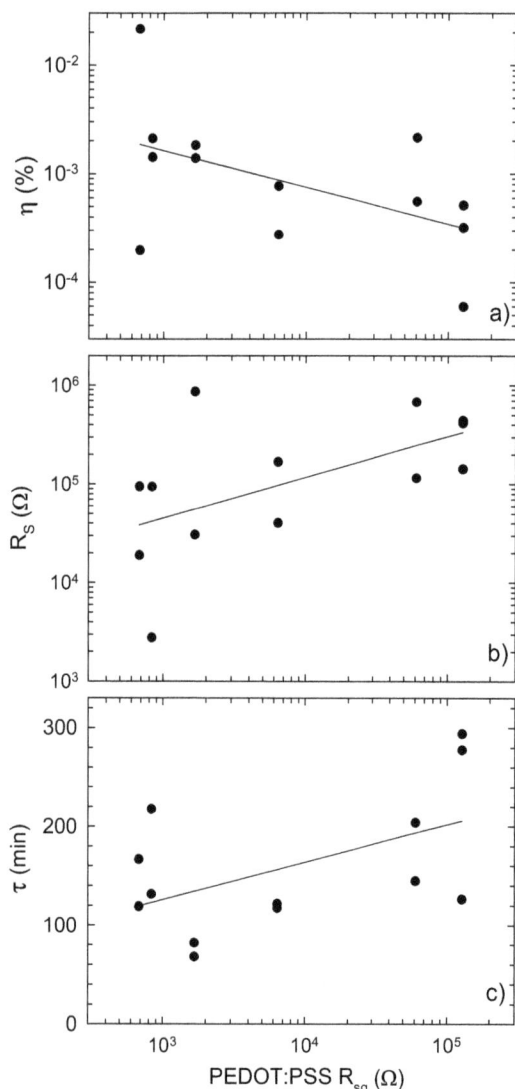

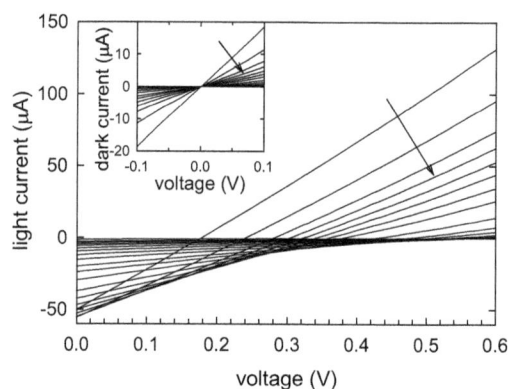

Fig. 4. Current-voltage characteristics for an illuminated solar cell showing an initial increase in efficiency. Inset: current-voltage characteristics near zero bias for the same solar cell.

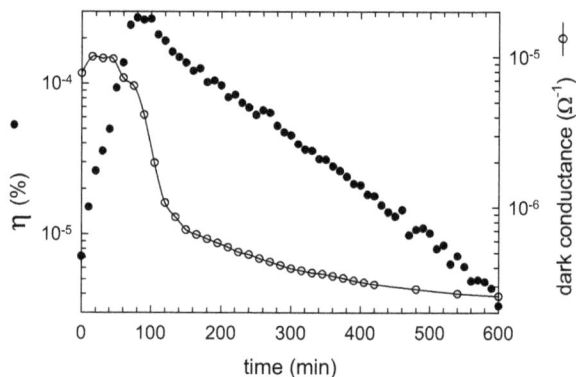

Fig. 5. Efficiency (left axis) and dark zero-bias conductance (right axis) for a solar cell that exhibits a large initial efficiency increase before its exponential decrease.

Fig. 3. Solar cell parameters of merit for PCBM/P3HT devices as dependent on the sheet resistance of the G-PEDOT:PSS films used as the transparent electrodes: (a) efficiency, (b) series resistance and (c) characteristic time (lifetime).

are consistent given that R_S directly affects the output current and hence the efficiency. Furthermore it is reasonable to expect that R_S will depend in part on the resistivity of the PEDOT:PSS contact since that should contribute to the device resistance. In Figure 3c, we observe that decreasing the G-PEDOT:PSS resistance *reduces* the lifetime of the devices albeit only about a factor of two over two decades of resistance change. We also point out that most of the *IV* curves recorded are nearly linear (for example see Fig. 4), yielding fill factors which are very close to 0.25 as would be expected.

Figure 4 shows a typical set of *IV* curves for an illuminated device as it ages (some traces have been removed for clarity). The arrow indicates the direction of increasing time. Besides the poor fill factors ($\sim$0.25) we note that both the initial short circuit current (current axis intercept) and the open circuit voltage (zero-current

crossing) *increase* from the initial values. In other words, the device efficiency is initially increasing before a decay process begins. This kind of behavior has been observed in previous samples from our laboratory [18], albeit as a much smaller effect. A clue for understanding this behavior is exhibited in the un-illuminated *IV* data for the same sample as displayed in the inset of Figure 4. These curves suggest a characteristic sample resistance which increases with time (again indicated by the arrow). We postulate that this initially "low" resistance represents a "leakage" or parallel conduction path that effectively shunts current directly through the device. This mechanism could potentially reduce both the current and voltage as long as the resistance is significantly low. To test this idea, we plot in Figure 5 $\eta(t)$ for a sample which shows an initial increase in η nearly two orders of magnitude before a decay mechanism becomes dominant. On the same figure (right axis) is the zero bias conductance for the same film taken in the dark. We note that the initial dark conductance drops precipitously at about 100 min, the same time that the maximum η is observed. Beyond that time we again see the typical exponential decay of the device efficiency [17]. Although we cannot speculate on the exact mechanism of this effect, it is certainly enhanced for samples that utilize G-PEDOT:PSS contacts.

4 Conclusions

In summary, we have characterized the resistivity PEDOT:PSS films as a function of glycerol doping. We have also measured the efficiency, series resistance and lifetime of PCBM/P3HT solar cells utilizing these G-PEDOT:PSS films as the transparent contact. These parameters of merit were found to vary with the glycerol doping, with the lifetime of cells negatively impacted by the addition of glycerol. We have also observed that these solar cells often exhibit efficiencies which initially increase before an exponential decay begins. We have proposed a simple leakage mechanism for this behavior based on the zero-bias resistance of the devices.

We acknowledge both G. Laskowski and G. Sloan for invaluable technical assistance. Funding was provided in part by a grant from IntelliVision Technologies and a Santa Clara University IBM Faculty Research Grant.

References

1. M. Jørgensen, K. Norrman, S.A. Gevorgyan, T. Tromholt, B. Andreasen, F.C. Krebs, Adv. Mater. **24**, 580 (2012)
2. F.C. Krebs, S.A. Gevorgyan, J. Alstrup, J. Mater. Chem. **19**, 5442 (2009)
3. F.C. Krebs, S.A. Gevorgyan, B. Gholamkhass, S. Holdcroft, C. Schlenker, M.E. Thompson, B.C. Thompson, D. Olson, D.S. Ginley, S.E. Shaheen, H.N. Alshareef, J.W. Murphy, W.J. Youngblood, N.C. Heston, J.R. Reynolds, S. Jia, D. Laird, S.M. Tuladhar, J.G.A. Dane, P. Atienzar, J. Nelson, J.M. Kroon, M.M. Wienk, R.A.J. Janssen, K. Tvingstedt, F. Zhang, M. Andersson, O. Inganäs, M. Lira-Cantu, R. de Bettignies, S. Guillerez, T. Aernouts, D. Cheyns, L. Lutsen, B. Zimmermann, U. Würfel, M. Niggemann, H.-F. Schleiermacher, P. Liska, M. Grätzel, P. Lianos, E.A. Katz, W. Lohwasser, B. Jannon, Sol. Energy Mater. Sol. Cells **93**, 1968 (2009)
4. Y.H. Kim, C. Sachse, M.L. Machala, C. May, L. Müller-Meskamp, K. Leo, Adv. Functional Mater. **21**, 1076 (2011)
5. S. Schwertheim, O. Grewe, I. Hamm, T. Mueller, R. Pichner, W.R. Fahrner, in *35th IEEE Photovoltaic Specialists Conference (PVSC), 2010*, pp. 001639-001642
6. F.C. Krebs, *Polymer photovoltaics a practical approach* (Wash.: SPIE Press, Bellingham, 2008)
7. X. Crispin, F.L.E. Jakobsson, A. Crispin, P.C.M. Grim, P. Andersson, A. Volodin, C. van Haesendonck, M. Van der Auweraer, W.R. Salaneck, M. Berggren, Chem. Mater. **18**, 4354 (2006)
8. J. Huang, P.F. Miller, J.S. Wilson, A.J. de Mello, J.C. de Mello, D.D.C. Bradley, Adv. Functional Mater. **15**, 290 (2005)
9. J. Ouyang, Q. Xu, C.-W. Chu, Y. Yang, G. Li, J. Shinar, Polymer **45**, 8443 (2004)
10. F. Zhang, M. Johansson, M.R. Andersson, J.C. Hummelen, O. Inganäs, Adv. Mater. **14**, 662 (2002)
11. K.-H. Tsai, S.-C. Shiu, C.-F. Lin, in *Proceedings SPIE, 2008*, Vol. 7052, pp. 70521B–70521B–8.
12. C.-J. Ko, Y.-K. Lin, F.-C. Chen, C.-W. Chu, Appl. Phys. Lett. **90**, 063509 (2007)
13. W. Zhang, B. Zhao, Z. He, X. Zhao, H. Wang, S. Yang, H. Wu, Y. Cao, Energy Environ. Sci. **6**, 1956 (2013)
14. F. Krebs, J. Carle, N. Cruysbagger, M. Andersen, M. Lilliedal, M. Hammond, S. Hvidt, Sol. Energy Mater. Sol. Cells **86**, 499 (2005)
15. F.C. Krebs, H. Spanggaard, Chem. Mater. **17**, 5235 (2005)
16. W. Ma, C. Yang, X. Gong, K. Lee, A.J. Heeger, Adv. Funct. Mater. **15**, 1617 (2005)
17. B.H. Johnson, E. Allagoa, R.L. Thomas, G. Stettler, M. Wallis, J.H. Peel, T. Adalsteinsson, B.J. McNelis, R.P. Barber Jr., Sol. Energy Mater. Sol. Cells **94**, 537 (2010)
18. E.L. Sena, J.H. Peel, D. Wesenberg, S. Nathan, M. Wallis, M.J. Giammona, T. Adalsteinsson, B.J. McNelis, R.P. Barber Jr., Sol. Energy Mater. Sol. Cells **100**, 192 (2012)

23

Route to enhance the efficiency of organic photovoltaic solar cells - by adding ferroelectric nanoparticles to P3HT/PCBM admixture

David Black, Iulia Salaoru, and Shashi Paul[a]

Emerging Technologies Research Centre, Hawthorn Building, De Montfort University, The Gateway, Leicester LE1 9BH, UK

Abstract We have demonstrated that by adding ferroelectric nanoparticles to poly(3-hexylthiophene) (P3HT) and [6,6]-phenyl-C_{61}-butyric acid methyl ester (PCBM) photovoltaic devices the relative efficiency can be increased compared to the same blend without these nanoparticles. In this work samples of 20 mg/ml concentrations of P3HT and PCBM were prepared in a 1:1 ratio and the samples prepared using ferroelectric barium titanate (BT) and strontium titanate (ST) nanoparticles in a 1:1:0.5 ratio. The samples were spin coated onto ITO coated glass with a layer of poly(3,4-ethylenedioxythiophene)poly(styrenesulfonate) (PEDOT:PSS). A top electrode of aluminium 1 cm^2 was deposited. The current-voltage characteristics of the devices were determined using a solar simulator and the absorption characteristics by UV-Vis spectroscopy. The samples with BT and ST exhibited increased absorption around 490 nm and increased open circuit voltage and short circuit current compared to the control P3HT/PCBM sample. The possible mechanism that helps to understand the increase in open circuit voltage and short circuit current is also proposed in this work.

1 Introduction

The most efficient bulk heterojunction photovoltaic devices comprise blends of poly(3-hexylthiophene) (P3HT) and [6,6]-phenyl-C_{61}-butyric acid methyl ester (PCBM) and the best P3HT/PCBM heterojunction devices currently have power conversion efficiencies of around 6% and quantum efficiencies approaching 100% [1]. In this work we have made an attempt, for the first time, to demonstrate a relative improvement in the efficiency of the devices by adding ferroelectric nanoparticles to the P3HT/PCBM blends. We have previously demonstrated that by adding ferroelectric nanoparticles, such as barium titanate into polymers, can have a significant effect on dielectric constant of the insulating polymers [2] and the photoconductivity of organic photoconductor materials [3, 4]. We are anticipating, but not yet fully demonstrated, that the increase in permittivity is also present in the photoconductive polymers doped with ferroelectric nanoparticles.

As it has already been established by a number of workers that organic photovoltaic (OPV) devices are excitonic in nature, which is to say that in conventional

semiconductor (CSC) cell the incident photon creates free carriers immediately once absorbed whereas the excitonic semiconductor (XSC) creates electrostatically bound charge carriers called excitons. Whether a material behaves conventionally or excitonically is strongly dependent upon the electrical permittivity of the material. It is possible to determine whether a material will behave as an XSC or a CSC by comparing the ratios of the Bohr radius with the Coulomb radius. Therefore let the ratio of r_c to r_B be called γ.

$$\gamma = \frac{r_c}{r_B} \approx \left(\frac{q^2}{4\pi\varepsilon_0 k_B r_0 m_e} \right) \left(\frac{m_{eff}}{\varepsilon^2 T} \right). \quad (1)$$

A value of $\gamma > 1$ indicates that the material will exhibit XSC behaviour and conversely a value of $\gamma < 1$ indicates a material that will behave as a CSC.

Where $r_0 = 0.53$ Å and is the distance between the electron and the nucleus for hydrogen like structures such as silicon and excitons, m_e is the mass of a free electron in a vacuum and m_{eff} is the effective mass of the electron in the semiconductor, q is the charge on the electron, ε is the relative permittivity of the material and r_c is the critical distance between the charges, k_B is Boltzmann's constant and T is the temperature of the system. The material will

[a] e-mail: spaul@dmu.ac.uk

behave in an excitonic manner if $r_c > r_B$ and where r_B is greater than the radius of the particle [5]. It is important to mention here two important caveats to this model; firstly m_{eff} is not well defined in XSC materials due to the fact that charge transport generally occurs due to a hopping mechanism rather than delocalised band transport, secondly ε is a bulk property of the material and tends towards the permittivity of free space at small scales [5].

Persson and Iganas [6] have suggested a seven step process of charge generation in XSC materials:

1. Photon incoupling.
2. Photon absorption.
3. Exciton formation.
4. Exciton migration.
5. Exciton dissociation.
6. Charge transport.
7. Charge collection.

Although it can be argued that processes 1 and 2 are essentially the same and this could be a six step process this description is essentially accurate and corresponds with the work of Moliton and Nunzi [7]. Once an exciton has formed (step 3) it is free to move within the polymer matrix, however the distance that an exciton can travel is limited to a few tens of nanometers before they recombine. It is essential therefore that they can dissociate before they recombine. *The object of this research is to attempt to maximize the dissociative region of the material and hence increase the overall electron harvest by maximizing the amount of excitons that dissociate into charge carriers.*

In bulk heterojunction devices charge separation occurs when an exciton travels to an interface between the photoconductive polymer where the exciton is produced and the electron accepting nanoparticles in the polymer matrix, in this case P3HT and PCBM, respectively. At the interface between the two materials an electrochemical potential is set up that is equal to the difference between the Fermi levels for each material [8]. This potential results in an electron flow from the PCBM to the polymer, which results in a depleted region surrounded by positively charged polymer, due to an excess of holes and a negatively charged PCBM molecules due to an excess of electrons. Once an equilibrium state is reached an electric field is set up which is essential for charge separation [8]. Exciton dissociation occurs if the Coulomb forces binding the exciton are less than the built in field of the junction $(q\Phi_{bi})$.

Therefore by increasing the relative permittivity of the bulk material we can influence at least two important factors in exciton dissociation and hence charge production, namely; increasing permittivity decreases the Coulomb potential in the exciton, and increases the Debye length of the material, which is related to the width of the depletion region [6], and hence to the dissociative region [8].

In this work we have added ferroelectric barium titanate (BT) and strontium titanate (ST) to blends of PCBM and P3HT. The commercially available BT and ST nanoparticles are cubic in nature and not ferroelectric in this state. By annealing these nanoparticles in

Table 1. Polymer-nanoparticle blends used in active layers.

Sample	P3HT (mg/ml)	PCBM (mg/ml)	BT (mg/ml)	ST (mg/ml)
1	20	20	0	0
2	30	30	0	0
3	40	40	0	0
4	20	20	0	10
5	20	20	10	0

air at 1000 °C the structure of the molecules was altered from cubic to tetragonal and hence to a ferroelectric state prior to inclusion in the blend [9–11]. The samples were electrically tested in our solar simulator and their absorption characteristics were measured using UV-Vis spectroscopy. The fill factor (FF) and power conversion efficiency (PCE) were calculated from the current-voltage (I-V) [12] characterisitics.

2 Experiment

A blend of P3HT with varying concentration ferroelectric nanoparticles was prepared to understand the effect on dielectric constant. The dielectric constant was deduced from capacitance-voltage measurement of metal-blend-metal structures.

Three sets of sample solar cells were fabricated with different concentrations of P3HT, PCBM and while keeping strontium titanate (average size <60 nm) and barium titanate nanoparticles (average size <70 nm) concentrations (10 mg/ml) the same. The base solution for each concentration was a blend of P3HT and PCBM, to which was added either no nanoparticles to form a control sample, ST or BT. The concentrations for the P3HT/PCBM samples were; 20, 30 and 40 mg/ml in a 1:1 ratio. TheP3HT/PCBM/nanoparticle blend was a 1:1:0.5 ratio. The concentrations of the samples are detailed in Table 1. Samples 1, 2 and 3 were compared initially to determine the ideal concentration, and then the samples containing the ferroelectric nanoparticles were prepared and compared with the control sample for that concentration.

The schematic of various structures of the samples is shown in Figure 1.

The devices were prepared by first spin coating a layer of poly(3,4-ethylenedioxythiophene)poly(styrenesulfonate) (PEDOT:PSS) (~30 nm) onto cleaned ITO (~80 nm) coated glass substrate. This followed by the deposition of P3HT/PCBM active layer (~100 nm) by spin coating. Aluminium top electrodes, 1 cm^2 area, were deposited by thermal evaporation at a chamber pressure of ~1 × 10^{-6} mbar.

Electrical tests were carried out on the samples using standard current-voltage (I-V) tests in both illuminated and non-illuminated conditions. The illumination used was an Oriel AM1.5 light source, equivalent to ~8.58 mW/cm^2.

UV-Vis measurements were carried out on samples deposited on ITO coated glass substrates using the Thermo Scientific Evolution 300 spectrometer.

3 Results and discussion

It has been known for some time that by adding high permittivity (high k) materials, particularly ceramics, to polymers and epoxy resins that the permittivity of the blend of materials will be higher than that of the original polymer or epoxy alone [13–18]. A number of models have been proposed which describe the effective permittivity (ε_{eff}) of the composite material and a few of which are presented and compared below. The simplest model is known as the volume fraction average model and is a summation of the relative fractions of each type of material by volume. This method is generally held to be inaccurate, but can be used for a quick first approximation [14].

$$\varepsilon_{eff} = \phi_1\varepsilon_1 + \phi_2\varepsilon_2 \qquad (2)$$

where; ϕ is the volume fraction, ε is the permittivity and the subscripts 1 and 2 represent the polymer and ceramic material, respectively. This model predicts a sharp increase in permittivity at a relatively small fraction of ceramic filler, which in practice does not occur [14]. More accurate models are based on "mean field theory" which reduces all interactions on a body to a single average interaction thus removing the uncertainty generally associated with a many body problem. The Maxwell-Garnett equation is one such model [14, 18] which can be written as:

$$\varepsilon_{eff} = \varepsilon_1 \frac{\varepsilon_2 + 2\varepsilon_1 - 2(1-\phi_1)(\varepsilon_1 - \varepsilon_2)}{\varepsilon_2 + 2\varepsilon_1 + (1-\phi_1)(\varepsilon_1 - \varepsilon_2)}. \qquad (3)$$

This is valid for a single ceramic or ferroelectric particle surrounded by polymer. This equation also has a drawback in that as it is valid only when the filler fraction is infinitesimally small.

The Bruggeman model is another mean field theory model that treats the polymer/ceramic matrix as a series of repeated units of spherical ceramics surrounded by polymer. This can be written as [14, 18]:

$$\phi_1\left(\frac{\varepsilon_1 - \varepsilon_{eff}}{\varepsilon_1 + 2\varepsilon_{eff}}\right) + \phi_2\left(\frac{\varepsilon_2 - \varepsilon_{eff}}{\varepsilon_2 + 2\varepsilon_{eff}}\right) = 0. \qquad (4)$$

In this model the value of the dielectric constant increases dramatically for ceramic filler volumes of 20% and above. The two final equations compared here are the logarithmic Lichtnecker equation:

$$\ln\varepsilon_{eff} = \phi_2\ln\varepsilon_2 + (1-\phi_2)\ln\varepsilon_1 \qquad (5)$$

and the "effective medium theory (EMT) model" that describes the permittivity of a two part system in terms of the relative fractions of the components:

$$\varepsilon_{eff} = \varepsilon_m\left[1 + \frac{\phi_2(\varepsilon_2 - \varepsilon_1)}{\varepsilon_1 + (n(1-\phi_2)(\varepsilon_2 - \varepsilon_1))}\right] \qquad (6)$$

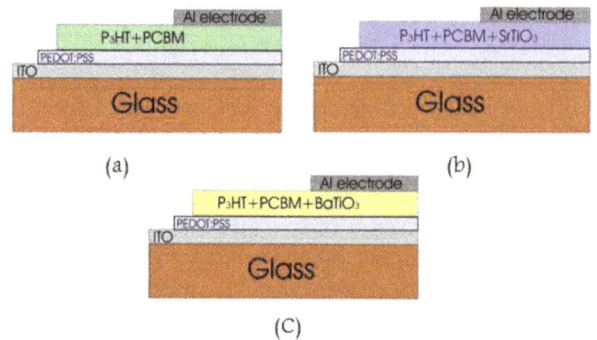

Fig. 1. Typical structure of organic heterojunction solar cells as used in this work, (a) control sample 1 of P3HT/PCBM, (b) sample 2, P3HT/PCBM plus strontium titanate and (c) P3HT/PCBM plus barium titanate.

where n is a correction factor which compensates for the non spherical nature of the ceramics and is generally less than 0.15 for a polymer/ceramic matrix.

To obtain comparative permittivity measurements for samples of P3HT containing varying quantities of barium and strontium titanate. The first set of samples used the capacitance measurements and after some optimisation produced reliable results. It is thought that the problems in making reliable measurements with this technique were twofold, firstly any minor imperfection in the layer such as pinholes could cause the device to fail and secondly the any clumping of nanoparticles could provide a path for electrical conduction. Also high leakage currents can lead to unusual results such as negative capacitances and unusually high or low results [19–21]. Once uniform films were produced after optimisation of various parameters, the majority of these issues disappeared and reliable measurements could be taken. In order to overcome the difficulties in making these measurements the polymer/nanoparticle solution was left in the ultrasonic bath for a minimum of 8 hours to maximise the dispersion of the nanoparticles in the polymer.

Given that barium titanate and strontium titanate caused an increase in photoconductivity both were investigated to determine what, if any, effect they would have on the permittivity of P3HT.

Figure 2 clearly demonstrates an almost linear increase in the permittivity of P3HT with increasing volume fraction of barium titanate. The upper limit of this effect has not been investigated as the concentration levels required to determine this would be far in excess of the levels required to fabricate functional photovoltaic devices. The EMT model fits the data well with an R^2 value of 0.925. The measured value of permittivity for the closest result to 20 mg/ml concentration is 8.24 compared to 7.74 as predicted by the model.

Figure 3 shows the results of the permittivity measurements for samples of P3HT with varying concentrations of strontium titanate. The increase in permittivity in this case is slightly higher than that produced by barium titanate with similar levels of concentration, which could be expected from the fact that strontium titanate is a

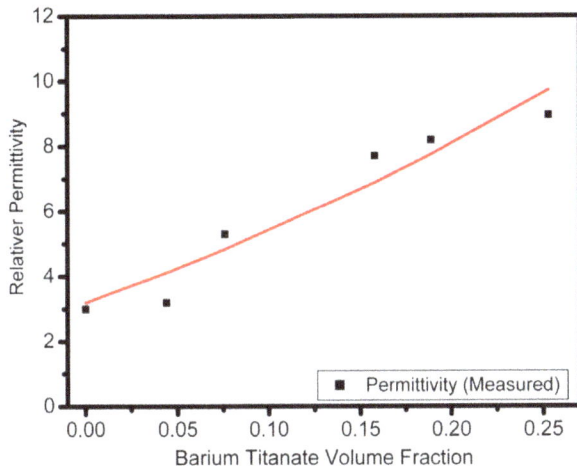

Fig. 2. Relative permittivity of P3HT with increasing volume fractions of barium titanate.

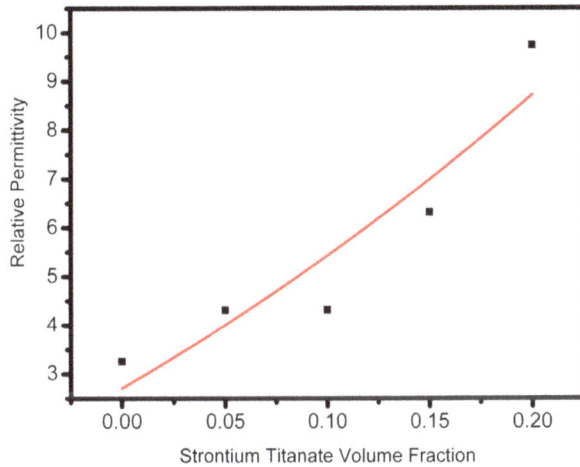

Fig. 3. Relative permittivity of P3HT with increasing volume fractions of Strontium Titanate.

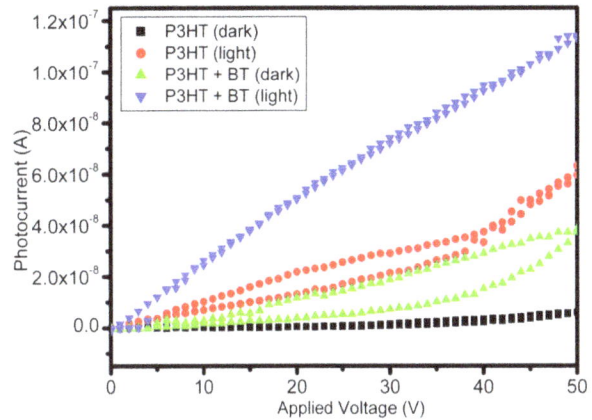

Fig. 4. Photoconductivity measurements for P3HT samples containing 20 mg/ml barium titanate.

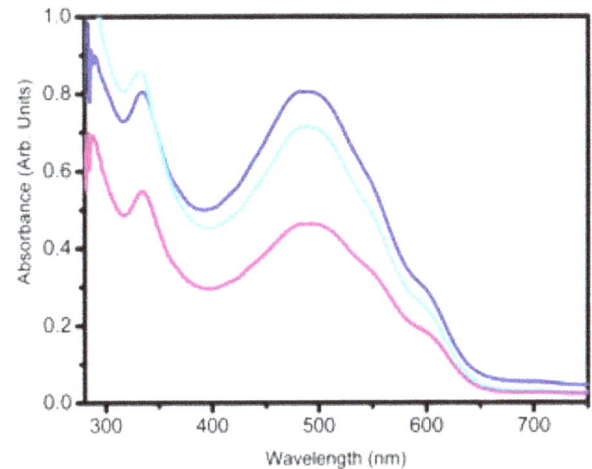

Fig. 5. UV-Vis spectrum of sample 1 (magenta), sample 4 (cyan) and sample 5 (blue) for comparison.

higher k value to begin with. These data confirm that both barium and strontium titanates are suitable candidate materials to be added to P3HT/PCBM blends to potentially increase the efficiency of this type of solar cell.

The first test of photoconductivity using P3HT films used only barium titanate, although subsequent tests also included strontium titanate. Figure 4 shows the results of the first test, with a base 20 mg/ml solution of P3HT was divided into two vials, one was kept as a control and the other was added to pre-measured barium titanate, which produced a solution with a concentration of BT of 20 mg/ml. The solution was ultrasonicated until the BT was in full suspension and then spin coated onto glass. Top metal electrodes were deposited to measure the photoconductivity.

The sample containing BT (blue line), clearly improves this to $\sim 1.1 \times 10^{-7}$ A and also reduces the hysteresis in the curve, producing an almost linear state. It is clearly evident from the data gather for dielectric constant of P3HT as a function of ferroelectric nanoparticles concentration and its effect on the photoconductivity may re-

sult in enhanced efficiency of organic photovoltaic solar cells. The clear role of ferroelectric nano-particles need further investigation, for example, measurement of charge carrier life time, etc. of thin film containing ferroelectric nano-particles.

The UV-Vis spectroscopy measurements shown in Figure 5 suggest that samples 1, 4 and 5 have absorption peaks at two different energy values; the first around 334 nm and the second at 490 nm. The first peak is slightly shifted to 331 nm in sample 4 containing strontium titanate but remains at 334 nm in the other two samples. The 490 nm peaks correspond in all three samples and agree with other works with regard to the location of the peaks in the green part of the spectrum. The absorption shoulders that are expected between 500 and 600 nm are not very pronounced in these samples. This may be a characteristic of the solvent which in this case was chloroform. Samples with chlorobenzene based solvents have been shown by Vanlaeke et al. to show flatter regions in this area [22]. It can clearly be seen that the relative absorbance of both polymer blends containing ferroelectric

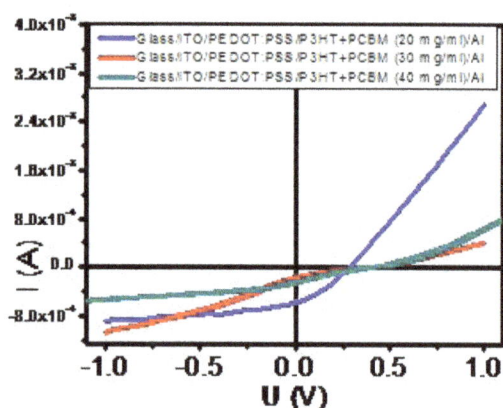

Fig. 6. Comparison of *I-V* characteristics of P3HT/PCBM samples with 20, 30 and 40 mg/ml concentrations in a 1:1 ratio.

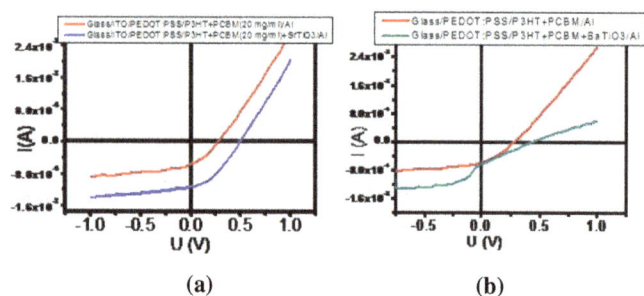

| | (a) | | (b) | |

Fig. 7. *I-V* characteristics of (a) sample 4 (blue) compared with sample 1 (red) and (b) sample 5 (green) compared with sample 1 (red).

Table 2. Results of electrical measurements for 1 cm^2 devices.

Sample	V_{OC} (V)	I_{SC} (mA)	FF	PCE %
1	0.27	0.50	0.38889	0.63
2	0.33	0.18	0.16111	0.11
3	0.40	0.27	0.15167	0.20
4	0.48	1.10	0.37652	2.37
5	0.43	0.70	0.19565	0.70

excitons that can dissociate into charge carriers, due to both a decrease in Coulomb forces and an increase in the volume of the dissociation region. The reason for the increase in open circuit voltage is less clear but is likely to be linked to the ferroelectric particles increasing the built in potential of the devices [23, 24]. We anticipate additional measurements in the near future to confirm the increase in permittivity of the material and to determine the variations in the built in potential.

Figure 7b shows the comparison between samples 1 and 5, as with sample 4 the open circuit voltage and short circuit current are increased, but the curve of sample 5 does not retain the expected ideal diode like shape compared to sample 1. In any solar cell, XSC or CSC, there are two resistances; a series resistance (r_s) related to the contact and bulk resistances and a shunt resistance (r_{sh}) arising from the contact alone, in an ideal device $r_s = 0$ and $r_{sh} = \infty$. As r_s increases and r_{sh} decreases the diode moves away from perfect diode like behaviour and the shape of the curve is flattened. For reasons not currently clear the resistances in the BT sample are obviously affecting the behaviour of the material, why this is not evident in the ST sample is also not known at present.

The V_{oc}, I_{sc}, fill factor and PCE of different samples are tabulated in Table 2.

While the overall efficiencies of these devices are not high compared to the best devices currently produced by other research groups, the relative efficiencies are compared in this work. Sample 4 presents an overall PCE of 2.37% which is almost a factor of four greater than the control sample. Even though the resistances have affected the diode characteristics of sample 5 it still has a higher PCE than the control. Strontium titanate, shows great promise as an additive to improve the performance of OPV devices as does barium titanate if the resistances in the devices can be overcome. It is hoped that by continuing to use ferroelectric nanoparticles in these devices that we can improve their efficiency compared to P3HT/PCBM devices alone.

nanoparticles is greater than those of P3HT/PCBM alone. We believe that this link to the increase in the dielectric constant of the P3HT/PCBM system by addition of ferroelectric nano-particles, which might increase the Debye length and longer availability of the charge carriers in admixture and further investigation is needed to confirm this.

The comparison of samples 1, 2 and 3 is presented in Figure 6. It is clear from both the shape of the curves and the fill factor and efficiency that sample 1 (20 mg/ml) has the highest overall efficiency. The open circuit voltage is higher for both samples 2 and 3, but the fill factor is reduced for these samples as there is an increased effect from resistances that is causing flattening of the curves. Form these initial results the decision was taken to use a base solution of 20 mg/ml and to add the ferroelectric nanoparticles to this solution. The solution used was from the same batch as solution 1 and the nanoparticles were added as described above.

The *I-V* characteristics of samples 1, 4 and 5 are presented and compared in Figure 7. Figure 7a shows a comparison between samples 1 and 4. Both samples exhibit curves of similar shape, presenting the expected diode like behaviour. Sample 4, containing ST clearly produces higher open circuit voltage and short circuit current than sample 1. From the theory stated above one explanation for the increase in current is the increased numbers of

4 Conclusions

We have demonstrated that out of three concentrations of P3HT/PCBM the most efficient and closest to an ideal diode is 20 mg/ml. We have also shown that it is possible to improve the efficiency of P3HT/PCBM OPV devices by adding ferroelectric nanoparticles such as strontium titanate or barium titanate to the active layer. The maximum efficiency obtained was by the sample containing strontium titanate, which was four times more efficient than the P3HT/PCBM control sample. The sample with barium titanate was more efficient than the control, but was limited by the fact that the resistances in the cell reduced the ideal diode like characteristics of this device. Ferroelectric nanoparticles may offer one route towards the efficient organic photovoltaic devices.

Authors would like to thank EPSRC for DTA student funding.

References

1. S.H. Park, A. Roy, S. Beaupre, S. Cho, N. Coates, J.-S. Moon, D. Moses, M. Leclerc, K. Lee, A.J. Heeger, Nat. Photon. **3**, 297 (2009)
2. D. Black, S. Paul, I. Salaoru, J. Nanosci. Nanotechnol. Lett. **2**, 1 (2010)
3. D. Black, S. Paul, Mater. Res. Soc. Symposium Proc. **1303**, 69 (2011)
4. S. Paul, D. Black, Organic photoconductive material (Patent application number GB2484743 – Filing date: 23 October 2010, Publication 25 April 2012)
5. B.A. Gregg, Coulomb Forces in Excitonic Solar Cells, in *Organic Photovoltaics: Mechanisms, Materials and Devices*, edited by S.S. Sun, N.S. Sariciftci (Taylor and Francis, Boca Raton, 2005), pp. 139–159
6. N.-K. Persson, O. Inganas, Simulations of Optical Processes in Organic Photovoltaic Devices, in *Organic Photovoltaics: Mechanisms, Materials and Devices*, edited by S.S. Sun, N.S. Sariciftci (Taylor and Francis, Boca Raton, 2005), pp. 107–138
7. A. Moliton, J.-M. Nunzi, Polymer Int. **55**, 583 (2006)
8. E. Kymakis, G.A.J. Amaratunga, Solar Cells Based on Composites of Donor Conjugated Polymers and Carbon Nanotubes, in *Organic Photovoltaics: Mechanisms, Materials and Devices*, edited by S.-S. Sun, N.S. Sariciftci (Taylor and Francis, Boca Raton, 2005)
9. L.M.B. Alldredge et al., Appl. Phys. Lett. **94**, 052904 (2009)
10. T.-I. Yang, P. Kofinas, Polymer **48**, 791 (2007)
11. I. Salaoru, S. Paul, Philos. Trans. Roy. Soc. A **367**, 4227 (2009)
12. J. Nelson, *The Physics of Solar Cells*, 1st edn. (Imperial College Press, 2003)
13. P. Kim et al., Adv. Mater. **19**, 1001 (2007)
14. P. Barber et al., Materials **2**, 1697 (2009)
15. Z.-M. Dang et al., Mater. Chem. Phys. **109**, 1 (2008)
16. S. George, M.T. Sebastian, Comput. Sci. Technol. **69**, 1298 (2009)
17. S. Ogitani, S.A. Bidstrup-Allen, P.A. Kohl, Adv. Packag. IEEE Trans. **23**, 313 (2000)
18. G. Subodh et al., Appl. Phys. Lett. **95**, 062903 (2009)
19. G. Arlt, D. Hennings, G. de With, J. Appl. Phys. **58**, 1619 (1985)
20. A.K. Jonscher, in *Dielectric Relaxation in Solids* (Chelsea Dielectrics Press, London, 1996), p. 380
21. B. Pradhan, A.J. Pal, Chem. Phys. Lett. **416**, 327 (2005)
22. P. Vanlaeke, G. Vanhoyland, T. Aernouts, D. Cheyns, C. Deibel, J. Manca, P. Heremans, J. Poortmans, Thin Solid Films **511-512**, 358 (2006)
23. C.J. Brabec, A. Cravino, D. Meissner, N.S. Sariciftci, T. Fromherz, M.T. Rispens, L. Sanchez, J.C. Hummelen, Adv. Funct. Mater. **11**, 374 (2001)
24. M.F. Lo, T.W. Ng, T.Z. Liu, V.A.L. Roy, S.L. Lai, M.K. Fung, C.S. Lee, S.T. Lee, Appl. Phys. Lett. **96**, 113303 (2010)

24

Towards 12% stabilised efficiency in single junction polymorphous silicon solar cells: experimental developments and model predictions

Sergey Abolmasov[1,2,a], Pere Roca i Cabarrocas[2], and Parsathi Chatterjee[2]

[1] R&D Center of Thin-Film Technologies in Energetics, Ioffe Institute 28 Polytekhnicheskaya, 194064 Saint Petersburg, Russia
[2] LPICM, CNRS, Ecole Polytechnique, Université Paris-Saclay, 91128 Palaiseau, France

Abstract We have combined recent experimental developments in our laboratory with modelling to devise ways of maximising the stabilised efficiency of hydrogenated amorphous silicon (a-Si:H) PIN solar cells. The cells were fabricated using the conventional plasma enhanced chemical vapour deposition (PECVD) technique at various temperatures, pressures and gas flow ratios. A detailed electrical-optical simulator was used to examine the effect of using wide band gap P-and N-doped μc-SiO$_x$:H layers, as well as a MgF$_2$ anti-reflection coating (ARC) on cell performance. We find that with the best quality a-Si:H so far produced in our laboratory and optimised deposition parameters for the corresponding solar cell, we could not attain a 10% stabilised efficiency due to the high stabilised defect density of a-Si:H, although this landmark has been achieved in some laboratories. On the other hand, a close cousin of a-Si:H, hydrogenated polymorphous silicon (pm-Si:H), a nano-structured silicon thin film produced by PECVD under conditions close to powder formation, has been developed in our laboratory. This material has been shown to have a lower initial and stabilised defect density as well as higher hole mobility than a-Si:H. Modelling indicates that it is possible to attain stabilised efficiencies of 12% when pm-Si:H is incorporated in a solar cell, deposited in a NIP configuration to reduce the P/I interface defects and combined with P- and N-doped μc-SiO$_x$:H layers and a MgF$_2$ ARC.

1 Introduction

Large-scale application of amorphous silicon based solar cells in the photovoltaic power generation industry requires improvement of their efficiency. Hydrogenated amorphous silicon (a-Si:H) based solar cells, in spite of their advantage of large area deposition, lag behind their crystalline counterparts with respect to conversion efficiency. Moreover they suffer from light-induced degradation, which can reduce cell efficiency by ∼25%. Therefore much efforts have been made to develop growth conditions leading to a-Si:H based materials having improved transport properties or stability e.g. [1, 2]. Also tandem or triple-junction solar cells, with thin, varying band gap materials in different sub-cells have been tried out and achieved an initial record efficiency of 16.3% [3]. This design improves light absorption over the entire visible spectrum and reduces light-induced degradation as it is possible here to use thinner sub-cells. However the design is complex and it would be ideal technologically to attain a high stabilised efficiency in single junction solar cells. Moreover, this is a requisite for high stabilized efficiency in tandem and triple junction solar cells.

To this end, we have been experimenting with an a-Si:H-like material: hydrogenated polymorphous silicon (pm-Si:H), which is produced by the standard RF glow discharge decomposition of silane highly diluted in hydrogen, under conditions close to powder formation [4–8]. Pm-Si:H is a nano-structured silicon material with medium range order which shows up in the superior hole transport properties as deduced from diffusion induced time resolved microwave conductivity [8] and time-of-flight measurements [9]. Moreover it exhibits lower initial [5] and stabilised defect densities than standard a-Si:H [1, 9, 10], with the $\eta\mu\tau$ product of electrons 200–700 times higher than that of standard a-Si:H for samples in the as-deposited state [1], and attaining values after light-soaking comparable to those of standard a-Si:H films before degradation [1, 4–7]. In reference [1], a comparative study has been made of the kinetics of defect creation in

---
[a] e-mail: s.abolmasov@hevelsolar.com

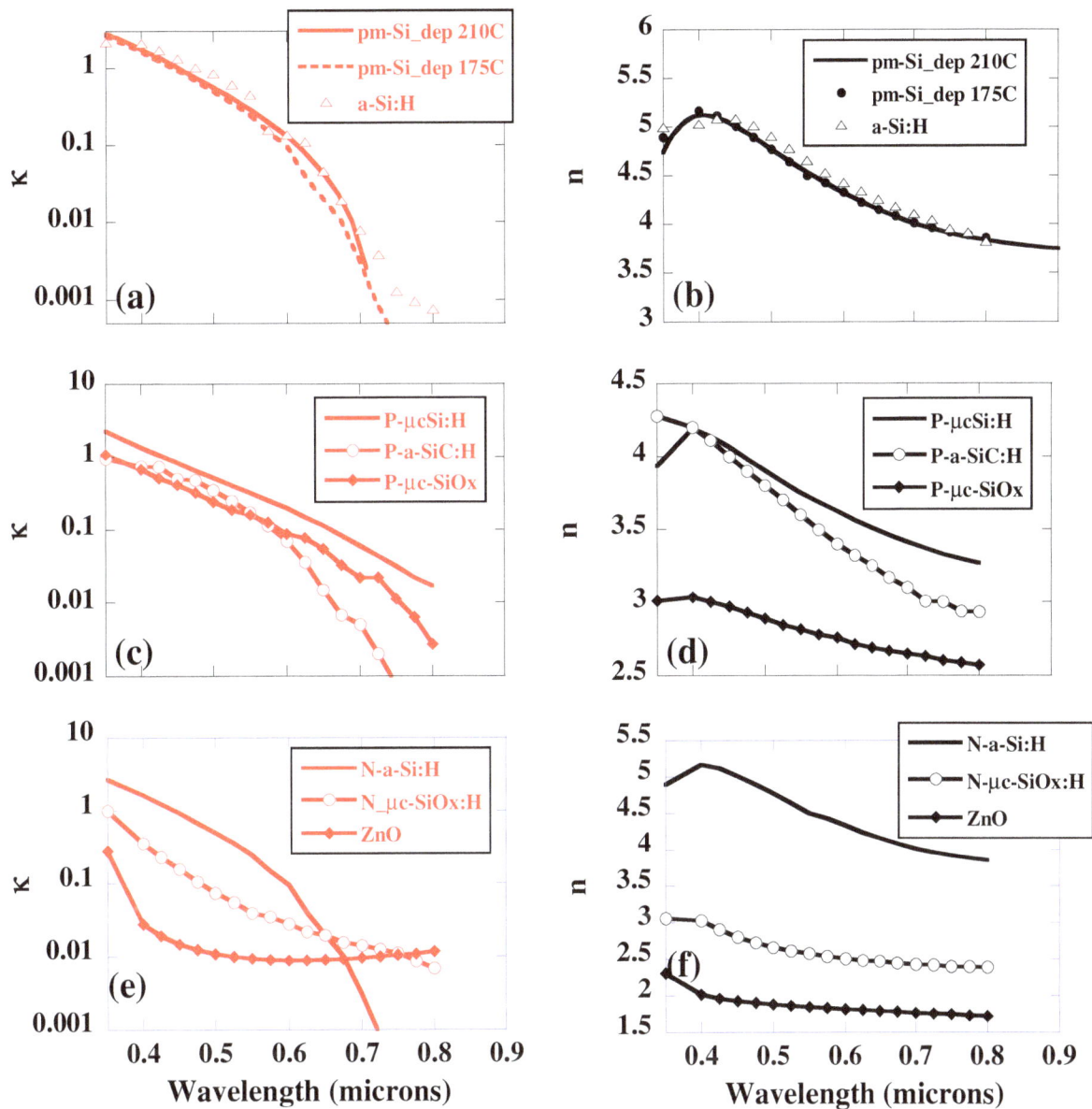

Fig. 1. Comparison of the complex refractive indices (κ, n) (a) and (b) of the intrinsic layers used in this study: a-Si:H, pm-Si:H deposited at 175 °C and 210 °C; (c) and (d) of the various P-layers studied: P-μc-Si:H, P-a-SiC:H and P-μc-SiO$_x$:H; and (e) and (f) of the different N-layers: N-a-Si:H, N-μc-SiO$_x$:H and the TCO: ZnO:Al.

standard a-Si:H and pm-Si:H films. It is found that the value of the absorption coefficient $\alpha(1.1\ \mathrm{eV})$, obtained at 'saturation' is lower for the pm-Si:H than standard a-Si:H films, indicating a lower stabilized density of states (DOS) for the former case.

Unfortunately the material has slightly lower absorption coefficients than a-Si:H, therefore lower coefficients of extinction (κ – Fig. 1a) over the visible spectrum [11]. The wavelength dependence of the real part of the complex refractive indices (CRINDs) of pm-Si:H and a-Si:H are given in Figure 1b. Also in the same article [11], using detailed modelling to simulate the measured characteristics of solar cells fabricated with this material as the intrinsic layer, we have shown that pm-Si:H develops a very defective P/I interface, probably due to bombardment by

charged silicon nanocrystals contributing to the deposition and due to the high flux of hydrogen required to develop the material. Moreover other experiments performed at our laboratory [12] have shown that light-soaking of pm-Si:H leads to the formation of hydrogen bubbles that introduce macroscopic defects. In this article we show by modelling how the above drawbacks of a pm-Si:H solar cell can be overcome to produce solar cells with a high stabilised efficiency.

We have also tried out different window structures using modelling, to see which combination leads to the highest efficiency. Two types of front window designs, each with two P-layers were examined: (a) P-μc-Si:H/P-a-SiC:H or (b) P-μc-Si:H/P-μc-SiO$_x$. In a previous paper [13] modelling the effects of P-μc-Si:H, P-a-SiC:H window layers and various combinations thereof, we have shown that a P-μc-Si:H/P-a-SiC:H stack is an efficient design. Indeed, the μc-Si:H P-layer has a lower surface band bending (sbb) and lower activation energy than P-a-SiC:H. These properties help to improve the built-in potential (V_{bi}), open-circuit voltage (V_{oc}) and fill factor (FF) of solar cells employing such a double window layer structure relative to a standard cell having a single P-a-SiC:H window layer [13]. These modelling results have received confirmation from the experimental work of Ma et al. [14]. The problem with a single P-μc-Si:H window is the large valence band discontinuity, since it has been shown by the IPE measurements of Xu et al. [15] that the P-μc-Si:H/a-Si:H band discontinuity is almost totally on the valence band side. This fact results in a large drop in potential at the P/I interface, leading to a collapse of the electric field inside the volume of the device and reducing the V_{oc} and the FF of a device employing such a single P-μc-Si:H window. Also considerable back diffusion of electrons takes place, reducing the blue-green part of the external quantum efficiency (EQE) and short-circuit current density (J_{sc}). In addition the P-μc-Si:H layer developed in our laboratory has a higher coefficient of extinction (hence higher absorption coefficient) compared to P-a-SiC:H or P-μc-SiO$_x$:H over the visible solar spectrum, as seen from Figure 1c. A thin P-a-SiC:H buffer sandwiched between the P-μc-Si:H and the a-Si:H layers solves all these problems, since the band discontinuity between a-Si:H and P-a-SiC:H is mainly on the conduction band side [16], preventing back diffusion of electrons and bringing up the short wavelength QE and J_{sc}. Also, the detrimental effect of the large valence band discontinuity is reduced by the presence of a thin P-a-SiC:H layer [13], leading to high values of V_{oc} and FF. Moreover the presence of P-a-SiC:H, allows to use a thinner P-μc-Si:H layer, so that the absorption of the shorter wavelengths in the P-layers is reduced. P-doped μc-SiO$_x$:H is also a promising P-layer, since it is more transparent (having a wider band gap) than P-a-SiC:H and its doping density can be adjusted to yield high conductivity. We have in this study therefore also tried out by modelling a P-μc-Si:H/P-μc-SiO$_x$:H double window structure, primarily because we know the TCO/P sbb for the case of P-μc-Si:H [14], but it is unknown in the case of P-μc-SiO$_x$:H. Nevertheless we have also experimented with a single P-μc-SiO$_x$:H window layer, assuming a range of values for the ZnO:Al/P-μc-SiO$_x$:H sbb. In all cases an ultra-thin intrinsic buffer layer follows the window P layers, both to prevent boron tailing into the intrinsic layer and to obtain a less defective P/I interface.

Different back contact designs were also investigated via modelling. This was either N-a-Si:H/ZnO:Al/Ag or simply N-μc-SiO$_x$:H/Ag with the ZnO:Al left out. The logic behind this latter design is that the band gap and doping density of N-μc-SiO$_x$:H can be adjusted in such a way that it can be used as a wide band gap TCO, while retaining a good enough conductivity to act as the N-layer of the cell [17]. Thus the deposition of an extra TCO layer can be avoided. However this is only possible in cases where the P-layer is deposited first on ZnO:Al coated glass. When a cell is deposited in a NIP configuration, a thin ZnO:Al layer between the silver and N-μc-SiO$_x$:H is mandatory to prevent silver diffusion into the latter. The CRIND's of N-μc-SiO$_x$:H are compared to those of standard N-a-Si:H and the TCO ZnO:Al in Figures 1e and 1f. Finally we have used modelling to study how a thin anti-reflection coating (ARC), deposited on the surface through which light enters the solar cell, can affect J_{sc} (and hence the conversion efficiency η).

In brief, the aim of this article is to use the measured values of the initial and stabilised state properties of the best a-Si:H and pm-Si:H materials developed in our laboratory so far, together with realistic P/I interface defect density and other properties extracted by modelling the corresponding solar cells in an earlier article [11], combined with different emitter and N-layer materials and designs, to predict using modelling the highest efficiency attainable in these structures. We find, using the material and deposition parameters of our a-Si:H solar cells, that it is difficult to attain 10% stabilised efficiency with this material, even though other laboratories e.g. [18] have actually developed a-Si:H single junction cells with stabilised efficiencies in excess of 10%. This is likely due to the superior nature and/or light absorption of the a-Si:H material developed, improved P/I interface properties in their deposited solar cells, etc. On the other hand we have been able to develop pm-Si:H [1, 4–8], a close cousin of a-Si:H, with lower initial and stabilised defect densities, higher hole mobility, but lower light absorption. Using modelling we show in this article that such a material when deposited in a solar cell in the NIP configuration, and in conjunction with suitable P- and N-layers and an ARC is capable of attaining a stabilised efficiency of around 12%, although our best experimental results so far show a lower stabilised efficiency [19].

Schematic diagrams of the a-Si:H solar cell deposited in a PIN configuration on TCO-coated glass, and of the pm-Si:H cell deposited in a NIP configuration on glass/Ag/ZnO:Al are shown in Figures 2a and 2b, respectively. These are the structures that have been modelled. It would have been better if the ARC were assumed to be textured also; however as will be pointed out in the section on the simulation model ASDMP, this latter at present is capable of handling only two textured interfaces. These have been taken at the TCO/P and N/back reflector (ZnO:Al/Ag or only Ag) interface to maximise light absorption in the intrinsic layer, except when μc-SiO$_x$:H is used as the N-layer. The refractive index of this N-layer is close to that of ZnO:Al, so that most of the light in reality suffers specular and diffused reflection at the I/N-μc-SiO$_x$:H interface. Hence it is this latter interface that is taken as the second textured interface in the modelling calculations. The layers between the top TCO and bottom reflector are thin and we assume that they almost exactly take up the structure of the preceding layer (conformal deposition). Therefore we have assumed that only specular reflection, obeying Fresnel's law, takes place at these interfaces.

(a)

(b)

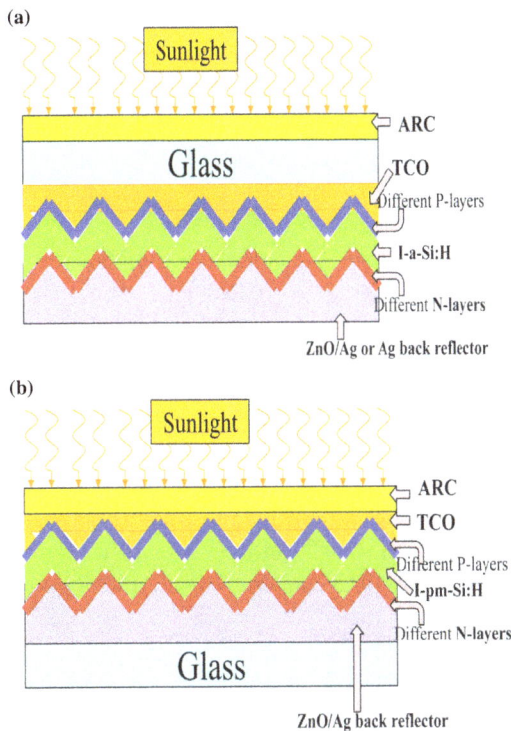

Fig. 2. Schematic representations of (a) an a-Si:H solar cell (deposited in the PIN configuration on TCO coated glass) and (b) a pm-Si:H solar cell deposited in the NIP configuration on TCO/Ag coated glass.

2 Experiments

The PIN devices were deposited at various temperatures ranging from 175 °C to 240 °C by the RF (13.56 MHz) glow discharge plasma enhanced chemical vapor deposition (PECVD) method in a multiplasma-monochamber reactor [20]. Standard a-Si:H films were produced by the dissociation of pure silane at 40 mTorr under an RF power of 10 mW/cm^2. On the other hand, pm-Si:H films were obtained by the dissociation of a 10% silane in hydrogen mixture at 2–4 Torr under an RF power of 90 mW/cm^2. A deposition rate of 9 Å/s could be achieved for the pm-Si:H films [21]. We experimented with two different deposition temperatures (175 °C and 210 °C), but were unable to obtain any significant increase in the absorption coefficients in the case of a-Si:H. However for pm-Si:H, a substrate temperature of 210 °C allowed to obtain films with a higher absorption over the visible wavelengths than for pm-Si:H deposited at 175 °C. The CRINDs of a-Si:H, and pm-Si:H deposited at two different temperatures are compared in Figures 1a and 1b.

The a-Si:H PIN cells were deposited in the conventional structure: glass/ZnO:Al/P/I/N/ZnO:Al/Ag (or simply Ag back eflector) (Fig. 2a). However it has been shown in a previous article [11] combining modelling with experiments, that a pm-Si:H intrinsic layer cell deposited in this order develops a very damaged P/I interface due to bombardment by the high dose of hydrogen required for developing pm-Si:H. Other experiments conducted in our

laboratory [12] further highlight that upon light-soaking treatments, hydrogen bubbles form and lead to macroscopic defects, such as blisters and delamination, that obviously cannot be reversed by annealing. Therefore, as we are interested in the stabilised efficiency, we have modelled pm-Si:H cells with the material deposited as follows: glass/silver/ZnO:Al/N/I/P/ZnO:Al/silver grid in that order (Fig. 2b). The high flux of hydrogen required for pm-Si:H deposition will then damage the N/I interface, to which, as will be shown later by modelling, the solar cell output is much less sensitive. Also macroscopic defects, if any, will now appear at the ZnO:Al/N contact and will therefore have far less repercussion on solar cell characteristics. The light of course should still enter through the P-layer during cell operation for improved efficiency, since the maximum number of holes needs to be created near the contact that is the collector of holes (the ZnO:Al/P contact in this case). This is because, even though the mobility of holes in pm-Si:H is a factor of 3 higher than that in a-Si:H [8,9], it is still less than the electron mobility. The a-Si:H PIN cell does not have a very damaged P/I interface [11], since this material can be deposited by the decomposition of SiH$_4$ without or very little hydrogen dilution. We therefore commence the investigations in this article on a-Si:H PIN cells, deposited in the normal manner with the P-layer first on ZnO:Al coated glass; but pm-Si:H cells deposited with N-layer first on glass/silver/ZnO:Al. For any deposition involving a rich hydrogen environment, ZnO is preferable to either ITO or SnO$_2$, as ZnO is more stable against reduction by atomic hydrogen. The RMS roughness of the LPCVD grown ZnO:Al is around 60 nm corresponding to a haze of ~23%, measured at a wavelength of 550 nm.

To attain the stabilised state, the samples were light soaked at 100 °C in order to accelerate the kinetics of creation of metastable defects [22]. They were exposed to a maximum intensity of 670 mW cm^{-2} unfiltered white light produced by a 1 kW halogen lamp and then quenched to 30 °C for their characterisation [1]. In some other light-soaking (LS) experiments [19], the cells were light soaked in the open-circuit condition using a mercury vapour lamp providing an intensity of 100 mW/cm^2. During LS, the cells were fan-cooled to reduce illumination-induced heating. The temperature of the cells was monitored using a PT100 thermocouple during LS, to verify that the device temperature remained under 50 °C. Light-soaking was continued for 500 h.

P-μc-Si:H, P-a-SiC:H and P-μc-SiO$_x$ layers have also been developed in our laboratory with properties given in Table 1. The values given in the table are representative of a series of each such film grown under the same conditions. The CRIND's of the P-layers are compared in Figures 1c and 1d. The conductivity of our P-μc-SiO$_x$ films is strongly dependent on the CO$_2$ flow rate [23], nevertheless, its conductivity compares well with that of optimised P-aSiC:H layers. The properties of the N-μc-SiO$_x$ have however been taken from reference [17] and its CRIND's are compared to those of standard N-a-Si:H and the TCO ZnO:Al in Figures 1e and 1f.

Table 1. The input parameters used to model different cases. The first 5 nm of the I-pm-Si:H layer at the N/I interface has a DB DOS of 5×10^{17} cm^{-3}, followed by a 30 nm intermediate layer with a DOS = 2×10^{16} cm^{-3}. The first 20 nm of the I-a-Si:H layer adjacent to the P/I interface is assumed to have a DOS of 2×10^{16} cm^{-3}. The values have been extracted from modelling previous experimental results [11]. The parameters of N-a-Si:H are not shown as these are standard.

Parameter	P-μc-Si:H	P-a-SiC:H	P-μc-SiO$_x$	N-μcSiO$_x$	I-pm-Si:H	I-a-Si:H
Layer thickness (nm)	6	4	10	10	250	200–300
Electron affinity (eV)	4.00	3.89	3.67	4.00	3.95	4.00
Mobility gap (eV)	1.6	2.0	2.5	2.4	1.96	1.84
Activation energy (eV)	0.12	0.46	0.47	0.09	0.92	0.86
Eff. DOS in bands (cm^{-3})	2×10^{20}	2×10^{20}	2×10^{20}	2×10^{20}	2×10^{20}	2×10^{20}
Ch Energy (VB tail) (eV)	0.035	0.120	0.120	0.025	0.050	0.050
Ch Energy (CB tail) (eV)	0.025	0.07	0.070	0.015	0.030	0.030
μ_n (μ_p) (cm^2/V s)	32 (8)	20 (4)	32 (8)	32 (8)	20 (12)	20 (4)
DOS (cm^{-3}) (annealed)	3×10^{18}	3×10^{18}	4×10^{18}	2×10^{18}	7×10^{14}	5×10^{15}
DOS (cm^{-3}) (stabilized)	3×10^{18}	3×10^{18}	4×10^{18}	2×10^{18}	2×10^{16}	10^{17}
σ_n (tails) (cm^2)	10^{-17}	10^{-17}	10^{-17}	10^{-17}	10^{-17}	10^{-16}
σ_c (tails) (cm^2)	10^{-16}	10^{-16}	10^{-16}	10^{-15}	10^{-15}	5×10^{-16}
σ_n (DB) (cm^2) (annealed)	10^{-16}	10^{-16}	10^{-16}	5×10^{-17}	10^{-15}	10^{-16}
σ_c (DB) (cm^2) (annealed)	5×10^{-16}	5×10^{-16}	5×10^{-16}	5×10^{-15}	5×10^{-15}	5×10^{-16}
σ_n (DB) (cm^2) (stabilized)	10^{-16}	10^{-16}	10^{-16}	5×10^{-17}	10^{-15}	2×10^{-16}
σ_c (DB) (cm^2) (stabilized)	5×10^{-16}	5×10^{-16}	5×10^{-16}	5×10^{-15}	2×10^{-14}	2×10^{-14}

3 Simulation model

Our one-dimensional electrical-optical model AS-DMP (Amorphous Semiconductor Device Modelling Programme) [24, 25], later extended to model crystalline silicon and HIT cells [26], solves the Poisson's equation and the two carrier continuity equations under steady state conditions for the given device structure and yields the dark and illuminated current density-voltage (J-V) characteristics and the QE. The electrical part of the modelling programme is described in references [27, 28]. The expressions for the free and trapped charges, the recombination term, the boundary conditions and the solution technique in ASDMP are similar to the AMPS computer programme [29], developed by Prof. Fonash's group.

The gap state model in ASDMP consists of two monovalent donor-like and acceptor-like tail states and two monovalent Gaussian distribution functions (one being of donor type and the other of acceptor type) to simulate the deep dangling bond (DB) states (as e.g. used in [29] (AMPS), [30, 31] (AFORS-HET)). A more realistic gap state distribution model in a-Si:H would consist of, besides the monovalent band tail states, a deep defect distribution of DBs determined from the defect pool model (DPM) [32, 33]. However in ASDMP, as in AMPS and AFORS-HET, since only monovalent states can be introduced, the deep DB distribution determined from DPM is replaced by two Gaussian distributions of monovalent states, donor like and acceptor like, separated by a correlation energy. Such a replacement has proved to be quite accurate in a-Si:H [34, 35].

The contact barrier heights for a cell with the P-layer in contact with a TCO at $x = 0$ and the N-layer in contact

with a TCO or metal at $x = L$, are given by:

$$\varphi_{B0} = E_\mu(P) - E_{ac}(P) - sbb \quad (1)$$

and

$$\varphi_{BL} = E_{ac}(N). \quad (2)$$

Here $E_\mu(P)$ and $E_{ac}(P)$ are the mobility gap and the activation energy of the P-layer, and $E_{ac}(N)$ is the activation energy of the N-layer, forming an ohmic contact at the back of the device. 'sbb' is the TCO/P surface band bending, which we have taken to be ~0.16 eV, the measured value [14] for this quantity for P-μc-Si:H, employed in all double window structures. A value of 0.2 eV has been assumed for P-μc-SiO$_x$:H, in simulations involving such a single P-layer. However, later in Section 4 we show the effect of varying the sbb (P-μc-SiO$_x$:H) on the solar cell output. With the activation energies and mobility band gaps of P-μc-Si:H and P-μc-SiO$_x$:H given in Table 1, φ_{B0} in the present PIN devices is 1.32 eV in the case of TCO/P-μc-Si:H and 1.83 eV for TCO/P-μc-SiO$_x$:H. φ_{BL} is 0.2 eV for the case of N-a-Si:H and 0.09 eV for N-μc-SiO$_x$:H [19].

The generation term in the continuity equations has been calculated using a semi-empirical model [36] that has been integrated into the modelling programme [24,25]. Both specular interference effects and diffused reflectances and transmittances due to the interface roughness are taken into account. As mentioned earlier, in the model it is possible to consider only two rough interfaces. The diffused reflection and transmission from the TCO (ZnO:Al) used in the simulations have both a wavelength and angular dependence. In some of the simulations the effect of using a MgF$_2$ ARC is given. This layer is assumed to have a thickness of 50 nm – however it was not optimised by changing its thickness. The complex refractive index for

Table 2. Sensitivity of a pm-Si:H solar cell output to a damaged N/I and a similarly damaged P/I interface (modelling results). The cells given here have a two P-layer structure with light entering the device on the P-layer side.

ZnO:Al/P-μc-Si/P-μc-SiO$_x$/I-pm-Si:H/ N-μc-SiO$_x$/ZnO:Al/Ag/glass		J_{sc} (mA cm^{-2})	V_{oc} (volts)	FF	Efficiency (%)
With a damaged N/I interface	Initial	15.29	1.140	0.772	13.45
	Stabilised	15.15	1.073	0.701	11.40
With a damaged P/I interface	Initial	15.41	1.063	0.665	10.90
	Stabilised	15.26	1.044	0.608	9.69

each layer of the structure is required as input and has been measured in-house by spectroscopic ellipsometry. In all cases studied in this article, experimentally or by modelling, light enters through the P-layer, the junction of the TCO with the P-layer being taken as $x = 0$ on the position scale in the modelling calculations. Voltage is also applied at $x = 0$, with the N/TCO or N/metal contact at $x = L$, at the end of the semiconductor layers of the device, being at ground potential.

The input material and device parameters of the model are given in Table 1. The values chosen for a-Si:H are the best average values for the material developed in our laboratory. The parameters for pm-Si:H (e.g., hole mobility, initial and stabilised DOS, etc.) are those deduced from measured data in references [1,5,8–10]. The superior hole transport properties of pm-Si:H, as already pointed out in the introductory section, have been deduced from diffusion induced time resolved microwave conductivity [8] and time-of-flight measurements [9]. The DOS at the Fermi level $N(E_F)$ in the initial state has been measured on pm-Si:H samples using both capacitance measurements on Schottky barriers and space-charge-limited current (SCLC) measurements on N$^+$/I/N$^+$ structures [5]. From both techniques $N(E_F)$ values of 7–8×10^{14} cm^{-3} eV^{-1} have been obtained which is almost an order of magnitude lower than that measured [5] for a-Si:H ($\sim 5 \times 10^{15}$ cm^{-3} eV^{-1}) or reported in the literature for standard a-Si:H. In reference [1], the evolution of the sub-band gap absorption during accelerated light-soaking at 100 °C was studied for a-Si:H and pm-Si:H films. It was observed that the value of $\alpha(1.1$ eV) at "saturation", namely α_{ss}, is about an order of magnitude lower for pm-Si:H relative to a-Si:H. All the above results are representative of the study of a large number of samples grown under the same conditions. The Tauc's gap was determined from standard optical transmission measurements [1]. For both a-Si:H and pm-Si:H we have assumed the mobility gap to be ~ 0.1 eV larger than the corresponding optical gap.

4 Results and discussion

In performing the modelling we have relied heavily on the input parameters derived from simulating a-Si:H and pm-Si:H solar cells described in reference [11]. The reason for using the P-μc-Si:H/P-a-SiC:H design is given in detail in reference [13] and for using the P-μc-Si:H/P-μc-SiO$_x$:H window structure described in the introductory section. However as already stated, we have also examined

the P-μc-SiO$_x$:H single window structure, assuming here a surface band bending of 0.2 eV, as for P-a-SiC:H – a value that is uncertain. The input parameters of the principal layers are given in Table 1 and their complex refractive indices shown in Figure 1. The P/I interface dangling bond DOS is 2×10^{16} cm^{-3} over 20 nm for the a-Si:H PIN cells, as inferred by modelling experimental results [11]. As the deposition of a pm-Si:H film involves a high flux of hydrogen to the substrate, we have preferred to study solar cells with this material deposited in a NIP configuration with the N-layer first on ZnO:Al/Ag/glass. The improvement in a pm-si:H solar cell output parameters when it is deposited in a NIP configuration has been demonstrated experimentally in reference [19]. This is because solar cell output is less sensitive to a damaged N/I interface than to a similarly damaged I/P interface. This fact is demonstrated in Table 2 by modelling. In actual experiments, for pm-Si:H solar cells deposited in PIN and NIP configurations in our laboratory [19], it has been found that cells deposited in the latter configuration degrade much less – from an initial efficiency of $\sim 9.3\%$ to about 8.2%; while those deposited in the PIN configuration degrade from a lower initial efficiency of $\sim 8.5\%$ to 5.8%, or even down to 3%, in the worst cases after 500 h of light-soaking. The reason for this has been attributed to the creation of macroscopic defects such as blisters and delamination after light-soaking [19]. Coming back to modelling, the N/I interface defect density for the pm-Si:H cells is taken $= 5 \times 10^{17}$ cm^{-3} over the first 5 nm, followed by an intermediate layer with a DOS of 2×10^{16} cm^{-3} over 30 nm, The existence of such an intermediate layer in pm-Si:H cells has been suggested by modelling in reference [11]. As already stated, in all cases an ultra-thin intrinsic buffer layer follows the window P-layers, both to prevent boron tailing into the main intrinsic layer and to achieve a lower interface defect density. As described in the introduction the back contact was either N-a-Si:H/ZnO:Al/Ag or N-μc-SiO$_x$:H/Ag with the ZnO:Al left out in the case of a-Si:H solar cells deposited in the PIN configuration, but retaining a thin ZnO:Al layer even for the latter design for the case of the pm-Si:H films, that are deposited in the NIP configuration. As mentioned earlier, this is to prevent silver diffusion into N-μc-SiO$_x$:H.

4.1 The intrinsic layer

The best solar cell output parameters of the a-Si:H solar cells deposited in our laboratory are given in the first

Table 3. Solar cell output parameters of a-Si:H PIN cells with different window layers and back contacts in the initial and light-stabilised states. Light enters through the glass/ZnO:Al/P-layer. a-Si:H layer thickness is 200 nm. ZnO in the table corresponds to aluminium-doped ZnO. The first row gives our best experimental results [23] with a 20 nm pm-Si:H P/I buffer, while the rest are model calculations. The last line shows the effect of using MgF_2 ARC on glass.

Cell structure		J_{sc} (mA cm^{-2})	V_{oc} (volts)	FF	Efficiency (%)
Expt: Glass/ZnO/P-μc-SiO$_x$/	Initial	15.17	0.87	0.729	9.63
pm-Si:H/I-a-Si:H/N-μc-Si:H/ZnO/Ag	Stabilised	14.65	0.843	0.636	7.85
Glass/ZnO/P-μc-Si/P-a-SiC/	Initial	15.55	0.988	0.782	12.01
I-a-Si:H/N-a-Si:H/ZnO/Ag	Stabilised	15.01	0.933	0.603	8.83
Glass/ZnO/P-μc-Si/P-μc-SiO$_x$	Initial	15.78	1.029	0.723	11.74
/I-a-Si:H/N-a-Si:H/ZnO/Ag	Stabilised	15.37	0.943	0.611	8.86
Glass/ZnO/P-μc-Si/P-μc-SiO$_x$/	Initial	16.07	1.032	0.729	12.09
I-a-Si:H/N-μc-SiO$_x$/Ag	Stabilised	15.53	0.944	0.612	8.97
MgF_2/Glass/ZnO/P-μc-Si/	Initial	16.29	1.032	0.729	12.25
P-μc-SiO$_x$ /I-a-Si:H/N-μc-SiO$_x$ /Ag	Stabilised	15.74	0.944	0.612	9.09

Table 4. Solar cell output of pm-Si:H solar cells, deposited in the NIP configuration, but with the light entering on the P-side. The first row gives our best experimental results [19], while the rest are model calculations. In the second row, properties of pm-Si:H deposited at 175 °C are used, while in the rest those of pm-Si:H deposited at 210 °C are used. ZnO in the table corresponds to aluminium-doped ZnO. The subsequent lines show the effect of different window layers and back contacts in the initial and light-stabilised states. The ZnO:Al/P-μc-SiO$_x$ sbb is taken =0.2 eV. The last two rows show the effect of using MgF_2 ARC on ZnO:Al. Thickness of the pm-Si:H layer is 250 nm except for the last case where it is 150 nm.

Cell type		J_{sc} (mA cm^{-2})	V_{oc} (volts)	FF	Efficiency (%)
Expt: ITO/P-μc-Si/P-μc-SiO$_x$/	Initial	15.7	0.95	0.63	9.3
I-pm-Si:H/N-a-Si:H/SnO$_2$:F/glass	Stabilised	15.0	0.93	0.60	8.2
ZnO/P-μc-Si/P-a-SiC/I-pm-Si:H	Initial	13.31	0.985	0.792	10.39
(175 °C)/N-a-Si:H/ZnO/Ag/glass	Stabilised	13.22	0.980	0.744	9.63
ZnO/P-μc-Si/P-a-SiC/I-pm-Si:H	Initial	14.48	0.987	0.792	11.32
(210 °C)/N-a-Si:H/ZnO/Ag/glass	Stabilised	14.38	0.982	0.743	10.5
ZnO/P-μc-Si/P-μc-SiO$_x$/I-pm-Si:H/	Initial	15.02	1.106	0.748	12.43
N-a-Si:H/ZnO/Ag/glass	Stabilised	14.89	1.061	0.694	10.97
ZnO/P-μc-Si/P-μc-SiO$_x$/I-pm-Si:H/	Initial	15.29	1.140	0.772	13.45
N-μc-SiO$_x$/ZnO/Ag/glass	Stabilised	15.15	1.073	0.701	11.40
ZnO/P-μc-SiO$_x$/I-pm-Si:H/	Initial	15.89	1.124	0.760	13.56
N-μc-SiO$_x$/ZnO/Ag/glass	Stabilised	15.74	1.057	0.687	11.43
MgF_2/ZnO/P-μc-SiO$_x$/I-pm-Si:H/	Initial	16.64	1.125	0,759	14.20
N-μc-SiO$_x$/ZnO/Ag/glass	Stabilised	16.49	1.059	0.686	11.98
MgF_2/ZnO/P-μc-SiO$_x$/	Initial	15.98	1.129	0.764	13.78
I-pm-Si:H/N-μc-SiO$_x$/ZnO/Ag/glass	Stabilised	15.92	1.080	0.714	12.27

row of Table 3 and those of the best pm-Si:H solar cells in the first row of Table 4. The remaining rows of the two tables give modelling results with improvements in the output parameters likely to occur when various emitter designs and N-layers are used. For both types of material, we studied the effect of the substrate temperature to see whether any improvement in the absorption coefficient, linked to the extinction coefficients (κ) could be achieved. No such increase was observed for a-Si:H over a range of temperatures around 200 °C, but the κ's of pm-Si:H increased somewhat when the deposition temperature was increased from 175 °C to 210 °C (Fig. 1a), resulting in superior values of J_{sc} (compare the results of the second and third rows of Tab. 4) and the EQE given in Figure 4a.

Comparing Tables 3 and 4 for the a-Si:H and pm-Si:H PIN cells respectively, we note that although in most cases, the initial J_{sc} in a-Si:H cells is higher, the stabilised efficiency is lower than in pm-Si:H cells. The reason for

higher current is the stronger absorption (higher coefficient of extinction – Fig. 1a) of long wavelength visible light in a-Si:H relative to pm-Si:H. On the other hand the stabilised dangling bond density in a-Si:H is one order of magnitude higher than in pm-Si:H (see Tab. 1) and this fact in conjunction with a lower hole mobility than in pm-Si:H brings down sharply the stabilised efficiency in a-Si:H PIN cells. In fact, we have been able to achieve 12% stabilised efficiency in pm-Si:H solar cells (Tab. 4), but not as yet in a-Si:H cells (Tab. 3) using parameters extracted by simulating our experimental results.

4.2 The window and emitter design

Figures 3a and 3b depict respectively typical band diagrams of a-Si:H and pm-Si:H PIN cells having a two-window P-μc-Si:H/P-μc-SiO$_x$:H structure and a

Table 5. Sensitivity of the pm-Si:H solar cell output to the band edge line-up between P-μc-SiO$_x$:H and pm-Si:H in cells employing a single window layer in the initial state. ΔE_c and ΔE_v are respectively the conduction band and valence band discontinuities and ΔE_μ the band gap discontinuity between the two. In Table 4 throughout we had assumed $\Delta E_c = \Delta E_v = \Delta E_\mu/2$.

ZnO:Al/P-μc-SiO$_x$/I-pm-Si:H/ N-μc-SiO$_x$/ZnO:Al/Ag/glass	V_{bi} (volts)	J_{sc} (mA cm^{-2})	V_{oc} (volts)	FF	Efficiency (%)
P-μc-SiO$_x$/pm-Si $\Delta E_c = \Delta E_v = \Delta E_\mu/2$	1.566	15.89	1.124	0.760	13.56
P-μc-SiO$_x$/pm-Si $\Delta E_c = 0$, $\Delta E_v = \Delta E_\mu = 0.54$ eV	1.814	16.26	1.162	0.198	3.74
P-μc-SiO$_x$/pm-Si $\Delta E_v = 0$, $\Delta E_c = \Delta E_\mu = 0.54$ eV	1.314	15.82	1.143	0.736	13.31

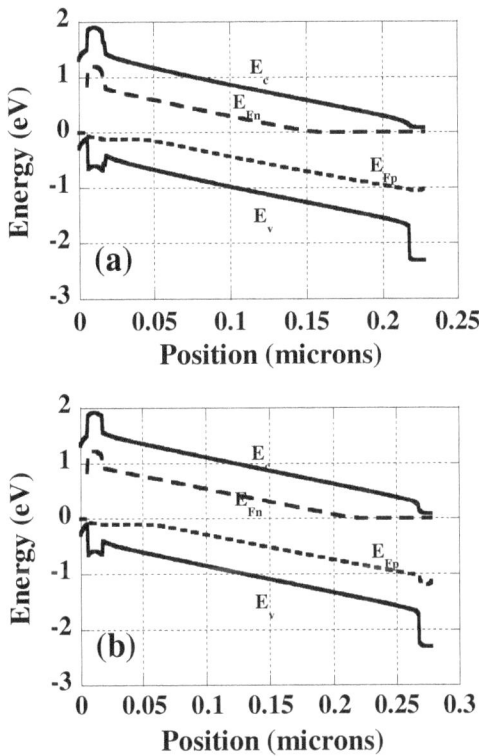

Fig. 3. Typical band diagrams under AM 1.5 light and short-circuit conditions in the annealed state of (i) a a-Si:H single junction solar cell having a P-μc-Si:H/P-μc-SiO$_x$:H double window design and a μc-SiO$_x$ N-layer; and (ii) a pm-Si:H solar cell with identical emitter and N-layer designs under the same conditions.

μc-SiO$_x$:H N-layer, under AM1.5 light illumination and short-circuit conditions, in the initial state. On the other hand, Figure 4b shows the EQE curves of a pm-Si:H single junction cell with the following doped layers: (i) P-μc-Si:H/P-a-SiC:H emitter layer and a-Si:H N-layer, (ii) P-μc-Si:H/P-μc-SiO$_x$:H and N-a-Si:H, (iii) P-μc-Si:H/P-μc-SiO$_x$:H and and N-μ-SiO$_x$:H and (iv) a single P-μc-SiO$_x$:H emitter with N-μ-SiO$_x$:H. Additionally, Figure 4c shows the effect of a MgF$_2$ anti-reflection coating (ARC) on the last case, while Figure 4d depicts the stabilised state EQE of Figure 4c. In Figure 5 we draw the EQE curves corresponding to the a-Si:H cells described in the second, third and fourth rows of Table 3 in the initial

(Fig. 5a) and the light-stabilised (Fig. 5b) states. A MgF$_2$ ARC on glass produces negligible improvement as seen by comparing the last two rows of Table 3; hence this case is not included in Figure 5.

In the results given in Tables 3 and 4, the P-a-SiC:H and I-a-Si:H or I-pm-Si:H band discontinuity is apportioned to the conduction band side, as inferred from reference [16]. Since there is no valence band discontinuity, the FF is better in this case than when P-μc-SiO$_x$:H is used. In the latter case, in the absence of any measured data (to our knowledge), we have apportioned half the band discontinuity to the valence band and this fact hinders hole collection and brings down the FF. Also the sbb is 0.16 eV for TCO/P-μc-Si:H [14]. As there are no experimental data for the TCO/P-μc-SiO$_x$:H band bending, we have taken the value of 0.2 eV (sixth row of Tab. 4), that is generally encountered at the TCO/P-a-SiC:H contact. The consequences on the solar cell output of changing the band edge line up and TCO/P-μc-SiO$_x$:H sbb are shown in Tables 5 and 6, respectively. Note that the effects of changing the band edge line-up and the sbb for the single P-μc-SiO$_x$:H design are shown only for the pm-Si:H PIN cell, since the variation is similar in a-Si:H cells, and, such a cell is more promising than an a-Si:H one, from the point of view of the stabilised efficiency. From Table 5, apportioning the entire band gap discontinuity to the valence band side leads to a higher built-in field and V_{oc}, but a strongly reduced FF, as the high ΔE_v hinders hole collection. However, in spite of ΔE_c being taken equal to zero for this case (a fact that may be expected to encourage electron back diffusion leading to a lower current), the current actually increases, indicating that the high V_{bi}, and consequently the higher field in this device, more than cancels the detrimental effect of $\Delta E_c = 0$ on J_{sc}. When the entire band gap discontinuity is on the conduction band side (row 3 of Tab. 5), no improvement in J_{sc} is seen over row 1 of the same table, indicating that even $\Delta E_c = \Delta E_\mu/2$ is sufficient to block the back diffusion of electrons. V_{bi} decreases for this case as expected, leading to a fall in the FF relative to row 1; surprisingly however V_{oc} increases. This fact may be explained as follows: as $\Delta E_v = 0$, and as we have assumed low I/P (P-layer deposited on top of I layer) interface defect density (our pm-Si:H cell has been deposited in a NIP configuration, so that the I/P interface is not damaged), negligible hole accumulation at this

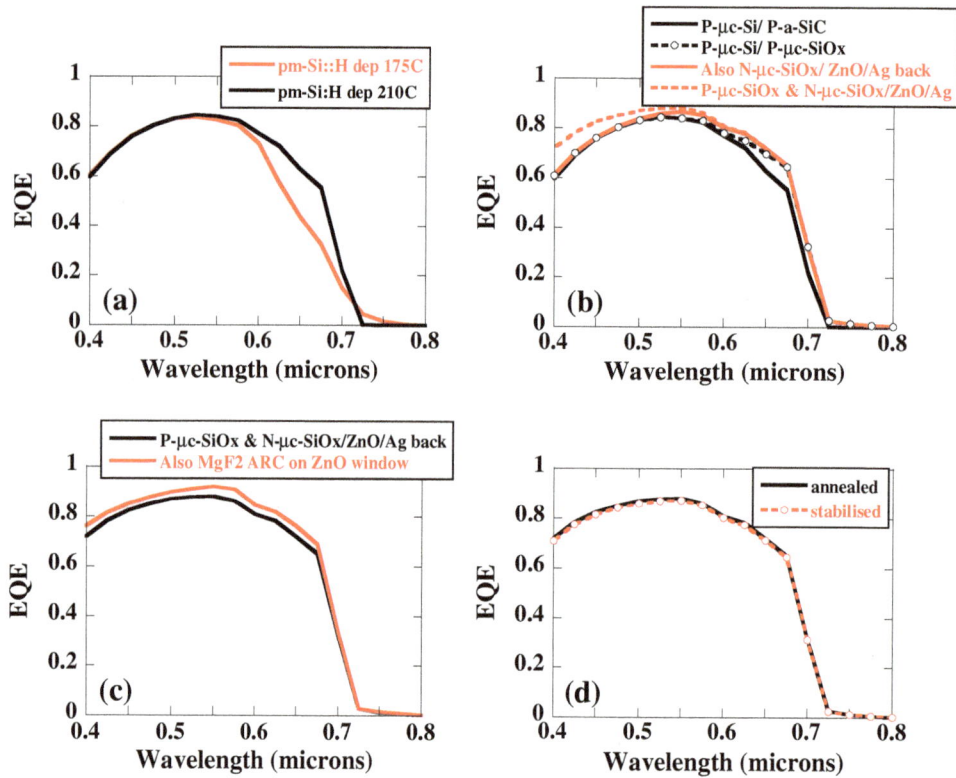

Fig. 4. Comparison of the external quantum efficiency (EQE) of pm-Si:H solar cells having (a) P-μc-Si:H/P-a-SiC:H double window design with an a-Si:H N-layer, but pm-Si:H intrinsic layers deposited at 175 °C and 210 °C in the initial state; (b) 4 types of solar cells in the initial state with P-μc-Si:H/P-a-SiC:H and P-μc-Si:H/P-μc-SiO$_x$ double window designs both with a-Si:H N-layers, the latter emitter design with N-a-Si:H replaced by N-μc-SiO$_x$ and the last design: with a single P-μc-SiO$_x$:H emitter; (c) the latter design in the initial state with and without a MgF$_2$ anti-reflection coating on the ZnO:Al window and (d) the ZnO:Al/P-μc-SiO$_x$/I-pm-Si:H/N-μc-SiO$_x$/ZnO:Al/Ag/glass cell in the initial and light-stabilised states.

Table 6. Sensitivity of the pm-Si:H solar cell output to the ZnO:Al/P-μc-SiO$_x$:H surface band bending. The case where the sbb = 0.2 eV (that is generally observed for ZnO:Al/P-a-SiC:H), is given in the first row of this table and assumed for the single window designs in Table 4. The front contact barrier height (φ_{b0}) for this case is 1.83 eV, since the band gap of P-μc-SiO$_x$ is taken to be 2.5 eV and its activation energy = 0.47 eV. Even for a slight increase of sbb – to 0.3 eV (φ_{b0} = 1.73 eV), V_{oc}, FF and the cell efficiency deteriorate appreciably (second line of table), as the V_{bi} falls. Any effort to increase the V_{bi} by increasing the P-μc-SiO$_x$:H thickness, brings down the J_{sc}. The third line indicates the effect of having φ_{b0} = 1.32 eV, the same as at the ZnO:Al/P-μc-Si:H interface of the double window design.

Cell type		J_{sc} (mA cm^{-2})	V_{oc} (volts)	FF	Efficiency (%)
pm-Si:H, ZnO:Al/P-μc-SiO$_x$:H sbb = 0.2 eV, φ_{b0} = 1.83 eV	Initial	15.89	1.124	0.760	13.56
pm-Si:H, ZnO:Al/P-μc-SiO$_x$:H sbb = 0.3 eV, φ_{b0} = 1.73 eV	Initial	15.88	1.039	0.732	12.07
pm-Si:H, ZnO:Al/P-μc-SiO$_x$:H sbb = 0.71 eV φ_{b0} = 1.32 eV	Initial	15.81	0.629	0.614	6.11

interface takes place for this case, leading to low interface field and more field in the bulk of the device that brings up V_{oc}.

Now comparing the results of Veneri et al. [17] to the solar cell output parameters of the a-Si:H cells given in Table 3, we note that our V_{oc}'s are considerably higher. Modelling also yields a higher V_{oc} than that achieved experimentally [19,23]. In such a heterojunction structure, where the P-μc-SiO$_x$ band gap is considerably higher than that of I-a-Si:H or I-pm-Si:H (Tab. 1), the band edge line

up has a strong influence on the FF (as seen from the modelling results in Tab. 5 for the case of a pm-Si:H cell, the effect on an a-Si:H cell is similar). Therefore, taking a comparison with the FF values given in reference [17], 19 and 23, no more than half the band gap discontinuity can lie on the valence band side, as assumed by us for all cases in Table 3. However the fact that our calculated V_{oc} is higher than the measured results [17,19,23], independently of this band edge line up (as shown in the example of pm-Si:H cells in Tab. 5), indicates that in the experimental cases,

Table 7. Optimisation of the thickness of the pm-Si:H absorber layer for a double P-layer window design: P-μc-Si:H/P-μc-SiO$_x$:H and N-μc-SiO$_x$/ZnO:Al/Ag back structure. Results indicate that any thickness between 250 nm and 150 nm is good with the thinnest pm-Si:H cell having an edge regarding the stabilised efficiency.

ZnO:Al/P-μc-Si/P-μc-SiO$_x$/I-pm-Si:H/N-μc-SiO$_x$/ZnO:Al/Ag/glass		J_{sc} (mA cm^{-2})	V_{oc}(volts)	FF	Efficiency (%)
Thickness = 250 nm	Initial	15.29	1.140	0.772	13.45
	Stabilised	15.15	1.073	0.701	11.40
Thickness = 300 nm	Initial	15.48	1.138	0.767	13.51
	Stabilised	15.29	1.065	0.685	11.16
Thickness = 200 nm	Initial	14.99	1.143	0.775	13.29
	Stabilised	14.90	1.082	0.715	11.53
Thickness = 150 nm	Initial	14.61	1.145	0.781	13.07
	Stabilised	14.56	1.094	0.731	11.64

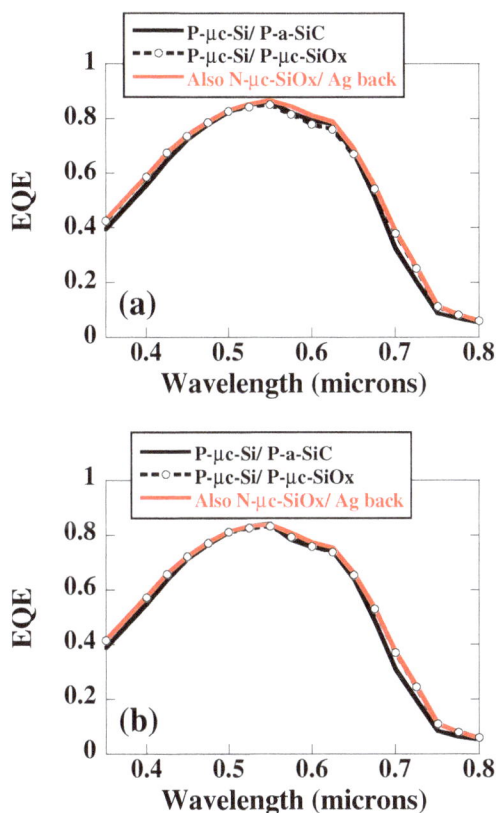

Fig. 5. Comparison of the EQE curves of a-Si:H solar cells in the (a) initial and (b) light stabilised states having different window designs and N-layer: with P-μc-Si:H/P-a-SiC:H and P-μc-Si:H/P-μc-SiO$_x$:H double window designs both with a-Si:H N-layers and the latter emitter design with N-a-Si:H replaced by N-μc-SiO$_x$.

the P/I interface is more defective than assumed in our calculations, and points to the fact that the V_{oc} and the cell efficiency may be considerably improved by improving the quality of the P/I interface.

We find from Tables 3 and 4 and Figures 4b and 5, that in both a-Si:H and pm-Si:H cells there is an increase in the current density when the P-a-SiC:H layer is substituted by P-μc-SiO$_x$:H. This fact is due to the lower absorption coefficients of P-μc-SiO$_x$ that is linked to the imaginary part

of the CRIND's input to the program – Figure 1c (the real part of the CRIND's of the different P-layers studied are plotted as a function of wavelength in Fig. 1d). The EQE curves indicate that it is mainly the increase in the long wave length response that contributes to the current increase. The absence of an appreciable increase in the short wavelength EQE is due to the fact that P-μc-SiO$_x$, being micro-crystalline, has been assumed thicker than the amorphous P-a-SiC:H layer, so that a possible increase in the blue response due to lower absorption coefficients of the former is annulled. Our EQE curves in Figure 5a for a PIN a-Si:H cell both having a P-μc-SiO$_x$ window layer but with the back contact design N-a-Si:H/ZnO:Al/Ag in one case (black dotted line) and N-μc-SiO$_x$/Ag in the other (red line) agree quite well with the EQE curves of Veneri et al. [17] ($R = 0$ N-μc-Si/ZnO/Ag and N-μc-SiO$_x$/Ag back contacts) except for wavelengths in the range of $0.65\,\mu m \leqslant \lambda \leqslant 0.7\mu m$, where our EQE curves are lower in both cases. This may be due to lower values of the imaginary part of the CRIND's for those wavelengths for the a-Si:H material developed in our laboratory.

Also from Table 4 and Figure 4b, we note that for the single window design with P-μc-SiO$_x$:H, the current is appreciably higher, since P-μc-Si:H is absent, but V_{oc} and FF are lower (because now we cannot take advantage of the lower E_{ac} and sbb of P-μc-Si:H) yielding nearly the same initial and stabilised efficiency. However in the case of the single P-μc-SiO$_x$:H window layer the uncertainty of the value of the TCO/P-μc-SiO$_x$:H sbb may be a crucial limiting factor as is evident from Table 6. Due to the large sensitivity to the TCO/P-μc-SiO$_x$:H sbb about which little is known at the moment and because, as seen from Table 4, there is negligible difference between the initial and stabilised efficiencies for the double window structure – ZnO:Al/P-μc-Si:H/P-μc-SiO$_x$:H, and the single window structure – ZnO:Al/P-μc-SiO$_x$:H (when this sbb is taken =0.2 eV), it was decided to choose the former design to optimise the thickness of the pm-Si:H film grown at 210 °C. This is because for this case we have a fairly good idea the ZnO:Al/P-μc-Si:H sbb [14]. The results of thickness optimisation are given in Table 7 and show that any thickness between 250 nm and 150 nm is acceptable, with the thinnest pm-Si:H cell having a small advantage regarding the stabilised efficiency.

Finally a MgF$_2$ ARC deposited on ZnO:Al, through which light enters the pmSi:H cell with the NIP configuration on glass (Fig. 2b), strongly enhances J_{sc} (Tab. 4) and the EQE (Fig. 4c). However, only a small improvement is observed in the case of the a-Si:H cell in the PIN configuration with MgF$_2$ deposited on glass (Tab. 3).

4.3 The N-side design

Tables 3 and 4 and the EQE curves of Figures 4b and 5 show the effect of substituting N-μc-SiO$_x$:H in place of N-a-Si:H. This brings up J_{sc} and the long wavelength EQE. For this layer the electrical and optical parameters have been taken from reference [17] in our simulation. This material with the real part n, of the complex refractive index considerably lower than that of N-a-Si:H and close to that of ZnO:Al (Fig. 1f), appreciably improves the long wavelength EQE. Indeed most of the light is now reflected at the I-layer/N-μc-SiO$_x$ interface and therefore does not need to cross the N-layer twice. Also it is nearly as transparent as ZnO:Al, with values of the extinction coefficient close to those of the latter (Fig. 1e). Since as well, its doping can be adjusted to make it highly conducting ($E_{ac} = 0.09$ eV, lower than that of N-a-Si:H – Tab. 1), it can effectively act both as the TCO and the N-layer of the structure [17]; also leading to improvements in V_{oc} and FF. Moreover, the deposition of an extra TCO layer may be avoided in this case. However the pm-Si:H cell has to be deposited in a NIP configuration, to avoid a very damaged I/P interface and macroscopic defects as already stated. Here a thin layer of ZnO:Al needs to be deposited on silver, to prevent the latter diffusing into the N-layer when the N-side of the junction is bombarded by a high flux of hydrogen required to form pm-Si:H. Thus in pm-Si:H cells, although the deposition of an extra TCO layer cannot be avoided, the use of N-μc-SiO$_x$:H instead of N-a-Si:H is still advantageous as can be seen from the 5th row of Table 4, and Figure 4b.

5 Conclusions

The possibility of single junction pm-Si:H solar cells with stabilised efficiencies around 12% has been demonstrated by modelling. We have shown that in order to achieve such efficiencies the pm-Si:H cell should be deposited in the NIP rather than the PIN configuration to avoid a heavily damaged P/I interface that is much more harmful for the solar cell efficiency than a similarly damaged N/I interface; and to avoid formation of macroscopic defects on the P-layer side. However light should enter the cell through the P-layer. A P-μc-SiO$_x$:H window layer, a suitably N-doped μc-SiO$_x$:H, as well as a MgF$_2$ ARC on ZnO:Al are required to achieve this goal. On the contrary, using the initial and stabilised state properties and interface parameters of the best a-Si:H material and solar cell developed in our laboratory so far, we could not achieve 10% stabilised efficiency in a-Si:H PIN solar cells, as the

stabilised DOS in this material is high and its hole mobility considerably lower than in pm-Si. However such a landmark has been achieved in a few other laboratories e.g. reference [18], probably due to better light absorption in their material, improved P/I interface properties in solar cells fabricated with this material, etc.

The calculated solar cells have a lower FF but a higher V_{oc} when P-μc-SiO$_x$:H forms part of the window layer than for the P-a-SiC:H case. The higher V_{oc} is due to a higher built-in field in the former case, while the lower FF for P-μc-SiO$_x$ is due to an assumed $\Delta E_v = 0.26$ eV, that hinders hole collection, while ΔE_v is $\sim$0 for P-a-SiC:H. The solar cells using P-μc-SiO$_x$:H show higher current density, due to the lower coefficient of extinction of this material compared to P-a-SiC:H. Also modelling shows that highly conducting and transparent N-μc-SiO$_x$:H can act both as the N-layer of the structure and the TCO. In fact this material with a refractive index considerably lower than that of N-a-Si:H and close to that of ZnO:Al, appreciably improves the long wavelength QE, since most of the light is now reflected at the I-layer/N-μc-SiO$_x$ interface and does not need to cross the N-layer twice. The improvements in the solar cell output parameters due to the use of the P- and N-layers described above and as deduced from modelling will be applicable independent of the absorber material used in the PIN solar cell.

In our calculations, there is uncertainty in the calculated solar cell output, due to the limited knowledge on the band edge line up of P-μc-SiO$_x$ with I-a-Si:H or I-pm-Si:H, and its surface band bending at the contact with ZnO:Al when used in the single window design. The effect of varying these parameters has been demonstrated. The latter uncertainty may be avoided by using the double window design. Also we have shown that ΔE_v cannot be higher than half the band gap discontinuity in order to attain the high experimental values of FF [17, 19, 23]. Moreover we infer that the V_{oc} in the deposited solar cells may be considerably increased by lowering the P/I interface defect density. We conclude, as Table 4 indicates, that $\sim$12% stabilised efficiency is achievable in thin pm-Si:H single junction solar cells (thickness of the I-layer $\leqslant$250 nm) with practical material and device parameters and a suitable anti-reflection coating.

This work was carried out in the framework of the FP7 project "Fast Track", funded by the EC under grant agreement No. 283501. The computer modelling code was developed (electrical part) by P. Chatterjee during projects funded by MNRE and DST, Government of India, and the optical part was added during her tenure as Marie Curie fellow at Ecole Polytechnique, Palaiseau, France.

References

1. R. Butté, R. Meaudre, M. Meaudre, S. Vignoli, C. Longeaud, J.P. Kleider, P. Roca i Cabarrocas, Phil. Mag. B **79**, 1079 (1999)
2. P.P. Ray, P. Choudhuri, P. Chatterjee, Thin Solid Films **403–4**, 275 (2002)

3. Baojie Yan, Guozhen Yue, L. Sivec, J. Yang, Subhendu Guha, C.-S. Jiang, Appl. Phys. Lett. **99**, 113512 (2011)

4. P. Roca i Cabarrocas, A. Fontcuberta i Morral, S. Lebib, Y. Poissant, Pure Appl. Chem. **74**, 359 (2002)

5. J.P. Kleider, C. Longeaud, M. Gauthier, M. Meaudre, R. Meaudre, R. Butté, S. Vignoli, P. Roca i Cabarrocas, Appl. Phys. Lett. **75**, 3351 (1999)

6. P. St'ahel, S. Hamma, P. Sladek, P. Roca i Cabarrocas, J. Non-Cryst. Solids **227–230**, 276 (1998)

7. P. Roca i Cabarrocas, Mater. Res. Soc. Symp. Proc. **507**, 855 (1998)

8. A. Fontcuberta i Morral, R. Brenot, E.A.G. Hamers, R. Vanderhaghen, P. Roca i Cabarrocas, J. Non-Cryst. Solids **266–269**, 48 (2000)

9. E.A. Schiff, J. Non-Cryst. Solids **352**, 1087 (2006)

10. M. Brinza, G. Adriaenssens, P. Roca i Cabarrocas, Thin Solid Films **427**, 123 (2003)

11. Y. Poissant, P. Chatterjee, P. Roca i Cabarrocas, J. Appl. Phys. **94**, 7305 (2003)

12. K.-H. Kim, E.V. Johnson, A. Abramov, P. Roca i Cabarrocas, Sol. Energy Mater. Sol. Cells **105**, 208 (2012)

13. N. Palit, P. Chatterjee, J. Appl. Phys. **86**, 6879 (1999)

14. W. Ma, T. Saida, C.C. Lim, S. Aoyama, H. Okamoto, Y. Hamakawa, in *Proceedings of the First World Conference on Photovoltaic Solar Energy Conversion, Hawaii* (IEEE, New York, 1994), p. 117

15. X. Xu, J. Yang, A. Banerjee, S. Guha, K. Vasanth, S. Wagner, Appl. Phys. Lett. **67**, 2323 (1995)

16. Z.Y. Wu, J.M. Siefert, B. Equer, in *Proc. 10th EU PVSEC, Lisbon, Portugal, 1991*, p. 953.

17. P.D. Veneri, L.V. Mercaldo, I. Usatii, Appl. Phys. Lett. **97**, 023512 (2010)

18. A. Lambertz, F. Finger, R.E.I. Schropp, U. Rau, V. Smirnov, Prog. Photovolt.: Res. Appl. **23**, 939 (2015)

19. K.-H. Kim, S. Kasouit, E.V. Johnson, P. Roca i Cabarrocas, Sol. Energy Mater. Sol. Cells **119**, 124 (2013)

20. P. Roca i Cabarrocas, J.B. Chévrier, J. Huc, A. Lloret, J.Y. Parey, J.P.M. Schmitt, J. Vac. Sci. Technol. A **9**, 2331 (1991)

21. Y.M. Soro, A. Abramov, M.E. Gueunier-Ferrat, E.V. Johnson, C. Longeaud, P. Roca i Cabarrocas, J.P. Kleider, J. Non-Cryst. Solids **354**, 2092 (2008)

22. S. Vignoli, R. Meaudre, M. Meaudre, Phil. Mag. B **73**, 261 (1996)

23. S.N. Abolmasov, H. Woo, R. Planques, J. Holovský, E.V. Johnson, A. Purkrt, P. Roca i Cabarrocas, Eur. Phys. J. Photovolt. **5**, 55206 (2014)

24. P. Chatterjee, M. Favre, F. Leblanc, J. Perrin, Mater. Res. Soc. Symp. Proc. **426**, 593 (1996)

25. N. Palit, P. Chatterjee, Sol. Energy Mater. Sol. Cells **53**, 235 (1998)

26. Madhumita Nath, P. Chatterjee, J. Damon-Lacoste, P. Roca i Cabarrocas, J. Appl. Phys. **103**, 034506 (2008)

27. P. Chatterjee, J. Appl. Phys. **76**, 1301 (1994)

28. P. Chatterjee, J. Appl. Phys. **79**, 7339 (1996)

29. P.J. McElheny, J.K. Arch, H.-S. Lin, S.J. Fonash, J. Appl. Phys. **64**, 1254 (1988)

30. R. Stangl, M. Kriegel, K. v. Maydell, L. Korte, M. Schmidt, W. Fuhs, in *Conference record 31st IEEE Photovoltaic Specialists Conference, 2005*, p. 1556

31. R. Varache, J.P. Kleider, W. Favre, L. Korte, J. Appl. Phys. **112**, 123717 (2012)

32. M.J. Powell, S.C. Deane, Phys. Rev. B **48**, 10815 (1993)

33. M.J. Powell, S.C. Deane, Phys. Rev. B **53**, 10121 (1996)

34. V. Halpern, Phil. Mag. B **54**, 473 (1986)

35. C. Longeaud, J.P. Kleider, Phys. Rev. B **48**, 8715 (1993)

36. F. Leblanc, J. Perrin, J. Schmitt, J. Appl. Phys. **75**, 1074 (1994)

25

Porous (001)-faceted anatase TiO$_2$ nanorice thin film for efficient dye-sensitized solar cell

Athar Ali Shah, Akrajas Ali Umar[a], and Muhamad Mat Salleh

Institute of Microengineering and Nanoelectronics, Universiti Kebangsaan Malaysia, 43600 UKM Bangi, Selangor, Malaysia

Abstract Anatase TiO$_2$ structures with nanorice-like morphology and high exposure of (001) facet has been successfully synthesized on an ITO surface using ammonium Hexafluoro Titanate and Hexamethylenetetramine as precursor and capping agent, respectively, under a microwave-assisted liquid-phase deposition method. These anatase TiO$_2$ nanoparticles were prepared within five minutes of reaction time by utilizing an inverter microwave system at a normal atmospheric pressure. The morphology and the size (approximately from 6 to 70 nm) of these nanostructures can be controlled. Homogenous, porous, 5.64 ± 0.002 μm thick layer of spongy-nanorice with facets (101) and (001) was grown on ITO substrate and used as a photo-anode in a dye-sensitized solar cell (DSSC). This solar cell device has emerged out with $4.05 \pm 0.10\%$ power conversion efficiency (PCE) and 72% of incident photon-to-current efficiency (IPCE) under AM1.5 G illumination.

1 Introduction

TiO$_2$ nanostructure with a larger surface area is ideal for solar cells [1–3], photolysis [4], sensors [5] and photocatalytic applications [4, 6, 7], as it improves the charge-transfer reaction, enhances the redox potential of photogenerated electrons and holes, and reduces the electron-hole recombination. For a solar cell application, TiO$_2$ with large surface area provides many active centers for reagent adsorption and reaction, improves dye molecules loading and facilitates facile electrolyte diffusion, leads to a facile electron transport in the device [8–12].

In a dye-sensitized solar cell (DSSC), anatase is the TiO$_2$ polymorph that shows an intriguing performance [13–17]. Since many surface reaction favours to occur at the high-energy site, such as defect, twinning or kinks [18, 19], to synthesize anatase TiO$_2$ nanostructures having such structural properties promises enhanced performance in applications. Moreover, anatase TiO$_2$ with high-energy plane, such as (001), and anisotropic-shape (such as nanorice) [20], containing high-surface area and high-defect further promotes active surface reaction and facile electron transfer in the device [5, 21]. Thus, high-performance solar cell or photocatalysis can be obtained from the structure.

In this paper, we present a straightforward method to prepare anatase TiO$_2$ nanorice with a large-area of high-energy plane of (001) containing high-surface defect via a microwave assisted liquid-phase deposition method. In typical procedure, TiO$_2$ nanorice (size in the range of 6 to 70 nm) with high-density (thickness of approximately 5.64 ± 0.002 μm) can be successfully grown on an ITO substrate surface via this method using a growth solution containing TiO$_2$ precursor and hexamethylenetetramine (HMT). The performance of the TiO$_2$ nanorice in DSSC has been examined. Power conversion efficiency and incident photon to current efficiency as high as $4.05 \pm 0.10\%$ and 72%, respectively, can be achieved so far. The performance of the device could be further enhanced via TiO$_2$ nanostructure properties as well as device properties improvements. The porous TiO$_2$ anatase should find a potential used in solar cell and photocatalysis applications.

2 Experimental

2.1 Synthesis and characterization of TiO$_2$ nanorice on an ITO substrate

The TiO$_2$ nanorices were synthesized on an ITO substrate by using a microwave-assisted liquid phase deposition method [22–24]. In typical process, the TiO$_2$ nanorices were prepared by immersing a cleaned ITO substrate (sheet resistance ca. 9–22 Ω/cm^2 purchased from VinKarola instuments USA) which was previously cleaned via an ultrasonication for 30 min in acetone

---

[a] e-mail: `akrajas@ukm.edu.my`

and ethanol into a 10 mL of aqueous solution containing equimolar (0.05 M) solution of ammonium hexafluoro titanate ($(NH_4)_2TiF_6$) (AHT) and hexamethylenetetramine (HMT). Both chemical reagents were purchased from Sigma-Aldrich, USA, and used directly without any purification process. To obtain an optimum porosity properties and high percentage of (001) lattice plane of anatase TiO_2, the HMT concentration was varied from 0.03 to 0.08 M. In this case, the TiO_2 precursor molarity was fixed at 0.05 M. The growth time was 5 min, meanwhile the microwave power used was 180 W. After a growth process, the sample was taken out from the solution and rinsed with copious amount of pure water and dried under a nitrogen gas flow. Finally, the substrate was annealed in air at 350 °C for an hour.

Field Emission Scanning Electron Microscope (FE-SEM) technique (ZeiSS SUPRA 55VP) was used for examining the surface morphology of the sample. Meanwhile, the crystallinity of the nanostructure was examined via a high resolution transmission electron microscopy (HRTEM) analysis using Ziess Libra 200FE HRTEM apparatus operating at 200 kV. The X-ray diffraction spectroscopy (BRUKER D8 Advance with $CuK\alpha$ radiation and scan step as low as 2°/min) and the UV/VIS spectrometer (Lambda 900 Perkin-Elmer) were used to confirm the structure and the phase, and the optical properties of the sample, respectively.

2.2 Fabrication of dye sensitized solar cell and characterization

DSSC with a structure of ITO|TiO_2:dye|electrolyte|Pt electrode was fabricated utilizing the TiO_2 nanorice as the photo active layer. Prior to the device fabrication, a TiO_2 nanostructures modified-ITO substrate was immersed into a 0.05 mM ethanolic solution of dye (N719, purchased from Sigma-Aldrich, USA) for 12 h. It was then gently rinsed with ethanol and dried using a flow of nitrogen gas. For simplicity, we called this structure as as photoanode. A counter electrode was prepared by depositing a platinum layer of approximately 150 nm thickness on glass substrate via a sputtering method. A DSSC was assembled by clamping a photoanode and a counter electrode together. An iodide/tridiode redox couple (Iodolyte AN-50, purchased from Solaronix Switzerland) was used as the electrolyte and injected into the space between the photoanode and Pt counter electrode. The active area of the DSSC device was controlled at 0.24 cm^2.

The photovoltaic responses (I-V and incident photon to current efficiency (IPCE)) of the DSSC device was evaluated using a Keithley high-voltage source-measure unit (SMU) model 237 under AM 1.5 simulated irradiation (100 mW/cm^2) provided by 150 W Newport low-cost solar simulator. The photovoltaic properties of the DSSC device was characterized via an electrochemical impedance spectroscopy method using Solartron 1260 under a frequency range of 0.01 to 1 MHz, bias voltage at 0.5 V, and alternating current amplitude of 50 mA. The current amplitude is required to be higher in this case to accelerate the response

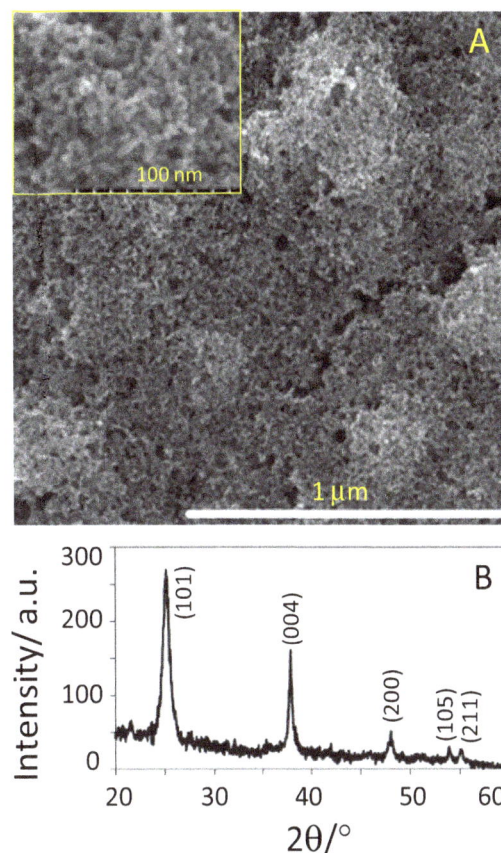

Fig. 1. (A) Typical FESEM image of TiO_2 nanorice grown on an ITO substrate prepared using a growth solution containing 0.05 M of ammonium hexafluoro titanate (AHT), and 0.08 M HMT. Inset shows detailed structure of nanorice. (B) XRD pattern of the TiO_2 nanorice on ITO substrate.

and to avoid electrolyte drying during the measurement. However, our device is stable at this high-current source.

The performance of the samples in the DSSC device was verified at least for five times and the uncorrected standard deviation of the measurement was used to validate the performance of the device.

3 Results and discussion

3.1 TiO$_2$ nanorice characterization

TiO_2 nanorices have been successfully grown directly on an ITO substrate using the present approach after following a growth process for 5 min in a growth solution containing ammonium hexafluoro titanate (AHT) and hexamethylenetetramine (HMT). In the typical process, TiO_2 nanorice with length-scale from 6 to 70 nm were formed on the surface of ITO with a thickness can be up to 6 μm. Figure 1A shows typical FESEM image of the TiO_2 nanorice prepared using equimolar, i.e. 0.05 M, solution of AHT and HMT. As Figure 1A shows, high-density networked-TiO_2 nanorice forms on the surface covering the entire

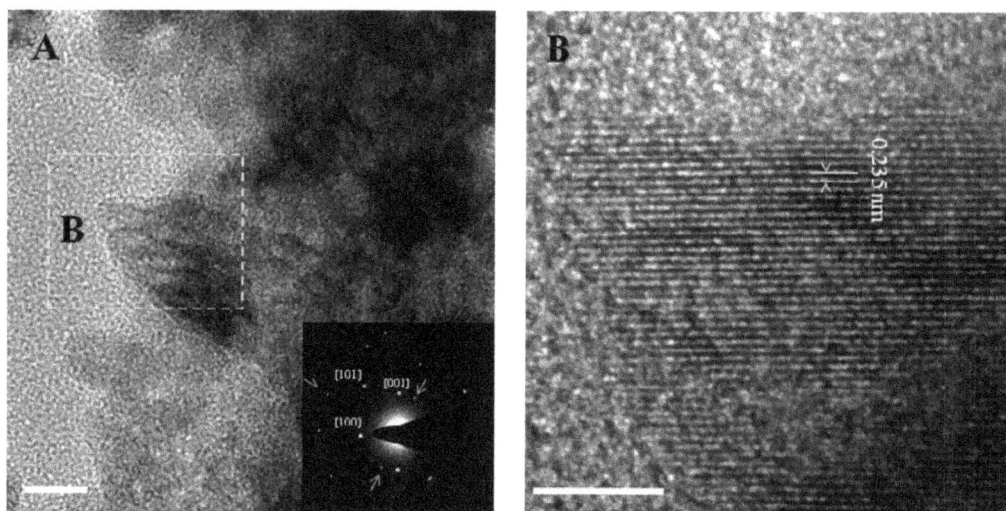

Fig. 2. (A) Low resolution, and (B) high resolution transmission electron microscope images of TiO_2 nanorice. The low resolution image verify nanorice morphology, and as well as SAED analysis (see inset in (A)), showing that nanorice are mono-crystalline. The high resolution image shows fringe spacing of 0.23 nm, reveals high exposure facet (001) with its growth along the [001] direction. Scale bars 10 nm in (A), and 5 nm in (B).

area of the substrate. Such networked-nanostructure produces a TiO_2 nanostructure films with a highly porous property (see inset in Fig. 1A), which is potential for solar cell application due to facilitating a high-dye loading and a facile diffusion of redox species on the surface of the nanostructure. As have been mentioned earlier, the networked-nanostructure is composed of nanoparticles with morphology resembles the rice shape. Probably due to a process of surface energy minimalization, they are connected each other, reflecting the individual TiO_2 nanorice bounded by a high-energy lattice plane. Figure 1B shows the corresponding X-ray diffraction spectrum of the samples. By comparing with the standard powder diffraction file for anatase TiO_2 (JCPDS file no. 21-1272), the obtained result is confirmed to be an anatase polymorph of TiO_2. By comparing with the JCPDS file, the TiO_2 nanorice's XRD spectrum peaks can be labeled as (101), (004), (200), (105) and (211) for peaks at 2θ of 25.5, 37.8, 48.2, 54.0 and 55.0°, respectively. One important fact that can be noted from the result is the peaks ratio between (004) and (101) is quite high, i.e. 0.7, which is much higher compared to normal anatase nanostructure (approximately ranging from 0.2 to 0.4). This reflects that the anatase TiO_2 nanorice is characterized by dominant (001) lattice plane, the second highest in the surface energy. Thus, we expect that enhanced performance in applications, such as solar cell and photocatalysis, can be obtained from this new TiO_2 nanostructure.

The TEM analysis of TiO_2 nanorice structure is presented in Figure 2. A low resolution TEM image shown in Figure 2A verifies the morphology attained by FESEM results. A high resolution TEM analysis highlights the defect-less, smooth, and twinning-less lattice fringes with a spacing approximately 0.235 nm (see Fig. 2B), which reveals that the nanorice are single crystalline in nature, with their unidirectional growth on ITO substrates. This

fringe spacing is corresponding to the facet (001), which is in a good agreement with the XRD results. The selected area electron diffraction (SAED) analysis of the nanorice (see inset in Fig. 2A) suggests an overlapping of two TiO_2 nanorice structures, which is depicted by two sets of bright spots, one of them is with high brightness (indexed diffraction pattern) and another with low brightness (pointed with arrows). The brighter set seems to be correspondent to the crystal at the top. The dimmer set can be correspondent to the crystal of TiO_2 nanorice placed at the bottom, as the image is due to diffraction of low energy scattered electrons or may be due to deviation from exact Bragg conditions, such as tilting of crystal (1–3°) and excitation errors. Nevertheless, it confirms that the nanorice is characterized by (001) high-energy lattice plane. A large exposure of high energy facet (001) can play a prominent role in the applications involving photolysis, catalysis and solar cells.

Under normal liquid-phase deposition method, which uses AHT and boric acid as the growth solution, continuous films of TiO_2 is obtained. In the present approach, while microwave energy applications only play a limited role in modifying nanocrystal growth morphology in the case of ZnO nanostructures [24, 25], in good agreement with the reported result by Parmar et al. [20] the microwave induces an anisotropic crystal growth in TiO_2, particularly nanorice-like morphology. By comparing the results obtained by them, which used acetylacetonate to decouple the hydrolysis and polycondensation of Ti ions with the result presented in this work, we remarked that the microwave energy likely induces an anisotropic stress and strain in the nanocrystallite and promotes the formation of anisotropic nanorice of TiO_2. And in the presence of the surfactant (HMT) here via an effective adhesion of its active amine functional onto the TiO_2 nanocrystallite, presumably on (001) plane, the nanorice of anatase

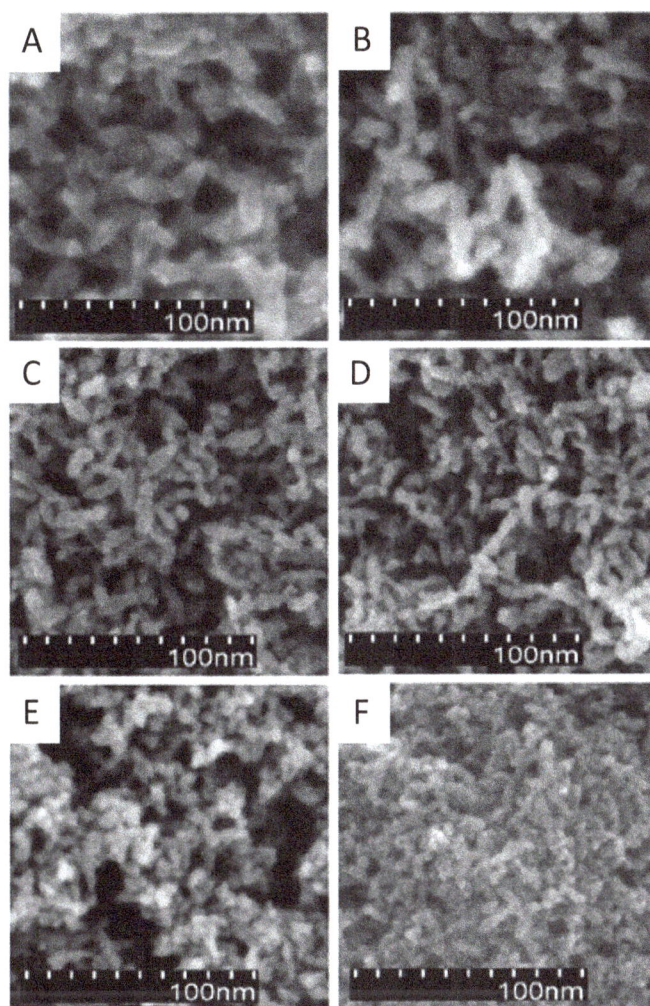

Fig. 3. FESEM of TiO_2 nanorice structures grown on ITO substrate prepared using different HMT concentrations, namely 0.03 (A), 0.04 (B), 0.05 (C), 0.06 (D), 0.07 (E) and 0.08 M (F). $(NH_4)_2TiF_6$ or (AHT) is fixed at 0.05 M. The growth time is 5 min.

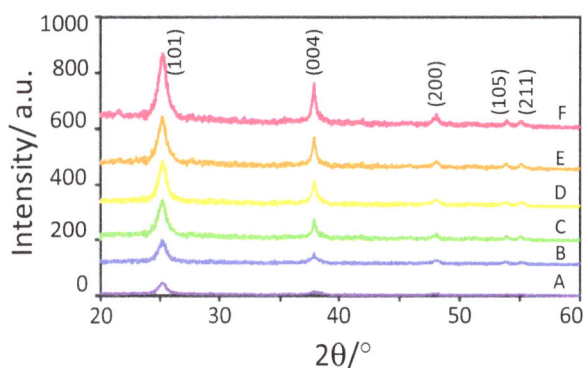

Fig. 4. XRD spectra of TiO_2 nanorice prepared using different HMT concentrations, namely 0.03 (A), 0.04 (B), 0.05 (C), 0.06 (D), 0.07 (E) and 0.08 M (F). AHT was fixed at 0.05 M.

Table 1. TiO_2 nanorice prepared using different HMT concentrations with AHT fixed at 0.05 M.

S. label	HMT(M)	Length (nm)	Width (nm)
A	0.03	70	27
B	0.04	45	18
C	0.05	27	11
D	0.06	24	9
E	0.07	20	7
F	0.08	16	6

TiO_2 that is bounded by this plane is realized. Such feature has also been found in the one-dimensional crystal growth in the case of ZnO [26, 27] and platelet [28, 29], brick-shape [30], nanofibrous [31, 32] and multipods [33] nanocrystal in the case of Au, Pt and Pd. Therefore, in order to obtain the extent role of HMT in the formation of (001)-faceted TiO_2 nanorice, we examined the nanocrystal growth properties under a different HMT concentration. The results are shown in Figure 3. As can be seen from the FESEM results, the length and the diameter of the nanorice decrease with the increasing of HMT concentration. This result reveals that the nature of nanorice packing and density can be controlled on the surface, which the density is increasing with the decreasing of nanorice dimension. Interestingly, from the figure, it was found that, although there is the change in the nanorice dimension, however, the aspect ratio; the length to diameter ratio, was relatively unchanged, namely 2.5. It was also observed that the morphology of the nanorice is unchanged with the

variation in the HMT concentration. Despite no morphological modification, however, the change in the dimension as well as the nature of nanorice assembly on the surface may have produced novel properties for enhanced-performance in solar cell application. Table 1 summarizes the dimension of the nanorice prepared from several HMT concentrations with AHT concentration was fixed at 0.05 M.

While the morphology of the nanorice is relatively unchanged upon variation of HMT concentration, the crystalline properties of the samples were also evaluated by using the XRD analysis. The result is shown in Figure 4. As Figure 3 reveals, the crystallographic orientation, i.e. the lattice plane preference, is also found to be unchanged with the variation of HMT in the growth solution. Nevertheless, it was found the peaks intensity of X-ray diffraction from prominent lattice plane increases with the decreasing of nanorice dimension (HMT increasing), while, the full-width at half-maximum (FWHM) decreases with the decreasing of dimension. This reflects that the shrinking in the nanorice dimension might have improved the surface area of particular lattice plane. Thus, novel and enhanced properties are expected to be produced from the nanostructures. Figure 5 shows the optical absorption spectra of the samples shown in Figure 3. In good agreement with the XRD results, the absorbance of the nanorice film effectively increases with the decreasing of the nanorice dimension. Judging from the FESEM results as shown in Figure 3, the increasing in the absorbance upon the decreasing in the nanorice dimension is resulted from the improvement of nanorice assembly, namely become more compact if the nanorice dimension reduces.

Table 2. Photovoltaic parameter of DSSCs device utilizing TiO$_2$ nanorice with nanograin size variation.

Device	Conc. AHT: HMT (M)	V_{oc} (V)	IPCE	J_{sc} (mA/cm^2)	R_{CT} (Ω)	R_S (Ω)	η (%)	FF
A	0.05:0.03	0.56 ± 0.023	9	1.58 ± 0.24	1014.52	42.39	0.32 ± 0.055	0.36 ± 0.004
B	0.05:0.04	0.62 ± 0.009	34	9.95 ± 0.18	587.15	42.46	2.41 ± 0.14	0.39 ± 0.004
C	0.05:0.05	0.64 ± 0.01	42	13.5 ± 0.22	405.36	40.62	3.24 ± 0.055	0.38 ± 0.004
D	0.05:0.06	0.68 ± 0.01	54	15.34 ± 0.27	307.74	38.9	3.73 ± 0.037	0.4 ± 0.004
E	0.05:0.07	0.68 ± 0.01	61	14.87 ± 0.20	268.81	33.88	3.81 ± 0.033	0.4 ± 0.006
F	0.05:0.08	0.64 ± 0.01	70	16.67 ± 0.265	224.27	34.76	4.05 ± 0.10	0.38 ± 0.004

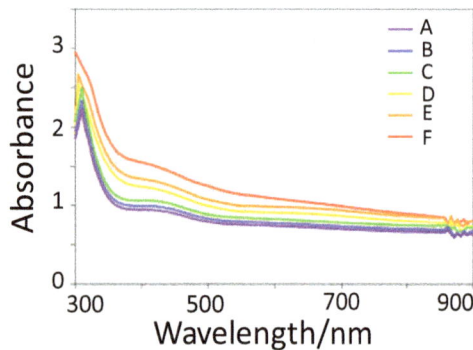

Fig. 5. Typical optical absorption spectra of TiO$_2$ nanorice prepare using different HMT concentrations.

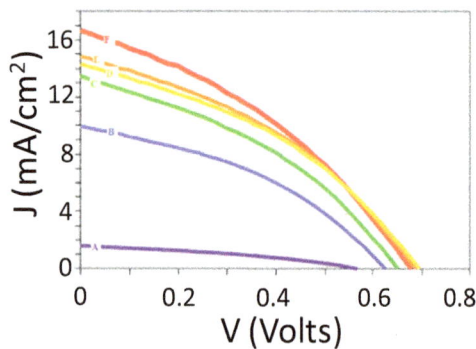

Fig. 6. J-V characteristic of the DSSCs utilizing photoanodes with different nanograin size, namely bigger grain size (A) to smaller (F), under A.M1.5, 100 W illumination.

A blue shift is also observed for the samples when the nanorice dimension reduced, which leads to the improvement of open-circuit voltage of the DSSC device [6]. Because of the nanorice assembly become more compact as the dimension reduced and considering the surface area of high-energy lattice plane increase, enhanced photoactivated surface reaction or charge-transfer [25] will be produced as the absorbance of the nanorice film increases with the decreasing of their dimension.

3.2 Solar cell characterization

A DSSC device with structure of ITO|TiO$_2$: dye (N719)|electrolyte (I^{3-}/I^{2-})|Pt was fabricated to evaluate the photovoltaic property of the new structure. Figure 6

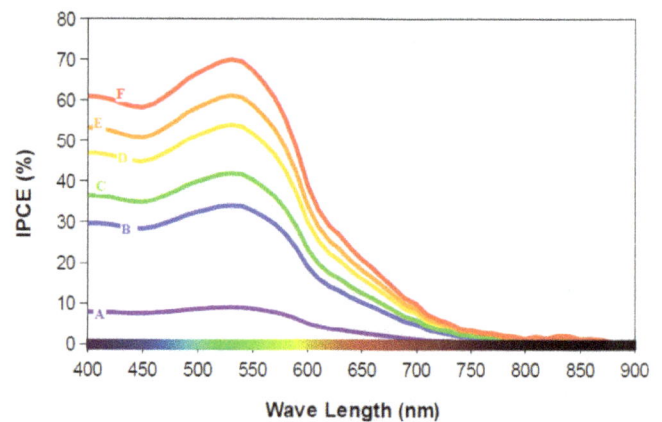

Fig. 7. Incident photon to current efficiency of the devices under A.M1.5, 100 W illumination.

shows typical J-V curve for the DSSC device that were fabricated using six different TiO$_2$ nanorice structures of which their images are shown in Figure 3. As can be seen from Figure 5, the DSSC performance increases with the decreasing of the nanorice dimension, for example, the short-circuit current density (J_{sc}) and the open circuit voltage (V_{oc}) of the device enhanced from 1.58 ± 0.24 mA/cm^2 and 0.56 ± 0.023 V for the high grain size nanorice (device A) to 16.67 ± 0.265 mA/cm^2 and 0.64 ± 0.00 V for small grain nanorice (dimension) (device F). The increase in the performance of the device with the decreasing of nanorice grain size can be attributed to the high photon absorption by the device and possible enhanced electron transport [34,35] as well as facile dye-TiO$_2$ charge transfer as the increase in the nanorice density and the surface are of high-energy (001) plane. The variation in the performance of the DSSC upon the variation of the nanorice size is unlikely related to the effect of surfactant because of the surfactant is seemed to be removed upon post-growth annealing at 350 °C for one hour. Therefore, it is clearly associated with the variation in the nanorice surface physico-chemistry. The photovoltaic parameters of the devices are summarized in Table 2.

IPCE responses of the devices as shown in Figure 7 further verifies such phenomenon. The increasing value of IPCE of the device with the reducing of nanograin size stamped the role of high energy facet (001) exposure to generate photo electrons, and facilitates facile electron transportation. Hence, J_{sc} is enhanced. As can be seen from Figure 5, the V_{oc} of the device increases

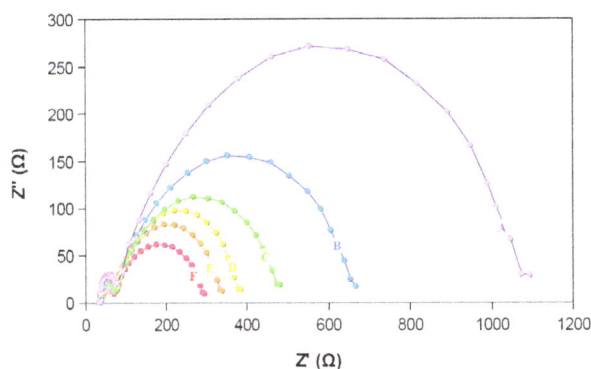

Fig. 8. Electrochemical impedance spectra (EIS) of the devices.

as the nanograin size decreases, reflecting the increasing of exciton lifetime or limited electron-hole recombination. Nevertheless, the device 'F' (device with the smallest nanograin size) shows a slight falls in V_{oc}, which can be attributed to the increasing of electron-hole recombination's rate [17, 36]. In spite of that fact, the device 'F' exhibits significant rise in J_{sc}, the result of enhanced exciton formation and facile electron transportation in the device probably due to greater exposure of high energy surface area.

Electrochemical impedance spectroscopy (EIS) study of the devices (A-F) has explained such transportation of electrons and electron-hole recombination's natures in the device. The results are shown in Figure 8. It is observed that device 'A' with high V_{oc} but low in fill factor, IPCE, and current density has higher value of charge transfer resistance (R_{CT}) in the region of Dye: TiO2|electrolyte interface, when compared with other devices 'B-F', rendering higher rate of recombination and weak charge transportation [2]. This is due to less reactive of (001) facet because of bulkier dimension. The R_{CT} decreased to a great extent when the nanograin size reduce, as per expectation due to more (001) facet exposure. This may improve dye adsorption and interconnection between these nanoparticles [37]. Thus, the PCE increased.

4 Conclusions

Thin films of anatase TiO_2 with nanorice morphology and rich of (001) facet has been successfully synthesized directly on an ITO substrate surface via a liquid-phase deposition method using a growth solution containing hexamethylenetetramine (HMT) and ammonium hexafluoro titanate (AHT) under a microwave irradiation. The size of the nanorice as well as the basal plane of the nanocrystal can be finely controlled by varying the concentration of HMT in the reaction. It was found that the performance of the dye-sensitized solar cell was improved if the fraction of (001) facet in the nanostructures was increased. It was found that the performance of the DSSC device increases with the decreasing of nanorice dimension. The optimum device demonstrates the power conversion efficiency as high as $4.05 \pm 0.10\%$ with internal quantum

efficiency (IPCE) as high as 72%. The decreasing in the nanorice dimension is predicted to increase the (001) facet exposure, enhancing the photoactivity, surface reactivity and electron transport in the device.

The authors would like to acknowledge the Ministry of Higher Education (MOHE), Malaysia for funding this work under research grants FRGS/1/2013/SG02/UKM/02/8 and HiCOE Project. The authors are also grateful for the financial support received from Ministry of Science, Technology and Innovation (MOSTI), Malaysia for the funding under Science Fund Grant (06-01-02-SF1157).

References

1. M. Graetzel, R.A.J. Janssen, D.B. Mitzi, E.H. Sargent, Nature **488**, 304 (2012)
2. M. Gratzel, Nature **414**, 338 (2001)
3. A. Ali Umar, S. Nafisah, S.K. Md Saad, S. Tee Tan, A. Balouch, M. Mat Salleh, M. Oyama, Sol. Energy Mater. Sol. C **122**, 174 (2014)
4. A.L. Linsebigler, G. Lu, J.T. Yates, Chem. Rev. **95**, 735 (1995)
5. W.-J. Ong, L.-L. Tan, S.-P. Chai, S.-T. Yong, A.R. Mohamed, Nanoscale **6**, 1946 (2014)
6. H. Ariga, T. Taniike, H. Morikawa, M. Tada, B.K. Min, K. Watanabe, Y. Matsumoto, S. Ikeda, K. Saiki, Y. Iwasawa, J. Am. Chem. Soc. **131**, 14670 (2009)
7. K. Lee, D. Kim, P. Roy, I. Paramasivam, B.I. Birajdar, E. Spiecker, P. Schmuki, J. Am. Chem. Soc. **132**, 1478 (2010)
8. J.R. Jennings, A. Ghicov, L.M. Peter, P. Schmuki, A.B. Walker, J. Am. Chem. Soc. **130**, 13364 (2008)
9. A.J. Frank, N. Kopidakis, J. van de Lagemaat, Coordin. Chem. Rev **248**, 1165 (2004)
10. E.J.W. Crossland, N. Noel, V. Sivaram, T. Leijtens, J.A. Alexander-Webber, H.J. Snaith, Nature **495**, 215 (2013)
11. C. Ducati, Nature **495**, 180 (2013)
12. E.L. Crepaldi, G.J.D.A.A. Soler-Illia, D. Grosso, F. Cagnol, F. Ribot, C. Sanchez, J. Am. Chem. Soc. **125**, 9770 (2003)
13. H. Zhang, Y. Wang, P. Liu, Y. Han, X. Yao, J. Zou, H. Cheng, H. Zhao, ACS. Appl. Mater. Interfaces **3**, 2472 (2011)
14. W.-Q. Wu, B.-X. Lei, H.-S. Rao, Y.-F. Xu, Y.-F. Wang, C.-Y. Su, D.-B. Kuang, Sci. Rep.-Uk **3**, 1352 (2013)
15. N. Wu, J. Wang, D.N. Tafen, H. Wang, J.-G. Zheng, J.P. Lewis, X. Liu, S.S. Leonard, A. Manivannan, J. Am. Chem. Soc. **132**, 6679 (2010)
16. A. Staniszewski, S. Ardo, Y. Sun, F.N. Castellano, G.J. Meyer, J. Am. Chem. Soc. **130**, 11586 (2008)
17. W. Shao, F. Gu, L. Gai, C. Li, Chem. Commun. **47**, 5046 (2011)
18. L.M. Falicov, G.A. Somorjai, Proc. Natl. Acad. Sci. **82**, 2207 (1985)
19. C.-Y. Chiu, P.-J. Chung, K.-U. Lao, C.-W. Liao, M.H. Huang, J. Phys. Chem. C **116**, 23757 (2012)
20. K.P.S. Parmar, E. Ramasamy, J. Lee, J.S. Lee, Chem. Commun. **47**, 8572 (2011)
21. C.-W. Peng, T.-Y. Ke, L. Brohan, M. Richard-Plouet, J.-C. Huang, E. Puzenat, H.-T. Chiu, C.-Y. Lee, Chem. Mater. **20**, 2426 (2008)

22. A.A. Umar, M.Y.A. Rahman, S.K.M. Saad, M.M. Salleh, M. Oyama, Appl. Surf. Sci. **270**, 109 (2013)

23. F.K.M. Alosfur, M.H.H. Jumali, S. Radiman, N.J. Ridha, M.A. Yarmo, A.A. Umar, Nanoscale Res. Lett. **8**, 1 (2013)

24. N.J. Ridha, A.A. Umar, F. Alosfur, M.H.H. Jumali, M.M. Salleh, J. Nanosci. Nanotechnol. **13**, 2667 (2013)

25. S.T. Tan, A.A. Umar, M. Yahaya, C.C. Yap, M.M. Salleh, J. Phys.: Conf. Ser. **431**, 012001 (2013)

26. L. Vayssieres, Adv. Mater. **15**, 464 (2003)

27. S.T. Tan, A.A. Umar, M. Yahaya, M.M. Salleh, C.C. Yap, H.-Q. Nguyen, C.-F. Dee, E.Y. Chang, M. Oyama, Sci. Adv. Mater. **5**, 803 (2013)

28. A.A. Umar, M. Oyama, M.M. Salleh, B.Y. Majlis, Cryst. Growth Design **9**, 2835 (2009)

29. A. Ali Umar, M. Oyama, M. Mat Salleh, B. Yeop Majlis, Cryst. Growth Design **10**, 3694 (2010)

30. A.A. Umar, M. Oyama, Cryst. Growth Design **8**, 1808 (2008)

31. A. Balouch, A. Ali Umar, A.A. Shah, M. Mat Salleh, M. Oyama, ACS Appl. Mater. Interfaces **5**, 9843 (2013)

32. A. Balouch, A.A. Umar, S.T. Tan, S. Nafisah, S.K. Md Saad, M.M. Salleh, M. Oyama, RSC Adv. **3**, 19789 (2013)

33. A.A. Umar, M. Oyama, Cryst. Growth Design **9**, 1146 (2009)

34. Q. Meng, T. Wang, E. Liu, X. Ma, Q. Ge, J. Gong, Phys. Chem. Chem. Phys. **15**, 9549 (2013)

35. M. Harb, P. Sautet, P. Raybaud, J. Phys. Chem. C **115**, 19394 (2011)

36. Z. He, C. Zhong, X. Huang, W.Y. Wong, H. Wu, L. Chen, S. Su, Y. Cao, Adv. Mater. **23**, 4636 (2011)

37. K. Zhu, N.R. Neale, A. Miedaner, A.J. Frank, Nano Lett. **7**, 69 (2007)

26

Highly transparent front electrodes with metal fingers for p-i-n thin-film silicon solar cells

Etienne Moulin[1,2,a,b,c], Thomas Christian Mathias Müller[2,c], Marek Warzecha[2,c], Andre Hoffmann[2], Ulrich Wilhelm Paetzold[2], and Urs Aeberhard[2,c]

[1] Ecole Polytechnique Fédérale de Lausanne (EPFL), Institute of Microengineering (IMT), Photovoltaics and Thin-Film Electronics Laboratory, Rue de la Maladière 71b, 2000 Neuchâtel, Switzerland

[2] IEK5-Photovoltaik, Forschungszentrum Jülich GmbH, 52425 Jülich, Germany

Abstract The optical and electrical properties of transparent conductive oxides (TCOs), traditionally used in thin-film silicon (TF-Si) solar cells as front-electrode materials, are interlinked, such that an increase in TCO transparency is generally achieved at the cost of reduced lateral conductance. Combining a highly transparent TCO front electrode of moderate conductance with metal fingers to support charge collection is a well-established technique in wafer-based technologies or for TF-Si solar cells in the substrate (n-i-p) configuration. Here, we extend this concept to TF-Si solar cells in the superstrate (p-i-n) configuration. The metal fingers are used in conjunction with a millimeter-scale textured foil, attached to the glass superstrate, which provides an antireflective and retroreflective effect; the latter effect mitigates the shadowing losses induced by the metal fingers. As a result, a substantial increase in power conversion efficiency, from 8.7% to 9.1%, is achieved for 1-μm-thick microcrystalline silicon solar cells deposited on a highly transparent thermally treated aluminum-doped zinc oxide layer combined with silver fingers, compared to cells deposited on a state-of-the-art zinc oxide layer.

1 Introduction

Optimized transparent conductive oxide (TCO) layers applied as front electrodes in thin-film silicon (TF-Si) solar cells should address two trade-offs in order to enable cells to perform well both optically and electrically. The first trade-off is related to the surface morphology of the TCO layers. To efficiently absorb light in the optically thin silicon (Si) layers, and consequently achieve high photocurrent values, an advanced light-trapping concept has to be implemented. Light trapping in TF-Si solar cells is conventionally realized by applying randomly textured TCO electrodes to scatter light and thereby elongate the light path within the absorber layers [1–6]; the compromise here resides in the fact that the TCO surface morphologies providing the most suitable light trapping in TF-Si solar cells often trigger the formation of defective porous regions in the subsequently grown Si layers, and these porous regions are detrimental for the electrical performance of the cells [7,8].

The second trade-off is related to the intrinsic properties of the applied TCO layers. TCOs should exhibit a high conductivity and a high transparency in the full spectral range of relevance for solar cells, i.e. between 300 nm and 1100 nm. As the optical and electrical properties of TCOs are interlinked via the charge-carrier density and mobility, a high transparency in the near infrared (NIR) range is generally associated with a high sheet resistance R_{sq} and vice versa [9,10]. To compare the suitability of different TCOs to be used as electrode materials, a commonly accepted figure of merit (FOM) is defined as $FOM = \sigma/\alpha$, where σ is the film conductivity in S cm^{-1} and α is its absorption coefficient in cm^{-1} at a given wavelength [11], the best TCO being the one with the highest conductivity and the lowest light absorption.

The most common TCOs used as front electrodes for TF-Si solar cells in the superstrate (p-i-n) configuration are aluminum-doped zinc oxide (ZnO:Al) deposited by sputtering [12,13], boron-doped zinc oxide (ZnO:B) grown by low-pressure chemical vapor deposition (LPCVD) [14] and fluorine-doped tin oxide (SnO$_2$:F) grown by atmospheric-pressure chemical vapor deposition (APCVD) [15,16].

Novel materials such as gallium-doped ZnO have been suggested as alternative compounds for use as front TCOs

---

[a] e-mail: `etienneantoine.moulin@epfl.ch`

[b] All the results presented in this work were obtained in IEK5 (Jülich). The author recently moved to EPFL (Neuchâtel).

[c] These authors contributed equally to this work.

in p-i-n solar cells [17]. Sputtered tin-doped indium oxide In_2O_3:SnO_2 (ITO) combined with a textured superstrate has also been proposed to fulfill this function [18], and more recently hydrogenated indium oxide In_2O_3:H (IOH) [19, 20]. Other approaches have also been introduced to improve the FOM of ZnO:Al front electrodes and, with it, cell performance: for instance, two-step post-deposition annealing treatments, comprising a thermal treatment under a protective layer, result in a decrease in sub-bandgap and free-carrier absorption combined with a remarkable gain in carrier mobility [9, 21, 22].

In wafer-based photovoltaic technologies or for TF-Si solar cells in the substrate (n-i-p) configuration, a metal grid is conventionally combined with the front TCO electrode to aid in lateral conduction [23, 24]; in this case, the front TCO layer usually exhibits relatively high R_{sq} (up to 100 Ω/sq) as its thickness is generally kept thin (around 70 nm) to fulfill the role of antireflective coating [25]. In the present study, we propose to extend this well-established concept to TF-Si solar cells in the superstrate (p-i-n) configuration to relax the constraint on conductance imposed on front TCOs. In particular, this approach was adopted in the framework of the "Nanospec" project, where TCOs with high transparency in the infrared spectral range (and thus moderate conductance) were required to take full advantage of up-converter systems placed behind the cell [26, 27].

Here, we show that single-junction microcrystalline silicon (μc-Si:H) solar cells with excellent electrical properties can be obtained on highly transparent TCO layers of relatively high R_{sq} (up to approximately 30 Ω/sq) fitted with silver (Ag) fingers with optimized density. We demonstrate that this optimum can be well predicted with the optoelectronic device simulator "Advanced Semiconductor Analysis" (ASA). Then, we show that the shadowing losses induced by the Ag fingers can be noticeably reduced by a transparent retroreflective textured foil placed at the air/glass interface. We highlight the relevance of the introduced concept by comparing the cell performance with that of a cell with a state-of-the-art ZnO:Al front electrode.

2 Experimental details

The p-i-n μc-Si:H solar cells were deposited on 1.1-mm-thick glass (Corning Inc.). Figure 1a illustrates the schematic cross section of a cell with Ag fingers. The Ag fingers were evaporated on glass through a stainless steel mask, which was patterned by laser scribing, providing Ag fingers with a width of approximately 100 μm or 200 μm and a thickness of around 700 nm. In this study, the height of the fingers was not varied. However, we believe that in practice, for the finger widths considered here ($\sim$100 μm), the thickness of the fingers could certainly be further reduced, while still providing sufficient conductivity and current-collection ability. The mask was engineered in a way that various finger geometries, i.e. with varying distances between the fingers (see Fig. 1b), could be obtained simultaneously on the same glass substrate. As a front

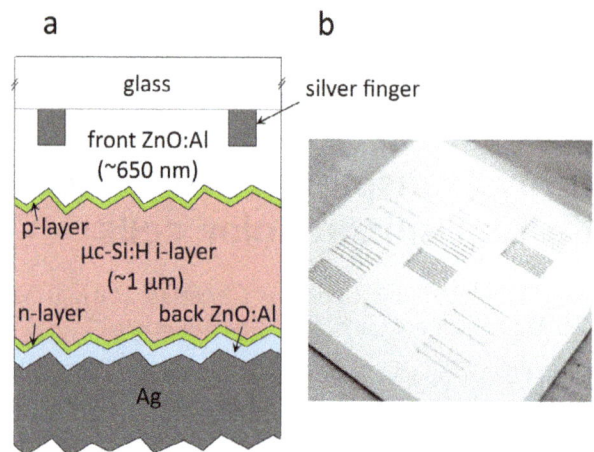

Fig. 1. (a) Schematic cross section of a μc-Si:H p-i-n solar cell deposited on a highly transparent front TCO layer with Ag fingers. (b) Evaporation mask employed to deposit the Ag fingers on the glass superstrate. The line width is varied between 100 μm and 200 μm.

electrode, a ZnO:Al bi-layer with a 1 wt.% aluminum doping concentration was deposited by radio-frequency magnetron sputtering, the first nanometers of ZnO:Al being deposited at room temperature to limit thermal stress in the Ag fingers, and the bulk ZnO:Al layer being deposited at a substrate temperature of around 300 °C. Note that, for cells with and without Ag fingers, the same front TCO was used (by co-deposition). Etching in a 0.5% w/w diluted hydrochloric acid (HCl) solution was carried out to texturize the initially smooth ZnO:Al surface, leading to a final average thickness of approximately 650 nm. The sheet resistance of the resulting ZnO:Al layers is typically around 8 Ω/sq.

In this study, a thermal treatment in vacuum (at 600 °C for 3.5 h to 5.5 h) was performed to increase the transparency of the front ZnO:Al layers fitted with the Ag fingers. A similar treatment at 400 °C did not show any noticeable modification of the opto-electrical properties of ZnO:Al. Annealing at 600 °C is challenging for industrial applications due to the severe cost issues associated to the use of Corning glass and to the thermal treatment itself. In principle, alternative TCO materials to ZnO:Al – with a higher transparency – could have been used in this study. However, one of the advantages of annealing is that it preserves the surface morphology of the front ZnO:Al [28]. It thus enables one to investigate the impact of the ZnO:Al intrinsic optical and electrical properties without considering eventual modifications of the light-trapping properties or of the Si growth conditions associated with a variation of the surface texture.

The Si layers were deposited by plasma-enhanced chemical vapor deposition (PECVD). The boron- and phosphorous-doped (p- and n-) layers and the intrinsic (i-) layer have thicknesses of approximately 20 nm and 1 μm, respectively. More details on the preparation of our baseline solar cells can be found elsewhere [29]. An 80-nm-thick sputtered ZnO:Al layer and a 700-nm-thick evaporated Ag layer were used together as a back electrode. The cell area (of 1 cm^2) is defined by the Ag back reflector, which serves as an etching mask for the underlying ZnO:Al layer.

An antireflective/retroreflective foil from the company DSM was applied to the front side of the device at the air/glass interface. Details on the specifications and working principle of this foil are given later in the text.

The reflectance R and transmittance T of TCOs – from which their absorptance was determined (as $A = 1 - R - T$) – were acquired using a LAMBDA 950 (Perkin Elmer) spectrophotometer equipped with an integrating sphere. Current-voltage (J-V) measurements were carried out using a dual-lamp solar simulator (Wacom) under standard test conditions (AM1.5G spectrum, 1000 W/m^2, 25 °C).

Simulations with ASA and model calibration

To support the measurements, full one-dimensional numerical device simulations were performed. For this purpose, we used the commercial optoelectronic device simulation software ASA [30]. The simulations account for two important aspects, namely the shadowing by the fingers and the resistive losses in the front electrode. The optical response of the solar cells was calibrated with absorptance measurements of a flat bifacial μc-Si:H p-i-n device, similar to the devices used here but without the back reflector [26]. Owing to the low coverage of the fingers (well below 5%), the additional light scattering presumably resulting from the surface modulation around the fingers has likely only a minor impact on the overall optical performance (and thus on J_{sc}). Also, due to the low height-to-width aspect ratio (<0.01) of the fingers, additional shadowing caused by illumination under oblique incidence is most probably negligible. Therefore, we assume that the 1-D model used here to account for the optical impact of the fingers already provides good estimation. The electrical parameters were derived from J-V characteristics of the device in the dark and under AM1.5G illumination. The illuminated J-V characteristics are therefore a superposition of J-V characteristics from areas in the dark (those under the fingers, J_d) and areas that receive illumination (everywhere else, J_i). The current is hence calculated as:

$$J = \frac{d_{Ag}}{d} J_d + \frac{d_{Ag-Ag}}{d} J_i,$$

where d_{Ag} is the width of a finger, d_{Ag-Ag} the gap between two adjacent fingers, and $1/d$ the finger density. Note that $d = d_{Ag-Ag} + d_{Ag}$.

Solving the differential equation for the electrostatic potential between two fingers in one dimension yields the contact series resistance

$$R_s = R_{s0} + \frac{R_{sq}^{ZnO} d_{Ag-Ag}^2}{8},$$

where R_{s0} is a constant offset (used here as a fitting parameter and set to a value of 0.08 Ω), and R_{sq}^{ZnO} the sheet resistance of the ZnO:Al front contact [26]. The variation of $1/d$ then yields – through R_s – the corresponding

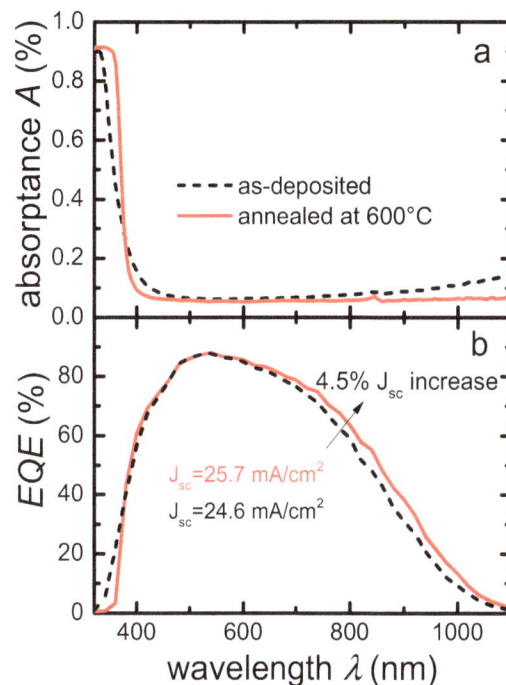

Fig. 2. (a) Absorptance A as a function of wavelength of glass/ZnO:Al superstrates with and without a post-deposition annealing treatment at 600 °C for 5.5 h. (b) External quantum efficiency EQE of the solar cells deposited on these superstrates. The corresponding photocurrent values (J_{sc}) are given in the graph.

change in fill factor (FF), short-circuit current density (J_{sc}) and efficiency of the solar cell.

3 Results and discussion

Figure 2 illustrates the impact of an annealing treatment on the optical properties of ZnO:Al and on the spectral performance of a complete solar cell. Below 380 nm, the ZnO:Al layer shows a higher absorptance (lower transparency) after annealing at 600 °C for 5.5 h (Fig. 2a). This effect is explained by the reduced band filling – and associated narrowing of the optical bandgap (Burstein-Moss effect) – caused by the decrease in free-carrier density upon thermal treatment [31, 32]. Above 700 nm, the reduction of the free-carrier density leads to a noticeable increase in layer transparency, which is mostly dictated by the shift of the free-electron plasma resonance towards longer wavelengths [33, 34]. Finally, sub-bandgap absorption in the spectral range below the fundamental absorption edge, i.e. from 380 nm to 500 nm, also decreases, as already reported in reference [9]. In accordance with the absorptance data of the glass/ZnO:Al superstrates, the external quantum efficiency (EQE) of the cell deposited on the annealed superstrate surpasses that of the reference cell in the near-infrared (NIR) wavelength region; this EQE gain compensates for the slight optical loss below 380 nm, such that the photocurrent increases from 24.6 mA/cm^2 to 25.7 mA/cm^2 (by 4.5%). The lower absorptance of the treated ZnO:Al between 380 nm and 500 nm does not translate into an *EQE* gain; the optical behaviour of single layers on glass strongly differs from that of more complex optical systems like complete solar cells and further investigation would be needed to understand this effect.

Annealing ZnO:Al in vacuum at 600 °C typically results in an increase in sheet resistance, due to the decreased carrier concentration and to a lower Hall mobility [10,33]. In the literature, the latter effect is attributed to residual oxygen gas in the vacuum chamber, leading to oxygen adsorption – and thus probably to an increased barrier height – at the grain boundaries of ZnO:Al [35,36].

After 3 h of annealing at 600 °C in vacuum, the ZnO:Al layer becomes highly transparent, even though R_{sq} remains below 25 Ω/sq. Figure 3 shows the *FF*, J_{sc} and efficiency of 1-cm^2 full-area μc-Si:H cells deposited on thermally treated front ZnO:Al layers (with a R_{sq} of around 20 Ω/sq) equipped with 200-μm-wide Ag fingers. The finger density is varied from 0 (i.e. with no finger) to 10 fingers per cm. The V_{oc} exhibits almost no variation upon modification of the finger density (not shown here). Owing to the relatively thick ZnO:Al layer (of approximately 650 nm), the abruptness of the transition step at the finger edges might be strongly attenuated, preventing the occurrence of too severe structural defects in the subsequent layers that would deteriorate cell performance. The *FF* first rapidly increases from 61% to 72% by adding a single finger (see Fig. 3a, solid black line) and then saturates at a value of 74% for higher finger densities. Note that the strong benefit observed when adding a single finger is explained by the rather long distance (of around 3 mm) separating the cell edges and the bus bar where the contact is made. As expected, the J_{sc} decreases linearly with increasing Ag finger density (Fig. 3b) due to increased shadowing. The highest $FF \times J_{sc}$ product, and consequently the highest efficiency, is obtained with two fingers per cm (Fig. 3c, solid black line). The existence of this optimal configuration is in good agreement with theoretical predictions obtained with ASA (Fig. 3, dashed lines).

To improve light coupling into solar cells, an antireflective coating is traditionally applied at the air/glass interface. Geometric textures with feature sizes well above the wavelength of the incoming light can be introduced at this interface to fulfill this function [37, 38]. The antireflective effect is based on the rebound of the incident light at the facets of the texture, as illustrated in Figure 4 (effect 1) for pyramidal features. One particularity of such textures is that they also permit some of the light reflected at the first interfaces (glass/ZnO:Al, ZnO:Al/Si) or at the Ag fingers to be redirected towards the absorber layer (see Fig. 4, effect 2). In the present study, the antireflective/retroreflective (AR-RR) texture is of particular importance since it mitigates the optical (shadowing) losses induced by the fingers by effectively retroreflecting the light back into the solar cell. Here, the textured cover consists of an adhesive flexible polymeric foil, from the company DSM, with a refractive index similar to that of glass, attached directly to the glass superstrate. The pos-

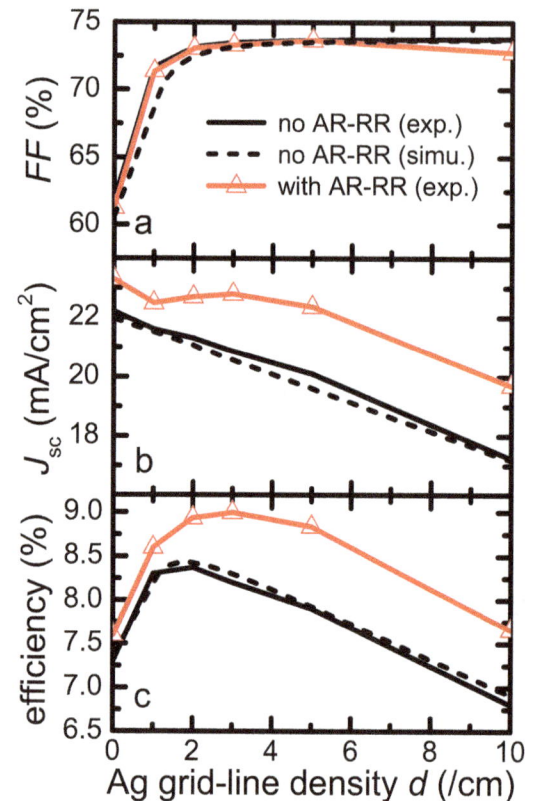

Fig. 3. Experimental and calculated parameters of μc-Si:H solar cells deposited on thermally treated front ZnO:Al layers fitted with Ag fingers of varying density. The curves with triangles represent parameters of cells with an antireflective/retroreflective (AR-RR) foil. Note that finger densities d of $0 < d < 1$ make sense only for the simulated data since the experimental parameters extracted here were measured on cells with an active area of 1×1 cm^2. The case $d = 0$ represents the cell with no Ag fingers.

itive impact of the AR-RR foil on the device performance is demonstrated in Figure 3 (lines with triangles): the AR-RR foil leads to a substantial gain in J_{sc} (Fig. 3b). Note that the relative gain in J_{sc} obtained with the foil increases with finger density, which underlines the effectiveness of its retroreflective ability. As a result of this effect, a new optimum of the finger geometry is found with the AR-RR foil, which gives a maximum power conversion efficiency of approximately 9% with three fingers (per cm).

By reducing the width of the fingers to 100 μm (i.e. by 50%) and by choosing a density of three fingers per cm, we were able to further increase the cell performance. Figure 5 depicts the J-V characteristics of a cell deposited on such a front electrode and of a reference cell, co-deposited on a bare, untreated ZnO:Al layer. The corresponding cell parameters with and without the AR-RR foil are listed in Table 1.

The major difference resides in the *FF*: The annealed ZnO:Al layer fitted with Ag fingers provides a higher *FF*, by around 2.5% (rel.), than the reference ZnO:Al electrode. This gain underlines the beneficial role of the metal fingers, which more than compensate for the increased

Table 1. *J-V* parameters of cells (average of the three best), with and without an AR-RR foil, deposited on a reference ZnO:Al layer (ref.) and on a ZnO:Al layer annealed at 600 °C for 3.5 h (ann.) fitted with Ag fingers.

	Efficiency (%)	FF (%)	J_{sc} (mA/cm^2)	V_{oc} (mV)
ref.	8.5	70.3	23.1	525
ref. + AR-RR	8.7	70.1	23.7	525
ann. + fingers	8.7	71.9	22.9	530
ann. + fingers + AR-RR	9.1	71.9	23.8	530

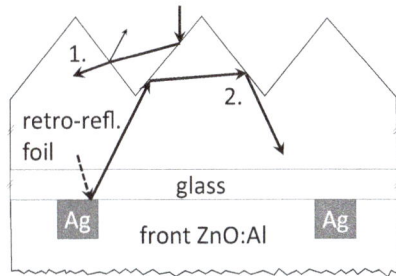

Fig. 4. Illustration of the geometric-optics interaction between a ray of light and the anti-reflective/retroreflective foil. Effect 1: the fraction of incoming light that is not transmitted into the foil upon the first incidence will hit a second facet, resulting in a higher fraction of transmitted light. Effect 2: light reflected out of the cell, in particular at the Ag fingers, is redirected back towards the active layers of the cell via a double rebound of the light beam at the textured-foil/air interface.

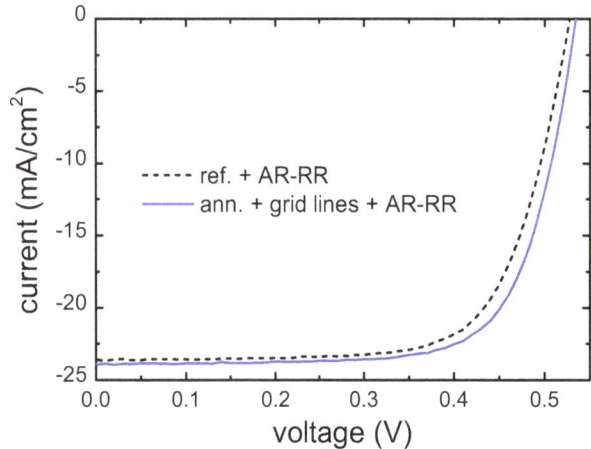

Fig. 5. *J-V* curves of 1-μm-thick μc-Si:H solar cells deposited on a reference ZnO:Al layer and on a ZnO:Al layer annealed at 600 °C for 3.5 h equipped with Ag fingers. The width of the fingers is approximately 100 μm and the density is 3 fingers per cm. Both cells are covered with an AR-RR foil.

Ohmic losses in the annealed ZnO:Al layer. For cells without an AR-RR foil, a slightly smaller J_{sc} is measured for the front electrode with Ag fingers, due to shadowing. The application of the AR-RR foil leads to a J_{sc} increase of around 2.6% for the reference cell and 4% for the cell with fingers. Notably, a slightly higher V_{oc} is measured for the cell with the annealed ZnO:Al layer (accounting for a relative efficiency gain of less than 1%). Overall, cells with the AR-RR foil are significantly improved by substituting the reference ZnO:Al layer with a highly transparent front electrode fitted with Ag fingers: The efficiency increases from 8.7% to 9.1%.

To conclude, we demonstrated that the combination of a highly transparent front TCO with metal fingers is a promising approach to obtain a better compromise between the optical and electrical constraints of TF-Si solar cell front electrodes. We highlighted the utility of this concept for 1-μm-thick μc-Si:H solar cells deposited on thermally treated ZnO:Al layers with Ag fingers, where a noticeable gain in FF has been measured as compared to reference cells deposited on bare, untreated ZnO:Al. By introducing a millimeter-scale textured foil providing an antireflective and retroreflective effect at the air/glass interface, we have been able to offset the impact of the shadowing losses induced by the fingers. By extending this approach to other TCO materials such as sputtered ZnO:Al with a lower doping concentration or to IOH or LPCVD-ZnO and by using optimized geometries for the metal fingers, we anticipate further advances in the effi-

ciency of TF-Si solar cells in the single- or multi-junction device configuration.

We gratefully acknowledge Jürgen Hüpkes, Stephan Haas, Carolin Ulbrich, and Juraj Hotovy for constructive discussions. This work was carried out in the framework of the project "Nanospec", funded by the European Community's Seventh Framework Programme (FP7/2007-2013) under grant agreement no 246200.

References

1. E. Yablonovitch, G.D. Cody, IEEE Trans. Electron Devices **29**, 300 (1982)
2. H.W. Deckman et al., Appl. Phys. Lett. **42**, 968 (1983)
3. O. Kluth et al., Thin Solid Films **351**, 247 (1999)
4. M. Ermes et al., J. Appl. Phys. **113**, 073104 (2013)
5. J.I. Owen et al., Phys. Stat. Sol. A **208**, 109 (2011)
6. J. Bailat et al., in *Proceedings of the 25th European Photovoltaic Solar Energy Conf./5th World Conf. Photovoltaic Energy Convers, Valencia, 2010*, p. 2720
7. M. Python et al., Prog. Photovolt.: Res. Appl. **18**, 491 (2010)
8. S. Hänni et al., IEEE J. Photovolt. **99**, 1 (2012)
9. F. Ruske et al., J. Appl. Phys. **107**, 013708 (2010)
10. T. Minami et al., J. Vacuum Sci. Technol. A **17**, 1822 (1999)

11. R.G. Gordon, MRS Bulletin **25**, 52 (2000)
12. A. Lambertz et al., Sol. Energy Mater. Sol. Cells **119**, 134 (2013)
13. S. Kim et al., Sol. Energy Mater. Sol. Cells **119**, 26 (2013)
14. M. Boccard et al., IEEE J. Photovolt. **2**, 229 (2012)
15. M. Kambe et al., in *Proceedings of the 3rd World Conf. Photovoltaic Energy Conversion, Osaka, 2003*, Vol. 2, p. 1812
16. J. Krc et al., Thin Solid Films **518**, 3054 (2010)
17. K.C. Lai et al., Sol. Energy Mater. Sol. Cells **94**, 397 (2010)
18. C. Battaglia et al., Appl. Phys. Lett. **96**, 213504 (2010)
19. C. Battaglia et al., J. Appl. Phys. **109**, 114501 (2011)
20. T. Koida et al., Jpn J. Appl. Phys. **28**, L685 (2007)
21. M. Wimmer et al., Thin Solid Films **520**, 4203 (2012)
22. S. Neubert et al., Prog. Photovolt.: Res. Appl. **22**, 1285 (2014)
23. J. Geissbühler et al., IEEE J. Photovolt. **4**, 1055 (2014)
24. H. Sai et al., in *Proceeding of the MRS conference, San Francisco, 2013*, Vol. 1536, p. 3
25. E. Kobayashi et al., in *Proceedings of the 28th European Photovoltaic Solar Energy Conference and Exhibition, Paris, 2013*, p. 691
26. T.C.M. Müller et al., Energy Procedia **10**, 76 (2011)
27. J.C. Goldschmidt et al., in *Renewable Energy and the environment* (Optical Society of America, 2013), paper PT3C.2
28. J. Owen, Ph.D. thesis, RWTH Aachen University, 2011
29. B. Rech et al., Thin Solid Films **511**, 548 (2006)
30. B.E. Pieters et al., J. Appl. Phys. **105**, 044502 (2009)
31. B.E. Sernelius et al., Phys. Rev. B **37**, 10244 (1988)
32. E. Burstein, Phys. Rev. **93**, 632 (1954)
33. M. Berginski et al., Thin Solid Films **516**, 5836 (2008)
34. A. Pflug et al., Thin Solid Films **455**, 201 (2004)
35. S. Takata et al., Thin Solid Films **135**, 183 (1986)
36. T. Minami, H. Sato, K. Ohashi, T. Tomofuji, S. Takata, J. Cryst. Growth **117**, 370 (1992)
37. C. Ulbrich, Prog. Photovolt.: Res. Appl. **21**, 1672 (2013)
38. J. Escarre et al., Sol. Energy Mat. Sol. Cells **98**, 185 (2012)

27

Improved electron collection in fullerene via caesium iodide or carbonate by means of annealing in inverted organic solar cells

Zouhair El Jouad[1,2], Guy Louarn[3], Thappily Praveen[4], Padmanabhan Predeep[4], Linda Cattin[3], Jean-Christian Bernède[1,a], Mohammed Addou[2], and Mustapha Morsli[5]

[1] L'UNAM, Université de Nantes, MOLTECH-Anjou, CNRS, UMR 6200, 2 rue de la Houssinière, BP 92208, 44000 Nantes, France
[2] Laboratoire Optoélectronique et Physico-chimie des Matériaux, Université Ibn Tofail, Faculté des Sciences, BP 133, 14000 Kenitra, Morocco
[3] Université de Nantes, Institut des Matériaux Jean Rouxel (IMN), CNRS, UMR 6502, 2 rue de la Houssinière, BP 32229, 44322 Nantes Cedex 3, France
[4] Laboratory for Unconventional Electronics and Photonics, Department of Physics, National Institute of Technology, 673 601 Calicut, Kerala, India
[5] L'UNAM, Université de Nantes, Faculté des Sciences et des Techniques, 2 rue de la Houssinière, BP 92208, 44000 Nantes, France

Abstract Inverted organic photovoltaic cells (IOPVCs), based on the planar heterojunction $C_{60}/CuPc$, were grown using MoO_3 as anode buffer layer and CsI or Cs_2CO_3 as cathode buffer layer (CBL), the cathode being an ITO coated glass. Work functions, Φ_f, of treated cathode were estimated using the cyclic voltammetry method. It is shown that Φ_f of ITO covered with a Cs compounds is decreased. This decrease is amplified by the annealing. It is shown that the thermal deposition under vacuum of the CBL induces a partial decomposition of the caesium compounds. In parallel, the formation of a compound with the In of ITO is put in evidence. This reaction is amplified by annealing, which allows obtaining IOPVCs with improved efficiency. The optimum annealing conditions is 150 °C for 5 min.

1 Introduction

Nowadays, organic photovoltaic cells (OPVCs) are devices the most studied in the field of the photovoltaic energy owing to their promising properties, lightness, flexibility, semi transparency. Until today they are still in need of power conversion efficiency and stability improvement. Conventional OPVCs consist of an organic active material, containing an electron donor (ED) and an electron acceptor (EA), sandwiched by a high workfunction, conductive and transparent electrode as the anode, such as indium tin oxide (ITO) and a low work function metal, such as Al, as the cathode. In this conventional architecture, ITO is the bottom electrode, deposited onto the substrate, while the cathode is the top electrode [1].

More recently, inverted OPVCs with modified ITO as the transparent cathode and a high work function metal as the anode were studied. In this architecture the anode is the top electrode. It allows the use of an air stable high work function material as top electrode to improve the air stability of the cells [2–4].

Efficient charge collections are usually achieved through electrode buffer layers. As a matter of fact, for efficient charge collection, work functions of cathode and anode must be matched to the lowest unoccupied molecular orbital (LUMO) of acceptor and the highest occupied molecular orbital (HOMO) of donor, respectively. Buffer layers (BF) are necessary in view of the difficulties in organic optoelectronic devices of the charge carrier transport between the organic materials and the electrodes. In the case of the anode/electron donor contact, a common solution is to introduce a thin anode buffer layer (ABL), which adjusts the electronic behaviour of the adjacent materials. We have shown that an ultra-thin (0.5 nm) Au film and or a thin (4 nm) MoO_3 film introduced between the anode and the organic material can be used to improve the devices performances [5–7] and we have used MoO_3 as ABL in the OPV studied here.

In the case of inverted OPVCs (IOPVCs) it was shown that the use of a MoO_3 layer thick of 6 nm allows obtaining

---
[a] e-mail: jean-christian.bernede@univ-nantes.fr

efficient hole collection whatever is the metal of the anode [8, 9]. The main aim of the present work being the study of the interface cathode/EA, we have used this classical MoO$_3$/anode hole collecting structure, with Al metal electrode. We have focused our interest on the effect of cathode buffer layer (CBL) based on caesium compounds, CsI and Cs$_2$CO$_3$. More precisely we have studied the effect of the annealing temperature of the ITO/caesium compound structure on the electron collection efficiency. We show that the improvement of the efficiency of the caesium compound as CBL is related to CBL/ITO chemical reaction during the annealing process.

2 Experimental details

The cells were fabricated onto ITO coated glass substrates with a sheet resistance of about 25 Ω/square. Here the ITO was used as cathode. The standard substrate dimensions were 25 mm by 25 mm. Since ITO covered the whole glass substrates, some ITO must be removed to obtain the under electrode. After masking a broad band of 25 mm by 20 mm, the ITO was etched by using Zn powder + HCl as etchant [8,9]. After scrubbing with soap, the ITO coated substrates were rinsed in running deionised water. Then the substrates were dried with an air flow and then loaded into a vacuum chamber (10^{-4} Pa). Deposition rate and film thickness were measured in situ by quartz monitor, after calibration for each material used. The organic donor/acceptor couple used is copper phthalocyanine (CuPc)/fullerene (C$_{60}$), the ABL is, as said above, MoO$_3$, while the CBL is either CsI or Cs$_2$CO$_3$. The Cs compound, C$_{60}$, CuPc, MoO$_3$ films were successively sublimated under vacuum and finally the metal anode was evaporated on the top of the device giving the following inverted OPVC:ITO/CBL/C$_{60}$ (40 nm)/CuPc (35 nm)/MoO$_3$ (6 nm)/Al (120 nm). The top electrode was deposited through a mask with 2×10 mm^2 active area.

Before deposition of the organic layers the ITO/Cs compound structures were annealed for 5 min at temperature between room temperature and 200 °C under argon flux. The effect of annealing temperature on the OPVCs performances, and on the properties of the bilayer ITO/Cs compound were studied.

The morphology of the different structures used as cathode was observed through scanning electron microscopy (SEM) with a JEOL 7600F. AFM images on different sites of the films were taken ex-situ at atmospheric pressure and room temperature. All measurements have been performed in tapping mode (Nanoscope IIIa, Veeco, Inc.). Classical silicon cantilevers were used (NCH, nanosensors). The average force constant and resonance were approximately 40 N/m and 300 kHz, respectively. The cantilever was excited at its resonance frequency. These structural characterizations were performed at the "Centre de micro-caractérisation de l'Université de Nantes".

XPS measurements were carried out at room temperature. An Axis Nova instrument from Kratos Analytical spectrometer with Al Kα line (1486.6 eV) as excitation source has been used. The core level spectra were acquired with an energy step of 0.1 eV and using a constant pass energy mode of 20 eV, to obtain data in a reasonable experimental time (energy resolution of 0.48 eV). Concerning the calibration, binding energy for the C1s hydrocarbon peak was set at 284.6 eV. The pressure in the analysis chamber was maintained lower than 10^{-7} Pa. The background spectra are considered as Shirley type.

For comparison, OPVCs with CsI, Cs$_2$CO$_3$ and without CBL were realized during the same deposition process. Successive (3–4) depositions were done for each configuration, which corresponds to 9 to 12 OPVCs, in order to check the reproducibility of the results.

The work function (Φ_f) of the ITO electrode and CBL are evaluated using cyclic voltammetry. Cyclic Voltammetry is an electrochemical method for determination of oxidation and reduction states. The working electrode (here ITO sample), dipped in an electrolyte solution (Acetonitrile containing 0.1 M tetrabutylammonium tetrafluoroborate), is biased with respect to the reference electrode (Ag/AgCl (3M NaCl)), which has a known potential. When the bias reaches the difference between the reference electrode and oxidation potential of the sample, the electrode is oxidized and the current is recorded at the counter electrode. Similarly, when the bias overcomes the reduction potential, relative to the reference, the electrode is reduced. A linear voltage ramp is used in the sweep. Linear ramping potential starting from 0 V up to a pre-defined limiting value with a scan rate of 50 mV/s and recorded the current. The optimised switching potential is 2 V for this experiment. At this potential, the direction of the potential scan is reversed and the current is again recorded. The sweep is repeated several times between the two limiting potentials. A three electrodes cell configuration with Pt as the counter electrode is used for this work. Three-electrodes configuration allows one electrode to be studied in isolation, without complications from the electrochemistry of the other electrodes. The instrument used is Zhaner make Electrochemical workstation (IM6ex).

The Ionization potential is calculated from the onset Oxidation potential using the equation

$$IP = Eox^{\text{onset}} + 4.4 \text{ eV}. \qquad (1)$$

The onset potentials are determined from the intersection of the two tangents drawn at the rising current and baseline charging current of the CV traces. The work function of ITO is taken as equal to ionization potential (IP), as it is assumed that there is no band bending at ITO interface and the injection barrier, $\Delta E_h \sim 0$.

3 Results and discussion

Figure 1 shows the photovoltaic response of typical IOPVCs using different cathode configurations. Table 1 shows that the post-deposition annealing temperature of the cathode buffer layer plays a major role in the device performances.

Table 1. Performances of the OPVCs with different cathode buffer layers and after different annealing.

Buffer layer	Annealing temperature (°C)	J_{sc} (mA/cm^2)	V_{oc} (V)	FF (%)	η (%)
CuI*	–	6.09	0.43	51	1.35
–	–	2.99	0.42	50	0.63
CsI	–	1.52	0.32	35	0.17
CsI	100	2.49	0.37	41	0.38
CsI	150	3.53	0.43	55	0.84
CsI	200	4.05	0.40	45	0.81
CsCO$_3$	–	1.63	0.28	35	0.16
CsCO$_3$	100	2.08	0.30	40	0.25
CsCO$_3$	150	3.12	0.40	51	0.63
CsCO$_3$	200	3.80	0.38	45	0.65

*Classical cell configuration: ITO/CuI/CuPc/C60/Alq$_3$/Al.

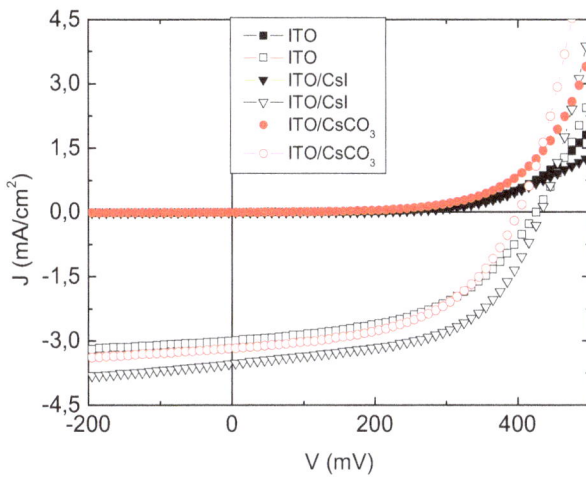

Fig. 1. J-V characteristics of OPVCs with different cathode buffer layers: ITO ($\blacksquare$), ITO/CsI ($\blacktriangledown$) and ITO/Cs$_2$O$_3$ ($\bullet$) after annealing at 150 °C of the structure ITO/Cs compound.

Firstly, we can see that a minimum annealing temperature is necessary for the buffer layer is effective. Below this temperature the power conversion efficiency (η) of the IOPVCs is smaller than that obtained without CBL. Similar results were already achieved, for instance in the case of Cs$_2$CO$_3$, it was attributed to its low conductivity [10]. Actually, when the annealing temperature is insufficient, the IOPVCs show poor short circuit current density, J_{sc}, open circuit voltage, V_{oc}, and fill factor, FF, implying that the Cs compound CBL provide poor function in terms of electron extraction and transport to the cathode [11]. Secondly, we can also note that the minimum annealing temperature required is 150 °C for 5 min. For higher annealing temperatures we note a tendency of stabilization of the IOPVCs performances. The variation in devices performances in not large in terms of V_{oc} and FF. However the variation in J_{sc} is obvious, due to the presence or not of a CBL and of its annealing temperature. Maximum J_{sc} of 3.53 mA/cm^2 and 3.80 mA/cm^2 are obtained after annealing 5 min at 150 °C for CsI and 200 °C for Cs$_2$CO$_3$, respectively.

It is known that the photocurrent of the IOPVCs depends on the light absorption, exciton dissociation, carrier

Table 2. Performances of the OPVCs with different CsI thickness after an annealing of 5 min at 150 °C.

CsI thickness (nm)	J_{sc} (mA/cm^2)	V_{oc} (V)	FF (%)	η (%)
1	3.01	0.42	53	0.67
1.5	3.53	0.43	55	0.84
3	2.40	0.41	32	0.32

transport and collection at electrodes. Here, due to the same ED/EA couple in each device, it can be supposed that exciton diffusion, dissociation and carrier transport are the same for all IOPVCs. Moreover the same anode configuration being used for all devices, the differences in the J_{sc} values should mainly due to cathode electron collection efficiency. So, in order to understand the origin of the IOPVCs improvement, the different CBL/ITO structures were investigated.

As the light is incident from the ITO electrode, a possible source of J_{sc} modification is the optical transmittance of the cathode. All samples, with or without CBL, show high transparency with transmittance above 90% in the visible range (not shown here). It shows that there is no contribution of the optical properties of the cathode to the variation of the J_{sc} value.

The electronic collection is also related to the passage of free electrons through the EA/CBL/ITO interface, before they are collected by the cathode. Therefore the conductivity of the CBL is significant for charge collection efficiency. The conductivity of the CBL used being very small, the electrons can cross this layer by tunnel effect. It is known that for thickness higher than 2 nm a fast decrease of the tunnelling current is observed, which justifies that the highest efficiency is obtained with CBL thickness of 1.5 nm. For thicker films, J_{sc} decreases due to limited tunnel effect. For thinner films, the CBL is strongly discontinuous which justifies its limited effect (Tab. 2).

Lastly, the energy alignment at the EA/cathode interface is decisive to the electron collection. In the case of poor band matching, the series resistance increases and V_{oc} is limited.

The series resistance of the different IOPVCs varies with the cathode structure. As expected, it decreases when J_{sc} increases. For instance it decreases from $Rs = 18$ Ω

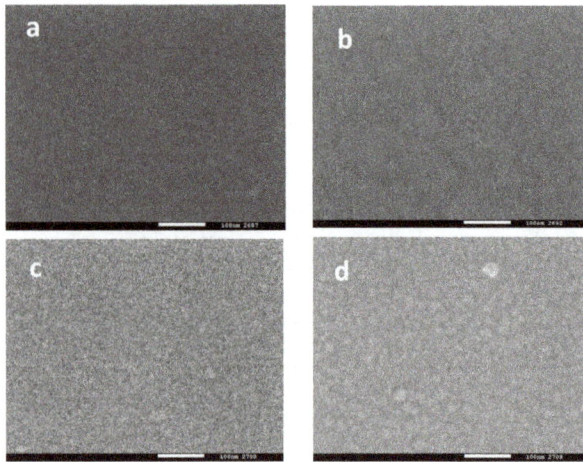

Fig. 2. Surface visualisation of ITO/CsI cathode not annealed (a) and annealed at 100 °C (b), 150 °C (c) and 200 °C (d).

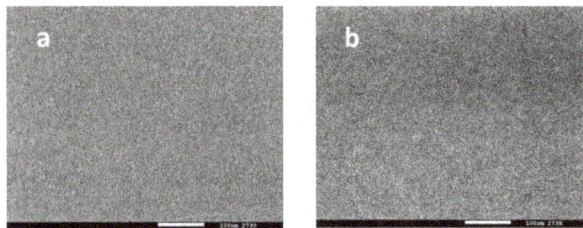

Fig. 3. Surface visualisation of ITO/Cs$_2$CO$_3$ cathode not annealed (a) and annealed at 150 °C (b).

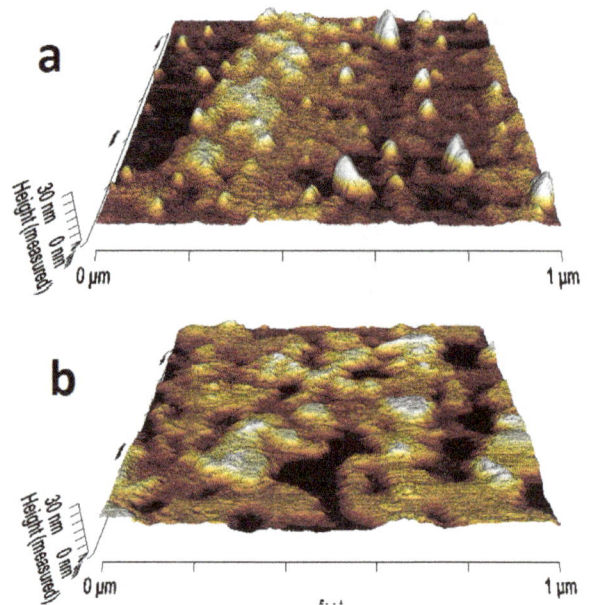

Fig. 4. AFM images of ITO/CsI cathode before (a) and after (b) annealing at 100 °C.

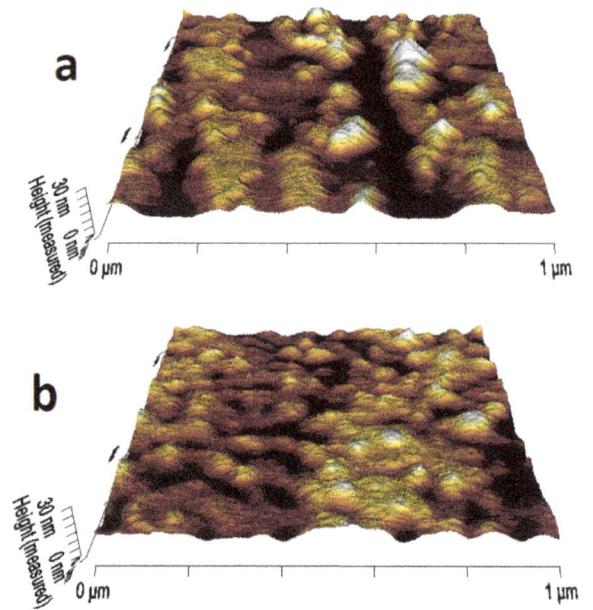

Fig. 5. AFM images of ITO/Cs$_2$CO$_3$ cathode before (a) and after (b) annealing at 100 °C.

without CBL to 8 Ω when the CsI CBL (1.5 nm) is annealed at 150 °C, indicating that the annealed structures are more favourable to electron collection. Also V_{oc} increases with the annealing temperature, at least up to the optimum temperature, which testifies of an improvement of the band matching.

In order to investigate more precisely the effect of the nature of the CBL and of the annealing treatment, on the ITO/CBL structure properties we proceed to some specific characterizations, SEM, AFM, XPS studies and Φ_M measurements.

We first consider the SEM images obtained for the surfaces of the ITO/CBL structures (Figs. 2 and 3) and the AFM topographic images obtained for the surfaces of these structures (Figs. 4 and 5).

Figures 2 and 3 show the SEM images of the CsI and the Cs$_2$CO$_3$ CBL onto the ITO electrode after annealing treatment or not. Without or with annealing the surfaces of the structures are fairly smooth and homogeneous even if some faint features seem appear when the annealing temperature increases.

The AFM topographic images of the ITO/CBL structures are shown in Figures 4 and 5. Before annealing the root mean square (rms) roughness is of 2.8 ± 0.2 nm in the case of CsI and 3.5 ± 0.2 nm in the case of Cs$_2$CO$_3$. After annealing, whatever the annealing temperature, the surface is smoothed. The rms value is 1.9 ± 0.2 nm in the case of CsI and 1.7 ± 0.2 nm for Cs$_2$CO$_3$. These results do not confirm the visual impression given by the SEM study.

Therefore the surface morphology studies of the ITO/CBL indicate that the nature of the CBL, CsI or Cs$_2$CO$_3$, and its annealing temperature have only a small effect on the surface morphology of the ITO/CBL structures. This suggests that, the surface morphology of the structures can hardly explain the different behaviours, so we preceded in XPS measurements.

About quantitative analysis, there is some uncertainty on the measures for the Cs$_2$CO$_3$ due to the transfer in room air from gloves box, where we proceed to the annealing, to the XPS apparatus, which induces some oxygen

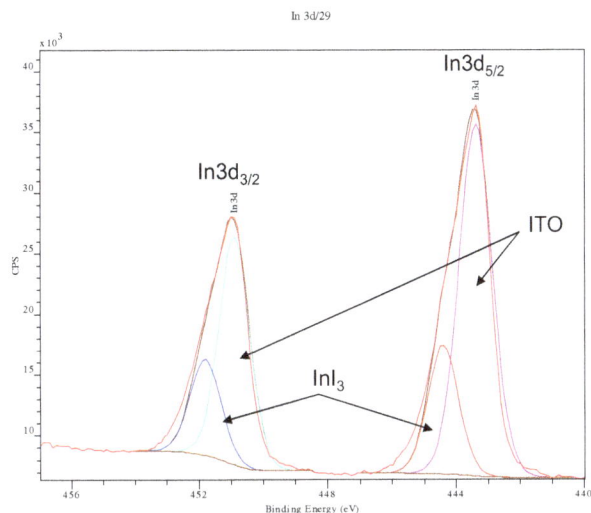

Fig. 6. Decomposition of the In3d XPS spectra of CsI/ITO structure annealed at 100 °C.

Table 3. Variation of the ratio In_{ITO}/In_x with the annealing temperature.

Annealing temperature (°C)	25	100	150	200
In_{ITO}/In_{InI_3}	4.0	3.0	2.15	2.0
$In_{ITO}/In_{In(OH)n}$	4.4	3.6	2.8	2.4

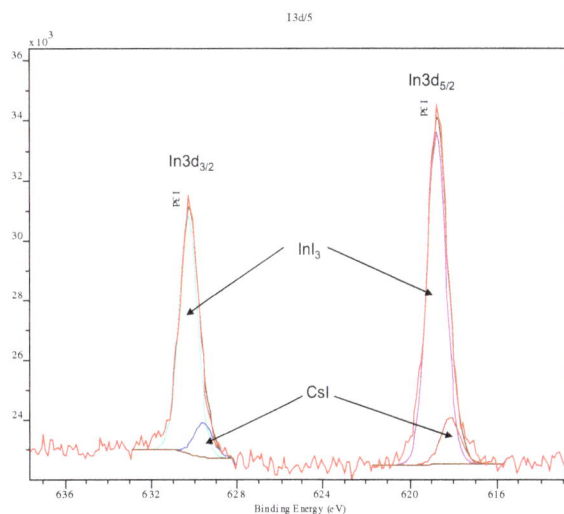

Fig. 7. Decomposition of the I3d XPS spectra of CsI/ITO structure annealed at 100 °C.

and carbon contamination. Moreover, the Cs_2CO_3 being very thin, the oxygen of ITO, the bottom layer, is also detected. This kind of problem does not exist in the case of the quantitative analysis of CsI and we found that the deposited films are systematically iodine deficient. Without annealing the relative atomic iodine concentration is only 15% and therefore that of Cs is 85 at.%. After annealing the iodine atomic concentration decreases progressively when the annealing temperature increases: 11 at.% after 5 min at 100 °C, 9 at.% at 150 °C and 7 at.% at 200 °C. It means that the vacuum deposition, and the annealing, of CsI induces, at least partly, its decomposition. This fact is in good agreement with earlier studies that show the tendency of Cs compounds to decompose during vacuum deposition for Cs_2CO_3 [12,13] and CsI [14]. Moreover, it should be noted that, the CBL layer being thin (1.5 nm), the elements of the bottom layer are systematically detected. For instance the doublet In3d of the indium of ITO is clearly visible. It can be seen in Figure 6 that the doublet In3d corresponds to two doublets. The first one, with $In3d5/2 = 444.5 \pm 0.2$ eV, corresponds to the indium bounded to oxygen in ITO, the second one, at $In3d5/2 = 445.5 \pm 0.2$ eV, can be attributed to some indium bounded to iodine in the form InI_3 [15]. Moreover the ratio In_{ITO}/In_{InI_3} decreases significantly when the annealing temperature increases from 4 before annealing to 2 after annealing at 200 °C (Tab. 3). The binding energy of the iodine doublet is $In3d5/2 = 618.2$ eV and 619.1 when it is bounded as CsI [16] and InI_3 [15], respectively. As a matter of fact the iodine signal corresponds to two doublets the main one, situated at 618.8 eV, can be attributed to the indium compound. The second one situated at around 618.3, can be attributed to CsI (Fig. 7). All this means that during the annealing there is InI_3 formation. In the case of Cs_2CO_3, the reaction is not so easy to put in evidence. However, here also, the In3d doublet can be decomposed into two doublets, the one situated at 444.3 eV corresponds at ITO and the second one,

which is smaller (Tab. 3) situated at 445 corresponds to some hydroxide compound such as $In(OH)_3$ [17]. During the annealing, as shown in Table 3, the contribution of the doublet corresponding to the ITO decreases for the benefit of the doublet corresponding to the hydroxide. This is confirmed by the evolution of the peak O1s of the oxygen. As it can be seen in Figure 8, before annealing and after annealing at 100 °C the O1s peak can be decomposed in three components, however their relative intensities are modified by the annealing. The peak situated at 531.6 eV corresponds to Cs_2CO_3 [18], that at 529.5 eV corresponds to ITO and that at 530.6 eV can be attributed to $In(OH)_3$. The relative intensity of this last contribution increases from 14 at.% before annealing to 27 at.% after annealing at 100 °C which corroborates the fact that the chemical interaction between the caesium compounds CBL with ITO is increased by the annealing process. This discussion is consolidated by the evolution of the binding energy Cs. Actually, the Cs3d doublets shift slightly toward higher binding energy upon annealing, which suggests that decomposition of Cs compounds occurs [18].

The cyclic voltammograms of ITO and all other samples with CBL layers are shown in Figures 9a–9g. The complete CV cycle is necessary to confirm the correctness of the experimental data and we got excellent CV traces for these samples as seen from Figure 9. To calculate the work function from the CV the onset oxidation potential is evaluated from the CV graph. This is usually done by noting the potential corresponding to the intersection of the two tangents drawn at the rising current and baseline charging current of the CV traces. But in this case,

Table 4. Variation of Φ_f with the nature of anode and annealing temperature.

No. in Figure 9	Sample	Annealing temperature ($^\circ$C)	Work function (eV)	Tolerance (eV)
1	ITO	–	4.51	± 0.01
2	ITO-CsCO$_3$	–	4.49	± 0.01
3	ITO-CsCO$_3$	100	4.42	± 0.02
4	ITO-CsCO$_3$	150	4.31	± 0.01
5	ITO-CsI	–	4.41	± 0.02
6	ITO-CsI	100	4.40	± 0.01
7	ITO-CsI	150	4.38	± 0.01

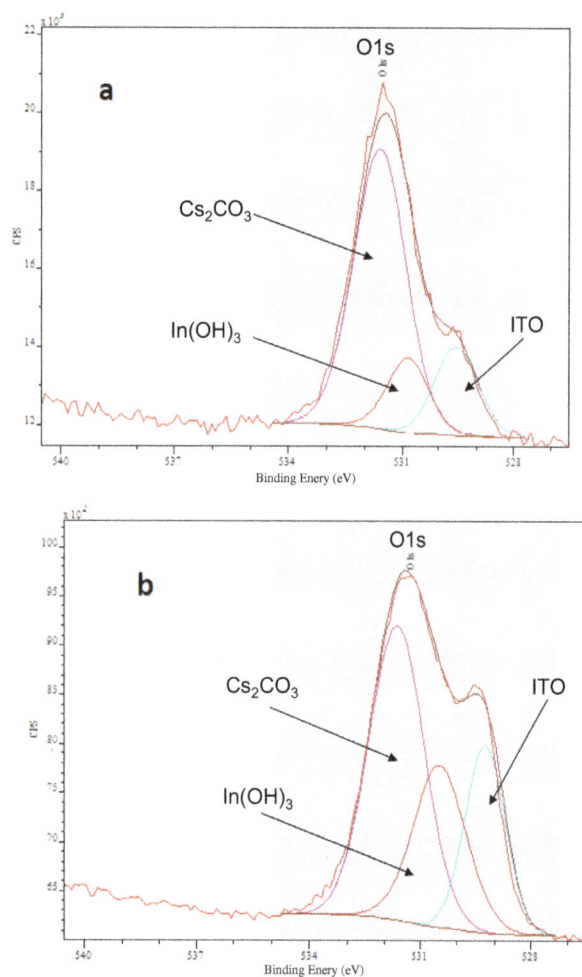

Fig. 8. Decomposition of the O1s XPS spectra of Cs$_2$CO$_3$/ITO structure before (a) and after (b) annealing at 100 $^\circ$C.

the required rising current peak is too small and thus CV graph is be magnified to see the correct appearance of the peak. Thus here the onset peak is found to be occurring between –0.5 V to +0.5 V. Here baseline is considered as the x-axis because the charging is started from 0 V.

Table 4 gives the deduced values of the work function. The measured values of the work function Φ_f of the different cathodes show that whatever the CBL used, it induces systematically a decrease of the value of Φ_f and that the more the temperature is raised, the higher is the reduction

in work function (Tab. 4). Of course, the technique used to measure Φ_f may not be one with extreme precision. However, it gives a relative variation of Φ_f. Band alignment and work functions in organic electronic materials are generally evaluated by three methods: cyclic voltammetry (CV), ultraviolet photoelectron spectroscopy (UPS) or Kelvin Probe techniques (KP). CV analysis employs the method of measurement of the reduction and oxidation potentials either for isolated molecules in solution or for thin films submerged in solvent, and gives the value of ionization potential. The working range of these measurements is limited [19] by the electrochemical stability of the solvent and values may not always correlate well with values observed at the solid state interfaces in actual devices. UPS analysis can provide both the ionization potential (IP), as well as its work function in ultrahigh vacuum (UHV). However, this method also has limitations [20–22] as the measurements made by UPS are extremely sensitive to surface conditions leading to a range of values reported for common materials due to variations in fabrication and handling history. Also it has been pointed out [19] that exposure of films to UHV conditions can also alter [23] the surface composition, particularly for metal oxide surfaces used in hybrid photovoltaics devices fabricated in ambient lab conditions. Often it has been found that the UPS determined value is about 0.3 eV lower than the KP value.

It has also been pointed out [19] that the discrepancy between KP and UPS values is well-documented [23–26]. Variation in the reference values will directly contribute to the differences observed by UPS and KP. Further fundamental differences exist [24] in the work function measuring principles of KP and UPS. UPS measures that of the lowest energy electrons that escape the film surface upon excitation with the UV source and invariably gives only lowest work function in the measured area. On the other hand KP measures [24] the average work function of the area just below the probe. This naturally ends up in higher work function values for KP than that of UPS as surfaces would not be uniform for all practical purposes. Another major problem with UPS is the degradation of the surface at UHV as it may reduce [19] the quantity of volatile surface adsorbates on the films, resulting in a work function difference. In the case of KP, it has been reported [24] discrepancies between KP values measured with three different instruments. A very important draw back of KP was reported [24] as though it is more sensitive than UPS it fails in detecting ITOs with different surface treatments. However, it has been shown [27] that

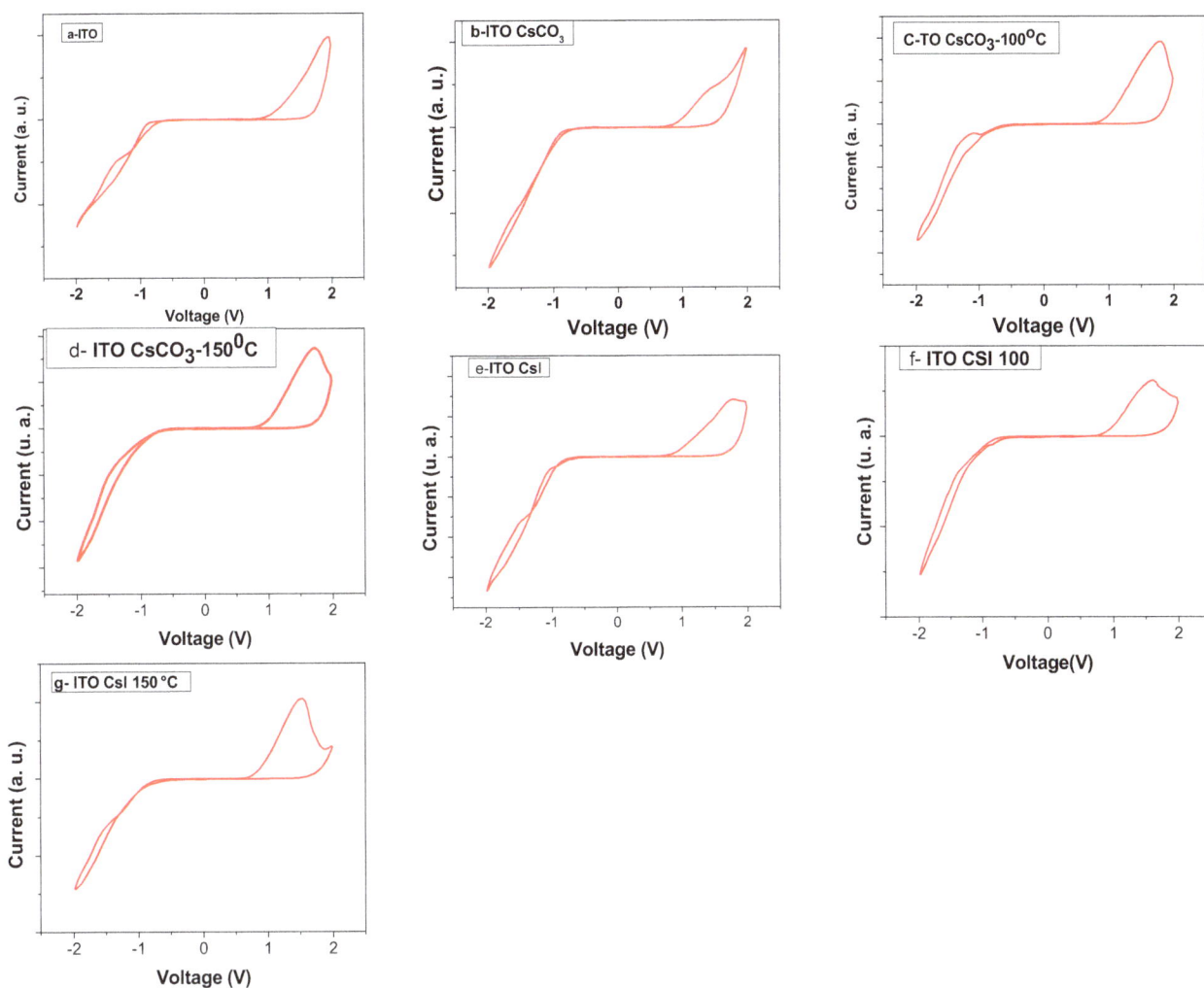

Fig. 9. Cyclic voltammograms of ITO and ITO coated with buffer layers.

CV does this nicely and convincingly. From the above discussion it is clear that none of these methods can be considered as absolute and all have distinct advantages and drawbacks. However the important thing is that the value of work function measured by all these three techniques differ only slightly in the range of 0.1 to 0.3 eV only and while studying the variation of work function with respect to a parameter all are capable of similar trends. Thus, the CV measurement of work function presented here gives a more or less accurate estimation of the degree of variation of work function on modifying the ITO electrodes with buffer layers.

As a matter of fact, it was already shown that the use of a caesium compound as CBL induces a decrease of the work function, Φ_f, of the cathode, which allows improving the band matching at the interface cathode electron acceptor [28, 29]. Xiao et al. [14] attribute this decrease of Φ_f to Cs-O bonds formation at the ITO surface. In the present work, we show that the indium of ITO reacts chemically with the caesium compound, which justifies this work function variation. The efficiency of the chemical reaction increases with the annealing temperature, which improve the decrease in Φ_f of the cathode.

4 Conclusion

We show that CsI is an efficient CBL between the ITO cathode and fullerene in inverted planar solar cells. It is as well, and even more, effective as Cs_2CO_3. When deposited under vacuum by thermal heating, the CsI is iodine deficient. This deficiency is amplified by the annealing to which it is necessary to submit the modified cathode to achieve efficient electron collection. Actually, we show by XPS analysis the presence of a chemical reaction between the indium of ITO and the anion of the caesium compound. This chemical reaction allows the decrease of the work function of the cathode. This decrease was checked through work function measurements using the cyclic voltammetry method. The efficiency of the IOPVCs was optimized by varying the annealing temperature and the CsI layer thickness. Obviously, the short circuit current J_{sc} is the more sensible parameter. The best IOPVCs were obtained after an annealing of 5 min at 150 °C, with a CsI thickness of 1.5 nm. IOPVCs with Cs_2CO_3 were studied for comparison with CsI. The results suggest that CsI is, at least, as efficient as Cs_2CO_3 as CBL in planar organic photovoltaic cells.

This work was supported by the France-Maroc contract: PHC Toubkal and the Hassan II Academy of Science and Technology (Morocco).

References

1. J.C. Bernède, J. Chil. Chem. Soc. **53**, 1549 (2008)
2. C. Tao, S.P. Ruan, X.D. Zhang, G.H. Xie, L. Shen, X.Z. Kong, W. Dong, C.X. Liu, W.Y. Chen, Appl. Phys. Lett. **93**, 193307 (2008)
3. C.Y. Jiang, X.W. Sun, D.W. Zhao, A.K.K. Kyaw, Y.N. Li, Sol. Energy Mater. Sol. Cells **94**, 1618 (2010)
4. M. Hösel, R.R. Sondergaard, M. Jorgensen, F.C. Krebs, Energy Technol. **1**, 102 (2013)
5. J.C. Bernède, L. Cattin, M. Morsli, Y. Berredjem, Sol. Energy Mater. Sol. Cells **92**, 1508 (2008)
6. L. Cattin, S. Tougaard, N. Stephant, S. Morsli, J.C. Bernède, Gold Bull. **44**, 199 (2011)
7. L. Cattin, F. Dahou, Y. Lare, M. Morsli, R. Tricot, K. Jondo, A. Khelil, K. Napo, J.C. Bernède, J. Appl. Phys. **105**, 034507 (2009)
8. F.-Z. Sun, A.-L. Shi, Z.-Q. Xu, H.-X. Wei, Y.-Q. Li, S.-T. Lee, J.-X. Tang, Appl. Phys. Lett. **102**, 133303 (2013)
9. A.S. Yapi, L. Toumi, Y. Lare, G.M. Soto, L. Cattin, K. Toubal, A. Djafri, M. Morsli, A. Khelil, M.A. Del Valle, J.-C. Bernède, Eur. Phys. J. Appl. Phys. **50**, 30403 (2010)
10. D.-W. Chou, K.-L. Chen, C.-J. Huang, Y.-J. Tsao, W.-R. Chen, T.-H. Meen, Thin Solid Films **536**, 235 (2013)
11. Z.-Q. Xu, J.-P. Yang, F.-Z. Sun, S.-T. Lee, Y.-Q. Li, Organic Electronics **13**, 697 (2012)
12. Y. Li, D.-Q. Zhang, L. Duan, R. Zhang, L.-D. Wang, Y. Qiu, Appl. Phys. Lett. **90**, 012119 (2007)
13. K. Morii, T. Kawase, S. Inoue, Appl. Phys. Lett. **92**, 213304 (2008)
14. T. Xiao, W. Cui, M. Cai, W. Leung, J.W. Anderegg, J. Shinar, R. Shinar, Org. Electron. **14**, 267 (2013)
15. B.H. Freeland, J.J. Habeed, D.G. Tuck, Can. J. Chem. **55**, 1528 (1977)
16. W.E. Morgan, J.R. Van Wazer, W.J. Stec, J. Am. Chem. Soc. **95**, 751 (1997)
17. M. Faur, M. Faur, D.T. Jayne, M. Goradia, C. Goradia, Surf. Interface Anal. **15**, 641 (1990)
18. H.-H. Liao, L.-M. Chen, Z. Xu, G. Li, Y. Yang, Appl. Phys. Lett. **92**, 173303 (2008)
19. R.J. Davis, M.T. Lloyd, S.R. Ferreira, M.J. Bruzek, S.E. Watkins, L. Lindell, P. Sehati, M. Fahlman, J.E. Anthony, J.W.P. Hsu, J. Mater. Chem. **21**, 1721 (2011)
20. Y. Park, V. Choong, Y. Gao, B.R. Hsieh, C.W. Tang, Appl. Phys. Lett. **68**, 2699 (1996)
21. C.C. Wu, C.I. Wu, J.C. Strum, A. Kahn, Appl. Phys. Lett. **70**, 1348 (1997)
22. K. Sugiyama, H. Ishii, Y. Ouchi, K. Seki, J. Appl. Phys. **87**, 295 (2000)
23. M. Lira-Cantu, F.C. Kreb, Sol. Energy Mater. Sol. Cells **90**, 2076 (2006)
24. J.S. Kim, B. Lagel, E. Moons, N. Johansson, I.D. Baikie, W.R. Salaneck, R.H. Friend, F. Cacialli, Synth. Met. **111-112**, 311 (2000)
25. W. Osikowicz, M.P. de Jong, S. Braun, C. Tengstedt, M. Fahlman, W.R. Salaneck, Appl. Phys. Lett. **88**, 193504 (2006)
26. E. Ito, H. Oji, N. Hayashi, H. Ishii, Y. Ouchi, K. Seki, Appl. Surf. Sci. **175-176**, 407 (2001)
27. H.D. Kwak, D.S. Choi, Y.K. Kim, B.C. Sohn, Mol. Cryst. Liq. Cryst. **370**, 47 (2001)
28. Y.-I. Lee, J.-H. Youn, M.-S. Ryu, J. Kim, H.-T. Moon, J. Jang, Sol. Energy Mater. Sol. Cells **99**, 3276 (2011)
29. B. Park, J.C. Shin, C.Y. Cho, Sol. Energy Mater. Sol. Cells **108**, 1 (2013)
30. M.M. Beerbom, B. Lagel, A.J. Cascio, B.V. Doran, R. Schlaf, J. Electron. Spectrosc. Relat. Phenom. **152**, 12 (2006)

The role of front and back electrodes in parasitic absorption in thin-film solar cells

Mathieu Boccard[a], Peter Cuony, Simon Hänni, Michael Stuckelberger, Franz-Josef Haug, Fanny Meillaud, Matthieu Despeisse, and Christophe Ballif

École Polytechnique Fédérale de Lausanne (EPFL), Institute of Microengineering (IMT), Photovoltaics and Thin Film Electronics Laboratory, Rue de la Maladière 71b, CP 526, 2002 Neuchâtel 2, Switzerland

Abstract When it comes to parasitic absorption in thin-film silicon solar cells, most studies focus on one electrode only, most of the time the substrate (in n-i-p configuration) or superstrate (in p-i-n configuration). We investigate here simultaneously the influence of the absorption in both front and back electrodes on the current density of tandem micromorph solar cells in p-i-n configuration. We compare four possible combinations of front and back electrodes with two different doping levels, but identical sheet resistance and identical light-scattering properties. In the infrared part of the spectrum, parasitic absorption in the front or back electrode is shown to have a similar effect on the current generation in the cell, which is confirmed by modeling. By combining highly transparent front and back ZnO electrodes and high-quality silicon layers, a micromorph device with a stabilized efficiency of 11.75% is obtained.

1 Introduction

Minimizing the absorption of non-photoactive layers is of crucial importance for most solar cell technologies. When a direct-band-gap photoactive material (such as gallium-arsenide) is used, parasitic absorption impinges on the photon recycling process and reduces the achievable voltage output [1]; whereas when indirect-band-gap photoactive material (such as silicon) is used, this parasitic absorption impinges on the current density output [2–5]. In the case of thin-film silicon technology, parasitic absorption due to non-active layer has been identified as a major limitation for present devices [6,7].

We focus in the following on micromorph tandem cells, which consist of a hydrogenated amorphous silicon (a-Si:H) top cell and a hydrogenated microcrystalline silicon (μc-Si:H) bottom cell. For this device, light trapping strategies are applied for a broad wavelength range of the solar spectrum, covering the near infrared (NIR) range between 600 nm and 1100 nm. The absorption coefficient of μc-Si:H silicon being weak in the NIR, efficient light trapping (typically via nanotextured interfaces) is mandatory to achieve large currents [7–14]. Yet, in the presence of a very efficient light-trapping scheme, parasitic absorption (A_P) due to the non-photoactive layers (such as the doped layers, the electrodes and the back

reflector) is also increased compared to a flat cell configuration [6,15–17]. For transparent conductive oxide (TCO) based front electrodes, decreasing the free-carrier density enables to greatly increase the transparency in the NIR [17–19].

In this work, we separate the influence of parasitic absorption in the front and back electrodes by using ZnO in both cases. Two different doping levels are compared for both electrodes, leading to four possible combinations of front and back electrodes. The sheet resistance is kept identical in all cases, and the light-scattering properties of all front electrodes are identical as well. Both electrodes are shown to have similar influence in the NIR spectrum range, which is confirmed by simulations.

2 Experimental details

The devices studied here are micromorph cells deposited by plasma enhanced chemical vapor deposition (PECVD). An in-situ 70-nm-thick silicon-oxide based intermediate reflecting layer (SiO-based IRL) [20,21] is inserted between the 240-nm-thick a-Si:H top cell and the 2.8-μm-thick μc-Si:H bottom cell. Front and back electrodes are made of ZnO deposited by low pressure chemical vapor deposition (LPCVD).

Figure 1 schematically presents the four micromorph devices compared here. Their front and back electrodes

---

[a] e-mail: `mathieu.boccard@epfl.ch`

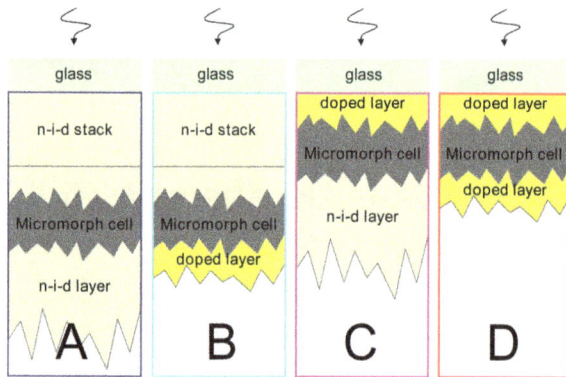

Fig. 1. Schematic view of the stacks composing the A, B, C and D micromorph devices. Only the front and back electrodes are changed. Light enters the devices from the top.

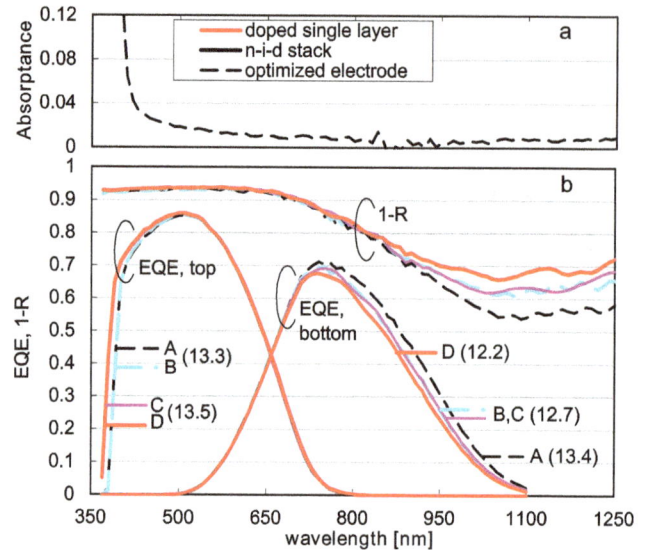

Fig. 2. (a) Absorptance of ZnO front electrodes used in this study as sketched in Figure 1, and of the optimized electrode used in the device of Figure 4. These measurements were made in air with an index-matching liquid (CH_2I_2). (b) EQE and 1-Reflection (1-R) of the four micromorph devices sketched in Figure 1 having only their ZnO electrodes changed. The number in parenthesis are current density values for the top and bottom cells, in mA/cm^2.

are the only changing parameters and two options are used: thick, non-intentionally-doped (n-i-d) LPCVD ZnO or thin, doped LPCVD ZnO. The free-carrier density in the n-i-d ZnO is around 1.2×10^{19} cm^{-3} and it is 1.0×10^{20} cm^{-3} in the doped one. The sheet resistance is kept similar for all electrodes (between 7 $\Omega/\square$ and 12 $\Omega/\square$) thanks to the much larger thickness in the n-i-d case. In order to keep the same morphology for both types of front electrodes, a stack of two layers is used in the n-i-d case. A 7-μm-thick layer is first deposited on glass and flattened by chemo-mechanical polishing (CMP) to erase the very large pyramidal features that naturally form on its surface during layer growth [22]. This polishing is made by immersing the sample in a diluted suspension of silica nanoparticles as typically used for CMP, and scanning the surface with a felt-covered drill-head. To inhibit epitaxial growth on this polished surface, an ultra-thin (<3 nm) n-doped μc-Si:H layer is subsequently deposited; a n-i-d ZnO film is then deposited on top, with thickness adjusted to match the surface morphology of the 2-μm-thick doped layer. Due to a reduction of the lateral growth rate of the grains when LPCVD ZnO is doped [22], this rough n-i-d layer is only 1.6-μm-thick.

For the back electrode, both electrodes are single-layers, 7-μm-thick for the n-i-d case and 2-μm-thick for the doped one. Though these back electrodes exhibit therefore different surface roughnesses, this was shown to have no noticeable impact on the light-trapping behavior of such devices, due to the efficient light scattering provided by the front electrode [23].

3 Results and discussion

Figure 2a compares absorptance measurements of the 8-μm-thick n-i-d ZnO stack and the 2-μm-thick doped ZnO layer. For wavelengths up to 550 nm the doped layer absorbs less than the n-i-d stack. This is due first to the Burstein-Moss effect, widening the band-gap in the case of the doped ZnO [24], but also to enhanced sub-gap absorption in the n-i-d stack because of the larger thickness. However, above 550 nm, the higher free-carrier absorption

(FCA) in the doped single layer makes the n-i-d stack more transparent despite its larger thickness.

Figure 2b shows the external quantum efficiency (EQE) spectrum of top and bottom cells for the four possible micromorph configurations employing these electrodes in front and back. The similarity of the morphology of all front electrodes is assessed by the perfect matching of the crossing of the top and bottom EQE curves for all substrates at the same wavelength (650 nm) [25]. The top cell current is only affected for such a thick cell by the front electrode (A, B and C, D are superimposed), and the higher optical band gap of the thin doped TCO layer enables a 0.2 mA cm^{-2} current gain. Then, the bottom cell is sensitive to the transparency of both electrodes, and in a very similar way, as deduced from the striking superimposition of the B and C curves and the identical current densities. This can also be seen in Figure 3a, which indicates the relative parasitic absorption for all cases compared to case A, evidencing the identical effect of FCA in the back or front electrode.

However, the current gained in the bottom cell by improving the transparency of the back electrode (i.e. going from B to A or from D to C) depends on the front electrode properties: 0.7 mA cm^{-2} are gained between B and A, whereas only 0.5 mA cm^{-2} is gained between D and C. This is illustrated with curves of relative spectral current density gain in Figure 3b. The same applies when improving the transparency of the front electrode: 0.7 mA cm^{-2} are gained in the bottom cell for C to A, but only 0.5 mA cm^{-2} for D to B. The current gain when making one electrode more transparent therefore scales with

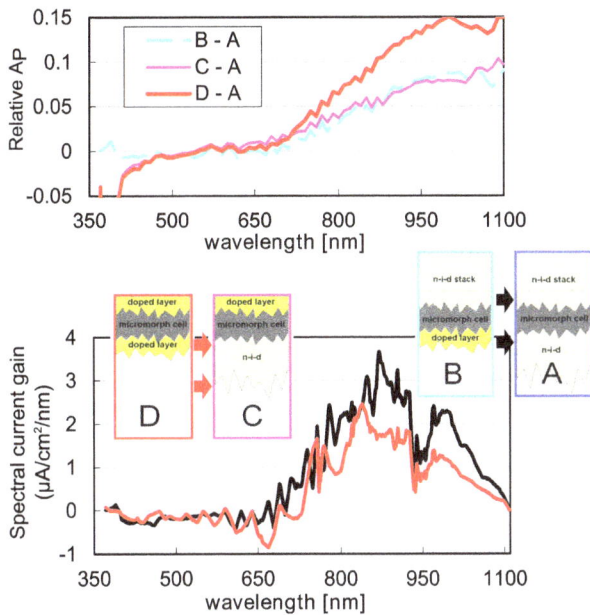

Fig. 3. (a) Relative parasitic absorption in the four micromorph devices sketched in Figure 1 compared to case A where both electrodes are non-intentionally-doped doped, and thus very transparent. (b) Spectral current density gain between configurations B and A, or D and C. In both cases, the same modification is made at the back electrode.

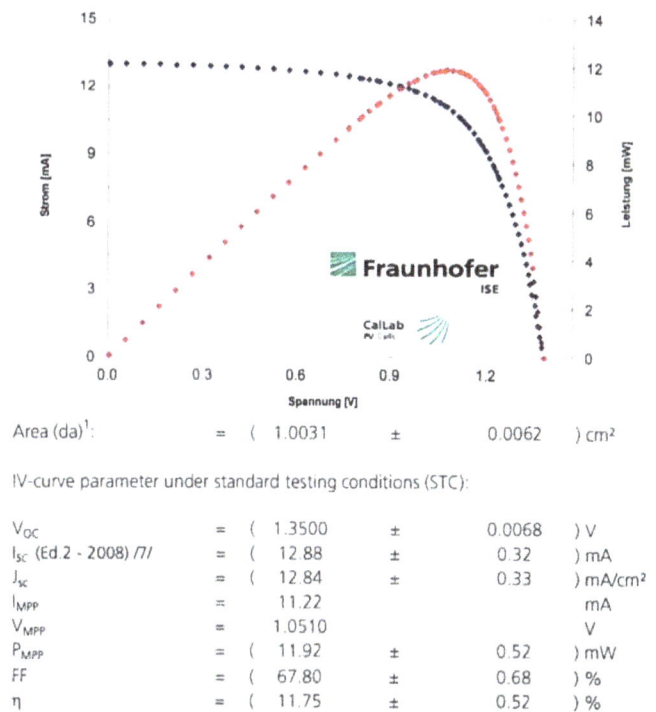

Fig. 4. IV curve of a micromorph cell with an efficiency of 11.75%, deposited using highly transparent ZnO electrodes.

the transparency of the other electrode, a more transparent unchanged electrode leading to a larger gain.

This difference comes from the fact that the light saved from parasitic absorption in one of the electrodes (by making it more transparent) will be distributed between valuable absorption (EQE), escape from the cell (R) and parasitic absorption in the remaining parasitically absorbing layers (A_P). The probability of the latter increases with the parasitic absorption of the unchanged electrode, as will be seen with modeling in the next section. This indicates that the largest gains from parasitic absorption reduction are expected when improving already well transparent electrodes. However, as evidenced by Battaglia et al. [2], doped layers also contribute to parasitic absorption. Their contribution to parasitic absorption might therefore at some point become larger than the contribution of the electrode, making further improving the electrode transparency of little effect.

By combining high-quality silicon layers and highly transparent electrodes, a micromorph cell with 11.75% efficiency is obtained, as shown in Figure 4. The front and back electrodes are 2.3-μm-thick LPCVD ZnO layers with a low doping (carrier density of 2×10^{19}) and a high mobility (over 50 cm^2/V/s), resulting in a sheet resistance below 30 $\Omega/\square$ for a high transparency as seen in Figure 2a [25]. The thickness of the intrinsic layer of the a-Si:H top cell is 230 nm, and the thickness of the intrinsic layer of the μc-Si:H bottom cell is 2.3 μm. An 80-nm-thick SiO-based IRL with a refractive index of 1.9 is used between the top and bottom cells, an intrinsic SiO buffer layer was used in the a-Si:H cell and a p-type SiO layer was used in the μc-Si:H cell [20,26]. The a-Si:H cell

and the SiO-based IRL were made in an Octopus I cluster tool made by INDEOtec SA, whereas the μc-Si:H cell was made in a research-scale reactor developed in-house [27].

4 Modeling

Figure 5 shows various spectra obtained when simulating this set of experimental data with the model presented in reference [7]. Only a μc-Si:H cell is used for the model, and the resulting simulated EQE curve is compared to the curve obtained by summing the top- and bottom-cell EQE curves. In absence of a correct AFM scan of the front-electrode surface, the light-trapping parameters of the model (a_0, b_1 and R_0) are fitted once for all four cases in absence of parasitic absorption using a "corrected EQE" (EQE$_c$) curve, calculated from experimental EQE and R curves of one of the devices [16]. Then, the measured absorptance spectra of the two different types of layers are implemented in A_{front} and A_{back}, and the p-layer thickness is kept identical for all curves.

An excellent agreement between experimental and simulated EQE and R curves is obtained in the infrared part of the spectrum. A notable discrepancy of the simulation compared to the measurement is observed, similarly for all devices, in the 600 nm–900 nm wavelength range. This is attributed to the many additional layers present in the real micromorph device compared to the simplified model considering only a μc-Si:H cell, such as the intermediate reflector and a-Si:H doped layers. These layers induce extra parasitic absorption and reflection, that is not taken into account in the model.

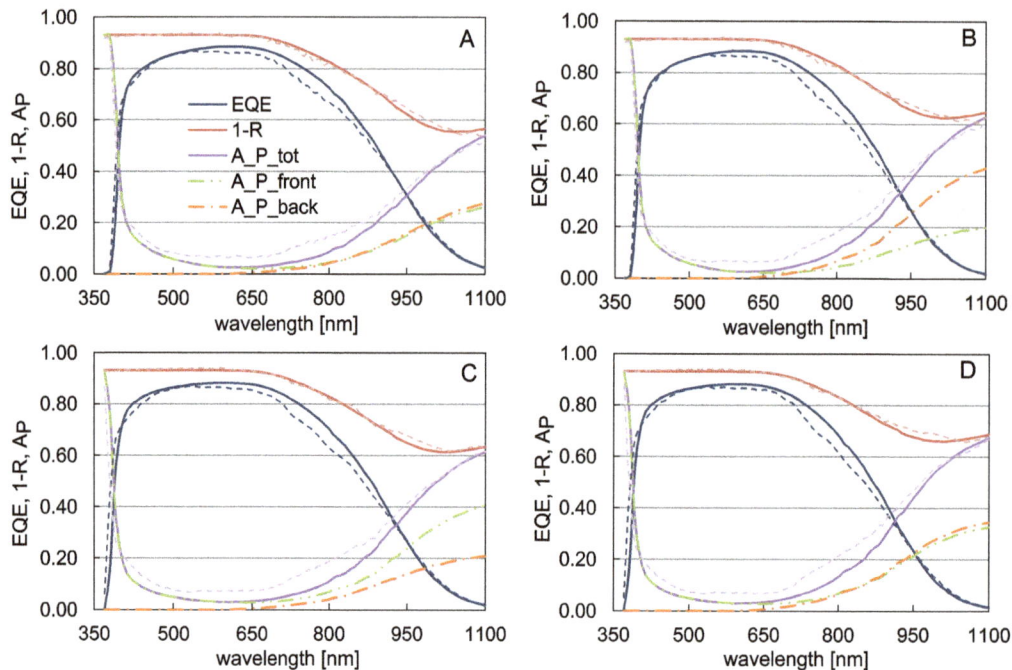

Fig. 5. Simulated (solid) and measured (dashed) EQE, total absorptance $(1-R)$ and front, back and total A_P curves simulated from experimental absorption data for the micromorph devices presented in Figure 1.

The parasitic absorption curve experimentally deduced from $A_P = 1 - \mathrm{EQE} - R$ is also plotted, as well as the curves representing calculated parasitic absorption which is attributed to the front side of the device $(A_P{}^{front})$, to the back side $(A_P{}^{back})$, and the total parasitic absorption $(A_P{}^{tot})$. It should be noted that possible collection issues are assimilated to A_P in the measurements, whereas a 98% collection efficiency was set for the whole spectrum in the model for all four cases – explaining the 2% higher A_P in all cases for the experimental case compared to the modeled case. As suggested by experimental measurements, $A_P{}^{front}$ (resp. $A_P{}^{back}$) is also influenced by the transparency of the layers at the back (resp. front) of the device: $A_P{}^{front}$ is different between cases A and B, even though all layers at the front of the cell are unchanged. Also, for wavelengths larger than 700 nm, very similar $A_P{}^{front}$ and $A_P{}^{back}$ are observed in symmetrical devices (A and D), which is in agreement with experimental findings.

5 Conclusion

We showed in this work that parasitic absorption in both electrodes of a solar cell has similar effect in the infra-red part of the spectrum. We assessed this result with simulations which reproduce well the observed trends. We also emphasized that a reduction of parasitic absorption in one electrode yields a larger EQE gain when the other electrode is made more transparent. Reducing the doping of TCO layers and compensating by thickness increase is shown to be an efficient way of reducing parasitic absorption in these layers. By using very transparent TCO layers and high-quality silicon layers, a micromorph device with 11.75% stable efficiency, independently certified by Fraunhofer ISE was obtained.

We thank the Swiss Federal Office for Energy for funding under grant SI/500750-01. A part of this work was carried out in the framework of the FP7 project "Fast Track", funded by the EC under Grant agreement No. 283501.

References

1. O.D. Miller, E. Yablonovitch, S.R. Kurtz, IEEE J. Photovolt. **2**, 303 (2012)
2. C. Battaglia, L. Erni, M. Boccard, L. Barraud, J. Escarre, K. Soederstroem, G. Bugnon, A. Billet, L. Ding, M. Despeisse, F.-J. Haug, S. De Wolf, C. Ballif, J. Appl. Phys. **109**, 114501 (2011)
3. S. Calnan, A.N. Tiwari, Thin Solid Films **518**, 1839 (2010)
4. T. Koida, H. Fujiwara, M. Kondo, Sol. Energy Mater. Sol. Cells **93**, 851 (2009)
5. Y. Tsunomura, Y. Yoshimine, M. Taguchi, T. Baba, T. Kinoshita, H. Kanno, H. Sakata, E. Maruyama, M. Tanaka, Sol. Energy Mater. Sol. Cells **93**, 670 (2009)
6. H.W. Deckman, C.R. Wronski, H. Witzke, E. Yablonovitch, Appl. Phys. Lett. **42**, 110968 (1983)
7. M. Boccard, C. Battaglia, F.-J. Haug, M. Despeisse, C. Ballif, Appl. Phys. Lett. **101**, 151105 (2012)
8. B. Stannowski, O. Gabriel, S. Calnan, T. Frijnts, A. Heidelberg, S. Neubert, S. Kirner, S. Ring, M. Zelt, B. Rau, J.-H. Zollondz, H. Bloess, R. Schlatmann, B. Rech, Sol. Energy Mater. Sol. Cells **119**, 196 (2013)
9. H. Sai, T. Koida, T. Matsui, I. Yoshida, K. Saito, M. Kondo, Appl. Phys. Express **6**, 104101 (2013)
10. A. Feltrin et al., Sol. Energy Mater. Sol. Cells **119**, 219 (2013)

11. H. Tan, R. Santbergen, A.H.M. Smets, M. Zeman, Nano Lett. **12**, 4070 (2012)

12. B. Lipovšek, J. Krč, O. Isabella, M. Zeman, M. Topič, Phys. Stat. Sol. (c) **7**, 1041 (2010)

13. J. Krc et al., J. Non-Cryst. Solids **352**, 1892 (2006)

14. J. Müller, B. Rech, J. Springer, M. Vanecek, Sol. Energy **77**, 917 (2004)

15. J.M. Gee, in *Photovoltaic Specialists Conference, Conference Record of the Twentieth IEEE, Las Vegas, USA, 1988*, pp. 549–554

16. M. Boccard, P. Cuony, C. Battaglia, M. Despeisse, C. Ballif, Phys. Stat. Sol. Rapid Res. Lett. **4**, 326 (2010)

17. M. Berginski, J. Hüpkes, M. Schulte, G. Schöpe, H. Stiebig, B. Rech, M. Wuttig, J. Appl. Phys. **101**, 074903 (2007)

18. Youn Ho Heo, Dong Joo You, Hyun Lee, Sungeun Lee, Heon-Min Lee, Sol. Energy Mater. Sol. Cells **122**, 107 (2014)

19. L. Ding, M. Boccard, G. Bugnon, M. Benkhaira, S. Nicolay, M. Despeisse, F. Meillaud, C. Ballif, Sol. Energy Mater. Sol. Cells **98**, 331 (2012)

20. P. Cuony, D.T.L. Alexander, I. Perez-Wurfl, M. Despeisse, G. Bugnon, M. Boccard, T. Soderstrom, A. Hessler-Wyser, C. Hebert, C. Ballif, Adv. Mater. **24**, 1182 (2012)

21. A. Lambertz, T. Grundler, F. Finger, J. Appl. Phys. **109**, 113109 (2011)

22. S. Faÿ, L. Feitknecht, R. Schlüchter, U. Kroll, E. Vallat-Sauvain, A. Shah, Sol. Energy Mater. Sol. Cells **90**, 2960 (2006)

23. P. Cuony, Ph.D. thesis, EPFL, 2012

24. J. Steinhauser, Ph.D. thesis, University of Neuchâtel, 2008

25. M. Boccard, T. Soderstrom, P. Cuony, C. Battaglia, S. Hanni, S. Nicolay, L. Ding, M. Benkhaira, G. Bugnon, A. Billet, M. Charriere, F. Meillaud, M. Despeisse, C. Ballif, IEEE J. Photovolt. **2**, 229 (2012)

26. G. Bugnon, G. Parascandolo, S. Hänni, M. Stuckelberger, M. Charrière, M. Despeisse, F. Meillaud, C. Ballif, Sol. Energy Mater. Sol. Cells **120**, 143 (2014)

27. S. Hänni, G. Bugnon, G. Parascandolo, M. Boccard, J. Escarré, M. Despeisse, F. Meillaud, C. Ballif, Prog. Photovolt.: Res. Appl. **21**, 821 (2013)

29

Recrystallized thin-film silicon solar cell on graphite substrate with laser single side contact and hydrogen passivation

Da Li[1,a], Stephan Wittmann[1], Thomas Kunz[1], Taimoor Ahmad[1], Nidia Gawehns[1], Maik T. Hessmann[1], Jan Ebser[2], Barbara Terheiden[2], Richard Auer[1], and Christoph J. Brabec[1,3]

[1] Bavarian Center for Applied Energy Research (ZAE Bayern), Haberstr. 2a, 91058 Erlangen, Germany
[2] Department of Physics, University of Konstanz, Box 676, 78457 Konstanz, Germany
[3] Institute of Materials for Electronics and Energy Technology (i-MEET), University of Erlangen-Nuremberg, Martensstr. 7, 91058 Erlangen, Germany

Abstract Laser single side contact formation (LSSC) and the hydrogen passivation process are studied and developed for crystalline silicon thin film (CSiTF) solar cells on graphite substrates. The results demonstrate that these two methods can improve cell performance by increasing the open circuit voltage and fill factor. In comparison with our previous work, we have achieved an increase of 3.4% absolute cell efficiency for a 40 μm thick 4 cm^2 aperture area silicon thin film solar cell on graphite substrate. Current density-voltage (J-V) measurement, quantum efficiency (QE) and light beam induced current (LBiC) are used as characterization methods.

For crystalline silicon thin film (CSiTF) solar cells on the foreign substrates, a recrystallization process plays an important role in enlarging the size of the silicon grains in order to reduce the density of the electrically active defects and increase the cell efficiencies, and hence the electron-beam recrystallization [1,2], zone melting recrystallization (ZMR) [3,4] and laser recrystallization [5–9] are widely developed [10–13]. Foreign substrate materials, such as ceramics [14–16] and graphite [17], are generally used for CSiTF cell fabrication with the high temperature approach. Schillinger et al. [18] described a CSiTF cell concept on zircon ceramic substrates using ZMR reached 8.1% conversion efficiency. Graphite substrates can be processed at high temperature. Moreover, they are available at high purity and with thermal expansion characteristics similar to silicon. While standard graphite types will not meet the cost requirements, developments for low-cost types such as biogenic substrates are in progress [19].

In previous work, we have reported a best laboratory cell efficiency of 6.8% with the 4 cm^2 aperture area on the graphite substrate [20]. That cell concept is shown in Figure 1a. The laser edge isolation (LEI) technique was applied to avoid the parasitic electrical connection between the front and back contacts. This technique was proved to be a convenient and accurate method instead of the plasma etching process. By using the LEI technique,

the best parallel resistance, approximately 1.7 Ω m^2, was achieved.

Much effort has been devoted to the metallization of CSiTF solar cells on foreign substrates. Stocks et al. [21] presented the concept of CSiTF solar cells with base front-contacts. Hebling et al. [22] realized the concept by using photolithography. Meanwhile, Lüdemann et al. [23] realized base front-contacts silicon solar cells on a SiC intermediate layer with a graphite substrate. Furthermore, Rachow et al. [24] named the concept as single side contact formation. Different microstructures of single side concepts can be formed using photolithography [21] or reactive ion etching (RIE) [25]. However, the photolithography is an expensive technology, whereas a typical RIE system consists of a vacuum chamber and a plasma generation system. Due to the development of laser techniques, laser processes are nowadays widely used in thin film solar cell fabric processing. Such techniques can now be used to complete a solar cell with single side contact formation in fewer steps.

In this paper, we developed the cell concept on graphite substrates as shown in Figure 1b. The current density-voltage (J-V) curves and quantum efficiency (QE) measurements show the improvements of open circuit voltage (V_{oc}), fill factor (FF), cell efficiency (η), and quantum efficiency due to the hydrogen bulk passivation process [26,27] and laser single side contact (LSSC) formation.

[a] e-mail: da.li@zae-bayern.de

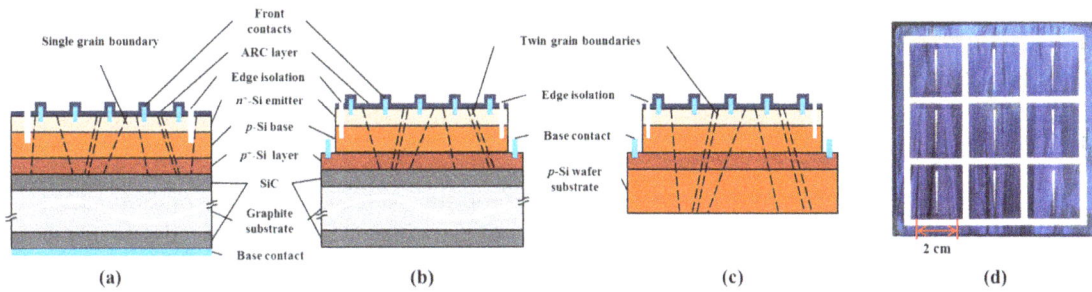

Fig. 1. The three different Si thin-film solar cell designs were studied. (a) The cell in previous work is based on front and back contacts. (b) The new design in this paper is based on laser single side front contacts (LSSC). (c) The reference cells are based on p-type Si wafer. The layers are not to scale. (d) Top view of batch Gra. C, on which are the nine cells with the graphite substrate. The cell No. 5 is in the center.

The combined solar cell process is carried out at a low temperature (<350 °C). The diffusion length was characterized using light beam induced current (LBiC) measurement and simulated by PC1D [28]. Here we report our 10.2% cell efficiency of a 40 μm thick and 4 cm^2 aperture area CSiTF solar cell on graphite substrate.

1 Experimental

1.1 Solar cell design and processing

Three different cell designs (Figs. 1a–1c) were studied. The designs illustrated in Figures 1a and 1b are based on the use of graphite as a foreign substrate, with different concepts of base contact formation. In designing Figure 1c, a multi-crystalline Si-wafer was chosen as a base material for reference purposes, because recrystallized silicon layers on graphite substrates have a similar grain size to typical multi-crystalline Si-wafers. Results from the design of Figure 1a have been described in a previous work [20], while the other two designs are investigated in this work.

For the foreign substrates (Figs. 1a and 1b), high purity graphite substrates ("FP479", Schunk Kohlenstofftechnik GmbH) with a size of 10 cm × 10 cm × 2 mm were used. The SiC layer with a thickness of 10 μm was deposited using a hot-wall chemical vapor deposition, which covered both surfaces of the substrate to prevent the diffusion of impurities. Then, a p^+-Si layer (acceptor concentration of 4×10^{18} cm^{-3}) approximately 20 μm thick was deposited on top of the SiC layer using a convection-assisted chemical vapor deposition (CoCVD) [29]. The p^+-Si layers served as a seed and back surface field layer. The ZMR process was applied to enlarge the size of the silicon grains from the micrometer to the millimeter range. Thus, a random texture surface was formed due to the various crystalline silicon directions [4].

For the reference substrate (Fig. 1c), the p^+-Si layer had the concentration of 4×10^{18} cm^{-3} and was epitaxially grown directly onto the wafer substrate.

Further processing was the same for both the foreign substrate and the Si-wafer substrate. A 20 μm thick epitaxial p-Si base layer was applied over the p^+-Si layer. This p-Si layer had a boron doping concentration of about

Table 1. Sample list. H$^+$ denotes the hydrogen passivation process.

Batches	Substrate	Contacts	H$^+$ [Yes/No]	Duration (min)
Gra. A	Graphite	Both sides	No	–
Gra. B	Graphite	LSSC	No	–
Gra. C	Graphite	LSSC	Yes	50
Waf. A	p-Si wafer	Both sides	No	–
Waf. B	p-Si wafer	LSSC	No	–
Waf. C	p-Si wafer	LSSC	Yes	50
Waf. D	p-Si wafer	LSSC	Yes	50
Waf. E	p-Si wafer	LSSC	Yes	50

2×10^{16} cm^{-3}. The n^+-Si emitter was formed by spin on doping (SOD) of a phosphorous solution followed by rapid thermal processing (RTP) in a furnace [30]. The sheet resistance is in the range from 80 to 120 Ω/square. The laser edge isolation process was applied directly after the removal of the phosphorous glass [20]. The metallization of the front contacts were formed by electron beam evaporation of Ti, Pd and Ag (30 nm, 30 nm, and 5 μm thick, respectively). Finally, using plasma-enhanced chemical vapor deposition (PECVD) we deposited a silicon nitride layer with a thickness of 75 nm, which served as an antireflection coating. The total thickness of the silicon layers was about 40 μm. As shown in Figure 1d, nine cells, each with a size of 2 cm × 2 cm, have been fabricated on the 10 cm × 10 cm graphite substrate.

1.2 Sample batches

As shown in Table 1, we prepared and studied eight different batches for the comparison of LSSC formation and hydrogen passivation process. The batches based on the graphite substrates are named Gra. A, B, and C, whereas the reference batches based on the p-type Si-wafers are named Waf. A, B, C, D, and E. The batches denoted by different capital letters correspond to the different processing sequences as shown in Figures 2 and 3:

– Gra. A and Waf. A are the samples with base contacts on the back of the solar cells. They are based on the design of Figure 1a. The base contact is formed by

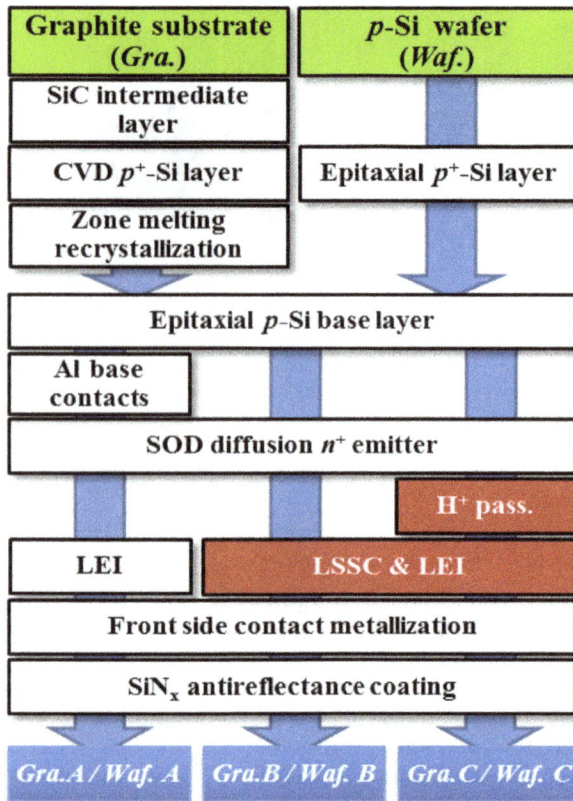

Fig. 2. Processing sequence of the cells. Eight solar cell batches are prepared. Gra. denotes the cells on graphite substrate and Waf. denotes the cells on wafers as reference. Laser edge isolation (LEI), laser single side contact (LSSC) and hydrogen bulk passivation (H$^+$ pass.) were applied in the cell process. Batches Gra. A–C on graphite substrates, reference batches Waf. A–C on Si wafers with the same processing after ZMR as Gra. A–C.

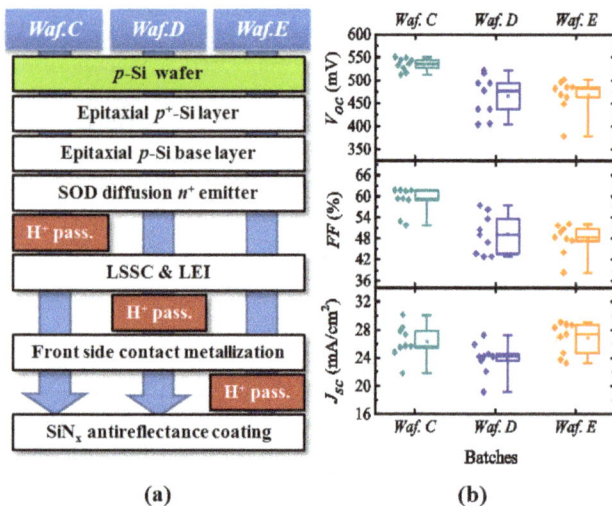

Fig. 3. Reference batches Waf. C–E on Si wafer with a different sequence of hydrogen passivation. (a) Processing sequence of the cells. (b) Open circuit voltage V_{oc} (top), fill factor FF (middle) and current density J_{sc} (bottom) due to the three different hydrogen passivation processes.

a 5 μm thick aluminum layer, which was deposited on the back of the substrate. Hydrogen passivation is not used in these batches.

- Gra. B and Waf. B's base contacts were formed from the front using LSSC. They are based on the design of Figure 1b, but without the hydrogen passivation process.
- Gra. C and Waf. C–E are with LSSC and the hydrogen passivation process. Gra. C is based on the design of Figure 1b. Waf. C–E are based on the design of Figure 1c.

1.3 Laser single side contact

Based on the development of laser technology, we fabricated the base contacts on the front using a laser process, i.e. laser single side contacts instead of the Al back base contacts. Each cell from batches Gra. A and Waf. A had an aluminum base contact, which was evaporated on the back. Here we used a Nd:YVO$_4$ laser (Rofin Power Line LP20, wavelength 1064 nm) to make approximately 30 μm deep trenches into the samples down to the p^+-Si layer and to obtain 3 mm wide contact stripes around the cells. After the laser trenching and LEI process, we evaporated the emitter contacts and the base contacts on the front at the same time and with the same metals Ti, Pd, and Ag.

1.4 Hydrogen passivation

The bulk recombination centers in the CSiTF solar cells can be restrained using a hydrogen passivation process, which is dependent on time and temperature. According to reference [27], the hydrogen passivation process needs less process time at high temperatures than at low temperatures. If the process temperature is 600 °C, the necessary reaction time is only about 5 min. In our case, the samples were put into the chamber at 350 °C for 50 min (microwave power: 1000 W, pressure: 0.04 mbar and hydrogen as the precursor). Experiments were carried out using a PECVD system of Roth and Rau AK1000 with microwave excitation frequency at 2.45 GHz.

The process sequence is critical for hydrogen passivation. As shown in Figure 3, in order to determine the best fabrication sequence of the hydrogen passivation, we prepared and compared three different process sequences. For this study, the reference system on wafer substrates was used in order to exclude influences from varying defect structures, as they may result from the recrystallization process. The hydrogen passivation process was included after

- the SOD diffusion processing (Waf. C);
- the laser trenching and LEI (Waf. D);
- the metallization processing (Waf. E).

1.5 Cell characterization

Internal quantum efficiency (IQE) and reflectance are measured using the pv-tools solar cell analysis system

Table 2. Cell parameters. Open circuit parallel resistance (R_{p}), series resistance (R_{s}), voltage (V_{oc}), current density (J_{sc}), fill factor (FF), and efficiency (η).

Batches	Avg. R_{p} ($\Omega\,\mathrm{m}^2$)	Avg. R_{s} ($\Omega\,\mathrm{cm}^2$)	Avg. V_{oc} (mV)	Avg. J_{sc} (mA/cm^2)	Avg. FF (%)	Avg. η (%)	Best η (%)
Gra. A	1.53	1.52	426	21.3	56.7	5.18	6.79
Gra. B	1.97	1.06	469	234	66.5	7.33	7.81
Gra. C	2.47	1.05	499	262	67.7	8.86	10.2
Waf. A	1.08	2.22	450	255	49.0	5.74	7.68
Waf. B	2.51	2.77	501	250	57.0	7.18	8.19
Waf. C	3.18	2.85	534	263	58.7	8.29	10.1
Waf. D	0.33	3.24	465	23.9	49.2	5.52	7.77
Waf. E	0.03	2.21	467	26.7	47.7	5.97	7.14

LOANA at the University of Konstanz and using the Enlitech solar cell analysis system QE-R at i-MEET. Light beam induced current (LBiC) measurements were carried out by using Semilab WT2000, which has four light sources (976 nm, 951 nm, 846 nm and 662 nm). The diffusion length was calculated from the LBiC measurements at various wavelengths.

2 Results and discussion

2.1 J-V characterization

Table 2 lists the parameters obtained from the illuminated and dark J-V measurements. We arranged the J_{sc}, V_{oc} and FF data of all the samples in the form of a box chart, as shown in Figure 3b. These results were used to identify the best hydrogen passivation sequence, which is described in Section 2.1.1. Furthermore, we observed significant improvement in V_{oc} and FF with hydrogen bulk passivation and LSSC formation in Figure 4, which is described in Section 2.1.2.

2.1.1 Hydrogen passivation sequence on reference batches

Comparing the results of these three batches (Waf. C–E) in Figure 3b and Table 2, we found the best hydrogen passivation sequence is the hydrogen bulk passivation process of Waf. C. This batch has the highest average values of R_{p}, V_{oc}, and FF among the three batches.

In the case of Waf. D, the hydrogen passivation was applied after the laser trenching and LEI process. This process sequence resulted in lower cell performance than Waf. C, which is most probably due to laser induced defects. Slaoui et al. [31] reported the hydrogen passivation of laser induced defects, which demonstrated that the quality of the hydrogen passivation highly depends on the temperature of the passivation process.

In the case of Waf. E, the contacts were already evaporated before the hydrogen passivation process. As shown in Table 2, this process sequence resulted in extremely lower average parallel resistance R_{p} ($\overline{R}_{\mathrm{p}}$) than Waf. C, which further resulted in a low average FF ($\overline{FF}$).

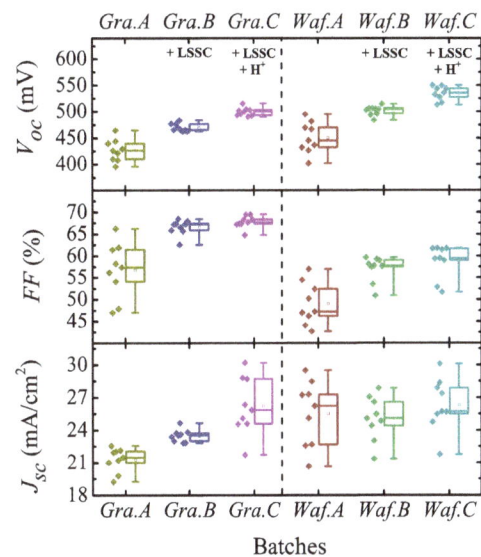

Fig. 4. V_{oc} (top), FF (middle) and J_{sc} (bottom) due to the both-side contact, LSSC formation (i.e. +LSSC), and hydrogen passivation processes (i.e. +H$^+$).

This was assumed due to the fact that the front contacts (Ag/Pd/Ti) penetrated through the thin emitter layer into the base layer during the passivation process at 350 °C [32].

Accordingly, the best hydrogen passivation sequence was made directly after the phosphor diffusion process, which was also applied to fabricate Gra. C.

2.1.2 Batches with hydrogen passivation and LSSC on graphite substrates

In order to confirm the improvement by LSSC formation and hydrogen passivation, we further compared the results of the graphite substrate batches (Gra. A–C) and the reference batches (Waf. A–C). There are three different formations as shown in Figure 2.

As shown in Figure 4, both the graphite-based (Gra.) and reference (Waf.) batches exhibit similar variations in

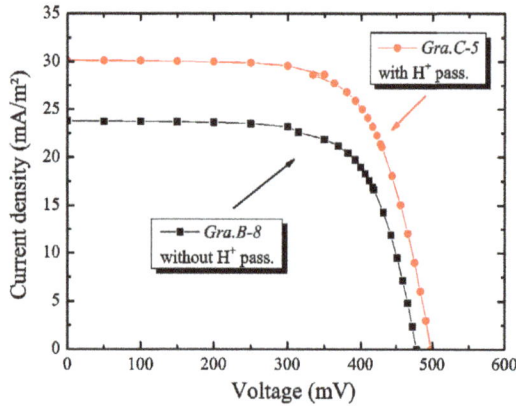

Fig. 5. *J-V* curves of the cells Gra. B-8 and Gra. C-5. (i.e. LSSC formation without (Gra. B) and with (Gra. C) hydrogen passivation).

$\overline{V}_{oc}$ and $\overline{FF}$, which can be described by:

$$\overline{V}_{oc} \text{(Gra. A)} < \overline{V}_{oc} \text{(Gra. B)} < \overline{V}_{oc} \text{(Gra. C)} \quad (1)$$

$$\overline{V}_{oc} \text{(Waf. A)} < \overline{V}_{oc} \text{(Waf. B)} < \overline{V}_{oc} \text{(Waf. C)} \quad (2)$$

$$\overline{FF} \text{(Gra. A)} < \overline{FF} \text{(Gra. B)} < \overline{FF} \text{(Gra. C)} \quad (3)$$

$$\overline{FF} \text{(Waf. A)} < \overline{FF} \text{(Waf. B)} < \overline{FF} \text{(Waf. C)}. \quad (4)$$

Those inequalities indicate that the batches with LSSC formation have better performance than the batches with both sides contact formation. The possible reason is that the LSSC formation had less recombination losses than the both-side contact formation, since the collected current bypassed the intermediate layer and the graphite substrate. Moreover, the hydrogen passivation reduced the bulk recombination, and therefore the batches, with both LSSC formation and hydrogen passivation applied have the best performance in $\overline{V}_{oc}$ and $\overline{FF}$. We also observed that the $\overline{J}_{sc}$ of the cells based on graphite substrates had the same trend:

$$\overline{J}_{sc} \text{(Gra. A)} < \overline{J}_{sc} \text{(Gra. B)} < \overline{J}_{sc} \text{(Gra. C)}. \quad (5)$$

We observed that the values of $\overline{J}_{sc}$ were improved using the LSSC formation by comparison $\overline{J}_{sc}$ (Gra. A) and $\overline{J}_{sc}$ (Gra. B). We also observed $\overline{J}_{sc}$ (Gra. C) was the highest due to the applied LSSC and hydrogen bulk passivation. However, the values of $\overline{J}_{sc}$ of the reference cells based on mc-Si wafers did not have the same trend. According to the process sequence, as shown in Figure 2, the reason can be that the reference cells were with neither an intermediate layer nor a ZMR process. Moreover, the back surfaces of the reference cells were not passivated. Therefore, the improvement using hydrogen bulk passivation cannot counteract the detrimental effect of the back surface recombination.

Figure 5 shows illuminated *J-V* curves of the best cells from Gra. B and C. The best cell in Gra. C (cell No. 5) i.e. Gra. C-5, has 3.17 Ω m^2 parallel resistance, 502 mV V_{oc} and 67.2% fill factor. In contrast, the best cell in Gra. B i.e. Gra. B-8, without the hydrogen bulk passivation has

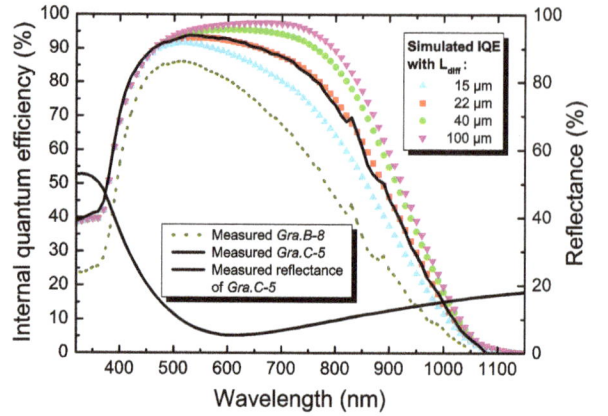

Fig. 6. The reflectance, measured and simulated IQE curves of the cell Gra. C-5. The measured IQE of the cell Gra. B-8 is plotted for comparison. The bulk lifetime was varied for fitting. The IQE is simulated by PC1D for diffusion lengths with 15, 22, 40, and 100 μm.

Table 3. PC1D parameters and simulation results. The parameters device area, base contact, thickness, p-type background doping and first rear diffusion were chosen according to cell Gra. C-5. *The parameters were varied for fitting IQE.

	Parameters	Results	Units
Input	Device area	4	cm^2
	Base contact	0.7125	Ω
	Thickness	40	μm
	p-type background doping	2×10^{16}	cm^{-3}
	Sheet Resistance	112	Ω/square
	1st rear diffusion	4×10^{18}	cm^{-3}
	Bulk recombination*	0.2	μs
	Front surface recombination*	8×10^5	cm/s
Output	I_{sc}	0.103	A
	V_{oc}	567.9	mV
	Efficiency	10.15	%

less V_{oc} and J_{sc} mainly due to the defects in the bulk. To date, we have achieved a cell efficiency of 10.2%, which is the highest efficiency of a 40 μm thick crystalline silicon solar cell on graphite substrate with a 4 cm^2 aperture area.

2.2 Quantum efficiency

Figure 6 shows the measured IQE of Gra. B-8 and Gra. C-5, the simulated IQE using PC1D and the reflectance of the Gra. C-5. Gra. B-8 had the shorter diffusion length and lower quantum efficiency than Gra. C-5 due to the surface and bulk recombination without the hydrogen bulk passivation. The simulation parameters are shown in Table 3. The simulation results are in very good agreement with the measured curve. The main difference is that the simulated V_{oc} is about 50 mV higher than the measured one. This may be a result of spatial non-uniformities due to various crystal grains, which cannot be accounted for in 1-dimensional simulations using PC1D.

The anti-reflection coating layers made up of silicon nitride 75 nm thick. The refractive index of the layer is approximately 2. Figure 6 shows that the minimum reflection of the reflectance curve is close to 600 nm, since we fabricated it close to the maximum of the number of incident photons. The SiN_x layers were formed at a low temperature without optimized passivation properties, which cannot reduce the surface recombination in the emitter. Therefore, as shown in the IQE curves, that was assumed to be the main reason why the cell has a spectral response below 90% in the range from 400 nm to 550 nm. In the range from 550 nm to 1000 nm, the IQE goes down gradually due to the bulk and rear recombination loss. In addition, the QE results also reveal that, there are some possibilities for improving the cell performance further, such as by lowering the surface recombination and by optimizing the bulk passivation process.

2.3 Diffusion length

The measured average diffusion length of Gra. C-5 is about 120 μm using the LBiC, however, the value is larger than the cell thickness, since these three wavelengths (976 nm, 951 nm and 846 nm) have the longer penetration lengths than the 40 μm thick active layers of the cells.

Therefore, we obtained the effective diffusion length of the CSiTF solar cell by using PC1D. The real thickness of the cell was included into the PC1D model. We used this model to fit the measured IQE. The simulated IQE characteristic curves for diffusion lengths were 15, 22, 40, and 100 μm, respectively. The effective diffusion length was varied for the fitting and is approximately 22 μm. This result again demonstrates that a surface passivation process and a higher layer quality are necessary for further increasing cell efficiency.

3 Conclusions

In this work, we have presented a new strategy to improve CSiTF solar cell on graphite substrates, using the laser single side contact formation and the hydrogen bulk passivation process at a low temperature. Both methods result in an increase of V_{oc} and FF. The recombination losses in the bulk and the surface are the main limitations on the cell performance. We achieved 10.2% cell efficiency on an aperture area of 4 cm^2.

By comparing different sequences with J-V measurement, we also found the best sequence for inclusion of the hydrogen bulk passivation. This significantly improved cell performance.

We found that one-side contact with LSSC formation is better than front and back contact formation for CSiTF solar cells on graphite substrates. Furthermore, the cell fabrication process can be simplified by making the metallization of all contacts at the same time. The successful application of the LSSC indicates a potential improvement: it is possible to obtain good cell performance at a cost-effective price, while all contacts are on the front side and the impact by substrates is reduced using the LSSC formation. Not only for the CSiTF solar cells on graphite substrates, but we also expect that the LSSC strategy and the best hydrogen bulk passivation process sequence can be applied for the cells based on other high temperature foreign substrates.

So far, a SiO_x intermediate layer is supposed to provide a better quality of surface passivation than a SiC layer [33]. In addition, the batches have not included optimum design of metal grids, light trapping, surface passivation or local diffusion. Thus, the results of both LBiC and QE show possibilities for further improvement.

The authors gratefully acknowledge funding by the German Federal Ministry for the Environment, Nature Conservation and Nuclear safety (BMU) under contract No. 0325031B. We thank Dr. Stefan Janz, Fraunhofer ISE Freiburg, for the recrystallization of the silicon layers.

References

1. D. Amkreutz, J. Müller, M. Schmidt, T. Hänel, T.F. Schulze, Prog. Photovolt.: Res. Appl. **19**, 937 (2011)
2. J. Haschke, D. Amkreutz, L. Korte, F. Ruske, B. Rech, Sol. Energy Mater. Sol. Cells **128**, 190 (2014)
3. A.M. Barnett, R.B. Hall, J.A. Rand, C.L. Kendall, D.H. Ford, Sol. Energy Mater. **23**, 164 (1991)
4. T. Kieliba. University of Konstanz, 2006
5. M. Weizman, H. Rhein, J. Dore, S. Gall, C. Klimm, G. Andrä, C. Schultz, F. Fink, B. Rau, R. Schlatmann, Sol. Energy Mater. Sol. Cells **120**, 521 (2014)
6. J. Dore, D. Ong, S. Varlamov, R. Egan, M.A. Green, IEEE J. Photovolt. **4**, 33 (2014)
7. J. Dore, R. Evans, U. Schubert, B.D. Eggleston, D. Ong, K. Kim, J. Huang, O. Kunz, M. Keevers, R. Egan, S. Varlamov, M.A. Green, Prog. Photovolt.: Res. Appl. **21**, 1377 (2013)
8. G. Andrä, F. Falk, Phys. Stat. Sol. C **5**, 3221 (2008)
9. S. Kühnapfel, N.H. Nickel, S. Gall, M. Klaus, C. Genzel, B. Rech, D. Amkreutz, Thin Solid Films **576**, 68 (2015)
10. K.R. Catchpole, M.J. McCann, K.J. Weber, A.W. Blakers, Sol. Energy Mater. Sol. Cells **68**, 173 (2001)
11. R. Brendel, *Thin-film Crystalline Silicon Solar Cells: Physics and Technology* (Wiley-Vch GmbH & Co. KGaA, Weinheim, 2003)
12. J. Poortmans, V. Arkhipov, *Thin Film Solar Cells: Fabrication, Characterization and Applications* (John Wiley & Sons, 2006)
13. C. Becker, D. Amkreutz, T. Sontheimer, V. Preidel, D. Lockau, J. Haschke, L. Jogschies, C. Klimm, J.J. Merkel, P. Plocica, S. Steffens, B. Rech, Sol. Energy Mater. Sol. Cells **119**, 112 (2013)
14. A. Focsa, I. Gordon, J.M. Auger, A. Slaoui, G. Beaucarne, J. Poortmans, C. Maurice, Renew. Energy **33**, 267 (2008)
15. T. Kunz, V. Gazuz, N. Gawehns, I. Burkert, M.T. Hessmann, R. Auer, in *Proceedings of 24th European Photovoltaic Solar Energy Conference and Exhibition, Hamburg, Germany, 2009*, p. 2553
16. S. Janz, P. Löper, M. Schnabel, Mater. Sci. Eng. B **178**, 542 (2013)

17. T. Kunz, M.T. Hessmann, R. Auer, A. Bochmann, S. Christiansen, C.J. Brabec, J. Cryst. Growth **357**, 20 (2012)

18. K. Schillinger, S. Janz, S. Reber, in *IEEE 39th Photovoltaic Specialists Conference, Tampa, USA, 2013*, p. 1784

19. T. Kunz, R. Auer, ZAE Bayern, Report No. FKZ 0325031B, 2013

20. T. Kunz, V. Gazuz, M.T. Hessmann, N. Gawehns, I. Burkert, C.J. Brabec, Sol. Energy Mater. Sol. Cells **95**, 2454 (2011)

21. M.J. Stocks, A. Cuevas, A.W. Blakers, Prog. Photovolt.: Res. Appl. **4**, 35 (1996)

22. C. Hebling, S.W. Glunz, C. Schetter, J. Knobloch, A. Räuber, Sol. Energy Mater. Sol. Cells **48**, 335 (1997)

23. R. Lüdemann, S. Schaefer, C. Schule, C. Hebling, in *Proceedings of the 26th IEEE Photovoltaics Specialist Conference, Anaheim, USA, 1997*, p. 159

24. T. Rachow, M. Ledinsky, S. Janz, S. Reber, A. Fejfar, in *Proceedings of 27th European Photovoltaic Conference and Exhibition, 2012*, p. 2386

25. S. Reber, W. Zimmermann, T. Kieliba, Sol. Energy Mater. Sol. Cells **65**, 409 (2001)

26. T. Kunz, I. Burkert, N. Gawehns, R. Auer, in *Proceedings of the 23rd European Photovoltaic Solar Energy Conference, Valencia, Spain, 2008*, p. 2202

27. B. Gorka, Ph.D. Thesis, Technical University of Berlin, 2010

28. D.A. Clugston, P.A. Basore, in *Proceedings of the 26th IEEE Photovoltaics Specialist Conference, Anaheim, USA, 1997*, p. 207

29. T. Kunz, I. Burkert, R. Auer, A.A. Lovtsus, R.A. Talalaev, Y.N. Makarov, J. Cryst. Growth **310**, 1112 (2008)

30. V. Gazuz, M. Muehlbauer, M. Scheffler, R. Weissmann, R. Auer, in *Proceedings of 21st European Photovoltaic Solar Energy Conference and Exhibition, Dresden, Germany, 2006*, p. 851

31. A. Slaoui, A. Barhdadi, J.C. Muller, P. Siffert, Appl. Phys. A **39**, 159 (1986)

32. F. Granek, A. Weeber, K. Tool, R. Kinderman, P. de Jong, in *Conference Record of the 2006 IEEE 4th World Conference on Photovoltaic Energy Conversion, 2006*, p. 1319

33. J. Dore, S. Varlamov, M.A. Green, IEEE J. Photovolt. **5**, 9 (2015)

PERMISSIONS

LIST OF CONTRIBUTORS

M. Morenoa and P. Roca i Cabarrocas
Laboratoire de Physique des Interfaces et des Couches Minces, ´Ecole Polytechnique, CNRS, Palaiseau, France

Piétrick Hudhommea
L'UNAM Université, Université d'Angers, Laboratoire MOLTECH-Anjou, CNRS UMR 6200, 2 boulevard Lavoisier, 49045 Angers, France

I. Ngo, M.E. Gueunier-Farreta, J. Alvarez and J.P. Kleider
LGEP, UMR 8507 CNRS, SUPELEC, UPMC, Université Paris-Sud 11, 11 rue Joliot-Curie, Plateau de Moulon, 91192 Gif-sur-Yvette Cedex, France

A. Chowdhury and A. Slaoui
ICube-University of Strasbourg and CNRS, Strasbourg, France

A. Bahouka and F. Mermet
IREPA Laser, Strasbourg, France

S. Steffens
HZB, Berlin, Germany

J. Schneider and J. Dore
SUNTECH, Thalheim, Germany

S. Krause and F. Steudel
Fraunhofer Center for Silicon Photovoltaics CSP, Walter-H¨ulse-Str. 1, 06120 Halle (Saale), Germany

P.T. Miclea
Fraunhofer Center for Silicon Photovoltaics CSP, Walter-H¨ulse-Str. 1, 06120 Halle (Saale), Germany

Institute of Physics, Martin Luther University of Halle-Wittenberg, Heinrich Damerow-Str. 4, 06120 Halle (Saale), Germany

S. Schweizer
Fraunhofer Center for Silicon Photovoltaics CSP, Walter-Hülse-Str. 1, 06120 Halle (Saale), Germany
Department of Electrical Engineering, South Westphalia University of Applied Sciences, L¨ubecker Ring 2, 59494 Soest, Germany

G. Seifert
Fraunhofer Center for Silicon Photovoltaics CSP, Walter-Hülse-Str. 1, 06120 Halle (Saale), Germany
Centre for Innovation Competence SiLi-nano®, Martin Luther University of Halle-Wittenberg, Karl-Freiherr-von-Fritsch-Str. 3, 06120 Halle (Saale), Germany

M. Nath, S. Chakraborty and P. Chatterjee
Energy Research Unit, Indian Association for the Cultivation of Science, 700 032 Kolkata, India

E.V. Johnson, A. Abramov and P. Roca i Cabarrocas
Laboratoire de Physique des Interfaces et des Couches Minces, ´Ecole Polytechnique, CNRS, 91128 Palaiseau, France

Matthias Meiera, Karsten Bittkau, Ulrich W. Paetzold, Jürgen Hüpkes, Stefan Muthmann, Ralf Schmitz, Andreas Mück, and Aad Gordijn
IEK5-Photovoltaik, Forschungszentrum Jülich GmbH, 52425 Jülich, Germany

Kitty Kumar and Jun Nogami
Department of Materials Science and Engineering, 184 College Street, Toronto, Ontario, M5S 3E4, Canada

Nazir P. Kherani
Department of Materials Science and Engineering, 184 College Street, Toronto, Ontario, M5S 3E4, Canada
Department of Electrical and Computer Engineering, 10 King's College Rd. Toronto, Ontario, M5S 3G4, Canada

Kenneth K.C. Lee and Peter R. Herman
Department of Electrical and Computer Engineering, 10 King's College Rd. Toronto, Ontario, M5S 3G4, Canada

S. Michard, V. Balmes, M. Meiera, A. Lambertz, T. Merdzhanova and F. Finger
Institute of Energy and Climate Research 5 – Photovoltaik, Forschungszentrum Jülich, 52425 Jülich, Germany

S. Delbos
EDF R&D, Institut de Recherche et Développement sur l'Énergie Photovoltaïque (IRDEP), 6 quai Watier, 78401 Chatou, France
CNRS, UMR 7174, 78401 Chatou, France
Chimie ParisTech, 75005 Paris, France

F. Pellé
Laboratoire de Chimie de la Matiére Condensée de Paris LCMCP UMR7574 CNRS/UPMC/Chimie ParisTech, Chimie ParisTech, 11 rue Pierre et marie Curie, 75235 Paris, France

S. Ivanova
Laboratoire de Chimie de la Matiére Condensée de Paris LCMCP UMR7574 CNRS/UPMC/Chimie ParisTech, Chimie ParisTech, 11 rue Pierre et marie Curie, 75235 Paris, France

Center for Information and Optics Technology, University of Information Technology, Mechanics and Optics, 199034, St. Petersburg, Russia

J.-F. Guillemoles
IRDEP, UMR 7174CNRS/EDF/Chimie ParisTech, 6 quai Watier, 78401 Chatou, France

F. Obereignera and R. Scheerb
Photovoltaics Group, Martin-Luther-University Halle-Wittenberg, 06120 Halle (Saale), Germany

X. Bril and P. Roca i Cabarrocas
LPICM, CNRS-École Polytechnique, 91128 Palaiseau Cedex, France

M. Labrune
LPICM, CNRS-École Polytechnique, 91128 2 TOTAL S.A., Gas & Power, R&D Division, Courbevoie, France Palaiseau Cedex, France

G. Patriarche, L. Largeau and O. Mauguin
Laboratoire de Photonique et de Nanostructures, CNRS, Marcoussis, France

P. Prod'homme
TOTAL SA – Gas & Power R&D Division Tour Lafayette, place des Vosges La Défense 6, 92400 Courbevoie, France

C. Charpentier
TOTAL SA – Gas & Power R&D Division Tour Lafayette, place des Vosges La Défense 6, 92400 Courbevoie, France
LPICM-CNRS – Laboratoire de Physique des Interfaces et Couches Minces, École Polytechnique, 91128 Palaiseau, France

M. Chaigneau and P. Roca i Cabarrocas
LPICM-CNRS – Laboratoire de Physique des Interfaces et Couches Minces, École Polytechnique, 91128 Palaiseau, France

I. Maurin
LPMC-CNRS – Laboratoire de Physique de la Matiére Condensée, École Polytechnique, 91128 Palaiseau, France

Thanh-Tuan Buia and Fabrice Goubardb
Laboratoire de Physicochimie des Polyméres et des Interfaces, Université de Cergy-Pontoise, 5 mail Gay Lussac, Neuville-sur-Oise, 95031 Cergy-Pontoise Cedex, France

L. Prönneke and G.C. Gläser
Institut für Photovoltaik, Universität Stuttgart, Pfaffenwaldring 47, 70569 Stuttgart, Germany

U. Rau
IEK5-Photovoltaik, Forschungszentrum Jülich, 52425 Jülich, Germany

T. Dimopoulos, A. Peić and S. Abermann
AIT-Austrian Institute of Technology, Energy Department, Giefinggasse 2, 1221 Vienna, Austria

M. Postl
NanoTecCenter Weiz Forschungsgesellschaft mbH, Franz-Pichler-Str. 32, 8160 Weiz, Austria

E.J.W. List-Kratochvil
NanoTecCenter Weiz Forschungsgesellschaft mbH, Franz-Pichler-Str. 32, 8160 Weiz, Austria
Graz University of Technology, Institute of Solid State Physics, Petergasse 16, 8010 Graz, Austria

R. Resel
Graz University of Technology, Institute of Solid State Physics, Petergasse 16, 8010 Graz, Austria

Lan Wang, Xianzhong Lin, Christian Wolf and Reiner Klenk
Helmholtz-Zentrum Berlin für Materialien und Energie, Hahn-Meitner-Platz 1, 14109 Berlin, Germany

Ahmed Ennaoui
Qatar Environment and Energy Research Institute and Hamad Bin Khalifa University, Education City, Doha, Qatar

Martha Ch. Lux-Steiner
Helmholtz-Zentrum Berlin für Materialien und Energie, Hahn-Meitner-Platz 1, 14109 Berlin, Germany
Freie Universität Berlin, Fachbereich Physik, Arnimallee 14, 14195 Berlin, Germany

Cordula Walder, Martin Kellermann, Karsten von Maydell and Carsten Agert
NEXT ENERGY · EWE Research Centre for Energy Technology at the University of Oldenburg, Carl-von-Ossietzky-Straße 15, 26129 Oldenburg, Germany

Elke Wendler and Jura Rensberg
Institut für Festkörperphysik, Friedrich-Schiller-Universität Jena, Helmholtzweg 3, 07743 Jena, Germany

Susanna Harndt, Christian A. Kaufmann, Martha C. Lux-Steiner and Reiner Klenk
Helmholtz-Zentrum Berlin für Materialien und Energie, Hahn-Meitner-Platz 1, 14109 Berlin, Germany

Reiner Nürnberg
Weierstraß-Institut für Angewandte Analysis und Stochastik, Mohrenstr. 39, 10117 Berlin, Germany

Maik Thomas Hessmann, Thomas Kunz, Taimoor Ahmad, Da Li, Stephan Wittmann, Arne Riecke, Michael Schmidt and Richard Auer
Bavarian Center for Applied Energy Research (ZAE Bayern), Haberstr. 2a, 91058 Erlangen, Germany

Jan Ebser and Barbara Terheiden
Department of Physics, University of Konstanz, Universitätsstr 10, Box 676, 78464 Konstanz, Germany

Kristian Cvecek
BLZ-Bavarian Laser Center, Konrad-Zuse-Str. 2-6, 91052 Erlangen, Germany

Chistoph J. Brabec
Bavarian Center for Applied Energy Research (ZAE Bayern), Haberstr. 2a, 91058 Erlangen, Germany
i-MEET: institute Materials for Electronics and Energy Technology, University of Erlangen-Nuremberg, Martensstr. 7, 91058 Erlangen, Germany

Emma Lewis and Richard P. Barber Jr.
Department of Physics and Center for Nanostructures, Santa Clara University, Santa Clara CA 95053, USA

Bhaskar Mantha
Department of Electrical Engineering, Santa Clara University, Santa Clara CA 95053, USA Abstract

David Black, Iulia Salaoru and Shashi Paula
Emerging Technologies Research Centre, Hawthorn Building, De Montfort University, The Gateway, Leicester LE1 9BH, UK

Sergey Abolmasov
R&D Center of Thin-Film Technologies in Energetics, Ioffe Institute 28 Polytekhnicheskaya, 194064 Saint Petersburg, Russia
LPICM, CNRS, Ecole Polytechnique, Université Paris-Saclay, 91128 Palaiseau, France

Pere Roca i Cabarrocas and Parsathi Chatterjee
LPICM, CNRS, Ecole Polytechnique, Université Paris-Saclay, 91128 Palaiseau, France

Athar Ali Shah, Akrajas Ali Umara and Muhamad Mat Salleh
Institute of Microengineering and Nanoelectronics, Universiti Kebangsaan Malaysia, 43600 UKM Bangi, Selangor, Malaysia

Etienne Moulin
Ecole Polytechnique Fédérale de Lausanne (EPFL), Institute of Microengineering (IMT), Photovoltaics and Thin-Film Electronics Laboratory, Rue de la Maladiére 71b, 2000 Neuchâtel, Switzerland

IEK5-Photovoltaik, Forschungszentrum Jülich GmbH, 52425 J¨ulich, Germany

Thomas Christian Mathias Müller, Marek Warzecha, Andre Hoffmann Ulrich Wilhelm Paetzold and Urs Aeberhard
IEK5-Photovoltaik, Forschungszentrum Jülich GmbH, 52425 Jülich, Germany

Jean-Christian Bernéde
L'UNAM, Université de Nantes, MOLTECH-Anjou, CNRS, UMR 6200, 2 rue de la Houssiniére, BP 92208, 44000 Nantes, France

Zouhair El Jouad
L'UNAM, Université de Nantes, MOLTECH-Anjou, CNRS, UMR 6200, 2 rue de la Houssiniére, BP 92208, 44000 Nantes, France
Laboratoire Optoélectronique et Physico-chimie des Matériaux, Université Ibn Tofail, Faculté des Sciences, BP 133, 14000 Kenitra, Morocco

Mohammed Addou
Laboratoire Optoélectronique et Physico-chimie des Matériaux, Université Ibn Tofail, Faculté des Sciences, BP 133, 14000 Kenitra, Morocco

Guy Louarn and Linda Cattin
Université de Nantes, Institut des Matériaux Jean Rouxel (IMN), CNRS, UMR 6502, 2 rue de la Houssiniére, BP 32229, 44322 Nantes Cedex 3, France

Thappily Praveen and Padmanabhan Predeep
Laboratory for Unconventional Electronics and Photonics, Department of Physics, National Institute of Technology, 673 601 Calicut, Kerala, India

Mustapha Morsli
L'UNAM, Université de Nantes, Faculté des Sciences et des Techniques, 2 rue de la Houssiniére, BP 92208, 44000 Nantes, France

Mathieu Boccarda, Peter Cuony, Simon Hänni, Michael Stuckelberger, Franz-Josef Haug, Fanny Meillaud, Matthieu Despeisse and Christophe Ballif
École Polytechnique Féd'erale de Lausanne (EPFL), Institute of Microengineering (IMT), Photovoltaics and Thin Film
Electronics Laboratory, Rue de la Maladiére 71b, CP 526, 2002 Neuchâtel 2, Switzerland

Da Li, Stephan Wittmann, Thomas Kunz, Taimoor Ahmad, Nidia Gawehns, Maik T. Hessmann and Richard Auer
Bavarian Center for Applied Energy Research (ZAE Bayern), Haberstr. 2a, 91058 Erlangen, Germany

Jan Ebser and Barbara Terheiden
Department of Physics, University of Konstanz, Box 676, 78457 Konstanz, Germany

Christoph J. Brabec
Institute of Materials for Electronics and Energy Technology (i-MEET), University of Erlangen-Nuremberg, Martensstr. 7, 91058 Erlangen, Germany

Index